Discrete Fourier Analysis and Wavelets

Discrete Fourier Analysis and Wavelets

Applications to Signal and Image Processing

Second Edition

S. Allen Broughton
Kurt Bryan

WILEY

This second edition first published 2018

Edition History
John Wiley & Sons, Inc. (1e, 2009)

Registered Office
John Wiley & Sons, Inc., 111 River Street, Hoboken, NJ 07030, USA

Editorial Office
111 River Street, Hoboken, NJ 07030, USA

For details of our global editorial offices, customer services, and more information about Wiley products visit us at www.wiley.com.

Wiley also publishes its books in a variety of electronic formats and by print-on-demand. Some content that appears in standard print versions of this book may not be available in other formats.

Library of Congress Cataloging-in-Publication Data:

Names: Broughton, S. Allen, 1951- author. | Bryan, Kurt, 1962- author.
Title: Discrete fourier analysis and wavelets : applications to signal and image processing / by S. Allen Broughton, Kurt Bryan.
Description: Second edition. | Hoboken, NJ : John Wiley & Sons, 2018. | Includes bibliographical references and index. |
Identifiers: LCCN 2017044668 (print) | LCCN 2017052511 (ebook) | ISBN 9781119258230 (pdf) | ISBN 9781119258247 (epub) | ISBN 9781119258223 (cloth)
Subjects: LCSH: Wavelets (Mathematics) | Signal processing–Mathematics. | Image processing–Mathematics.
Classification: LCC QA403.5 (ebook) | LCC QA403.5 .B77 2018 (print) | DDC 515/.2433–dc23
LC record available at https://lccn.loc.gov/2017044668

Cover design by Wiley
Cover image: Courtesy of Frances Silta

Set in 10/12pt WarnockPro by SPi Global, Chennai, India

Printed in the United States of America

10 9 8 7 6 5 4 3 2 1

To our wives, Judith Broughton and Frances Silta, and our families.

Contents

Preface

Philosophy of the Text

In this text, we provide an introduction to the mathematics that underlies much of modern signal and image processing. In particular, we

- develop the mathematical framework in which signal and image processing takes place, specifically, vector and inner product spaces,
- develop traditional Fourier-based transform techniques, primarily in the discrete case, but also to some extent in the continuous setting,
- provide entry-level material on filtering, convolution, filter banks, lifting schemes, frames, and wavelets,
- make extensive use of computer-based explorations for concept development.

These topics are extremely beautiful and important areas of classical and modern applied mathematics. They also form the foundation for many techniques in science and engineering. However, while they are usually taught (sometimes in bits and pieces) in the engineering and science curricula, they are often overlooked in mathematics courses, or addressed only as brief asides. We hope to change this.

Throughout the text, we often use image compression as a concrete and motivational hook for the mathematics, but our goal here is not a technical manual on how to program a JPEG encoder. Rather, we include this topic so that the reader can begin applying the theory to interesting problems relatively quickly. We also touch upon a number of other important applications in addition to compression, for example, progressive transmission of images, image denoising, spectrographic analysis, and edge detection. We do not discuss the very large topic of image restoration, although the mathematics we develop underlies much of that subject as well. Because audio signals are one-dimensional, they are somewhat simpler to deal with than images, so we often use audio data to illustrate the mathematics of interest before considering images.

In this text, we assume that the student is familiar with calculus and elementary matrix algebra, and possesses the "mathematical maturity" typical of a second or third year undergraduate. We do not assume any background in image or signal processing. Neither do we assume that the reader knows Matlab, though some experience with computer programming, however modest, is helpful. This text might also serve as a useful supplement for more advanced image processing texts such as [14], which is written at a graduate level and has a rather terse mathematical development.

Infinite-dimensional vector and inner product spaces form a natural mathematical framework for signal and image processing. Although we do develop a fair amount of general theory in this abstract framework, we focus a bit more on the discrete and finite-dimensional cases. The use of function spaces and the associated analysis is, especially at the beginning, an unnecessary complication for readers learning the basics. Moreover, the infinite-dimensional and functional settings do not lend themselves as well to computer-based exercises and experimentation. However, we return to a more general functional framework and some associated analysis in the last chapter, in which we discuss wavelets.

Outline of the Text

In Chapters 1 through 3, we develop the vector space and matrix framework for analyzing signals and images, as well as the discrete Fourier transform and the discrete cosine transform. We also develop some general theory concerning inner product spaces and orthogonal bases, and consider Fourier series. A principal motivating application throughout these chapters is to develop an understanding of the traditional JPEG compression algorithm. Some easily developed ideas such as noise removal in signals and images are also included. In Chapters 4 and 5, we consider convolution, filtering, and windowing techniques for signals and images. In particular, in Chapter 5, we develop some traditional windowing techniques for "localizing" Fourier-like transforms. Chapter 6 examines the relatively new topic of "frames," and its application to redundancy and oversampling. In Chapter 7, we develop the theory of filter banks, as a means of understanding the multiscale localized analysis that underlies the JPEG 2000 compression standard. Chapter 8 examines "lifting schemes," the modern approach used by most software for implementing filter banks and the discrete wavelet transforms (as well as designing them). In Chapter 9, we build on Chapter 7 to give a brief account of how filter banks give rise to wavelets.

The interdependence of the chapters is summarized in Figure 1. For a one-semester course, Chapters 1–5 and 7 could form the core topics, depending on student preparation, with part or all of chapter 9 covered as well. If time is an issue Chapter 5 can be skipped, since none of the material there is

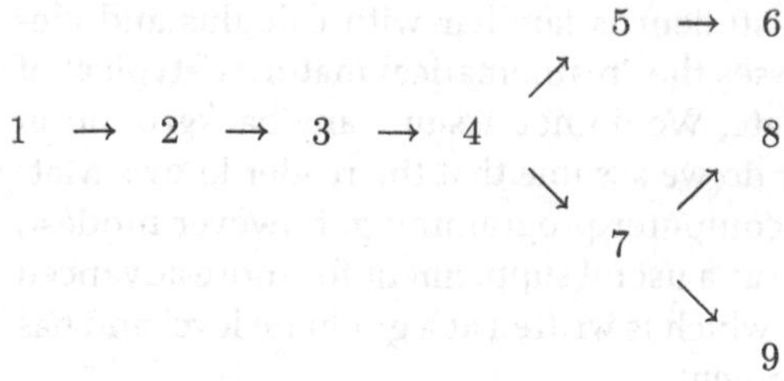

Figure 1 Chapter dependence.

essential for Chapters 7 or 9. Section 7.7 is also not essential for Chapter 9, but is very helpful in seeing the connection between filter banks and wavelets. Chapter 6 is not essential for any of the material that follows it, but merely builds on Chapter 5 and the preceding chapters. Similarly, Chapter 8 is not necessary for the material in Chapter 9.

The entire book could be done comfortably in a two-semester sequence.

Computer Use, Supporting Routines, and Text Web Page

It is an easy and accurate guess that many concrete problems in signal and image processing are very computationally intensive. As a result, both the examples in the text and the exercises make frequent use of Matlab; a computer algebra system such as Maple or Mathematica may also be useful. Various Matlab routines used in the text are available on the textbook web page [32]. In addition, on the web page, a large portion of the Matlab material has been migrated to Scientific Python for those readers who do not have access to Matlab.

Exercises and Projects

At the end of each chapter, there is a Matlab project and a number of exercises. The exercises are generally organized by theme. Some of the exercises are simple "by hand" computations, others are fairly large or theoretical, and others are Matlab programming or discovery exercises designed to refine intuition about the material or illustrate a mathematical concept. Many exercises contain hints, and we provide additional hints and answers to selected exercises in an appendix. The projects are essentially guided explorations that make use of Matlab. Many are suitable for use in a "class laboratory" setting, in which students work alone or in groups.

Most of the exercises and projects make use of standard Matlab commands, but in a few places we make use of commands from various Matlab Toolboxes. In Chapter 3, for example, we make use of the "`dct`" command from Matlab's signal processing toolbox. For those readers who do not have access to this toolbox, we have made available on the textbook web page [32] a substitute "`kdct`." It is not as efficient as Matlab's dedicated DCT command, but it works. The Matlab project in Chapter 7 makes heavy use of the Wavelet Toolbox, but again, if the reader does not have this toolbox, we have alternate routines and even an alternate Matlab project.

August, 2017
Terre Haute, Indiana

S. Allen Broughton
Kurt Bryan

Acknowledgments

We would like to acknowledge Rose-Hulman professors Ed Doering and Roger Lautzenheiser for their suggested improvements to the original set of notes upon which this text is based. We also thank the many Rose-Hulman students who gave us feedback concerning these notes and exercises, and the students and colleagues from around the world who gave us feedback concerning the first edition of this text.

We also thank Frances Silta for supplying the images used to construct the cover (McDonald Creek in Glacier National Park in Montana) as well as the images that form the basis for Figures 1.5, 1.12, 2.13, 3.11–3.17, and 7.16–7.19.

S. Allen Broughton
Kurt Bryan

About the Companion Website

This book is accompanied by a companion website:

www.wiley.com/go/Broughton/Discrete_Fourier_Analysis_and_Wavelets

The companion website (BCS Standard: Student and Instructor—two sites) contains:

- Student site: all of revised and new software content (in Matlab code) produced by the authors during the rewrite of the manuscript. This content is referred to in the text.
- Student site: Scientific Python code that parallels the Matlab code referred to in the text. The SciPy code is not referred to in the text but is provided for those instructors wishing to use Scientific Python for exploratory software support.
- Student site: list of post–publication errata discovered by our readers.
- Instructor site (password protected): complete solution set to all problems in the book, organized by chapter, and SciPy partial solution files for the chapter projects.

1

Vector Spaces, Signals, and Images

1.1 Overview

In this chapter, we introduce the mathematical framework of vector spaces, matrices, and inner products. We motivate the study of these topics by using these tools to analyze signals and images, both outside the computer (the *analog* signal as it exists in the "real world") and inside the computer (the *digitized* signal, suitable for computer storage and processing). In either case the signal or image may be viewed as an element of a vector space, so we define and develop some essential concepts concerning these spaces. In particular, to analyze signals and images, we will decompose them into linear combinations of basic sinusoidal or complex exponential waveforms. This is the essence of the discrete Fourier and cosine transforms.

The process of sampling the analog signal and converting it to digital form causes an essential loss of information, called "aliasing" and "quantization error." We examine these errors rather closely. Analysis of these errors motivates the development of methods to quantify the distortion introduced by an image compression technique, which leads naturally to the concept of the "energy" of a signal and a deeper analysis of inner products and orthogonality.

1.2 Some Common Image Processing Problems

To open this chapter, we discuss a few common challenges in image processing, to try to give the reader some perspective on the subject, and why mathematics is such an essential tool. In particular, we take a very short look at the following:

- image compression,
- image restoration and denoising,
- edge and defect detection.

Discrete Fourier Analysis and Wavelets: Applications to Signal and Image Processing, Second Edition.
S. Allen Broughton and Kurt Bryan.

Companion Website: www.wiley.com/go/Broughton/Discrete_Fourier_Analysis_and_Wavelets

We also briefly discuss the "transform" paradigm that forms the basis of so much signal and image processing, and indeed, much of mathematics.

1.2.1 Applications

1.2.1.1 Compression

Digitized images are everywhere. The Internet provides obvious examples, but we also work with digitized images when we take pictures, scan documents into computers, send faxes, photocopy documents, and read books on CD. Digitized images underlie video games, and television has gone entirely digital. In each case the memory requirements for storing and transmitting digitized images are an important issue. For example, in a digital camera or cell phone we want to pack as many pictures as we can into the available memory, and we all want web pages to load large images with little delay. Minimizing the memory requirements for digitized images is thus important, and this task is what motivates much of the mathematics in this text.

Without going into too much detail, let us calculate the memory requirement for a typical photograph taken with a cell phone or digital camera. Assume that we have full 24-bit color, so that one byte of memory is required for each of the red, green, and blue (RGB) components of each pixel. With a 3264×2448 pixel image (a typical size) there will be $3264 \times 2448 \times 3 = 23,970,816$ bytes or 24 MB of memory required, if no compression is used. On a phone, with all the competition for memory from apps, music, and so on, there may not be a lot of room to store images. Even with 5 GB of free memory for photos you can only store about 200 such large, gorgeous pictures, which will not get you through a week-long vacation. More sophisticated storage is needed to reduce memory requirements by a substantial factor.

However, the compression algorithm we devise cannot sacrifice significant image quality. Even casual users of digital cameras frequently enlarge and print portions of their photographs, so any degradation of the original image will rapidly become apparent. Besides, more than aesthetics may be at stake: medical images (e.g., X-rays) may be compressed and stored digitally, and any corruption of the images could have disastrous consequences. The FBI has also digitized and compressed its database of fingerprints, where similar considerations apply [4].

At this point the reader is encouraged to do Exercise 1.

1.2.1.2 Restoration

Images can be of poor quality for a variety of reasons: low-quality image capture (e.g., security video cameras), blurring when the picture is taken, physical damage to an actual photo or negative, or noise contamination during the image capture process. Restoration seeks to return the image to its original quality or even "better." Some of this technology is embedded into image capture devices such as scanners. A very interesting and mathematically sophisticated area of research involves *inpainting*, in which one tries to recover missing portions of an image, perhaps because a film negative was scratched, or a photograph written on.

1.2.1.3 Edge Detection

Sometimes the features of essential interest in an image are the edges, areas of sharp transition that indicate the end of one object and the start of another. Situations such as this may arise in industrial processing, for automatic detection of defects, or in automated vision and robotic manipulation.

1.2.1.4 Registration

Image registration refers to the process of aligning two or more images into a common frame of reference, possibly to compare the images, or amalgamate data from the images. This is often required when one images a single object using more than one imaging system (e.g., at different wavelengths). Registration is often a prelude to other processing, for example, facial recognition.

1.2.2 Transform-Based Methods

The use of transforms is ubiquitous in mathematics. The general idea is to take a problem posed in one setting, transform to a new domain where the problem is more easily solved, then inverse transform the solution back to the original setting. For example, if you have taken a course in differential equations you may have encountered the Laplace transform, which turns linear differential equations into algebra problems that are more easily solved.

Many image processing procedures begin with some type of transform T that is applied to the original image. The transform T takes the image data from its original format in the "image domain" to an altered format in the "frequency domain." Operations like compression, denoising, or other restoration are sometimes more easily performed in this frequency domain. The modified frequency domain version of the image can then be converted back to the original format in the image domain by applying the inverse of T.

The transform operator T is almost always linear, and for finite-sized signals and images such linear operators are implemented with matrix algebra. The matrix approach thus constitutes a good portion of the mathematical development of the text. Other processes, such as quantization and clipping (discussed later), are nonlinear. These are usually the lossy parts of the computation; that is, they cause irreversible but (we hope!) acceptable loss of data.

1.3 Signals and Images

Before beginning a general discussion of vector spaces, it will be helpful to look at a few specific examples that provide physical realizations of the mathematical objects of interest. We will begin with one-dimensional signals, then move on to two-dimensional images.

1.3.1 Signals

A signal may be modeled as a real-valued function of a real independent variable t, which is usually time. More specifically, consider a physical process that is dependent on time. Suppose that at each time t within some interval $a \leq t \leq b$ we perform a measurement on some aspect of this process, and this measurement yields a real number that may assume any value in a given range. In this case, our measurements are naturally represented by a real-valued function $x(t)$ with domain $a \leq t \leq b$. We will refer to $x(t)$ as an *analog signal*. The function $x(t)$ might represent the intensity of sound at a given location (an audio signal), the current through a wire, the speed of an object, and so on.

For example, a signal might be given by the function

$$x(t) = 0.75\sin(3t) + 0.5\sin(7t)$$

over the range $0 \leq t \leq 4\pi$. The graph of this function is shown in Figure 1.1. This signal is somewhat unrealistic, however, for it is a linear combination or superposition of a small number of simple sinusoidal functions with no noise. In general, in signal processing we can depend on being vexed by a few persistent annoyances:

- We almost never have an explicit formula for $x(t)$.
- Most signals are more complicated.
- Most signals have noise.

Despite the difficulty of writing out an analog description in any specific instance, many physical processes are naturally modeled by analog signals. Analog models also have the advantage of being amenable to analysis using methods from calculus and differential equations. However, most modern signal processing takes place in computers where the computations can be

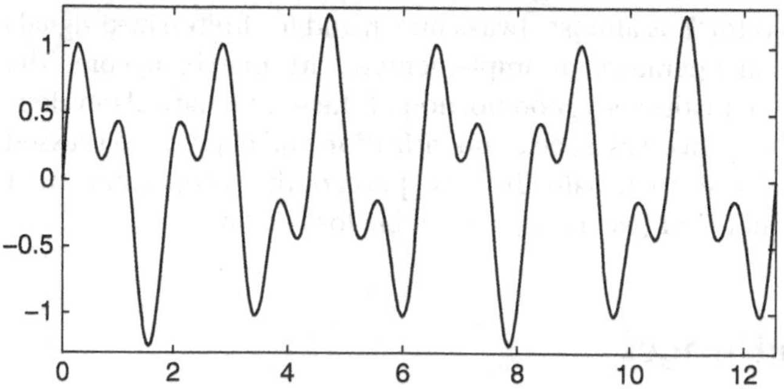

Figure 1.1 Analog or continuous model $x(t) = 0.75\sin(3t) + 0.5\sin(7t)$.

done quickly and flexibly. Unfortunately, analog signals generally cannot be stored in a computer in any meaningful way.

1.3.2 Sampling, Quantization Error, and Noise

To store a signal in a computer we must first *digitize* the signal. The first step in digitization consists of measuring the signal's instantaneous value at specific times over a finite interval of interest. This process is called *sampling*. For the moment let us assume that these measurements can be carried out with "infinite precision." The process of sampling the signal converts it from an analog form to a finite list of real numbers, and is usually carried out by hardware known as an *analog-to-digital* ("A-to-D") converter.

More explicitly, suppose that the signal $x(t)$ is defined on the time interval $a \leq t \leq b$. Choose an integer $N \geq 1$ and define the *sampling interval* $\Delta t = (b-a)/N$. We then measure $x(t)$ at times $t = a, a+\Delta t, a+2\Delta t, \ldots$, to obtain samples

$$x_n = x(a + n\Delta t), \quad n = 0, 1, \ldots, N.$$

Define

$$\mathbf{x} = (x_0, x_1, \ldots, x_N) \in \mathbb{R}^{N+1}$$

with the given indexing $x_0 = x(a)$ and $x_N = x(b)$. The vector $\mathbf{x}$ is the sampled version of the signal $x(t)$. The quantity $1/\Delta t$ is the number of samples taken during each time period, and so is called the *sampling rate*.

In Figure 1.2 we have a graphical representation of the sampled signal from Figure 1.1. It should be intuitively clear that sampling causes a loss of information. That is, if we know only the sampled signal, then we have no idea what the underlying analog signal did between the samples. The nature of this information loss can be more carefully quantified, and this gives rise to the concept of *aliasing*, which we examine later.

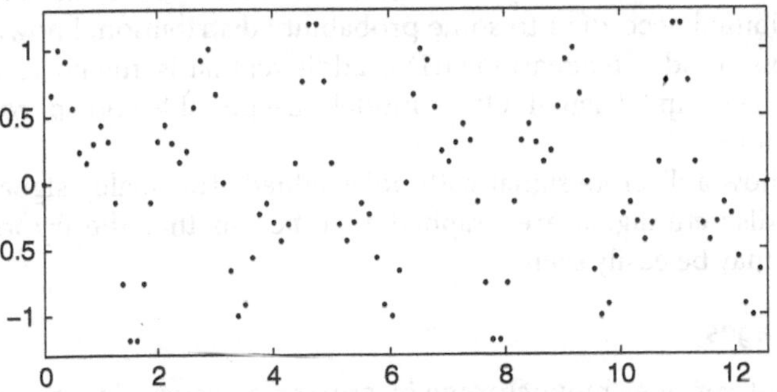

Figure 1.2 Discrete or sampled model, $x(t) = 0.75\sin(3t) + 0.5\sin(7t)$.

The sampling of the signal in the independent variable t is not the only source of error in our A-to-D conversion. In reality, we cannot measure the analog signal's value at any given time with infinite precision, for the computer has only a finite amount of memory. Consider, for example, an analog voltage signal that ranges from 0 to 1 V. An A-to-D converter might divide up this 1 V range into $2^8 = 256$ equally sized intervals, say with the kth interval given by $k\Delta x \leq x < (k+1)\Delta x$ where $\Delta x = 1/256$ and $0 \leq k \leq 255$. If a measurement of the analog signal at an instant in time falls within the kth interval, then the A-to-D converter might simply record the voltage at this time as $k\Delta x$. This is the *quantization* step, in which a continuously varying quantity is truncated or rounded to the nearest of a finite set of values. An A-to-D converter as above would be said to be "8-bit," because each analog measurement is converted into an 8-bit quantity. The error so introduced is called the *quantization error*.

Unfortunately, quantization is a nonlinear process that corrupts the algebraic structure afforded by the vector space model (see Exercise 6). In addition quantization introduces irreversible, though usually acceptable, loss of information. This issue is explored further in Section 1.9.

An additional potential source of error is *clipping*, in which the signal $x(t)$ goes outside the range that we can discretize, for example, 0–1 V in the above discussion. In this case, signals that exceed 1 V might simply be recorded as 1 V, and similarly signals that fall below 0 V will be recorded as 0 V. We assume that our analog signals are scaled so that clipping does not occur.

The combination of sampling and quantization allows us to *digitize* a signal or image, and thereby convert it into a form suitable for computer storage and processing.

One last source of error is random noise in the sampled signal; this may come from the transmission channel over which the image was sent, or the imaging sensor used to collect image. If the noiseless samples are given by x_n as above, the noisy sample values y_n might be modeled as

$$y_n = x_n + \epsilon_n, \tag{1.1}$$

where ϵ_n represents the noise in the nth measurement. The errors ϵ_n are usually assumed to be distributed according to some probability distribution, known or unknown. The noise model in equation (1.1) is additive; that is, the noise is merely added onto the sampled signal. Other models are possible and appropriate in some situations.

In Figure 1.3 we show a discrete signal with noise added. The analog signal and the corrupted discrete signal are graphed together so that the errors introduced by noise may be easily seen.

1.3.3 Grayscale Images

For simplicity, we first consider monochrome or *grayscale* images. An analog grayscale image is modeled as a real-valued function $f(x, y)$ defined on a

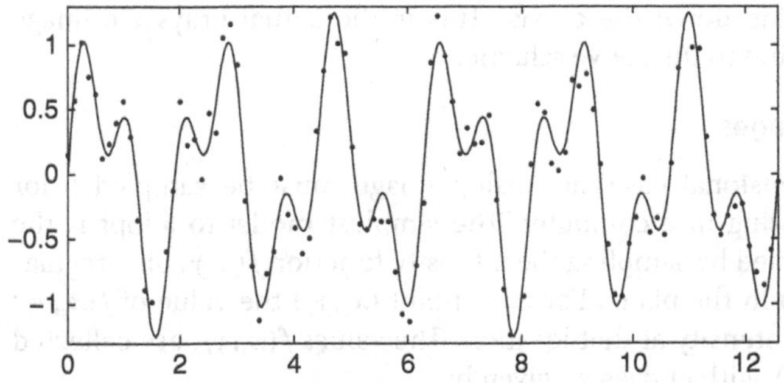

Figure 1.3 Analog and discrete models with noise, $x(t) = 0.75\sin(3t) + 0.5\sin(7t)$.

two-dimensional region Ω. Usually Ω is a rectangle, defined in xy coordinates by $a \leq x \leq b,\ c \leq y \leq d$. The value $f(x, y)$ represents the "intensity" of the image at the point (x, y) in Ω. Grayscale images are typically displayed visually so that smaller values of f correspond to darker shades of gray (down to black) and higher values to lighter shades (up to white).

For natural images $f(x, y)$ would never be a simple function. Nonetheless, to illustrate let us consider the image defined by the function

$$\begin{aligned} f(x, y) = {} & 1.5\ \cos(2x)\cos(7y) + 0.75\ \cos(5x)\sin(3x) \\ & -1.3\ \sin(9x)\cos(15y) + 1.1\ \sin(13x)\sin(11y) \end{aligned}$$

on the domain $\Omega = \{(x, y); 0 \leq x \leq 1, 0 \leq y \leq 1\}$. The situation is illustrated in Figure 1.4. The plot on the left is a conventional $z = f(x, y)$ plot with the surface "gray-coded" according to height, where $f = -5$ corresponds to black and $f = 5$ to white. The plot on the right is the same surface but viewed from

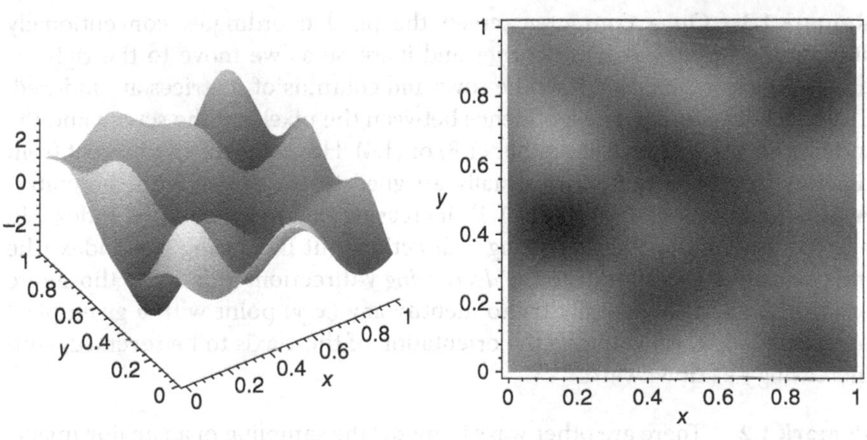

Figure 1.4 Grayscale image from two perspectives.

directly above, looking down the z axis. This is the actual grayscale image encoded by f according to the above scheme.

1.3.4 Sampling Images

As in the one-dimensional case, an analog image must be sampled prior to storage or processing in a computer. The simplest model to adopt is the discrete model obtained by sampling the intensity function $f(x, y)$ on a regular grid of points (x_s, y_r) in the plane. For each point (x_s, y_r) the value of $f(x_s, y_r)$ is the "graylevel" or intensity at that location. The values $f(x_s, y_r)$ are collected into a $m \times n$ matrix $\mathbf{A}$ with entries a_{rs} given by

$$a_{rs} = f(x_s, y_r). \tag{1.2}$$

If you are wondering why it is $f(x_s, y_r)$ instead of $f(x_r, y_s)$, see Remark 1.1.

The sampling points (x_s, y_r) can be chosen in many ways. One approach is as follows: subdivide the rectangle Ω into mn identical subrectangles, with m equal vertical (y) subdivisions and n equal horizontal (x) subdivisions, of length $\Delta x = (b - a)/n$ and height $\Delta y = (d - c)/m$. We may take the points (x_s, y_r) as the centers of these subrectangles so that

$$(x_s, y_r) = \left(a + \left(s - \frac{1}{2}\right)\Delta x, d - \left(r - \frac{1}{2}\right)\Delta y\right), \tag{1.3}$$

or alternatively as the lower right corners of these rectangles,

$$(x_s, y_r) = (a + s\Delta x, d - r\Delta y), \tag{1.4}$$

where in either case $1 \leq r \leq m, 1 \leq s \leq n$.

Note that in either case $x_1 < x_2 < \cdots < x_n$ and $y_1 > y_2 > \cdots > y_m$. Using the centers seems more natural, although the method (1.4) is a bit cleaner mathematically. For images of reasonable size, however, it will make little difference.

Remark 1.1 On a computer screen the pixel coordinates conventionally start in the upper left-hand corner and increase as we move to the right or down, which is precisely how the rows and columns of matrices are indexed. This yields a natural correspondence between the pixels on the screen and the indexes for the matrix $\mathbf{A}$ for either (1.3) or (1.4). However, this is different from the way that coordinates are usually assigned to the plane: with the matrix $\mathbf{A}$ as defined by either (1.3) or (1.4), increasing column index (the index s in a_{rs}) corresponds to the increasing x-direction, but increasing row index (the index r in a_{rs}) corresponds to the *decreasing* y-direction. Indeed, on those rare occasions when we actually try to identify any (x, y) point with a given pixel or matrix index, we will take the orientation of the y axis to be reversed, with increasing y as downward.

Remark 1.2 There are other ways to model the sampling of an analog image. For example, we may take a_{rs} as some kind of integral or weighted average of f near the point (x_s, y_r). These approaches can more accurately model the physical process of sampling an analog image, but the function evaluation model in equation (1.2) has reasonable accuracy and is a simple conceptual model. For almost all of our work in this text, we will assume that the sampling has been done and the input image matrix or signal is already in hand.

Remark 1.3 The values of m and n are often decided by the application in mind, or perhaps storage restrictions. It is useful and commonplace to have both m and n to be divisible by some high power of 2.

1.3.5 Color

There are a variety of approaches to modeling color images. One of the simplest is the "RGB" model in which a color image is described using three functions $r(x, y)$, $g(x, y)$, and $b(x, y)$, appropriately scaled, that correspond to the intensities of these three additive primary colors at the point (x, y) in the image domain. For example, if the color components are each scaled in the range 0–1, then $r = g = b = 1$ (equal amounts of all colors, full intensity) at a given point in the image would correspond to pure white, while $r = g = b = 0$ is black. The choice $r = 1, g = b = 0$ would be pure red, and so on. See [23] for a general discussion of the theory of color perception and other models of color such as HSI (hue, saturation, and intensity) and CMY (cyan, magenta, and yellow) models.

For simplicity's sake, we are only going to consider grayscale and RGB models, given that computer screens are based on RGB. In fact, we will be working almost exclusively with grayscale images in order to keep the discussion simple and focused on the mathematical essentials. An example where the CMY model needs to be considered is in color laser printers that use CMY toner. The printer software automatically makes the translation from RGB to CMY. It is worth noting that the actual JPEG compression standard specifies color images with a slightly different scheme, the luminance–chrominance or "YCbCr" scheme. Images can easily be converted back and forth from this scheme to RGB.

When we consider RGB images, we will assume the sampling has already been done at points (x_s, y_r) as described above for grayscale images. In the sampled image at a given pixel location on the display device the three colors are mixed according to the intensities $r(x_s, y_r)$, $g(x_s, y_r)$, and $b(x_s, y_r)$ to produce the desired color. Thus, a sampled m by n pixel image consists of three m by n arrays, one array for each color component.

1.3.6 Quantization and Noise for Images

Just as for one-dimensional signals, quantization error is introduced when an image is digitized. In general, we will structure our grayscale images so that each pixel is assigned an integer value from 0 to 255 (2^8 values) and displayed with 0 as black, 255 as white, and intermediate values as shades of gray. Thus,

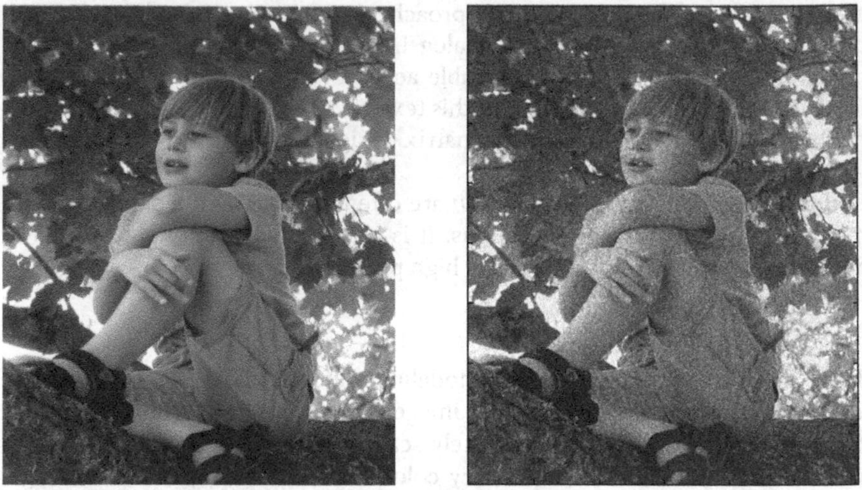

(a) (b)

Figure 1.5 Image without (a) and with additive noise (b).

the range is quantized with 8-bit precision. Similarly each color component in an RGB image will be assigned value in the 0–255 range, so each pixel needs three bytes to determine its color. Some applications require more than 8-bit quantization. For example, medical images are often 12-bit grayscale, offering 4096 shades of gray.

Like one-dimensional signals, images may have noise. For example, let $\mathbf{T}$ be the matrix with entries t_{sr} representing the image on the Figure 1.5a and let $\mathbf{A}$ be the matrix with entries a_{sr} representing the noisy image, shown on the panel (b). Analogous to audio signals we can posit an additive noise model

$$a_{sr} = t_{sr} + \epsilon_{sr}, \tag{1.5}$$

where $\mathbf{E}$ has entries ϵ_{sr} and represents the noise. The visual effect is to give the image a kind of "grainy" appearance.

1.4 Vector Space Models for Signals and Images

We now develop a natural mathematical framework for signal and image analysis. At the core of this framework lies the concept of a *vector space.*

Definition 1.1 A vector space over the real numbers $\mathbb{R}$ is a set V with two operations, *vector addition* and *scalar multiplication,* with the properties that

1. *for all vectors* $\mathbf{u}, \mathbf{v} \in V$ *the vector sum* $\mathbf{u} + \mathbf{v}$ *is defined and lies in* V *(closure under addition);*

2. *for all* $\mathbf{u} \in V$ *and scalars* $a \in \mathbb{R}$ *the scalar multiple* $a\mathbf{u}$ *is defined and lies in* V *(closure under scalar multiplication);*
3. *the "familiar" rules of arithmetic apply, specifically, for all scalars* a, b *and* $\mathbf{u}, \mathbf{v}, \mathbf{w} \in V$:
   (a) $\mathbf{u} + \mathbf{v} = \mathbf{v} + \mathbf{u}$, *(addition is commutative),*
   (b) $(\mathbf{u} + \mathbf{v}) + \mathbf{w} = \mathbf{u} + (\mathbf{v} + \mathbf{w})$ *(addition is associative),*
   (c) *there is a "zero vector"* $\mathbf{0}$ *such that* $\mathbf{u} + \mathbf{0} = \mathbf{0} + \mathbf{u} = \mathbf{u}$ *(additive identity),*
   (d) *for each* $\mathbf{u} \in V$ *there is an additive inverse vector* $\mathbf{w}$ *such that* $\mathbf{u} + \mathbf{w} = \mathbf{0}$; *we conventionally write* $-\mathbf{u}$ *for the additive inverse of* $\mathbf{u}$,
   (e) $(ab)\mathbf{u} = a(b\mathbf{u})$,
   (f) $(a + b)\mathbf{u} = a\mathbf{u} + b\mathbf{u}$,
   (g) $a(\mathbf{u} + \mathbf{v}) = a\mathbf{u} + a\mathbf{v}$,
   (h) $1\mathbf{u} = \mathbf{u}$.

If we replace $\mathbb{R}$ above by the field of complex numbers $\mathbb{C}$, then we obtain the definition of a *vector space over the complex numbers.*

We will also make frequent use of *subspaces*:

Definition 1.2 A nonempty subset W of a vector space V is called a *"subspace" of* V if W is itself closed under addition and scalar multiplication (as defined for V).

Let us look at a few examples of vector spaces and subspaces, especially those useful in signal and image processing.

1.4.1 Examples—Discrete Spaces

We will first consider examples appropriate for sampled signals or images.

Example 1.1 *The vector space* $\mathbb{R}^N$ *consists of vectors* $\mathbf{x}$ *of the form*

$$\mathbf{x} = (x_1, x_2, \ldots, x_N), \tag{1.6}$$

where the x_k *are all real numbers. Vector addition and scalar multiplication are defined component by component as*

$$\mathbf{x} + \mathbf{y} = (x_1 + y_1, x_2 + y_2, \ldots, x_N + y_N), \quad c\mathbf{x} = (cx_1, cx_2, \ldots, cx_N),$$

where $\mathbf{y} = (y_1, y_2, \ldots, y_N)$ *and* $c \in \mathbb{R}$. *The space* $\mathbb{R}^N$ *is appropriate when we work with sampled audio or other one-dimensional signals. If we allow the* x_k *in (1.6) and scalar* c *to be complex numbers, then we obtain the vector space* $\mathbb{C}^N$. *That* $\mathbb{R}^N$ *or* $\mathbb{C}^N$ *satisfy the properties of a vector space (with addition and scalar multiplication as defined) follows easily, with zero vector* $\mathbf{0} = (0, 0, \ldots, 0)$ *and additive inverse* $(-x_1, -x_2, \ldots, -x_n)$ *for any vector* $\mathbf{x}$.

As we will see, use of the space $\mathbb{C}^N$ *can simplify much analysis, even when the signals we work with are real-valued.*

Remark 1.4 *Warning*: In later work we will almost always find it convenient to index the components of vectors in $\mathbb{R}^N$ or $\mathbb{C}^N$ starting with index 0, that is, as $\mathbf{x} = (x_0, x_1, \ldots, x_{N-1})$, rather than the more traditional range 1 to N.

Example 1.2 *The sets $M_{m,n}(\mathbb{R})$ or $M_{m,n}(\mathbb{C})$, $m \times n$ matrices with real or complex entries, respectively, form vector spaces. Addition is defined in the usual way for matrices, entry by entry, as is multiplication by scalars. The vector $\mathbf{0}$ is just the matrix with all zero entries, and the additive inverse for a matrix $\mathbf{M}$ with entries m_{jk} is the matrix with entries $-m_{jk}$. Any multiplicative properties of matrices are irrelevant in this context. On closer examination it should be clear that these vector spaces are nothing more than $\mathbb{R}^{mn}$ or $\mathbb{C}^{mn}$, but spaces where we choose to display the "vectors" as m rows of n components rather than a single row or column with mn entries.*

The vector space $M_{m,n}(\mathbb{R})$ is an appropriate model for the discretization of images on a rectangle. As in the one-dimensional case, analysis of images is often facilitated by viewing them as members of space $M_{m,n}(\mathbb{C})$.

Example 1.3 *On occasion it is useful to think of an analog signal $f(t)$ as beginning at some time $t = a$ and continuing "indefinitely." If we sample such a signal at intervals of Δt starting at time $t = a$ without stopping, we obtain a vector*

$$\mathbf{x} = (x_0, x_1, x_2, \ldots) \tag{1.7}$$

with real components $x_k = f(a + k\Delta t)$, $k \geq 0$. Given another vector $\mathbf{y} = (y_0, y_1, y_2, \ldots)$, we define vector addition and scalar multiplication as

$$c\mathbf{x} = (cx_0, cx_1, cx_2, \ldots), \quad \mathbf{x} + \mathbf{y} = (x_0 + y_0, x_1 + y_1, \ldots).$$

Let V denote the resulting set with these operations. It is an easy algebra problem to verify that V is a vector space over the real numbers with zero vector $\mathbf{0} = (0, 0, 0, \ldots)$; the additive inverse of $\mathbf{x}$ above is $(-x_0, -x_1, -x_2, \ldots)$. And though it may seem painfully obvious, to say that "$\mathbf{x} = \mathbf{y}$" in V means precisely that $x_k = y_k$ for each $k \geq 0$. We will later encounter vector spaces where we have to be quite careful about what is meant by "$\mathbf{x} = \mathbf{y}$."

A simple variant of this vector space is the bi-infinite space of vectors

$$\mathbf{x} = (\ldots, x_{-2}, x_{-1}, x_0, x_1, x_2, \ldots) \tag{1.8}$$

with the analogous vector space structure. A space like this would be appropriate for modeling a physical process with a past and future of indefinite duration.

Example 1.4 *As defined, the set V in the previous example lacks sufficient structure for the kinds of analysis we usually want to do, so we typically impose additional conditions on the components of $\mathbf{x}$. For example, let us impose the additional condition that for each $\mathbf{x}$ as defined in equation (1.7) there is some number M (which may depend on $\mathbf{x}$) such that $|x_k| \leq M$ for all $k \geq 0$.*

In this case, the resulting set (with addition and scalar multiplication as defined above for V) is a vector space called $L^\infty(\mathbb{N})$ *(here* $\mathbb{N} = \{0, 1, 2, \ldots\}$ *denotes the set of natural numbers), or often just* ℓ^∞*. This would be an appropriate space for analyzing the class of sampled signals in which the magnitude of any particular signal remains bounded for all* $t \geq 0$.

The verification that $L^\infty(\mathbb{N})$ *is a vector space over* $\mathbb{R}$ *is fairly straightforward. The algebraic properties of item 3 in Definition 1.1 are verified exactly as for V in the previous example, where again the zero vector is* $(0, 0, 0, \ldots)$ *and the additive inverse of* $\mathbf{x}$ *is* $(-x_0, -x_1, -x_2, \ldots)$*. To show closure under vector addition, consider vectors* $\mathbf{x}$ *and* $\mathbf{y}$ *with* $|x_k| \leq M_x$ *and* $|y_k| \leq M_y$ *for all* $k \geq 0$*. From the triangle inequality for real numbers*

$$|x_k + y_k| \leq |x_k| + |y_k| \leq M_x + M_y,$$

so the components of $\mathbf{x} + \mathbf{y}$ *are bounded in magnitude by* $M_x + M_y$*. Thus,* $\mathbf{x} + \mathbf{y} \in L^\infty(\mathbb{N})$*, and the set is closed under addition. Similarly for any k the kth component* cx_k *of* $c\mathbf{x}$ *is bounded by* $|c|M_x$*, and the set is closed under scalar multiplication. This makes* $L^\infty(\mathbb{N})$ *a subspace of the vector space V from the previous example.*

If we consider bi-infinite vectors as defined in equation (1.8) with the condition that for each $\mathbf{x}$ *there is some number M such that* $|x_k| \leq M$ *for all* $k \in \mathbb{Z}$*, then we obtain the vector space* $L^\infty(\mathbb{Z})$.

Example 1.5 *We may impose the condition that for each sequence of real numbers* $\mathbf{x}$ *of the form in (1.7) we have*

$$\sum_{k=0}^{\infty} |x_k|^2 < \infty, \tag{1.9}$$

in which case the resulting set is called $L^2(\mathbb{N})$*, or often just* ℓ^2*. This is even more stringent than the condition for* $L^\infty(\mathbb{N})$*; verification of this assertion and that* $L^2(\mathbb{N})$ *is a vector space is left for Exercise 12. We may also let the components* x_k *be complex numbers, and the result is still a vector space.*

Conditions like (1.9) that bound the "squared value" of some object are common in applied mathematics and usually correspond to finite energy in an underlying physical process.

A very common variant of $L^2(\mathbb{N})$ *is the space* $L^2(\mathbb{Z})$*, consisting of vectors of the form in equation (1.8) that satisfy*

$$\sum_{k=-\infty}^{\infty} |x_k|^2 < \infty.$$

The space $L^2(\mathbb{Z})$ *will play an important role throughout Chapters 7 and 9.*

Variations on the spaces above are possible, and common. Which vector space we work in depends on our model of the underlying physical process and the analysis we hope to carry out.

1.4.2 Examples—Function Spaces

In the above examples, the spaces all consist of vectors that are lists or arrays, finite or infinite, of real or complex numbers. Functions can also be interpreted as elements of vector spaces, and this is the appropriate setting when dealing with analog signals or images. The mathematics in this case can be more complicated, especially when dealing with issues concerning approximation, limits, and convergence (about which we have said little so far). We will have limited need to work in this setting, at least until Chapter 9. Here are some relevant examples.

Example 1.6 *Consider the set of all real-valued functions f that are defined and continuous at every point in a closed interval $[a, b]$ of the real line. This means that for any $t_0 \in [a, b]$,*

$$\lim_{t \to t_0} f(t) = f(t_0),$$

where t approaches from the right only in the case that $t_0 = a$ and from the left only in the case that $t_0 = b$. The sum $f + g$ of two functions f and g is the function defined by $(f + g)(t) = f(t) + g(t)$, and the scalar multiple cf is defined via $(cf)(t) = cf(t)$. With these operations this is a vector space over $\mathbb{R}$, for it is closed under addition since the sum of two continuous functions is continuous. It is also closed under scalar multiplication, since a scalar multiple of a continuous function is continuous. The algebraic properties of item 3 in Definition 1.1 are easily verified with the "zero function" as the additive identity and $-f$ as the additive inverse of f. The resulting space is denoted $C[a, b]$.

The closed interval $[a, b]$ can be replaced by the open interval (a, b) to obtain the vector space $C(a, b)$. The spaces $C[a, b]$ and $C(a, b)$ do not coincide, for example, $f(t) = 1/t$ lies in $C(0, 1)$ but not $C[0, 1]$. In this case, $1/t$ is not defined at $t = 0$, and moreover this function cannot even be extended to $t = 0$ in a continuous manner.

Example 1.7 *Consider the set of all real-valued functions f that are piecewise continuous on the interval $[a, b]$; that is, f is defined and continuous at all but finitely many points in $[a, b]$. With addition and scalar multiplication as defined in the last example this is a vector space over $\mathbb{R}$. The requisite algebraic properties are verified in precisely the same manner. To show closure under addition, just note that any point of discontinuity for $f + g$ must be a point of discontinuity for f or g; hence $f + g$ can have only finitely many points of discontinuity. The discontinuities for cf are precisely those for f.*

Both $C(a, b)$ and $C[a, b]$ are subspaces of this vector space (which does not have any standard name).

Example 1.8 *Let V denote those functions f in C(a, b) for which*

$$\int_a^b f^2(t)\,dt < \infty. \tag{1.10}$$

A function f that is continuous on (a, b) can have no vertical asymptotes in the open interval, but may be unbounded as t approaches the endpoint t = a or t = b. Thus, the integral above (and all integrals in this example) should be interpreted as improper integrals, that is,

$$\int_a^b f^2(t)\,dt = \lim_{p\to a^+}\int_p^r f^2(t)\,dt + \lim_{q\to b^-}\int_r^b f^2(t)\,dt,$$

where r is any point in (a, b).

To show that V is closed under scalar multiplication, note that

$$\int_a^b (cf)^2(t)\,dt = c^2\int_a^b f^2(t)\,dt < \infty,$$

since f satisfies the inequality (1.10). To show closure under vector addition, first note that for any real numbers p and q, we have $(p+q)^2 \le 2p^2 + 2q^2$ *(this follows easily from* $0 \le (p-q)^2$*). As a consequence, for any two functions f and g in V and any* $t \in (a, b)$*, we have*

$$(f(t)+g(t))^2 \le 2f^2(t) + 2g^2(t).$$

Integrate both sides above from t = a to t = b (as improper integrals) to obtain

$$\int_a^b (f(t)+g(t))^2\,dt \le 2\int_a^b f^2(t)\,dt + 2\int_a^b g^2(t)\,dt < \infty,$$

so f + g is in V. The algebraic properties in Definition 1.1 follow as before, so that V is a vector space over $\mathbb{R}$*.*

The space V as defined above does not have any standard name, but it is "almost" the vector space commonly termed $L^2(a, b)$*, also called "the space of square integrable functions on (a, b)." More precisely, the space defined above is the intersection* $C(a, b) \cap L^2(a, b)$*. Nonetheless, we will generally refer to it as "*$L^2(a, b)$*," and say more about it in Section 1.10.5. This space will be discussed more in the text, especially in Chapter 9.*

Similar to the inequality (1.9), the condition (1.10) comes up fairly often in applied mathematics and usually corresponds to signals of finite energy.

Example 1.9 *Consider the set of functions f(x, y) defined on some rectangular region* $\Omega = \{(x, y); a \le x \le b, c \le y \le d\}$*. We make no particular assumptions about the continuity or other nature of the functions. Addition and scalar multiplication are defined in the usual way, as* $(f+g)(x, y) = f(x, y) + g(x, y)$ *and* $(cf)(x, y) = cf(x, y)$*. This is a vector space over* $\mathbb{R}$*. The proof is in fact the same as in the case of functions of a single variable. This space would be useful* *for image analysis, with the functions representing graylevel intensities and* Ω *the image domain.*

Of course, we can narrow the class of functions, for example, by considering only those that are continuous on Ω*; this space is denoted* $C(\Omega)$*. Or we can impose the further restriction that*

$$\int_a^b \int_c^d f^2(x,y)\,dy\,dx < \infty,$$

which, in analogy to the one-dimensional case, we denote by $L^2(\Omega)$*. There are many other important and potentially useful vector spaces of functions.*

We could also choose the domain Ω *to be infinite, for example, a half-plane or the whole plane. The region is selected to give a good tractable vector space model and to be relevant to the physical situation of interest, though unbounded domains are not generally necessary in image processing.*

In addition to the eight basic arithmetic properties listed in Definition 1.1, certain other arithmetic properties of vector spaces are worth noting.

Proposition 1.1 *If* V *is a vector space over* $\mathbb{R}$ *or* $\mathbb{C}$*, then*

1. the vector $\mathbf{0}$ is unique;
2. $0\mathbf{u} = \mathbf{0}$ for any vector $\mathbf{u}$;
3. the additive inverse of any vector $\mathbf{u}$ is unique, and is given by $(-1)\mathbf{u}$.

These properties look rather obvious and are usually easy to verify in any specific vector space as in Examples 1.1–1.9. But in fact they must hold in any vector space, and can be shown directly from the eight arithmetic properties for a vector space. The careful proofs can be a bit tricky though! (see Exercise 13).

We have already started to use the additive vector space structure when we modeled noise in signals and images with equations (1.1) and (1.5). The vector space structure will be indispensable when we discuss the decomposition of signals and images into linear combinations of more "basic" components.

Tables 1.1 and 1.2 give a brief summary of some of the important spaces of interest, as well as when each space might be used. As mentioned, the analog models are used when we consider the actual physical processes that underlie signals and images, but for computation we always consider the discrete version.

1.5 Basic Waveforms—The Analog Case

1.5.1 The One-Dimensional Waveforms

To analyze signals and images, it can be extremely useful to decompose them into a sum of more elementary pieces or patterns, and then operate on the

Table 1.1 Discrete signal models and uses.

Notation	Vector space description
$\mathbb{R}^N$	$\{\mathbf{x} = (x_1, \ldots, x_N) : x_i \in \mathbb{R}\}$, finite sampled signals
$\mathbb{C}^N$	$\{\mathbf{x} = (x_1, \ldots, x_N) : x_i \in \mathbb{C}\}$, analysis of sampled signals
$L^\infty(\mathbb{N})$ or ℓ^∞	$\{\mathbf{x} = (x_0, x_1, \ldots) : x_i \in \mathbb{R}, \lvert x_i\rvert \leq M$ for all $i \geq 0\}$ bounded, sampled signals, infinite time
$L^2(\mathbb{N})$ or ℓ^2	$\{\mathbf{x} = (x_0, x_1, \ldots) : x_i \in \mathbb{R}$ or $x_i \in \mathbb{C}, \sum_k \lvert x_k\rvert^2 < \infty\}$ sampled signals, finite energy, infinite time
$L^2(\mathbb{Z})$	$\{\mathbf{x} = (\ldots, x_{-1}, x_0, x_1, \ldots) : x_i \in \mathbb{R}$ or $x_i \in \mathbb{C}, \sum_k \lvert x_k\rvert^2 < \infty\}$ sampled signals, finite energy, bi-infinite time
$M_{m,n}(\mathbb{R})$	Real $m \times n$ matrices, sampled rectangular images
$M_{m,n}(\mathbb{C})$	Complex $m \times n$ matrices, analysis of images

Table 1.2 Analog signal models and uses.

Notation	Vector space description
$C(a,b)$ or $C[a,b]$	Continuous functions on (a,b) or $[a,b]$, continuous analog signal
$L^2(a,b)$	f Riemann integrable and $\int_a^b f^2(x)\,dx < \infty$, analog signal with finite energy
$L^2(\Omega)$ $(\Omega = [a,b] \times [c,d])$	f Riemann integrable and $\int_a^b \int_c^d f^2(x,y)\,dy\,dx < \infty$, analog image

decomposed version, piece by piece. We will call these simpler pieces the *basic waveforms*. They serve as the essential building blocks for signals and images. In the context of Fourier analysis of analog signals, these basic waveforms are simply sines and cosines, or equivalently, complex exponentials. Specifically, the two basic waveforms of interest are $\cos(\omega t)$ and $\sin(\omega t)$, or their complex exponential equivalent $e^{i\omega t}$, where ω acts as a frequency parameter.

The complex exponential basic waveforms will be our preferred approach. Recall Euler's identity,

$$e^{i\theta} = \cos(\theta) + i\sin(\theta). \tag{1.11}$$

From this we have (with $\theta = \omega t$ and $\theta = -\omega t$)

$$\begin{aligned} e^{i\omega t} &= \cos(\omega t) + i\ \sin(\omega t), \\ e^{-i\omega t} &= \cos(\omega t) - i\ \sin(\omega t), \end{aligned} \tag{1.12}$$

which can be solved for $\cos(\omega t)$ and $\sin(\omega t)$ as

$$\begin{aligned}\cos(\omega t) &= \frac{e^{i\omega t} + e^{-i\omega t}}{2},\\ \sin(\omega t) &= \frac{e^{i\omega t} - e^{-i\omega t}}{2i}.\end{aligned} \tag{1.13}$$

If we can decompose a given signal $x(t)$ into a linear combination of waveforms $\cos(\omega t)$ and $\sin(\omega t)$, then equation (1.13) makes it clear that we can also decompose $x(t)$ into a linear combination of appropriate complex exponentials. Similarly equation (1.12) can be used to convert any complex exponential decomposition into sines and cosines. Thus, we also consider the complex exponential functions $e^{i\omega t}$ as basic waveforms.

Remark 1.5 In the real-valued sine/cosine case we only need to work with $\omega \geq 0$, since $\cos(-\omega t) = \cos(\omega t)$ and $\sin(-\omega t) = -\sin(\omega t)$. Any function that can be constructed as a sum using negative values of ω has an equivalent expression with positive ω.

Example 1.10 *Consider the signal* $x(t) = \sin(t) + 3\sin(-2t) - 2\cos(-5t)$. *From Remark 1.5 we can express* $x(t)$ *as* $x(t) = \sin(t) - 3\sin(2t) - 2\cos(5t)$, *using only positive values of* ω *in the expressions* $\sin(\omega t)$ *and* $\cos(\omega t)$. *Equation (1.13) also yields*

$$x(t) = \frac{1}{2i}e^{it} - \frac{1}{2i}e^{-it} - \frac{3}{2i}e^{2it} + \frac{3}{2i}e^{-2it} - e^{5it} - e^{-5it},$$

a sum of basic complex exponential waveforms. Whether we work in trigonometric functions or complex exponentials matters little from the mathematical perspective. The trigonometric functions, because they are real-valued and familiar, have a natural appeal, but the complex exponentials often yield much cleaner mathematical formulas. As such, we will usually prefer to work with the complex exponential waveforms.

We can visualize $e^{i\omega t}$ by simultaneously graphing the real and imaginary parts as functions of t, as in Figure 1.6, with real parts solid and imaginary parts dashed. Note that

$$\begin{aligned}\cos(\omega t) &= \mathrm{Re}(e^{i\omega t}),\\ \sin(\omega t) &= \mathrm{Im}(e^{i\omega t}).\end{aligned} \tag{1.14}$$

Of course the real and imaginary parts, and $e^{i\omega t}$ itself, are periodic and ω controls the frequency of oscillation. The parameter ω is called the *radial frequency* of the waveform.

The period of $e^{i\omega t}$ can be found by considering those values of λ for which $e^{i\omega(t+\lambda)} = e^{i\omega t}$ for all t, which yields

$$e^{i\omega(t+\lambda)} = e^{i\omega t}e^{i\omega\lambda} = e^{i\omega t}(\cos(\omega\lambda) + i\sin(\omega\lambda)), \tag{1.15}$$

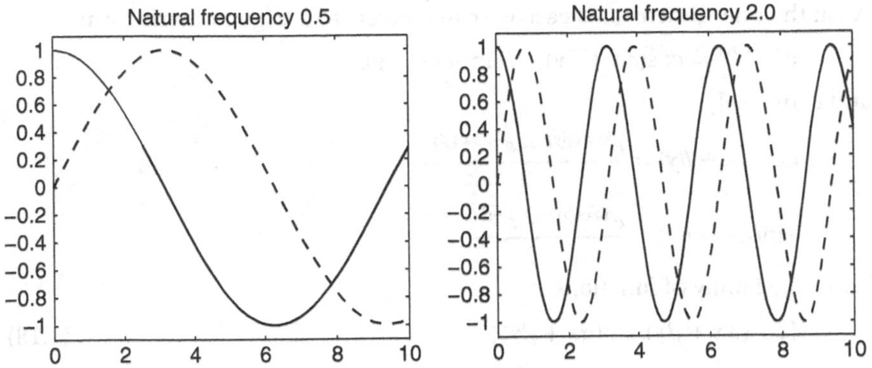

Figure 1.6 Real (*solid*) and imaginary (*dashed*) parts of complex exponentials.

so that $e^{i\omega(t+\lambda)} = e^{i\omega t}$ forces $\cos(\omega\lambda) + i\sin(\omega\lambda) = 1$. The smallest positive value of λ for which this holds satisfies $\lambda|\omega| = 2\pi$. Thus, $\lambda = 2\pi/|\omega|$, which is the *period* of $e^{i\omega t}$ (or the *wavelength* if t is a spatial variable).

The quantity $q = l/\lambda = \omega/2\pi$ (so $\omega = 2\pi q$) is the number of oscillations made by the waveform in a unit time interval and is called the *frequency* of the waveform. If t denotes time in seconds, then q has units of *hertz*, or cycles per second. It is often useful to write the basic waveform $e^{i\omega t}$ as $e^{2\pi iqt}$, to explicitly note the frequency q of oscillation. More precisely, the frequency (in Hz) of the waveform $e^{2\pi iqt}$ is $|q|$ hertz, since frequency is by convention nonnegative.

In real-valued terms, we can use $\cos(2\pi qt)$ and $\sin(2\pi qt)$ with $q \geq 0$ in place of $\cos(\omega t)$ and $\sin(\omega t)$ with $\omega \geq 0$.

As we will see later, any "reasonable" (e.g., bounded and piecewise continuous) function $x(t)$ defined on an interval $[-T, T]$ can be written as an infinite sum of basic waveforms $e^{i\omega t}$, as

$$x(t) = \sum_{k=-\infty}^{\infty} c_k e^{\pi ikt/T} \tag{1.16}$$

for an appropriate choice of the constants c_k. The radial frequency parameter ω assumes the values $\pi k/T$ for $k \in \mathbb{Z}$, or equivalently the frequency q assumes values $k/2T$. An expansion analogous to (1.16) also exists using the sine/cosine waveforms.

1.5.2 2D Basic Waveforms

The 2D basic waveforms are governed by a pair of frequency parameters α and β. Let (x, y) denote coordinates in the plane. The basic waveforms are products of complex exponentials and can be written in either additive form (left side) or a product form (right side),

$$e^{i(\alpha x+\beta y)} = e^{i\alpha x}e^{i\beta y}. \tag{1.17}$$

As in the one-dimensional case we can convert to a trigonometric form,

$$e^{i(\alpha x+\beta y)} = \cos(\alpha x + \beta y) + i\sin(\alpha x + \beta y),$$

and conversely,

$$\cos(\alpha x + \beta y) = \frac{e^{i(\alpha x+\beta y)} + e^{-i(\alpha x+\beta y)}}{2},$$

$$\sin(\alpha x + \beta y) = \frac{e^{i(\alpha x+\beta y)} - e^{-i(\alpha x+\beta y)}}{2i}.$$

Thus, the family of functions

$$\{\cos(\alpha x + \beta y), \sin(\alpha x + \beta y)\} \tag{1.18}$$

is an alternate set of basic waveforms.

Sometimes a third class of basic waveforms is useful. An application of Euler's formula to both exponentials on the right in equation (1.17) shows that

$$e^{i(\alpha x+\beta y)} = \cos(\alpha x)\cos(\beta y) - \sin(\alpha x)\sin(\beta y) + i(\cos(\alpha x)\sin(\beta y) + \sin(\alpha x)\cos(\beta y)),$$

so these complex exponential basic waveforms can be expanded into linear combinations of functions from the family

$$\{\cos(\alpha x)\cos(\beta y), \sin(\alpha x)\sin(\beta y), \cos(\alpha x)\sin(\beta y), \sin(\alpha x)\cos(\beta y)\}. \tag{1.19}$$

Conversely, each of these functions can be written in terms of complex exponentials (see Exercise 16). The functions in (1.19) also form a basic set of waveforms.

Just as in the one-dimensional case, for the real-valued basic waveforms (1.18) or (1.19) we can limit our attention to the cases $\alpha, \beta \geq 0$.

We almost always use the complex exponential waveforms in our analysis, however, except when graphing. As in the one-dimensional case these exponential waveforms can be written in a frequency format as

$$e^{i(\alpha x+\beta y)} = e^{2\pi i(px+qy)},$$

where p and q are frequencies in the x- and y-directions. In Figure 1.7, we show the real parts of the basic exponential waveforms for several values of p and q, as grayscale images on the unit square $0 \leq x, y \leq 1$, with y downward as per Remark 1.1. The waves seem to have a direction and wavelength (see Exercise 19).

1.6 Sampling and Aliasing

1.6.1 Introduction

As remarked prior to equation (1.16), an analog signal or function on an interval $[-T, T]$ can be decomposed into a linear combination of basic

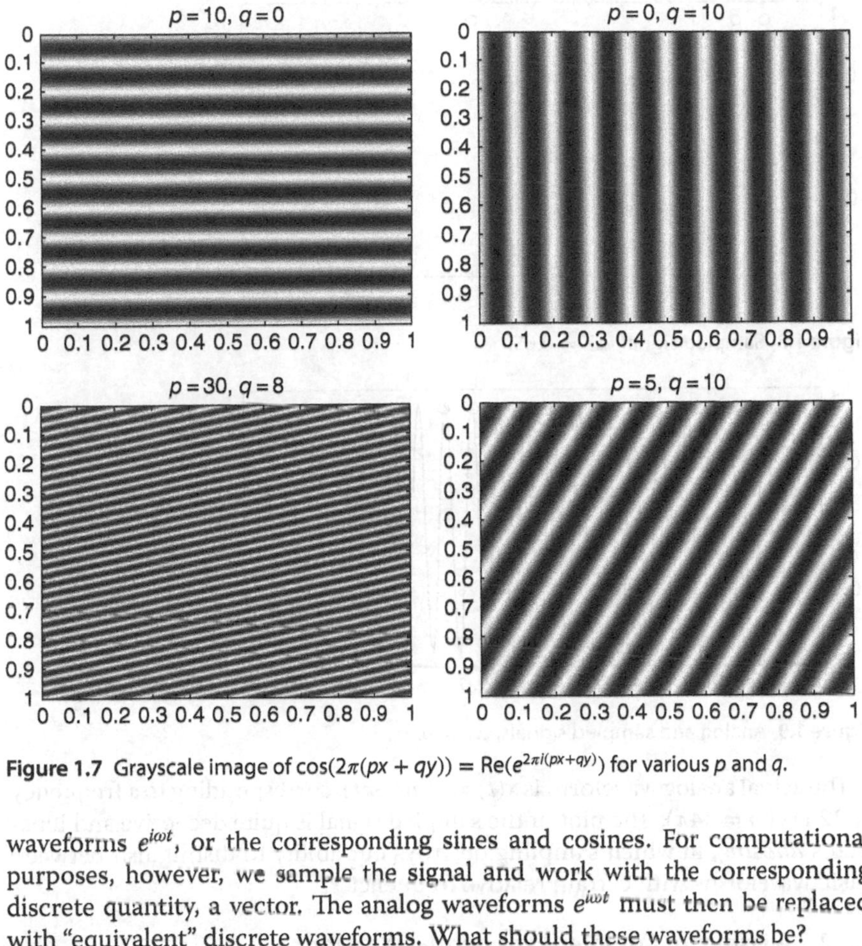

Figure 1.7 Grayscale image of $\cos(2\pi(px + qy)) = \mathrm{Re}(e^{2\pi i(px+qy)})$ for various p and q.

waveforms $e^{i\omega t}$, or the corresponding sines and cosines. For computational purposes, however, we sample the signal and work with the corresponding discrete quantity, a vector. The analog waveforms $e^{i\omega t}$ must then be replaced with "equivalent" discrete waveforms. What should these waveforms be?

The obvious answer is to use the sampled analog waveforms, but an interesting phenomenon called *aliasing* shows up. It should be intuitively clear that sampling destroys information about the analog signal. It is in the case where the sampling is done on basic waveforms that this loss of information is especially easy to quantify. Thus, we will take a short detour to discuss aliasing, and then proceed to the discrete model waveforms in the next section.

To illustrate aliasing, consider the sampled signal graphed in Figure 1.8, obtained by sampling the basic waveform $\sin(\omega t)$ for some "unknown" ω on the interval $0 \le t \le 1$ at intervals of $\Delta T = 0.05$. The sampled waveform appears to make exactly two full cycles in the time interval $[0, 1]$, corresponding to a frequency of 2 Hz and the basic waveform $\sin(4\pi t)$. However, Figure 1.9 shows a plot of the "true" analog signal, superimposed on the sampled signal!

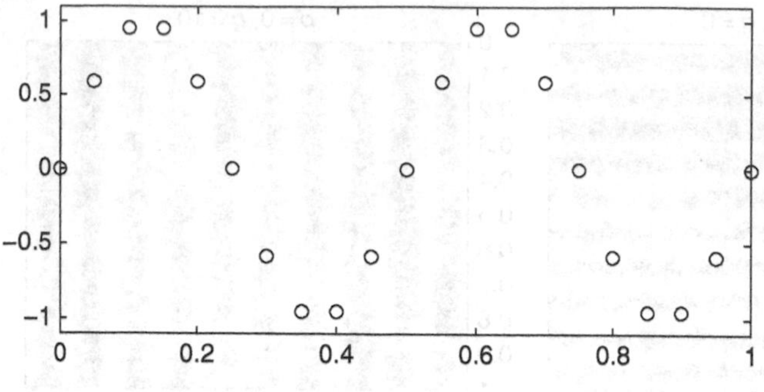

Figure 1.8 Sampled signal, $\Delta T = 0.05$.

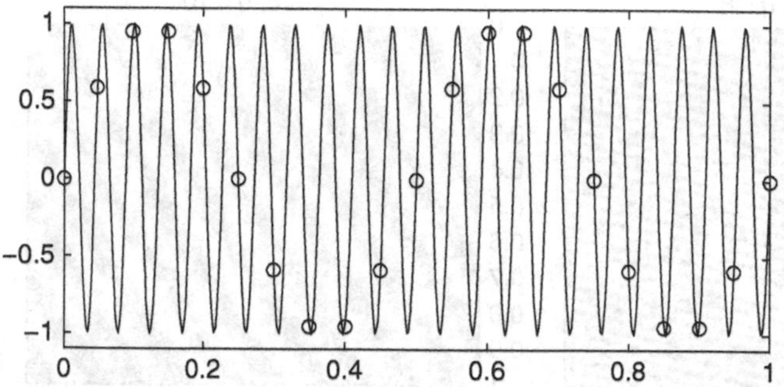

Figure 1.9 Analog and sampled signals, $\Delta T = 0.05$.

The actual analog waveform is $x(t) = \sin(44\pi t)$, corresponding to a frequency of 22 Hz ($\omega = 44\pi$). The plot of the sampled signal is quite deceptive and illustrates *aliasing*, in which sampling destroys our ability to distinguish between basic waveforms with certain relative frequencies.

1.6.2 Aliasing for Complex Exponential Waveforms

To quantify this phenomenon, let us first look at the situation for complex waveforms $e^{i\omega t}$. For simplicity, we write these in frequency form $e^{2\pi iqt}$ with $q = \omega/2\pi$, so q is in hertz; note q is not required to be an integer. Suppose that we sample such a waveform N times per second, at times $t = k/N$ for $k = 0, 1, 2, \ldots$. The sampled waveform yields values $e^{2\pi iqk/N}$. Under what circumstances will another analog waveform $e^{2\pi i\tilde{q}t}$ at frequency $\tilde{q}$ hertz yield the same sampled values at times $t = k/N$? Stated quantitatively, this means that

$$e^{2\pi iqk/N} = e^{2\pi i\tilde{q}k/N}$$

for all integers $k \geq 0$. Divide both sides above by $e^{2\pi iqk/N}$ to obtain $1 = e^{2\pi i(\tilde{q}-q)k/N}$, or equivalently,

$$1 = (e^{2\pi i(\tilde{q}-q)/N})^k \tag{1.20}$$

for all $k \geq 0$. Now if a complex number z satisfies $z^k = 1$ for all integers k then $z = 1$ (consider the case $k = 1$). From equation (1.20) we conclude that $e^{2\pi i(\tilde{q}-q)/N} = 1$. Since $e^x = 1$ only when $x = 2\pi i m$ where $m \in \mathbb{Z}$, it follows that $2\pi i(\tilde{q} - q)/N = 2\pi i m$, or $\tilde{q} - q = mN$. Thus, the waveforms $e^{2\pi i q t}$ and $e^{2\pi i \tilde{q} t}$ sampled at times $t = k/N$ yield identical values exactly when

$$\tilde{q} - q = mN \tag{1.21}$$

for some integer m.

Equation (1.21) quantifies the phenomenon of aliasing: When sampled with sampling interval $\Delta T = 1/N$ (frequency N Hz) the two waveforms $e^{2\pi i q t}$ and $e^{2\pi i \tilde{q} t}$ will be *aliased* (yield the same sampled values) whenever $\tilde{q}$ and q differ by any multiple of the sampling rate N. Equivalently, $e^{i\omega t}$ and $e^{i\tilde{\omega} t}$ yield exactly the same sampled values when $\tilde{\omega} - \omega = 2\pi m N$.

Aliasing has two implications, one "physical" and one "mathematical." The physical implication is that if an analog signal consists of a superposition of basic waveforms $e^{2\pi i q t}$ and is sampled at N samples per second, then for any particular frequency q_0 the waveforms

$$\ldots, e^{2\pi i(q_0-2N)t}, e^{2\pi i(q_0-N)t}, e^{2\pi i q_0 t}, e^{2\pi i(q_0+N)t}, e^{2\pi i(q_0+2N)t} \ldots$$

are all aliased. Any information concerning their individual characteristics (amplitudes and phases) is lost. The only exception is if we know a priori that the signal consists only of waveforms in a specific and sufficiently small frequency range. For example, if we know that the signal consists only of waveforms $e^{2\pi i q t}$ with $-N/2 < q \leq N/2$ (i.e., frequencies $|q|$ between 0 and $N/2$), then no aliasing will occur because $q \pm N, q \pm 2N$, and so on, do not lie in this range. This might be the case if the signal has been *low-pass filtered* prior to being sampled, to remove (by analog means) all frequencies greater than $N/2$. In this case, sampling at frequency N would produce no aliasing.

The mathematical implication of aliasing is this: when analyzing a signal sampled at frequency N, we need only use the sampled waveforms $e^{2\pi i q k/N}$ with $-N/2 < q \leq N/2$. Any discrete basic waveform with frequency outside this range is aliased with, and hence identical to, a basic waveform within this range.

1.6.3 Aliasing for Sines and Cosines

Similar considerations apply when using the sine/cosine waveforms. From equation (1.13) it is easy to see that when sampled at frequency N the functions $\sin(2\pi q t)$ or $\cos(2\pi q t)$ will be aliased with waveforms $\sin(2\pi \tilde{q} t)$ or $\cos(2\pi \tilde{q} t)$ if $\tilde{q} - q = mN$ for any integer m. Indeed, one can see directly that if $\tilde{q} = q + mN$, then

$$\sin(2\pi \tilde{q} k/N) = \sin(2\pi(q + mN)k/N) = \sin(2\pi q k/N + 2\pi k m) = \sin(2\pi q k/N),$$

since $\sin(t + 2\pi k m) = \sin(t)$ for any t, where k and m are integers. A similar computation holds for the cosine. Thus, as in the complex exponential case we may restrict our attention to a frequency interval in q of length N, for example, $-N/2 < q \leq N/2$.

However, in the case of the sine/cosine waveforms the range for q (or ω) can be narrowed a bit further. In the light of Remark 1.5 we need not consider $q < 0$ if our main interest is the decomposition of a signal into a superposition of sine or cosine waveforms, for $\cos(-2\pi qt) = \cos(2\pi qt)$ and $\sin(-2\pi qt) = -\sin(2\pi qt)$. In the case of the sine/cosine waveforms we need only consider the range $0 \leq q \leq N/2$. This is actually identical to the restriction for the complex exponential case, where the frequency $|q|$ of $e^{2\pi iqt}$ is restricted to $0 \leq |q| \leq N/2$.

1.6.4 The Nyquist Sampling Rate

For both complex exponential waveforms $e^{2\pi iqt}$ and the basic trigonometric waveforms $\sin(2\pi qt), \cos(2\pi qt)$, sampling at N samples per second results in frequencies greater than $N/2$ being aliased with frequencies between 0 and $N/2$. Thus, if an analog signal is known to contain only frequencies of magnitude F and lower, sampling at a frequency $N \geq 2F$ (so $F \leq N/2$) results in no aliasing. This is one form of what is called the *Nyquist sampling rate* or *Nyquist sampling criterion*: to avoid aliasing, we must sample at twice the highest frequency present in an analog signal. This means at least two samples per cycle for the highest frequency. One typically samples at a slightly greater rate to ensure greater fidelity. Thus, commercial CD's use a sample rate of 44.1 kHz, so that 44.1/2 kHz, or 22.05 kHz, is slightly greater than the generally accepted maximum audible frequency of 20 kHz. A CD for dogs would require a higher sampling rate!

A closely related result, the Shannon sampling theorem, states that if an analog signal $x(t)$ contains only frequencies in the range 0 to $N/2$ and is sampled at sampling rate N, then $x(t)$ can be perfectly recovered for all t [19 , p. 87].

1.6.5 Aliasing in Images

Aliasing also occurs when images are sampled. Consider the simple grayscale image embodied by the function

$$f(x, y) = 128 + 128\sin(2\pi(50x + 70y))$$

on the domain $0 \leq x, y \leq 1$. In Figure 1.10 are images based on sampling f on n by n grids for $n = 60, 100, 300$, and 1000, and displayed with 0 as black, 255 as white (rounded down). To avoid aliasing, we expect to need a sampling frequency of at least 100 samples per unit distance in the x-direction, 140 in the y-direction. The 1000 by 1000 image comes closest to the "true" analog image, while the $n = 60$ and $n = 100$ images completely misrepresent the nature of the underlying analog signal. When $n = 60$, the stripes actually go the wrong way.

In general, however, aliasing in images is difficult to convey consistently via the printed page, or even on a computer screen, because the effect is

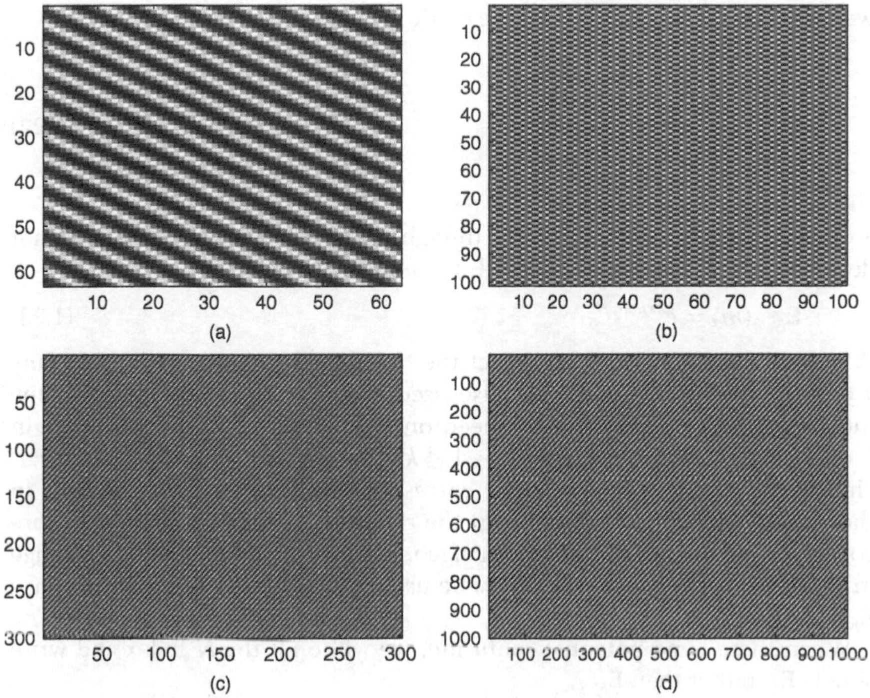

Figure 1.10 Aliasing in a 2D region, $n = 60$ (a), $n = 100$ (b), $n = 300$ (c), and $n = 1000$ (d).

highly dependent on printer and screen resolution. See Section 1.11, in which you can construct your own aliasing examples in Matlab, as well as audio examples.

1.7 Basic Waveforms—The Discrete Case

1.7.1 Discrete Basic Waveforms for Finite Signals

Consider a continuous signal $x(t)$ defined on a time interval $[0, T]$, sampled at the N times $t = nT/N$ for $n = 0, 1, 2, \dots, N-1$; note we do not sample $x(t)$ at $t = T$. This yields discretized signal $\mathbf{x} = (x_0, x_1, \dots, x_{N-1})$, where $x_n = nT/N$, a vector in $\mathbb{R}^N$. In all that follows we will index vectors in $\mathbb{R}^N$ from index 0 to index $N-1$, as per Remark 1.4.

As we show in a later section, the analog signal $x(t)$ can be decomposed into an infinite linear combination of basic analog waveforms, in this case of the form $e^{2\pi ikt/T}$ for $k \in \mathbb{Z}$. As discussed in the previous section, the appropriate basic waveforms are then the discretized versions of the waveforms $e^{2\pi ikt/T}$, obtained by sampling at times $t = nT/N$. This yields a sequence of basic waveform vectors, which we denote by $\mathbf{E}_{N,k}$, indexed by k, of the form

$$\mathbf{E}_{N,k} = \begin{bmatrix} e^{2\pi ik0/N} \\ e^{2\pi ik1/N} \\ \vdots \\ e^{2\pi ik(N-1)/N} \end{bmatrix}, \tag{1.22}$$

a discrete version of $e^{2\pi ikt/T}$. Note though that the waveform vectors do not depend on T. The mth component $\mathbf{E}_{N,k}(m)$ of $\mathbf{E}_{N,k}$ is given by

$$\mathbf{E}_{N,k}(m) = e^{2\pi ikm/N}. \tag{1.23}$$

For any fixed N we can construct the basic waveform vector $\mathbf{E}_{N,k}$ for any $k \in \mathbb{Z}$, but as shown when we discussed aliasing, $\mathbf{E}_{N,k} = \mathbf{E}_{N,k+mN}$ for any integer m. As a consequence we need only consider the $\mathbf{E}_{N,k}$ for a range in k of length N, say of the form $k_0 + 1 \le k \le k_0 + N$ for some k_0. A "natural" choice is $k_0 = -N/2$ (if N is even) corresponding to $-N/2 < k \le N/2$ as in the previous aliasing discussion, but the range $0 \le k \le N-1$ is usually more convenient for indexing and matrix algebra. However, no matter which range in k we use to index, we will always be using the same set of N vectors since $\mathbf{E}_{N,k} = \mathbf{E}_{N,k-N}$.

When there is no potential confusion, we will omit the N index and write simply $\mathbf{E}_k$, rather than $\mathbf{E}_{N,k}$.

Example 1.11 *As an illustration, here are the vectors $\mathbf{E}_{4,k}$ for $k = -2$ to $k = 3$:*

$$\mathbf{E}_{4,-2} = \begin{bmatrix} 1 \\ -1 \\ 1 \\ -1 \end{bmatrix}, \quad \mathbf{E}_{4,-1} = \begin{bmatrix} 1 \\ -i \\ -1 \\ i \end{bmatrix}, \quad \mathbf{E}_{4,0} = \begin{bmatrix} 1 \\ 1 \\ 1 \\ 1 \end{bmatrix},$$

$$\mathbf{E}_{4,1} = \begin{bmatrix} 1 \\ i \\ -1 \\ -i \end{bmatrix}, \quad \mathbf{E}_{4,2} = \begin{bmatrix} 1 \\ -1 \\ 1 \\ -1 \end{bmatrix}, \quad \mathbf{E}_{4,3} = \begin{bmatrix} 1 \\ -i \\ -1 \\ i \end{bmatrix}.$$

Note the aliasing relations $\mathbf{E}_{4,-2} = \mathbf{E}_{4,2}$ and $\mathbf{E}_{4,-1} = \mathbf{E}_{4,3}$. In particular, the sets $\{\mathbf{E}_{4,-1}, \mathbf{E}_{4,0}, \mathbf{E}_{4,1}, \mathbf{E}_{4,2}\}$ and $\{\mathbf{E}_{4,0}, \mathbf{E}_{4,1}, \mathbf{E}_{4,2}, \mathbf{E}_{4,3}\}$ (corresponding to $\mathbf{E}_{N,k}$ on the range $-N/2 < k \le N/2$ or $0 \le k \le N-1$) are identical.

It is also worth noting the relation

$$\overline{\mathbf{E}_{N,k}} = \mathbf{E}_{N,N-k}, \tag{1.24}$$

where the overline denotes complex conjugation. Equation (1.24) is sometimes called "conjugate aliasing" (see Exercise 21).

Remark 1.6 There are a lot of periodic functions in the discussion above. For example, the basic waveform $e^{2\pi ikt/T}$ is periodic in t with period T/k. The quantity $\mathbf{E}_{N,k}(m)$ in equation (1.23) is defined for all k and m and periodic in both. As a consequence $\mathbf{E}_{N,k}$ is defined for all k, and periodic with period N. The one entity that is not manifestly periodic is the analog time signal $x(t)$, or its sampled version $\mathbf{x} = (x_0, x_1, \dots, x_{N-1})$. At times it will be useful, at least conceptually, to extend either periodically. The sampled signal $\mathbf{x} = (x_0, \dots, x_{N-1})$ can be extended periodically with period N in its index by defining

$$x_m = x_{m \bmod N}$$

for all m outside the range $0 \le m \le N-1$. We can also extend the analog signal $x(t)$ periodically to all real t by setting $x(t) = x(t \bmod P)$, where P denotes the period of $x(t)$.

1.7.2 Discrete Basic Waveforms for Images

As with the one-dimensional waveforms, the appropriate discrete waveforms in the two-dimensional case are the sampled basic waveforms. These discrete waveforms are naturally rectangular arrays or matrices.

To be more precise, consider a rectangular domain or image defined by $0 \le x \le S, 0 \le y \le R$, but recall Remark 1.1; here increasing y is downward. The sampling will take place on an m (in the y-direction) by n (x-direction) rectangular grid. The basic two-dimensional waveforms were given in equation (1.17). As we will see later, the parameters α and β are most conveniently taken to be of the form $\alpha = 2\pi l/S$ and $\beta = 2\pi k/R$ for integers k and l. Thus, the analog basic waveforms to be sampled are the functions $e^{2\pi i(lx/S+ky/R)}$. Each such waveform is sampled at points of the form $x_s = sS/n, y_r = rR/m$, with $0 \le s \le n-1, 0 \le r \le m-1$. The result is an $m \times n$ matrix $\mathcal{E}_{m,n,k,l}$ with row r, column s entry

$$\mathcal{E}_{m,n,k,l}(r,s) = e^{2\pi i(kr/m+ls/n)}. \tag{1.25}$$

Note that $\mathcal{E}_{m,n,k,l}$ does not depend on the image dimensions R or S. The indexes may seem a bit confusing, but recall that m and n are fixed by the discretization size (an m by n pixel image); l and k denote the frequency of the underlying analog waveform in the x- and y-directions, respectively. The parameters s and r correspond to the x and y coordinates of the sample point. These $m \times n$ matrices $\mathcal{E}_{m,n,k,l}$ constitute the basic waveforms in the discrete setting.

In Exercise 18 you are asked to show that $\mathcal{E}_{m,n,k,l}$ can be factored into a product

$$\mathcal{E}_{m,n,k,l} = \mathbf{E}_{m,k}\mathbf{E}_{n,l}^T, \tag{1.26}$$

where the superscript T denotes the matrix transpose operation and where the vectors $\mathbf{E}_{m,k}$ and $\mathbf{E}_{n,l}$ are the discrete basic waveforms in one dimension, as defined in equation (1.22) (as column vectors). For example,

$$\mathcal{E}_{4,4,1,2} = E_{4,1}E_{4,2}^T = \begin{bmatrix} 1 \\ i \\ -1 \\ -i \end{bmatrix} \begin{bmatrix} 1 & -1 & 1 & -1 \end{bmatrix} = \begin{bmatrix} 1 & -1 & 1 & -1 \\ i & -i & i & -i \\ -1 & 1 & -1 & 1 \\ -i & i & -i & i \end{bmatrix}$$

As in the one-dimensional case we will write $\mathcal{E}_{k,l}$ instead of $\mathcal{E}_{m,n,k,l}$ when there is no possibility for confusion.

A variety of aliasing relations for the waveforms $\mathcal{E}_{m,n,k,l}$ follow from equation (1.26) and those for the $\mathbf{E}_{m,k}$ (see Exercise 22). Also the $\mathcal{E}_{m,n,k,l}$ are periodic in k with period m, and periodic in l with period n. If we confine our attention to the ranges $0 \le k < m$, $0 \le l < n$, there are exactly mn distinct $\mathcal{E}_{m,n,k,l}$ waveforms. For any index pair (l, k) outside this range the corresponding waveform is identical to one of the mn basic waveforms in this range.

The 2D images of the discrete 2D waveforms look pretty much the same as the analog ones do for low values of k and l. For larger values of k and l the effects of aliasing begin to take over and the waveforms are difficult to accurately graph.

Remark 1.7 As in the one-dimensional case, it may occasionally be convenient to extend an image matrix with entries $a_{r,s}$, $0 \le r \le m-1$ and $0 \le s \le n-1$, periodically to the whole plane. We can do this as

$$a_{r,s} = a_{r \bmod m,\, s \bmod n}.$$

1.8 Inner Product Spaces and Orthogonality

1.8.1 Inner Products and Norms

Vector spaces provide a convenient framework for analyzing signals and images. However, it is helpful to have a bit more mathematical structure to carry out the analysis, specifically some ideas from geometry. Most of the vector spaces we will be concerned with can be endowed with geometric notions such as "length" and "angle." Of special importance is the idea of "orthogonality." All of these notions can be quantified by adopting an *inner product* on the vector space of interest.

1.8.1.1 Inner Products

The inner product is a generalization of the familiar dot product from basic multivariable calculus. In the definition below a "function on $V \times V$" means a function whose domain consists of ordered pairs of vectors from V.

Definition 1.3 Let V be a vector space over $\mathbb{C}$ (resp., $\mathbb{R}$). An *inner product* (or *scalar product*) on V is a function from $V \times V$ to $\mathbb{C}$ (resp., $\mathbb{R}$). We use $(\mathbf{v}, \mathbf{w})$ to denote the inner product of vectors $\mathbf{v}$ and $\mathbf{w}$ and require that for all vectors $\mathbf{u}, \mathbf{v}, \mathbf{w} \in V$ and scalars a, b in $\mathbb{C}$ (resp., $\mathbb{R}$).

1. $(\mathbf{v}, \mathbf{w}) = \overline{(\mathbf{w}, \mathbf{v})}$ *(conjugate symmetry)*
2. $(a\mathbf{u} + b\mathbf{v}, \mathbf{w}) = a(\mathbf{u}, \mathbf{w}) + b(\mathbf{v}, \mathbf{w})$ *(linearity in the first argument)*
3. $(\mathbf{v}, \mathbf{v}) \geq 0$, *and* $(\mathbf{v}, \mathbf{v}) = 0$ *if and only if* $\mathbf{v} = \mathbf{0}$ *(positive definiteness).*

In the case where V is a vector space over $\mathbb{R}$, condition 1 is simply $(\mathbf{v}, \mathbf{w}) = (\mathbf{w}, \mathbf{v})$. If V is over $\mathbb{C}$, then condition 1 also immediately implies that $(\mathbf{v}, \mathbf{v}) = \overline{(\mathbf{v}, \mathbf{v})}$ so that $(\mathbf{v}, \mathbf{v})$ is always real-valued, and hence condition 3 makes sense. The linearity with respect to the first variable in property 2 easily extends to any finite linear combination.

One additional fact worth noting is that the inner product is conjugate-linear in the second argument, that is,

$$\begin{aligned}(\mathbf{w}, a\mathbf{u} + b\mathbf{v}) &= \overline{(a\mathbf{u} + b\mathbf{v}, \mathbf{w})} && \text{by property 1,}\\ &= \overline{a}\,\overline{(\mathbf{u}, \mathbf{w})} + \overline{b}\,\overline{(\mathbf{v}, \mathbf{w})} && \text{by property 2,}\\ &= \overline{a}\,(\mathbf{w}, \mathbf{u}) + \overline{b}\,(\mathbf{w}, \mathbf{v}) && \text{by property 1.}\end{aligned} \tag{1.27}$$

A vector space equipped with an inner product is called an *inner product space.*

1.8.1.2 Norms

Another useful geometric notion on a vector space V is that of a *norm,* a way of quantifying the size or length of vectors in V. This also allows us to quantify the distance between elements of V.

Definition 1.4 A *norm* on a vector space V (over $\mathbb{R}$ or $\mathbb{C}$) is a function $\|\mathbf{v}\|$ from V to $\mathbb{R}$ with the properties that

1. $\|\mathbf{v}\| \geq 0$, and $\|\mathbf{v}\| = 0$ if and only if $\mathbf{v} = \mathbf{0}$
2. $\|a\mathbf{v}\| = |a|\,\|\mathbf{v}\|$
3. $\|\mathbf{v} + \mathbf{w}\| \leq \|\mathbf{v}\| + \|\mathbf{w}\|$ *(the triangle inequality)*

for all vectors $\mathbf{v}, \mathbf{w} \in V$, and scalars a.

A vector space equipped with a norm is called (not surprisingly) a *normed vector space,* or sometimes a *normed linear space.*

An inner product $(\mathbf{v}, \mathbf{w})$ on a vector space V always induces a corresponding norm via the relation

$$\|\mathbf{v}\| = \sqrt{(\mathbf{v}, \mathbf{v})} \tag{1.28}$$

(see Exercise 28). Thus, every inner product space is a normed vector space, but the converse is not true (see Example 1.16 and Exercise 30). We will make frequent use of equation (1.28) in the form $\|\mathbf{v}\|^2 = (\mathbf{v}, \mathbf{v})$.

Remark 1.8 It is also useful to define the distance between two vectors $\mathbf{v}$ and $\mathbf{w}$ in a normed vector space as $\|\mathbf{v}-\mathbf{w}\|$. For example, in $\mathbb{R}^n$,

$$\|\mathbf{v}-\mathbf{w}\| = ((v_1-w_1)^2+\cdots+(v_n-w_n)^2)^{1/2},$$

which is the usual distance formula.

Generally, the function $\|\mathbf{v}-\mathbf{w}\|$ on $V\times V$ defines a *metric* on V, a way to measure the distance between the elements of the space, and turns V into a *metric space*. The study of metrics and metric spaces is a large area of mathematics, but in this text we will not need this much generality.

Remark 1.9 If the normed vector space V in question has any kind of physical interpretation, then the quantity $\|\mathbf{v}\|^2$ frequently turns out to be proportional to some natural measure of "energy." The expression for the energy of most physical systems is quadratic in nature with respect to the variables that characterize the state of the system. For example, the kinetic energy of a particle with mass m and speed v is $\frac{1}{2}mv^2$, quadratic in v. The energy dissipated by a resistor is V^2/R, quadratic in V, where R is the resistance and V the potential drop across the resistor. The concept of energy is also important in signal and image analysis, and the quantification of the energy in these settings is quadratic in nature. We will say more on this later.

1.8.2 Examples

Here are some specific, useful inner product spaces, and the corresponding norms.

Example 1.12 $\mathbb{R}^n$ *(real Euclidian space): The most common inner product on* $\mathbb{R}^n$ *is the dot product, defined by*

$$(\mathbf{x},\mathbf{y}) = x_1y_1+x_2y_2+\cdots+x_ny_n$$

for vectors $\mathbf{x}=(x_1,\dots,x_n), \mathbf{y}=(y_1,\dots,y_n)$ *in* $\mathbb{R}^n$. *Properties 1–3 for inner products are easily verified.*

The corresponding norm from equation (1.28) is

$$\|\mathbf{x}\| = (x_1^2+x_2^2+\cdots+x_n^2)^{1/2},$$

the usual Euclidean norm.

Example 1.13 $\mathbb{C}^n$ *(complex Euclidean space): On* $\mathbb{C}^n$ *the usual inner product is*

$$(\mathbf{x},\mathbf{y}) = x_1\overline{y_1}+x_2\overline{y_2}+\cdots+x_n\overline{y_n}$$

for vectors $\mathbf{x}=(x_1,\dots,x_n), \mathbf{y}=(y_1,\dots,y_n)$ *in* $\mathbb{C}^n$. *Conjugation of the second vector's components is important! The corresponding norm from equation (1.28) is*

$$\|\mathbf{x}\| = (|x_1|^2 + |x_2|^2 + \cdots + |x_n|^2)^{1/2},$$

where we have made use of the fact that $z\overline{z} = |z|^2$ for any complex number z.

Example 1.14 $M_{m,n}(\mathbb{C})$ *($m \times n$ matrices with complex entries): On $M_{m,n}(\mathbb{C})$ an inner product is given by*

$$(\mathbf{A}, \mathbf{B}) = \sum_{j=1}^{m} \sum_{k=1}^{n} a_{j,k}\overline{b_{j,k}}$$

for matrices **A** *and* **B** *with entries $a_{j,k}$ and $b_{j,k}$, respectively. The corresponding norm from equation (1.28) is*

$$\|\mathbf{A}\| = \left(\sum_{j=1}^{m} \sum_{k=1}^{n} |a_{j,k}|^2 \right)^{1/2},$$

called the Frobenius *norm. As in Example 1.2, as an inner product space $M_{m,n}(\mathbb{C})$ is really "identical" to $\mathbb{C}^{mn}$.*

Example 1.15 *$C[a, b]$ (continuous functions on $[a, b]$): Consider the vector space $C[a, b]$, and suppose that the functions can assume complex values. An inner product on this space is given by*

$$(f, g) = \int_a^b f(t)\overline{g(t)}\, dt.$$

It is not hard to see that the integral is well defined, for f and g are continuous functions on a closed interval, hence bounded, so that the product $f\overline{g}$ is continuous and bounded. The integral therefore converges. Of course, if the functions are real-valued, the conjugation is unnecessary.

Properties 1 and 2 for the inner product follow easily from properties of the Riemann integral. Only property 3 for inner products needs comment. First,

$$(f, f) = \int_a^b |f(t)|^2\, dt \geq 0, \tag{1.29}$$

since the integrand is nonnegative and the integral of a nonnegative function is nonnegative. However, the second assertion in property 3 for an inner product needs some thought—if the integral in (1.29) actually equals zero, must f be the zero function?

We can prove that this is so by contradiction: suppose that f is not identically zero, say $f(t_0) \neq 0$ for some $t_0 \in [a, b]$. Then $|f(t_0)|^2 > 0$. Moreover, since f is continuous, so is $|f(t)|^2$. Thus, we can find some small interval $(t_0 - \delta, t_0 + \delta)$ with $\delta > 0$ on which $|f(t)|^2 \geq |f(t_0)|^2/2$. Then

$$(f, f) = \int_a^{t_0-\delta} |f(t)|^2\, dt + \int_{t_0-\delta}^{t_0+\delta} |f(t)|^2\, dt + \int_{t_0+\delta}^{b} |f(t)|^2\, dt.$$

Since $|f(t)|^2 \geq |f(t_0)|^2/2$ for $t \in (t_0 - \delta, t_0 + \delta)$ the middle integral on the right above is positive and greater than $\delta|f(t_0)|^2$ (the area of a 2δ width by $|f(t_0)|^2/2$ tall rectangle under the graph of $|f(t)|^2$). The other two integrals are at least non-negative. We conclude that if $f \in C[a,b]$ is not the zero function, then $(f,f) > 0$. Equivalently $(f,f) = 0$ only if $f \equiv 0$, so that the second assertion in property 3 for an inner product holds.

The corresponding norm for this inner product is

$$\|f\| = \left(\int_a^b |f(t)|^2\, dt\right)^{1/2}.$$

In light of the discussion above, $\|f\| = 0$ if and only if $f \equiv 0$.

Example 1.16 *Another commonly used norm on the space $C[a,b]$ is the* supremum *norm, defined by*

$$\|f\|_\infty = \sup_{x\in[a,b]} |f(x)|.$$

Recall that the supremum of a set $A \subset \mathbb{R}$ is the smallest real number M such $a \leq M$ for every $a \in A$, meaning M is the "least upper bound" for the elements of A. It is a fact that each subset of $\mathbb{R}$ that is bounded above has a unique supremum. If f is continuous, then we can replace "sup" in the definition of $\|f\|_\infty$ with "max," since a continuous function on a closed bounded interval $[a,b]$ must attain its supremum.

The supremum norm does not come from any inner product in equation (1.28) (see Exercise 30).

Example 1.17 *$C(\Omega)$ (the set of continuous complex-valued functions on a closed rectangle $\Omega = \{(x,y); a \leq x \leq b, c \leq y \leq d\}$): An inner product on this space is given by*

$$(f,g) = \int_a^b \int_c^d f(x,y)\overline{g(x,y)}\, dy\, dx.$$

As in the one-dimensional case, the integral is well defined since f and g must be bounded; hence the product $f\overline{g}$ is continuous and bounded. The integral therefore converges. An argument similar to that of Example 1.15 shows that property 3 for inner products holds.

The corresponding norm is

$$\|f\| = \left(\int_a^b \int_c^d |f(x,y)|^2\, dy\, dx\right)^{1/2}.$$

This space can be considerably enlarged, to include many discontinuous functions that satisfy $\|f\| < \infty$.

When we work in a function space like $C[a,b]$ we will sometimes use the notation $\|f\|_2$ (rather than just $\|f\|$) to indicate the Euclidean norm that stems from the inner product, and so avoid confusion with the supremum norm (or other norms).

1.8.3 Orthogonality

Recall from elementary vector calculus that the dot product $(\mathbf{v}, \mathbf{w})$ of two vectors $\mathbf{v}$ and $\mathbf{w}$ in $\mathbb{R}^2$ or $\mathbb{R}^3$ satisfies the relation

$$(\mathbf{v}, \mathbf{w}) = \|\mathbf{v}\|\|\mathbf{w}\|\cos(\theta), \tag{1.30}$$

where $\|\mathbf{v}\|$ is the length of $\mathbf{v}$, $\|\mathbf{w}\|$ the length of $\mathbf{w}$, and θ is the angle between $\mathbf{v}$ and $\mathbf{w}$. In particular, it is easy to see that $(\mathbf{v}, \mathbf{w}) = 0$ exactly when $\theta = \pi/2$ radians, so $\mathbf{v}$ and $\mathbf{w}$ are orthogonal to each other. The notion of orthogonality can be an incredibly powerful tool. This motivates the following general definition:

Definition 1.5 Two vectors $\mathbf{v}$ and $\mathbf{w}$ in an inner product space V are "*orthogonal*" if $(\mathbf{v}, \mathbf{w}) = 0$.

The notion of orthogonality depends on not only the vectors but also the inner product we are using (there may be more than one!). We will say that a subset S (finite or infinite) of vectors in an inner product space V is *pairwise orthogonal*, or more commonly just *orthogonal*, if $(\mathbf{v}, \mathbf{w}) = 0$ for any pair of distinct $(\mathbf{v} \neq \mathbf{w})$ vectors $\mathbf{v}, \mathbf{w} \in S$.

Example 1.18 *Let $S = \{\mathbf{e}_1, \mathbf{e}_2, \ldots, \mathbf{e}_n\}$ denote the standard basis vectors in $\mathbb{R}^n$ ($\mathbf{e}_k$ has a "1" in the kth position, zeros elsewhere), with the usual Euclidean inner product (and indexing from 1 to n). The set S is orthogonal since $(\mathbf{e}_j, \mathbf{e}_k) = 0$ when $j \neq k$.*

Example 1.19 *Let S denote the set of functions $e^{\pi ikt/T}$, $k \in \mathbb{Z}$, in the vector space $C[-T, T]$, with the inner product as defined in Example 1.15. The set S is orthogonal, for if $k \neq m$, then*

$$\begin{aligned}
(e^{\pi ikt/T}, e^{\pi imt/T}) &= \int_{-T}^{T} e^{\pi ikt/T}\overline{e^{\pi imt/T}}\,dt \\
&= \int_{-T}^{T} e^{\pi ikt/T}e^{-\pi imt/T}\,dt \\
&= \int_{-T}^{T} e^{\pi i(k-m)t/T} \\
&= \frac{T(e^{\pi i(k-m)} - e^{-\pi i(k-m)})}{\pi i(k-m)}\,dt \\
&= 0,
\end{aligned}$$

since $e^{\pi in} - e^{-\pi in} = 0$ for any integer n.

Example 1.20 *Let S denote the set of basic discrete waveforms* $\mathbf{E}_{N,k} \in \mathbb{C}^N$, $-N/2 < k \leq N/2$, *as defined in equation (1.22). With the inner product on* $\mathbb{C}^N$ *as defined in Example 1.13, but on the index range 0 to* $N-1$ *instead of 1 to N, the set S is orthogonal. The proof is based on the surprisingly versatile algebraic identity*

$$1 + z + z^2 + \cdots + z^{N-1} = \frac{1 - z^N}{1 - z}, \qquad \text{if } z \neq 1. \tag{1.31}$$

If $k \neq l$, *then*

$$\begin{aligned}
(\mathbf{E}_k, \mathbf{E}_l) &= \sum_{r=0}^{N-1} e^{2\pi ikr/N} \overline{e^{2\pi ilr/N}} \\
&= \sum_{r=0}^{N-1} e^{2\pi ikr/N} e^{-2\pi ilr/N} \\
&= \sum_{r=0}^{N-1} e^{2\pi i(k-l)r/N} \\
&= \sum_{r=0}^{N-1} (e^{2\pi i(k-l)/N})^r.
\end{aligned} \tag{1.32}$$

Let $z = e^{2\pi i(k-l)/N}$ *in (1.31) and equation (1.32) becomes*

$$\begin{aligned}
(\mathbf{E}_k, \mathbf{E}_l) &= \frac{1 - (e^{2\pi i(k-l)/N})^N}{1 - e^{2\pi i(k-l)/N}} \\
&= \frac{1 - e^{2\pi i(k-l)}}{1 - e^{2\pi i(k-l)/N}} \\
&= 0,
\end{aligned}$$

since $e^{2\pi i(k-l)} = 1$. *Moreover the denominator above cannot equal zero for* $-N/2 < k, l \leq N/2$ *if* $k \neq l$. *Note the similarity of this computation to the computation in Example 1.19.*

Remark 1.10 A very similar computation to that of Example 1.20 shows that the waveforms or matrices $\mathcal{E}_{m,n,k,l}$ in $M_{m,n}(\mathbb{C})$ (with the inner product from Example 1.14 but indexing from 0) are also orthogonal.

1.8.4 The Cauchy–Schwarz Inequality

The following inequality will be extremely useful. It has wide-ranging application and is one of the most famous inequalities in mathematics.

Theorem 1.1 ***Cauchy–Schwarz*** *For any vectors* **v** *and* **w** *in an inner product space V over* $\mathbb{C}$ *or* $\mathbb{R}$,

$$|(\mathbf{v}, \mathbf{w})| \leq \|\mathbf{v}\| \|\mathbf{w}\|,$$

where $\|\cdot\|$ *is the norm induced by the inner product via equation (1.28).*

Proof: Note that $0 \le (\mathbf{v} - c\mathbf{w}, \mathbf{v} - c\mathbf{w})$ for any scalar c, from property 3 for inner products. If we expand this out by using the properties of the inner product (including equation (1.27)), we find that

$$\begin{aligned} 0 &\le (\mathbf{v} - c\mathbf{w}, \mathbf{v} - c\mathbf{w}) \\ &= (\mathbf{v}, \mathbf{v}) - c(\mathbf{w}, \mathbf{v}) - \bar{c}(\mathbf{v}, \mathbf{w}) + (c\mathbf{w}, c\mathbf{w}) \\ &= \|\mathbf{v}\|^2 - c(\mathbf{w}, \mathbf{v}) - \bar{c}(\mathbf{v}, \mathbf{w}) + |c|^2\|\mathbf{w}\|^2. \end{aligned}$$

Let us suppose that $\mathbf{w} \neq \mathbf{0}$, for otherwise, Cauchy–Schwarz is obvious. Choose $c = (\mathbf{v}, \mathbf{w})/(\mathbf{w}, \mathbf{w})$ so that $\bar{c} = (\mathbf{w}, \mathbf{v})/(\mathbf{w}, \mathbf{w})$; note that $(\mathbf{w}, \mathbf{w})$ is real and positive. Then $c(\mathbf{w}, \mathbf{v}) = \bar{c}(\mathbf{v}, \mathbf{w}) = |(\mathbf{v}, \mathbf{w})|^2/\|\mathbf{w}\|^2$ and

$$0 \le \|\mathbf{v}\|^2 - 2\frac{|(\mathbf{v}, \mathbf{w})|^2}{\|\mathbf{w}\|^2} + \frac{|(\mathbf{v}, \mathbf{w})|^2}{\|\mathbf{w}\|^2} = \|\mathbf{v}\|^2 - \frac{|(\mathbf{v}, \mathbf{w})|^2}{\|\mathbf{w}\|^2}$$

from which the Cauchy–Schwarz inequality follows. Note that the choice $c = (\mathbf{v}, \mathbf{w})/(\mathbf{w}, \mathbf{w})$ minimizes the value of $\|\mathbf{v} - c\mathbf{w}\|^2$ and makes $\mathbf{v} - c\mathbf{w}$ orthogonal to $\mathbf{w}$ (see Exercise 29).

1.8.5 Bases and Orthogonal Decomposition

1.8.5.1 Bases

Recall that the set of vectors $S = \{\mathbf{e}_1, \mathbf{e}_2, \dots, \mathbf{e}_n\}$ in $\mathbb{R}^n$ is called the *standard basis*. The reason is that any vector $\mathbf{x} = (x_1, x_2, \dots, x_n)$ in $\mathbb{R}^n$ can be written as a linear combination

$$\mathbf{x} = x_1\mathbf{e}_1 + x_2\mathbf{e}_2 + \cdots + x_n\mathbf{e}_n$$

of elements from S, in one and only one way. Thus, the $\mathbf{e}_k$ form a convenient set of building blocks for $\mathbb{R}^n$. The set is "minimal" in the sense that any vector $\mathbf{x}$ can be built from the elements of S in only one way.

The following concepts may be familiar from elementary linear algebra in $\mathbb{R}^N$, but they are useful in any vector space.

Definition 1.6 A set S (finite or infinite) in a vector space V over $\mathbb{C}$ (resp., $\mathbb{R}$) is said to "*span* V" if every vector $\mathbf{v} \in V$ can be constructed as a finite linear combination of elements of S,

$$\mathbf{v} = \alpha_1\mathbf{v}_1 + \alpha_2\mathbf{v}_2 + \cdots + \alpha_n\mathbf{v}_n,$$

for suitable scalars α_k in $\mathbb{C}$ (resp., $\mathbb{R}$) and vectors $\mathbf{v}_k \in S$.

Definition 1.7 A set S (finite or infinite) in a vector space V over $\mathbb{C}$ (resp., $\mathbb{R}$) is said to be "*linearly independent*" if, for any finite set of vectors $\mathbf{v}_1, \dots, \mathbf{v}_n$ in S, the only solution to

$$\alpha_1\mathbf{v}_1 + \alpha_2\mathbf{v}_2 + \cdots + \alpha_n\mathbf{v}_n = \mathbf{0}$$

is $\alpha_k = 0$ for all $1 \le k \le n$.

A set S that spans V is thus sufficient to build any vector in V by superposition. Linear independence ensures that no vector can be built in more than one way. A set S that both spans V and is linearly independent is especially useful, for each vector in V can be built from elements of S in a unique way.

Definition 1.8 A linearly independent set S that spans a vector space V is called a basis for V.

Be careful; a basis S for V may have infinitely many elements, but according to the above definition we must be able to construct any *specific* vector in V using only a *finite* linear combination of vectors in S. The word "basis" has a variety of meanings in mathematics, and the more accurate term for the type of basis defined above, in which only finite combinations are allowed, is a "Hamel basis." If infinite linear combinations of basis vectors are allowed (as in Section 1.10), then issues concerning limits and convergence arise. In either case, however, we will continue to use the term "basis," and no confusion should arise.

It is worth noting that no linearly independent set and hence no basis can contain the zero vector.

The standard basis in $\mathbb{R}^n$ or $\mathbb{C}^n$, of course, provides an example of a basis. Here are a couple slightly more interesting examples.

Example 1.21 *Consider the space $M_{m,n}(\mathbb{C})$ of $m \times n$ complex matrices, and define mn distinct elements $\mathbf{A}_{p,q} \in M_{m,n}(\mathbb{C})$ as follows: let the row p, column q entry of $\mathbf{A}_{p,q}$ equal 1, and set all other entries of $\mathbf{A}_{p,q}$ equal to zero (quite analogous to the standard basis of $\mathbb{R}^n$ or $\mathbb{C}^n$). The set $S = \{\mathbf{A}_{p,q}; 1 \le p \le m, 1 \le q \le n\}$ forms a basis for $M_{m,n}(\mathbb{C})$.*

Example 1.22 *Let P denote the vector space consisting of all polynomials in the variable x,*

$$p(x) = a_0 + a_1 x + \cdots + a_n x^n$$

with real coefficients a_k and no condition on the degree n. You should convince yourself that this is indeed a vector space over $\mathbb{R}$, with the obvious operations. And note that a polynomial has a highest degree term; we are not allowing expressions like $1 + x + x^2 + \cdots$, that is, power series. One basis for P is given by the infinite set

$$S = \{1, x, x^2, x^3, \ldots\}.$$

It is not hard to see that any polynomial can be expressed as a finite linear combination of elements of S, and in only one way.

A vector space typically has many different bases. In fact each vector space that will interest us in this text has infinitely many different bases. Which basis we use depends on what we are trying to do.

If a vector space V has a finite basis $S = \{\mathbf{v}_1, \ldots, \mathbf{v}_n\}$, then V is said to be *finite-dimensional.* It is a fact from elementary linear algebra that any other basis for V must also contain exactly n vectors. In this case, V is called an *n-dimensional* vector space. In light of the remarks above we can see that $\mathbb{R}^n$ really is an n-dimensional vector space over $\mathbb{R}$ (surprise!), while $\mathbb{C}^n$ is n-dimensional over $\mathbb{C}$. Based on Example 1.21 the spaces $M_{m,n}(\mathbb{R})$ and $M_{m,n}(\mathbb{C})$ are both mn-dimensional vector spaces over $\mathbb{R}$ or $\mathbb{C}$, respectively. The space P in Example 1.22 is infinite-dimensional.

1.8.5.2 Orthogonal and Orthonormal Bases

A lot of vector algebra becomes ridiculously easy when the vectors involved are orthogonal. In particular, finding an orthogonal set of basis vectors for a vector space V can greatly aid analysis and facilitate certain computations.

One very useful observation is the following theorem.

Theorem 1.2 *If a set $S \subset V$ of non-zero vectors is orthogonal then S is linearly independent.*

Proof: Consider the equation

$$\alpha_1 \mathbf{v}_1 + \alpha_2 \mathbf{v}_2 + \cdots + \alpha_n \mathbf{v}_n = \mathbf{0},$$

where the $\mathbf{v}_k$ are elements of S and the α_k scalars. Form the inner product of both sides above with any one of the vectors $\mathbf{v}_m$, $1 \le m \le n$, and use the linearity of the inner product in the first variable to obtain

$$\sum_{k=1}^{n} \alpha_k (\mathbf{v}_k, \mathbf{v}_m) = (\mathbf{0}, \mathbf{v}_m).$$

The right side above is, of course, 0. Since S is orthogonal, $(\mathbf{v}_k, \mathbf{v}_m) = 0$ unless $k = m$, so the above equation degenerates to $\alpha_m(\mathbf{v}_m, \mathbf{v}_m) = 0$. Because $(\mathbf{v}_m, \mathbf{v}_m) > 0$ (each $\mathbf{v}_m$ is nonzero by hypothesis), we obtain $\alpha_m = 0$ for $1 \le m \le n$.

In the remainder of this section, we assume that V is a finite-dimensional vector space over either $\mathbb{R}$ or $\mathbb{C}$. Of special interest are bases for V that are orthogonal so that $(\mathbf{v}_k, \mathbf{v}_m) = 0$ for any two distinct basis vectors. In this case, it is easy to explicitly write any $\mathbf{v} \in V$ as a linear combination of basis vectors.

Theorem 1.3 *Let $S = \{\mathbf{v}_1, \mathbf{v}_2, \ldots, \mathbf{v}_n\}$ be an orthogonal basis for a vector space V. Then any $\mathbf{v} \in V$ can be expressed as*

$$\mathbf{v} = \sum_{k=1}^{n} \alpha_k \mathbf{v}_k, \tag{1.33}$$

where $\alpha_k = (\mathbf{v}, \mathbf{v}_k)/(\mathbf{v}_k, \mathbf{v}_k)$.

Proof: This proof is very similar to that of Theorem 1.2. First, since S is a basis, there is some set of α_k that work in equation (1.33). Form the inner product of both sides of equation (1.33) with any $\mathbf{v}_m$, $1 \le m \le n$, and use the linearity of the inner product in the first variable to obtain

$$(\mathbf{v}, \mathbf{v}_m) = \sum_{k=1}^{n} \alpha_k (\mathbf{v}_k, \mathbf{v}_m).$$

Since S is orthogonal $(\mathbf{v}_k, \mathbf{v}_m) = 0$ unless $k = m$, in which case the equation above becomes $(\mathbf{v}, \mathbf{v}_m) = \alpha_m (\mathbf{v}_m, \mathbf{v}_m)$. Thus, $\alpha_m = (\mathbf{v}, \mathbf{v}_m)/(\mathbf{v}_m, \mathbf{v}_m)$ is uniquely determined. The denominator $(\mathbf{v}_m, \mathbf{v}_m)$ cannot be zero, since $\mathbf{v}_m$ cannot be the zero vector (because $\mathbf{v}_m$ is part of a linearly independent set).

Definition 1.9 An orthogonal set S is *orthonormal* if $\|\mathbf{v}\| = 1$ for each $\mathbf{v} \in S$.

In the case where a basis S for V forms an orthonormal set, the expansion in Theorem 1.3 becomes a bit simpler since $(\mathbf{v}_k, \mathbf{v}_k) = \|\mathbf{v}_k\|^2 = 1$, so we can take $\alpha_k = (\mathbf{v}, \mathbf{v}_k)$ in equation (1.33).

Remark 1.11 Any orthogonal basis S can be replaced by a "rescaled" basis that is orthonormal. Specifically, if S is an orthogonal basis for a vector space V, let S' denote the set obtained by replacing each vector $\mathbf{x} \in S$ by the rescaled vector $\mathbf{x}' = \mathbf{x}/\|\mathbf{x}\|$ of length 1. If a vector $\mathbf{v} \in V$ can be expanded according to Theorem 1.3, then $\mathbf{v}$ can also be written as a superposition of elements of S', as

$$\mathbf{v} = \sum_{k=1}^{n} (\alpha_k \|\mathbf{v}_k\|) \mathbf{v}'_k,$$

where $\mathbf{v}'_k = \mathbf{v}_k / \|\mathbf{v}_k\|$ has norm one.

Example 1.23 *The standard basis for $\mathbb{C}^N$ certainly has its place, but for many types of analysis the basic waveforms $\mathbf{E}_{N,k}$ are often more useful. In fact they also form an orthogonal basis for $\mathbb{C}^N$. We have already shown this to be true when in Example 1.20 we showed that the N vectors $\mathbf{E}_{N,k}$ for $0 \le k \le N-1$ (or $-N/2 < k \le N/2$) are mutually orthogonal. By Theorem 1.2, the vectors are necessarily linearly independent, and since a set of N linearly independent vectors in $\mathbb{C}^N$ must span $\mathbb{C}^N$, the $\mathbf{E}_{N,k}$ form a basis for $\mathbb{C}^N$.*

From equation (1.33) we then obtain a simple decomposition formula

$$\begin{aligned} \mathbf{x} &= \sum_{k=0}^{N-1} \frac{(\mathbf{x}, \mathbf{E}_{N,k})}{(\mathbf{E}_{N,k}, \mathbf{E}_{N,k})} \mathbf{E}_{N,k} \\ &= \frac{1}{N} \sum_{k=0}^{N-1} (\mathbf{x}, \mathbf{E}_{N,k}) \mathbf{E}_{N,k} \qquad (1.34) \end{aligned}$$

for any vector $\mathbf{x} \in \mathbb{C}^N$*, where we have made use of* $(\mathbf{E}_{N,k}, \mathbf{E}_{N,k}) = N$ *for each* k *(see Exercise 32). Equation (1.34) will be of paramount importance later; indeed most of Chapter 3 is devoted to the study of equation (1.34)!*

Example 1.24 *An entirely analogous argument shows that the matrices* $\mathcal{E}_{m,n,k,l}$ *form a basis for* $M_{m,n}(\mathbb{C})$*, and for any matrix* $\mathbf{A} \in M_{m,n}(\mathbb{C})$

$$\begin{aligned}\mathbf{A} &= \sum_{k=0}^{m-1}\sum_{l=0}^{n-1} \frac{(\mathbf{A}, \mathcal{E}_{m,n,k,l})}{(\mathcal{E}_{m,n,k,l}, \mathcal{E}_{m,n,k,l})}\mathcal{E}_{m,n,k,l} \\ &= \frac{1}{mn}\sum_{k=0}^{m-1}\sum_{l=0}^{n-1} (\mathbf{A}, \mathcal{E}_{m,n,k,l})\mathcal{E}_{m,n,k,l},\end{aligned} \tag{1.35}$$

where we have made use of $(\mathcal{E}_{m,n,k,l}, \mathcal{E}_{m,n,k,l}) = mn$ *(see Exercise 33).*

1.8.5.3 Parseval's Identity

Suppose that S is an orthonormal basis for an n-dimensional vector space V. Let $\mathbf{v} \in V$ be expanded according to equation (1.33). Then

$$\begin{aligned}\|\mathbf{v}\|^2 &= (\mathbf{v}, \mathbf{v}) \\ &= \left(\sum_j \alpha_j \mathbf{v}_j, \sum_k \alpha_k \mathbf{v}_k\right) \\ &= \sum_{j,k=1}^{n} \alpha_j \overline{\alpha_k}(\mathbf{v}_j, \mathbf{v}_k) \\ &= \sum_{k=1}^{n} |\alpha_k|^2,\end{aligned} \tag{1.36}$$

where we have used the properties of the inner product (including equation (1.27)), $\alpha_k\overline{\alpha_k} = |\alpha_k|^2$, and the fact that S is orthonormal so $(\mathbf{v}_k, \mathbf{v}_k) = 1$. Equation (1.36) is called *Parseval's identity*.

As noted in Remark 1.9, $\|\mathbf{v}\|^2$ is often interpreted as the energy of the discretized signal. If the basis set S is orthonormal, then each vector $\mathbf{v}_k$ represents a basis signal that is scaled to have energy equal to 1, and it is easy to see that the signal $\alpha_k \mathbf{v}_k$ has energy $|\alpha_k|^2$. In this case, Parseval's identity asserts that the "energy of the sum is the sum of the energies."

1.9 Signal and Image Digitization

As discussed earlier in the chapter, general analog signals and images cannot be meaningfully stored in a computer and so must be converted to digital form. As noted in Section 1.3.2, this introduces a quantization error. It is tempting to minimize this error by storing the underlying real numbers as high-precision floating point values, but this would be expensive in terms of storage. In Matlab a grayscale image stored as double precision floating point numbers requires eight bytes per pixel, compared to one byte per pixel for the 8-bit quantization scheme discussed Section 1.3.6. Furthermore, if very fast processing is required, it is usually better to use integer arithmetic chips than floating point hardware. Thus, we must balance quantization error with the storage and computational costs associated with more accurate digitization. In order to better understand this issue, we now take a closer look at quantization and a more general scheme than that presented in Section 1.3.2.

1.9.1 Quantization and Dequantization

Let us start with a simple but representative example.

Example 1.25 *Consider an analog signal $x(t)$ that can assume "any" real value at any particular time t. The signal would, of course, be sampled to produce a string of real numbers x_k for k in some range, say $0 \leq k \leq n$. This still does not suffice for computer storage though, since we cannot store even a single real number x_k to infinite precision. What we will do is this: divide the real line into "bins," say the disjoint intervals $(-\infty, -5], (-5, 3], (3, 7]$, and $(7, \infty)$. Note that these are chosen arbitrarily here, solely for the sake of example. Thus, we have four* quantization intervals. *Every real number falls into exactly one of these intervals. We will refer to the interval $(-\infty, -5]$ as "interval 0," $(-5, 3]$ as "interval 1," $(3, 7]$ as "interval 2," and $(7, \infty)$ as "interval 3." In this manner, any real number z can be associated with an integer in the range 0–3, according to the quantization interval in which z lies. This defines a* quantization map *q from $\mathbb{R}$ to the set $\{0, 1, 2, 3\}$. Rather than storing z we store (with some obvious loss of information) $q(z)$. Indeed, since $q(z)$ can assume only four distinct values, it can be stored with just two bits. We can store the entire discretized signal $x(t)$ with just $2(n+1)$ bits, for example, as "00" for a sample x_k in interval 0, "01" for interval 1, "10" for interval 2, "11" for interval 3.*

To reconstruct an approximation to any given sample x_k, we proceed as follows: for each quantization interval we choose a representative value z_k for quantization interval k. For example, we can take $z_0 = -10, z_1 = -1, z_2 = 5$, and $z_3 = 10$. Here z_1 and z_2 are chosen as the midpoints of the corresponding interval, z_0 and z_3 as "representative" of their intervals. If a sample x_k falls in quantization interval 0 (i.e., was stored as the bit sequence "00"), we reconstruct it approximately as $\tilde{x}_k = z_0$. A similar computation is performed for the other intervals. This yields a dequantization map *$\tilde{q}$ from the set $\{0, 1, 2, 3\}$ back to $\mathbb{R}$.*

As a specific example, consider the sampled signal $\mathbf{x} \in \mathbb{R}^5$ with components $x_0 = -1.2, x_1 = 2.3, x_2 = 4.4, x_3 = 8.8$, and $x_4 = -2.8$. The quantization map yields $q(\mathbf{x}) = (1, 1, 2, 3, 1)$ when q is applied component-by-component to $\mathbf{x}$. The reconstructed version of $\mathbf{x}$ is $\tilde{q}(q(\mathbf{x})) = (-1, -1, 5, 10, -1)$.

1.9.1.1 The General Quantization Scheme

The quantization scheme above generalizes as follows. Let r be the number of quantization levels ($r = 4$ in the previous example). Choose $r - 1$ distinct quantization "jump points" $\{y_1, \ldots, y_{r-1}\}$, real numbers that satisfy $-\infty < y_1 < y_2 < \cdots < y_{r-1} < \infty$ (we had $y_1 = -5, y_2 = 3, y_3 = 7$ above). Let $y_0 = -\infty$ and $y_r = \infty$. We call the interval $(y_k, y_{k+1}]$ the *kth quantization interval*, where $0 \le k \le r - 1$ (we should really use an open interval (y_{r-1}, ∞) for the last interval). The leftmost and rightmost intervals are unbounded. Each real number belongs to exactly one of the r quantization intervals.

The *quantization map* $q : \mathbb{R} \to \{0, 1, \ldots, r - 1\}$ assigns an integer quantization level $q(x)$ to each $x \in \mathbb{R}$ as follows: $q(x)$ is the index k such that $y_k < x \le y_{k+1}$, meaning, x belongs to the kth quantization interval. The function q can be written more explicitly if we define the Heaviside function

$$H(x) = \begin{cases} 0, & x \le 0, \\ 1, & x > 0, \end{cases}$$

in which case

$$q(x) = \sum_{k=1}^{r-1} H(x - y_k).$$

A sample graph of a quantization function is given in Figure 1.11, in which we have chosen $y_1 = -3, y_2 = -1.5, y_3 = 1.5, y_4 = 3$, and $y_0 = -\infty, y_5 = \infty$. Notice that the quantization interval around zero is bigger than the rest. This is not uncommon in practice.

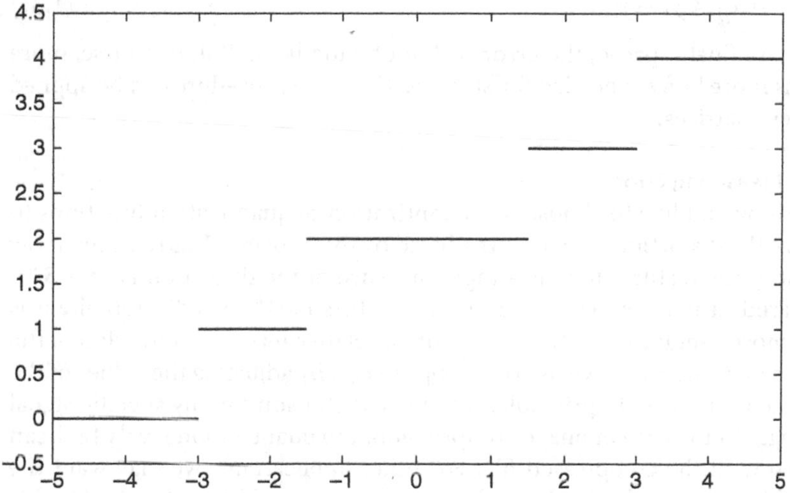

Figure 1.11 Quantization function $H(x + 3) + H(x + 1.5) + H(x - 1.5) + H(x - 3)$.

The quantization map is used as follows: Let $\mathbf{x}$ denote a sampled signal, so each component x_j of $\mathbf{x}$ is a real number and has not yet been quantized. The quantized version of $\mathbf{x}$ is just $q(\mathbf{x})$, in which q is applied component-by-component to $\mathbf{x}$. Each x_j is thus assigned to one of the r quantization intervals. If $r = 2^b$ for some $b > 0$, then each quantized x_j can be stored using b bits, and we refer to this as "b-bit quantization." A similar procedure would be applied to images.

1.9.1.2 Dequantization

Once quantized, the vector $\mathbf{x}$ cannot be exactly recovered because q is not invertible. If we need to approximately reconstruct $\mathbf{x}$ after quantization we do the following: pick real numbers $z_0, z_1, \dots, z_{r-1}$ such that z_k lies in the kth quantization interval, that is $y_{k-1} < z_{k-1} \le y_k$. Ideally the value of z_k should be a good approximation to the average value of the entries of $\mathbf{x}$ that fall in the kth quantization interval. A simple choice is to take z_k as the midpoint of the kth quantization interval, as in the example above. The set $\{z_k : 0 \le k \le r-1\}$ is called the *codebook* and the z_k are called the *codewords*. Define the *dequantization map* $\tilde{q} : \{0, \dots, r-1\} \to \mathbb{R}$ that takes k to z_k. Then define the approximate reconstruction of $\mathbf{x}$ as the vector $\tilde{\mathbf{x}}$ with components $\tilde{x}_j$ where

$$\tilde{x}_j = \tilde{q}(q(x_j)) = z_{q(x_j)}.$$

If $y_{k-1} < x_j \le y_k$, then x_j is mapped to the kth quantization interval (i.e., $q(x_j) = k$) and $\tilde{x}_j = \tilde{q}(q(x_j)) = z_k$ where $y_{k-1} < z_j \le y_k$. In other words, x_j and the quantized/dequantized quantity $\tilde{x}_j$ both lie in the interval $y_{k-1} < x_j \le y_k$, so at worst the discrepancy is

$$|x_j - \tilde{x}_j| \le y_k - y_{k-1}. \tag{1.37}$$

If the y_k are finely spaced, the error will not be too large. But, of course, more y_k means more bits are needed for storage. The same procedure can be applied to images/matrices.

1.9.1.3 Measuring Error

Ideally we would like to choose our quantization/dequantization functions to minimize the distortion or error introduced by this process. This requires a way to quantify the distortion. One simple measure of the distortion is $\|\mathbf{x} - \tilde{\mathbf{x}}\|^2$, the squared distance between $\mathbf{x}$ and $\tilde{\mathbf{x}}$ as vectors in $\mathbb{R}^n$ (or $\mathbb{C}^n$). Actually it is slightly more useful to quantify distortion in relative terms, as a fraction of the original signal energy, so we use $(\|\mathbf{x} - \tilde{\mathbf{x}}\|^2)/\|\mathbf{x}\|^2$. By adjusting the values of the y_k and the z_k, we can, in principle, minimize distortion for any specific signal $\mathbf{x}$ or image. In the case of image compression, the quantization levels $\{z_k\}$ can be stored with the compressed file. For other applications, we may want the quantization levels to be fixed ahead of time, say in an audio application. In this case, it is sensible to minimize the distortion over an entire class of signals or images. We would also like the quantization and dequantization computations to be simple. The following example gives a scheme that works fairly well.

Example 1.26 *In this example, we will quantize an image, but the same principles apply to a one-dimensional signal. Let us assume that the intensity values of a class of grayscale images of interest satisfy* $m \leq a(x,y) \leq M$ *on some rectangle* Ω*, where* $a(x,y)$ *is the analog image intensity. Let* $\mathbf{A}$ *denote the matrix with components* a_{jk} *obtained by sampling (but not quantizing) the analog image* $a(x,y)$*. Select the* y_k *so that they split up the interval* $[m, M]$ *into* r *subintervals of equal length, and let* z_k *be the mid-point of each interval. If we define* $h = (M - m)/r$*, then we obtain the following formulas for the* y_k*s, the* z_k*s,* q*, and* $\tilde{q}$*:*

$$\begin{aligned} y_k &= m + kh, \quad k = 1, \dots, r-1, \; y_0 = -\infty, \; y_r = \infty, \\ z_k &= m + \left(k + \frac{1}{2}\right) h, \quad k = 0, \dots, r-1, \\ q(x) &= \mathrm{ceil}\left(r\frac{x-m}{M-m}\right) - 1 \quad \text{for } x > m, \quad q(m) = 0, \\ \tilde{q}(k) &= m + \left(k + \frac{1}{2}\right) h. \end{aligned}$$

The ceiling function "ceil" from $\mathbb{R}$ *to* $\mathbb{Z}$ *is defined by taking* $\mathrm{ceil}(x)$ *as the smallest integer greater than or equal to* x*.*

To illustrate, the image at the Figure 1.12a has a sample matrix $\mathbf{A}$ *(stored as double precision floating point) with limits* $0 \leq a_{ij} \leq 255$*. The quantization method above at 5 bpp (32 quantization intervals) or greater gives no measurable distortion. In Figure 1.12 we illustrate quantization at each of* $b = 4, 2, 1$ *bits per pixel (bpp) in order to see the distortion. The measure of the distortion* mD *is reported as a percentage of the total image energy,*

$$mD = 100\frac{\|\mathbf{A} - \tilde{\mathbf{A}}\|^2}{\|\mathbf{A}\|^2},$$

where $\|\cdot\|$ *denotes the Frobenius norm of Example 1.14. The resulting errors are* 0.2%, 3.6%, *and* 15.2% *for the 4, 2, and 1 bit quantizations, respectively.*

Example 1.26 illustrates uniform quantization with midpoint codewords. It also yields an improvement over the error estimate (1.37), namely

$$|a_{ij} - \tilde{a}_{ij}| \leq \frac{h}{2}. \tag{1.38}$$

1.9.2 Quantifying Signal and Image Distortion More Generally

Suppose that we have signal that we want to compress or denoise to produce a processed approximation. The quantization discussion above provides a concrete example, but other similar situations will arise later. In general, what is a reasonable way to quantify the accuracy of the approximation?

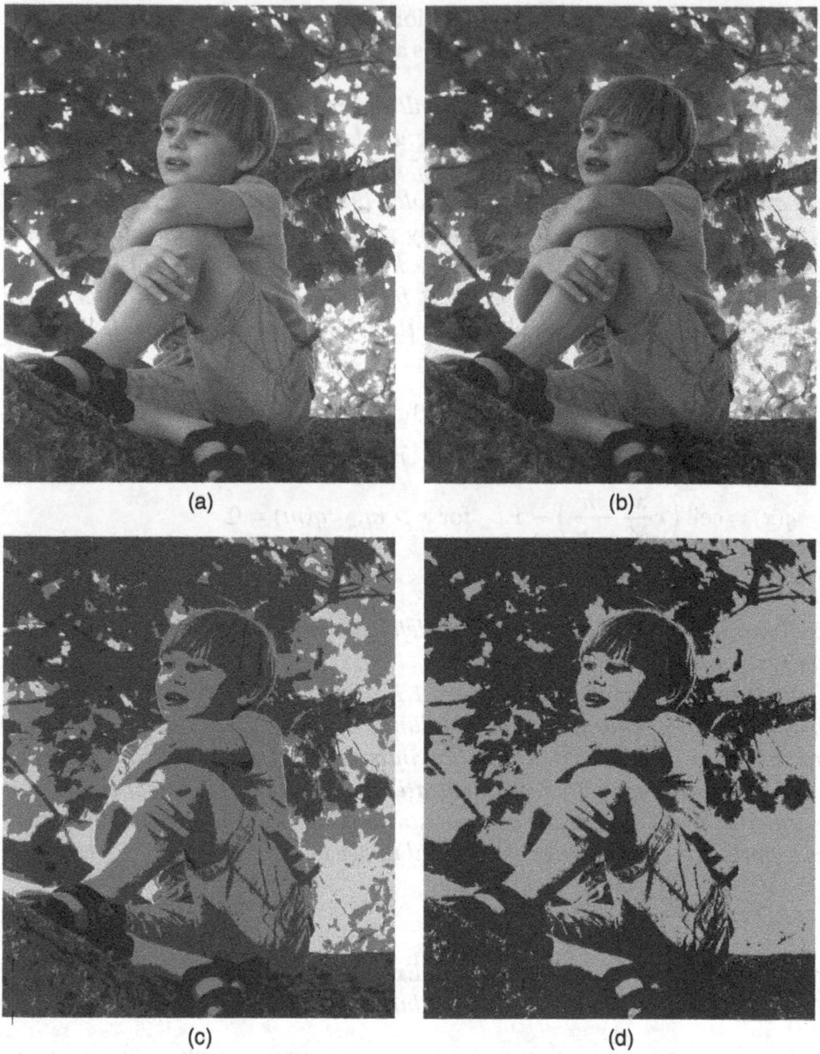

Figure 1.12 Original image (a) and quantization at 4 bits (b), 2 bits (c), and 1 bit (d).

Many approaches are possible, but we will do essentially as we did for the image in Example 1.26. For a discretized image (or signal) $\mathbf{A}$ approximated by $\tilde{\mathbf{A}}$, write

$$\tilde{\mathbf{A}} = \mathbf{A} + \mathbf{E},$$

where $\mathbf{E} = \tilde{\mathbf{A}} - \mathbf{A}$ is the error introduced by using the approximation $\tilde{\mathbf{A}}$ for $\mathbf{A}$. We could also consider $\mathbf{E}$ as some kind of random noise. Our measure of distortion (or noise level, if appropriate) is

$$mD = \frac{\|\tilde{\mathbf{A}} - \mathbf{A}\|^2}{\|\mathbf{A}\|^2} = \frac{\|\mathbf{E}\|^2}{\|\mathbf{A}\|^2}, \tag{1.39}$$

the relative error as is typically done in any physical application. We will often report this error as a percentage as we did with the images above, by multiplying by 100.

1.10 Infinite-Dimensional Inner Product Spaces

Analog signals and images are naturally modeled by functions of one or more real variables, and as such the proper setting for the analysis of these objects is a function space such as $C[a, b]$. However, vector spaces of functions are infinite-dimensional, and some of the techniques developed for finite-dimensional vector spaces need a bit of adjustment. In this section, we give an outline of the mathematics necessary to carry out orthogonal expansions in these function spaces, especially orthogonal expansions with regard to complex exponentials. The ideas in this section play a huge role in applied mathematics. They also provide a nice parallel to the discrete ideas.

1.10.1 Example: An Infinite-Dimensional Space

Let us focus on the vector space $C[a, b]$ for the moment. This space is not finite-dimensional. This can be shown by demonstrating the existence of m linearly independent functions in $C[a, b]$ for any integer $m > 0$. To do this, let $h = (b - a)/m$ and set $x_k = a + kh$ for $0 \le k \le m$; the points x_k partition $[a, b]$ into m equal subintervals, each of length h (with $x_0 = a, x_m = b$). Let $I_k = [x_{k-1}, x_k]$ for $1 \le k \le m$, and define m functions

$$\phi_k(x) = \begin{cases} 0, & \text{if } x \text{ is not in } I_k, \\ \frac{2}{h}(x - x_{k-1}), & x_{k-1} \le x \le (x_{k-1} + x_k)/2, \\ \frac{2}{h}(x_k - x), & (x_{k-1} + x_k)/2 < x \le x_k, \end{cases}$$

for $1 \le k \le m$. Each function ϕ_k is continuous and piecewise linear, identically zero outside I_k, with $\phi_k = 1$ at the midpoint of I_k (such a function is sometimes called a "tent" function—draw a picture). It is easy to show that the functions ϕ_k are linearly independent, for if

$$\sum_{k=1}^{m} c_k \phi_k(x) = 0$$

for $a \leq x \leq b$, then evaluating the left-hand side above at the midpoint of I_j immediately yields $c_j = 0$. Thus, the ϕ_k thus form a linearly independent set. If $C[a, b]$ were finite-dimensional, say of dimension n, then we would not be able to find a set of $m > n$ linearly independent functions. Thus, $C[a, b]$ is not finite-dimensional.

A similar argument can be used to show that any of the vector spaces of functions from Section 1.4.2 are infinite-dimensional.

1.10.2 Orthogonal Bases in Inner Product Spaces

It can be shown that any vector space, even an infinite-dimensional space, has a basis in the sense of Definition 1.8 (a Hamel basis), in which only finite combinations of the basis vectors are allowed. However, such bases are usually difficult to exhibit explicitly and of little use for computation. As such, we are going to expand our notion of basis to allow infinite combinations of the basis vectors. Of course, the word "infinite" always means limits are involved, and if so, we need some measure of distance or length, since limits involve some quantity "getting close" to another.

In the light of this criterion, it is helpful to restrict our attention to vector spaces in which we have some notion of distance, such as a normed vector space. But orthogonality, especially in the infinite-dimensional case, is such a valuable asset that we are going to restrict our attention to inner product spaces. The norm will be that associated with the inner product via equation (1.28).

Let V be an inner product space, over either $\mathbb{R}$ or $\mathbb{C}$. We seek a set S of vectors in V,

$$S = \{\mathbf{v}_1, \mathbf{v}_2, \mathbf{v}_3, \ldots\}, \tag{1.40}$$

that can act as a basis for V in some reasonable sense. Note that we are assuming that the elements of S can be listed, meaning put into a one-to-one correspondence with the positive integers. Such a set is said to be *countable* (many infinite sets are not!).

Remark 1.12 In what follows it is not important that the elements of the set S be indexed as $1, 2, 3, \ldots$. Indeed the elements can be indexed from any subset of $\mathbb{Z}$, $\mathbb{Z}$ itself, or even $\mathbb{Z} \times \mathbb{Z}$. The main point is that we must be able to sum over the elements of S using traditional summation notation, $\sum$. We index the elements starting from 1 in the discussion that follows solely to fix notation.

In the finite-dimensional case the basis vectors must be linearly independent, and this was an automatic consequence of orthogonality. In the infinite-dimensional case, we will cut straight to the chase: in this text, we will only consider as prospective bases those sets that are orthogonal (though nonorthogonal bases can be constructed).

In the finite-dimensional case, a basis must also span the vector space. In the present case, we are allowing infinite linear combinations of basis vectors, and want to be able to write

$$\mathbf{v} = \sum_{k=1}^{\infty} \alpha_k \mathbf{v}_k \tag{1.41}$$

for any $\mathbf{v} \in V$ by choosing the coefficients α_k suitably. But infinite linear combinations have no meaning in a general vector space. How should equation (1.41) be interpreted?

Recall that in elementary calculus the precise definition of an infinite sum

$$\sum_{k=1}^{\infty} a_k = A$$

is that $\lim_{n\to\infty}(\sum_{k=1}^{n} a_k) = A$ (the sequence of partial sums converges to A). This is equivalent to

$$\lim_{n\to\infty} \left| \sum_{k=1}^{n} a_k - A \right| = 0.$$

This motivates our interpretation of equation (1.41) and the definition of what it means for the set S in (1.40) to span V. However, in an infinite-dimensional inner product space the term "span" is replaced by "complete."

Definition 1.10 An (orthogonal) set S as in (1.40) is "*complete*" if for each $\mathbf{v} \in V$ there are scalars α_k, $k \geq 1$, such that

$$\lim_{n\to\infty} \left\| \sum_{k=1}^{n} \alpha_k \mathbf{v}_k - \mathbf{v} \right\| = 0. \tag{1.42}$$

The limit in (1.42) is just an ordinary limit for a sequence of real numbers. The norm on the inner product space takes the place of absolute value in $\mathbb{R}$.

Definition 1.11 A set S as in (1.40) is called an "*orthogonal basis*" for V if S is complete and orthogonal. If S is orthonormal, then S is called an "*orthonormal basis.*"

The existence of an orthogonal basis as in Definition 1.11 is not assured but depends on the particular inner product space. However, all of the function spaces of interest from Section 1.8.2 have such bases. In the case of spaces consisting of functions of a single real variable (e.g., $C[a,b]$), we can write out a basis explicitly, and also for function spaces defined on a rectangle in the plane. We will do this shortly.

1.10.3 The Cauchy–Schwarz Inequality and Orthogonal Expansions

For now let us assume that an orthogonal basis $S = \{\mathbf{v}_1, \mathbf{v}_2, \dots\}$ exists. How can we compute the α_k in the expansion (1.41)? Are they uniquely determined? It is tempting to mimic the procedure used in the finite-dimensional case: take the inner product of both sides of (1.41) with a specific basis vector $\mathbf{v}_m$ to obtain $(\mathbf{v}, \mathbf{v}_m) = (\sum_k \alpha_k \mathbf{v}_k, \mathbf{v}_m)$ then use linearity of the inner product in the first argument to obtain $(\mathbf{v}, \mathbf{v}_m) = \sum_k \alpha_k (\mathbf{v}_k, \mathbf{v}_m) = \alpha_m (\mathbf{v}_m, \mathbf{v}_m)$. This immediately yields $\alpha_m = (\mathbf{v}, \mathbf{v}_m)/(\mathbf{v}_m, \mathbf{v}_m)$, just as in the finite-dimensional case. This reasoning is a bit suspect, though, because it requires us to invoke linearity for the inner product with respect to an infinite sum. Unfortunately, the definition of the inner product makes no statements concerning infinite sums. We need to be a bit more careful (though the answer for α_m is correct!).

To demonstrate the validity of the conclusion above more carefully we will use the Cauchy–Schwarz inequality in Theorem 1.1. Specifically, suppose that $S = \{\mathbf{v}_1, \mathbf{v}_2, \dots\}$ is an orthogonal basis, so that for any $\mathbf{v} \in V$ there is some choice of scalars α_k for which equation (1.42) holds. For some fixed m consider the inner product $(\mathbf{v} - \sum_{k=1}^{n} \alpha_k \mathbf{v}_k, \mathbf{v}_m)$. If we expand this inner product and suppose $n \geq m$ while taking absolute values throughout, we find that

$$\left| \left(\mathbf{v} - \sum_{k=1}^{n} \alpha_k \mathbf{v}_k, \mathbf{v}_m \right) \right| = \left| (\mathbf{v}, \mathbf{v}_m) - \sum_{k=1}^{n} \alpha_k (\mathbf{v}_k, \mathbf{v}_m) \right|$$
$$= |(\mathbf{v}, \mathbf{v}_m) - \alpha_m (\mathbf{v}_m, \mathbf{v}_m)|. \tag{1.43}$$

Note that all sums above are finite. On the other hand, the Cauchy–Schwarz inequality yields

$$\left| \left(\mathbf{v} - \sum_{k=1}^{n} \alpha_k \mathbf{v}_k, \mathbf{v}_m \right) \right| \leq \left\| \mathbf{v} - \sum_{k=1}^{n} \alpha_k \mathbf{v}_k \right\| \|\mathbf{v}_m\|. \tag{1.44}$$

Combine (1.43) and (1.44) to find that for $n \geq m$

$$|(\mathbf{v}, \mathbf{v}_m) - \alpha_m (\mathbf{v}_m, \mathbf{v}_m)| \leq \left\| \mathbf{v} - \sum_{k=1}^{n} \alpha_k \mathbf{v}_k \right\| \|\mathbf{v}_m\|. \tag{1.45}$$

The completeness assumption forces $\|\mathbf{v} - \sum_{k=1}^{n} \alpha_k \mathbf{v}_k\| \to 0$ as $n \to \infty$ on the right-hand side of (1.45) (and $\|\mathbf{v}_m\|$ remains fixed, since m is fixed). The left-hand side of (1.45) must also approach zero, but the left-hand side does not depend on n. We must conclude that $(\mathbf{v}, \mathbf{v}_m) - \alpha_m (\mathbf{v}_m, \mathbf{v}_m) = 0$, so $\alpha_m = (\mathbf{v}, \mathbf{v}_m)/(\mathbf{v}_m, \mathbf{v}_m)$. Of course, if S is orthonormal, this becomes just $\alpha_m = (\mathbf{v}, \mathbf{v}_m)$.

Note that we assumed an expansion as in (1.42) exists. What we have shown above is that IF such an expansion exists (i.e., if S is complete), THEN the α_k are uniquely determined and given by the formula derived. Let us state this as a theorem.

Theorem 1.4 *If S is an orthogonal basis for an inner product space V, then for any* $\mathbf{v} \in V$, *equation (1.42) holds where* $\alpha_k = (\mathbf{v}, \mathbf{v}_k)/(\mathbf{v}_k, \mathbf{v}_k)$.

It is conventional to write the expansion of **v** in the shorthand form of (1.41), with the understanding that the precise meaning is the limit in (1.42).

Interestingly, Parseval's identity still holds. Note that if S is an orthonormal basis and $\alpha_k = (\mathbf{v}, \mathbf{v}_k)$, then

$$\left\| \mathbf{v} - \sum_{k=1}^{n} \alpha_k \mathbf{v}_k \right\|^2 = \left(\mathbf{v} - \sum_{k=1}^{n} \alpha_k \mathbf{v}_k, \mathbf{v} - \sum_{k=1}^{n} \alpha_k \mathbf{v}_k \right)$$
$$= (\mathbf{v}, \mathbf{v}) - \sum_{k=1}^{n} \alpha_k (\mathbf{v}_k, \mathbf{v}) - \sum_{k=1}^{n} \overline{\alpha_k} (\mathbf{v}, \mathbf{v}_k) + \sum_{k=1}^{n} |\alpha_k|^2$$
$$= \|\mathbf{v}\|^2 - \sum_{k=1}^{n} |\alpha_k|^2.$$

As $n \to \infty$ the left-hand side approaches zero; hence so does the right-hand side, and we obtain

$$\sum_{k=1}^{\infty} |\alpha_k|^2 = \|\mathbf{v}\|^2, \tag{1.46}$$

which is Parseval's identity in the infinite-dimensional case.

There are a variety of equivalent characterizations or definitions of what it means for a set S to be complete (e.g., S is complete if Parseval's identity holds for all **v**), and some may make it easier or harder to verify that any given set S is complete. Proving that a given set of vectors in a specific inner product space is complete always involves some analysis ("analysis" in the sense of limits, inequalities, and estimates). We will not go into these issues in much detail in this text. We will simply exhibit, without proof, some orthogonal bases for common spaces of interest.

1.10.4 The Basic Waveforms and Fourier Series

For the moment let us focus on the space $C[-T, T]$ of continuous complex-valued functions on the closed interval $[-T, T]$, where $T > 0$. This, of course, includes the real-valued functions on the interval. The vector space $C[-T, T]$ becomes an inner product space if we define the inner product as in Example 1.15.

1.10.4.1 Complex Exponential Fourier Series

Let S denote the set of basic analog waveforms $\phi_k(t) = e^{ik\pi t/T}$ for $k \in \mathbb{Z}$ introduced in Example 1.19. In that example it was shown that this set is orthogonal. It can be shown with a bit of nontrivial analysis that this set is complete in $C[-T,T]$ or $L^2(-T,T)$ (see, e.g., [25] or [9], Sec. II.4.4). As a consequence S is a basis.

Note that the basis ϕ_k here is indexed in k, which ranges over all of $\mathbb{Z}$, but according to Remark 1.12 this makes no essential difference.

Consider a typical function $f(t)$ in $C[-T,T]$. From Theorem 1.4 we have the expansion

$$f(t) = \sum_{k=-\infty}^{\infty} \alpha_k e^{ik\pi t/T}, \tag{1.47}$$

where

$$\alpha_k = \frac{(f, e^{ik\pi t/T})}{(e^{ik\pi t/T}, e^{\pi ikt/T})} = \frac{1}{2T}\int_{-T}^{T} f(t)e^{-ik\pi t/T}\,dt, \tag{1.48}$$

since $\int_{-T}^{T} e^{ik\pi t/T}e^{-ik\pi t/T}\,dt = 2T$. The right-hand side of equation (1.47) is called the Fourier series for f, and the α_k of equation (1.48) are called the *Fourier coefficients.*

Example 1.27 *Let $T = 1$, and consider the function $f(t) = t^2$ in $C[-1,1]$. The Fourier coefficients from equation (1.48) are given by*

$$\begin{aligned}\alpha_k &= \frac{1}{2}\int_{-1}^{1} t^2 e^{-ik\pi t}\,dt\\ &= \frac{2(-1)^k}{\pi^2 k^2}\end{aligned}$$

for $k \neq 0$ (integrate by parts twice), while $\alpha_0 = \frac{1}{3}$. We can write out the Fourier series for this function as

$$f(t) = \frac{1}{3} + \frac{2}{\pi^2}\sum_{k=-\infty, k\neq 0}^{\infty} \frac{(-1)^k e^{ik\pi t}}{k^2}.$$

It should be emphasized that the series on the right "equals" $f(t)$ only in the sense defined by (1.42). As such, it is instructive to plot $f(t)$ and the right-hand side above summed from $k = -n$ to n for a few values of n, as in Figure 1.13. In the case where $n = 0$ the approximation is just $f(t) \approx a_0 = (\int_{-T}^{T} f(t)\,dt)/2T$, the average value of f over the interval. In general, there is no guarantee that the Fourier series for any specific value $t = t_0$ converges to the value $f(t_0)$, unless one knows something more about f. However, the Fourier series will converge in the sense of equation (1.42) for any function $f \in L^2(-T,T)$.

It is also worth noting that the Fourier series itself continues periodically outside the interval $[-T,T]$ with period $2T$, since each of the basic waveforms are periodic with period $2T$. This occurs even though f does not need to be defined outside this interval. This case is shown in Figure 1.14.

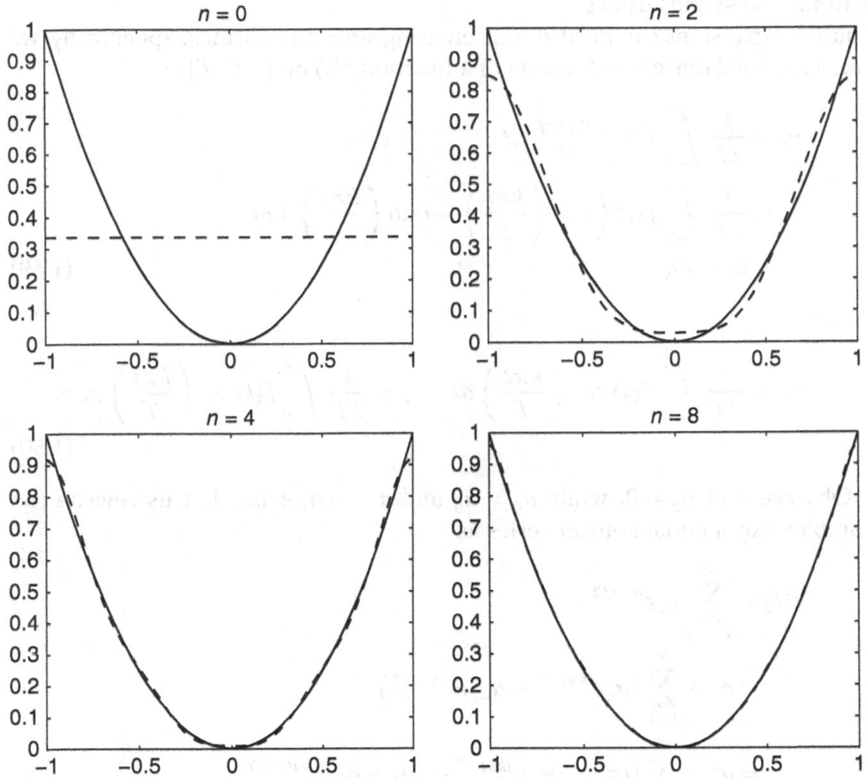

Figure 1.13 Function $f(t) = t^2$ (solid) and Fourier series approximations (dashed), $n = 0, 2, 4, 8$.

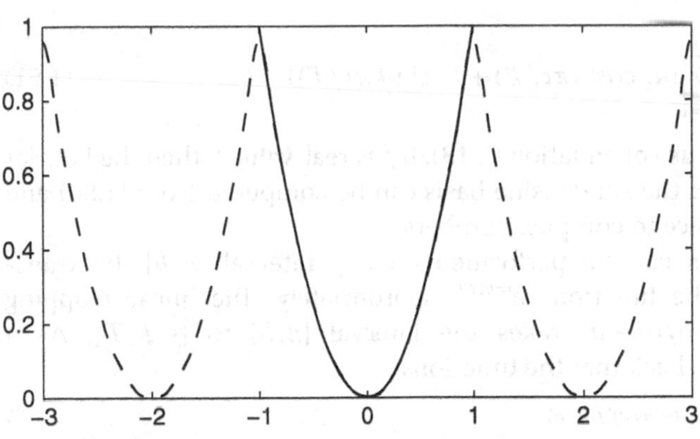

Figure 1.14 Function $f(t) = t^2, -1 \le t \le 1$ and eight-term Fourier series approximation extended to $[-3, 3]$.

1.10.4.2 Sines and Cosines

Fourier expansions can be also written using sines and cosines. Specifically, we can write the Fourier coefficients of a function $f(t)$ on $[-T, T]$ as

$$\begin{aligned}\alpha_k &= \frac{1}{2T}\int_{-T}^{T} f(t)e^{-ik\pi t/T}\,dt \\ &= \frac{1}{2T}\int_{-T}^{T} f(t)\left(\cos\left(\frac{k\pi t}{T}\right) - i\sin\left(\frac{k\pi t}{T}\right)\right)dt \\ &= a_k - ib_k,\end{aligned} \tag{1.49}$$

where

$$a_k = \frac{1}{2T}\int_{-T}^{T} f(t)\cos\left(\frac{k\pi t}{T}\right)dt,\quad b_k = \frac{1}{2T}\int_{-T}^{T} f(t)\sin\left(\frac{k\pi t}{T}\right)dt. \tag{1.50}$$

Observe that $b_0 = 0$, while $a_0 = \alpha_0$ and $\alpha_{-k} = a_k + ib_k$. Let us rewrite the complex exponential Fourier series as

$$\begin{aligned}f(t) &= \sum_{k=-\infty}^{\infty} \alpha_k e^{ik\pi t/T} \\ &= \alpha_0 + \sum_{k=1}^{\infty}\left(\alpha_k e^{ik\pi t/T} + \alpha_{-k}e^{-ik\pi t/T}\right) \\ &= a_0 + \sum_{k=1}^{\infty}\left((a_k - ib_k)e^{ik\pi t/T} + (a_k + ib_k)e^{-ik\pi t/T}\right) \\ &= a_0 + \sum_{k=1}^{\infty}\left(a_k\left(e^{ik\pi t/T} + e^{-ik\pi t/T}\right) - ib_k\left(e^{ik\pi t/T} - e^{-ik\pi t/T}\right)\right) \\ &= a_0 + 2\sum_{k=1}^{\infty}(a_k\cos(k\pi t/T) + b_k\sin(k\pi t/T)),\end{aligned} \tag{1.51}$$

where we have made use of equation (1.13). If f is real-valued, then the Fourier series with respect to the sine/cosine basis can be computed from (1.50) and (1.51) with no reference to complex numbers.

Fourier expansions can be performed on any interval $[a, b]$, by translating and scaling the functions $e^{i\pi kt/T}$ appropriately. The linear mapping $\phi : t \to T(2t - (a + b))/(b - a)$ takes the interval $[a, b]$ to $[-T, T]$. As a consequence we can check that the functions

$$e^{i\pi k\phi(t)/T} = e^{ik\pi(2t-(a+b))/(b-a)}$$

are orthogonal on $[a, b]$ with respect to the $L^2(a, b)$ inner product, and so form a basis.

1.10.4.3 Fourier Series on Rectangles

The set of functions $\phi_{k,m}(x,y)$, $k \in \mathbb{Z}$, and $m \in \mathbb{Z}$ defined by

$$\phi_{k,m}(x,y) = e^{2\pi i(kx/a+my/b)}$$

forms an orthogonal basis for the set $C(\Omega)$ where $\Omega = \{(x,y); 0 \le x \le a, 0 \le y \le b\}$. As in the one-dimensional case, we can also write out an equivalent basis of sines and cosines.

Orthogonal bases can be proved to exist for more general two-dimensional regions, or regions in higher dimensions, but they cannot usually be written out explicitly.

1.10.5 Hilbert Spaces and $L^2(a,b)$

The inner product space of continuous, square-integrable functions on an interval (a,b) was defined in Example 1.15. This space is satisfactory for some purposes, but in many cases it is helpful to enlarge the space to contain discontinuous functions. The inclusion of discontinuous functions brings with it certain technical difficulties that, in full generality and rigor, require some sophisticated analysis to resolve. This analysis is beyond the scope of this text. Nonetheless, we give below a brief sketch of the difficulties, the resolution, and a useful fact concerning orthogonal bases.

1.10.5.1 Expanding the Space of Functions

In our above Fourier series examples we worked in inner product spaces that consist of continuous functions. However, the definition of the inner product

$$(f,g) = \int_a^b f(t)\overline{g(t)}\,dt$$

requires much less of the functions involved. Indeed we might want to write out Fourier expansions for functions that have discontinuities. Such expansions can be very useful, so it makes sense to enlarge this inner product space to include more than just continuous functions.

The Cauchy–Schwarz inequality indicates the direction we should move. The inner product (f,g) of two functions is guaranteed to be finite if the L^2 norms $\|f\|$ and $\|g\|$ are both finite. This was exactly the condition imposed in Example 1.8. Thus, we will enlarge our inner product space to include all functions f that satisfy

$$\int_a^b |f(t)|^2\,dt < \infty. \tag{1.52}$$

This guarantees that the inner product (f,g) is defined for all pairs $f,g \in V$. However, equation (1.52) presumes that the function $|f(t)|^2$ can be meaningfully integrated using the Riemann integral. This may seem like a technicality, but while continuous functions are automatically Riemann integrable, arbitrary functions are not. Thus, we require Riemann integrability of f (which ensures that $|f|^2$ is Riemann integrable). Our notion of $L^2(a,b)$ is thus enlarged to include many types of discontinuous functions.

1.10.5.2 Complications

This enlargement of the space creates a problem of its own though. Consider a function f that is zero at all but one point in $[a,b]$. Such a function, as well as $|f(t)|^2$, is Riemann integrable, and indeed the integral of $|f(t)|^2$ will be zero. In short, $\|f\| = 0$ and also $(f,f) = 0$, even though f is not identically zero. This violates property 1 for the norm and property 3 for the inner product. More generally, functions f and g that differ from each other at only finitely many (perhaps even more) points satisfy $\|f - g\| = 0$, even though $f \neq g$. Enlarging our inner product space has destroyed the inner product structure and notion of distance! This was not a problem when the functions were assumed to be continuous; recall Example 1.15.

The fix for this problem requires us to redefine what we mean by "$f = g$" in $L^2(a,b)$; recall the remark at the end of Example 1.3. Functions f_1 and f_2 that differ at sufficiently few points are considered identical under an equivalence relation "$\sim$," where "$f_1 \sim f_2$" means $\int |f_1 - f_2|^2\,dt = 0$. The elements of the space of $L^2(a,b)$ thus consist of equivalence classes of functions that differ at sufficiently few points. This is similar to the rational numbers, which consist of "fractions," but with a_1/b_1 and a_2/b_2 identified under the equivalence relation $a_1/b_1 \sim a_2/b_2$ if $a_1b_2 = a_2b_1$. Thus, for example, $\frac{1}{2}$, $\frac{2}{4}$, and $\frac{3}{6}$ are all the same rational number. As a consequence functions in $L^2(a,b)$ do not have well-defined point values. When we carry out computations in $L^2(a,b)$, it is almost always via integration or inner products.

These changes also require some technical modifications to our notion of integration. It turns out that Riemann integration is not sufficient for the task, especially when we need to deal with sequences or series of functions in $L^2(a,b)$. The usual approach requires replacing the Riemann integral with the Lebesgue integral, a slightly more versatile approach to integration. The modifications also have the happy consequence of "completing" the inner product space by guaranteeing that Cauchy sequences in the space converge to a limit within the space, which greatly simplifies the analysis. A complete inner product space is known as a *Hilbert space*. The Hilbert space obtained in this situation is known as $L^2(a,b)$. These spaces play an important role in applied mathematics and physics.

Despite all of these complications, we will use the notation $L^2(a,b)$ for the set of all Riemann integrable functions that satisfy inequality (1.52). Indeed, if we limit our attention to the subset of piecewise continuous functions, then no real difficulties arise. For the details of the full definition of $L^2(a,b)$ the interested reader can refer to [24] or any advanced analysis text.

1.10.5.3 A Converse to Parseval

Parseval's identity (1.46) shows that if ϕ_k, $k \geq 0$, is an orthonormal basis for $L^2(a,b)$ and $f \in L^2(a,b)$, then f can be expanded as $f = \sum_k c_k \phi_k$ and $\sum_k |c_k|^2 = \|f\|^2 < \infty$. That is, every function in $L^2(a,b)$ generates a sequence $\mathbf{c} = (c_0, c_1, \ldots)$ in $L^2(\mathbb{N})$. In our new and improved space $L^2(a,b)$ it turns out that the converse is true: if we choose any sequence $\mathbf{c} = (c_0, c_1, \ldots)$ with $\sum_k c_k^2 < \infty$, then the sum

$$\sum_{k=0}^{\infty} c_k \phi_k \tag{1.53}$$

converges to some function $f \in L^2(a,b)$, in the sense that

$$\lim_{n\to\infty} \left\| f - \sum_{k=0}^{n} c_k \phi_k \right\| = 0.$$

In short, the functions in $L^2(a,b)$ and sequences in $L^2(\mathbb{N})$ can be matched up one to one (the correspondence depends on the basis ϕ_k). This is worth stating as a theorem.

Theorem 1.5 *Let ϕ_k, $k \geq 0$, be an orthonormal basis for $L^2(a,b)$. There is an invertible linear mapping $\Phi : L^2(a,b) \to L^2(\mathbb{N})$ defined by*

$$\Phi(f) = \mathbf{c},$$

where $\mathbf{c} = (c_0, c_1, \ldots)$ with $c_k = (f, \phi_k)$. Moreover, $\|f\|_{L^2(a,b)} = \|\mathbf{c}\|_{L^2(\mathbb{N})}$ (the mapping is an "isometry," that is, length-preserving).

1.11 Matlab Project

This section is designed as a series of Matlab explorations that allow the reader to play with some of the ideas in the text. It also introduces a few Matlab commands and techniques we will find useful later. Matlab commands that should executed are in a bold `typewriter font` and usually displayed prominently.

1. Start Matlab. Most of the computation in the text is based on indexing vectors beginning with index 0, but Matlab indexes vectors from 1. This will not usually cause any headaches, but keep it in mind.
2. Consider sampling the function $f(t) = \sin(2\pi(440)t)$ on the interval $0 \leq t < 1$, at 8192 points (sampling interval $\Delta T = 1/8192$) to obtain samples $f_k = f(k\Delta T) = \sin(2\pi(440)k/8192)$ for $0 \leq k \leq 8191$. The samples can be arranged in a vector `f`. You can do this in Matlab with

```
f = sin(2*pi*440/8192*(0:8191));
```

Do not forget the semicolon or Matlab will print out `f`!

The sample vector `f` is stored in double precision floating point, about 15 significant figures. However, we will consider `f` as not yet quantized. That is, the individual components f_k of `f` can be thought of as real numbers that vary continuously, since 15 digits is pretty close to continuous for our purposes.

(a) What is the frequency of the sine wave $\sin(2\pi(440)t)$, in hertz?

(b) Plot the sampled signal with the command `plot(f)`. It probably does not look too good, as it goes up and down 440 times in the plot range. You can plot a smaller range, say the first 100 samples, with `plot(f(1:100))`.

(c) At the sampling rate 8192 Hz, what is the Nyquist frequency? Is the frequency of $f(t)$ above or below the Nyquist frequency?

(d) Type `sound(f)` to play the sound out of the computer speaker. By default, Matlab plays all sound files at 8192 samples per second, and assumes the sampled audio signal is in the range −1 to 1. Our signal satisfies these conditions.

(e) As an example of aliasing, consider a second signal $g(t) = \sin(2\pi(440 + 8192)t)$. Repeat parts (a) through (d) with sampled signal

```
g = sin(2*pi*(440+8192)/8192*(0:8191));
```

The analog signal $g(t)$ oscillates much faster than $f(t)$, and we could expect it to yield a higher pitch. However, when sampled at frequency 8192 Hz, $f(t)$ and $g(t)$ are aliased and yield precisely the same sampled vectors **f** and **g**. They should sound the same too.

(f) To illustrate the effect of quantization error, let us construct a 2-bit (four quantization levels) version of the audio signal $f(t)$ as in the scheme of Example 1.26. With that notation we have minimum value $m = -1$ and maximum value $M = 1$ for our signal, with $r = 4$. The command

```
qf = ceil(2*(f+1))-1;
```

produces the quantized signal $q(\mathbf{f})$. Sample values of $f(t)$ in the ranges $(-1, -0.5]$, $(-0.5, 0]$, $(0, 0.5]$, and $(0.5, 1]$ are mapped to the integers 0, 1, 2, and 3, respectively.
To approximately reconstruct the quantized signal, we apply the dequantization formula to construct $\tilde{\text{f}}$ as

```
ftilde = -1 + 0.5*(qf+0.5);
```

This maps the integers 0, 1, 2, and 3 to values −0.75, −0.25, 0.25, and 0.75, respectively (the codewords in this scheme).

(g) Plot the first hundred values of `ftilde` with `plot(ftilde(1:100));`. Play the quantized signal with `sound(ftilde);`. It should sound harsh compared to `f`.

(h) Compute the distortion (as a percentage) of the quantized/dequantized signal using equation (1.39). In Matlab this is implemented as

```
100*norm(f-ftilde).^2/norm(f).^2
```

The `norm` command computes the standard Euclidean norm of a vector.

(i) Repeat parts (f) through (h) using 3-,4-,5-,6-,7-, and 8-bit quantization. For example, 5-bit quantization is accomplished with `qf=ceil (16*(f+1))-1`, dequantization with `ftilde=-1+(qf+0.5)/16`. Here $16 = 2^{5-1}$. Make sure to play the sound in each case. Make up a table showing the number of bits in the quantization scheme, the corresponding distortion, and your subjective rating of the sound quality. At what point can your ear no longer distinguish the original audio signal from the quantized version?

3. Type `load('splat')`. This loads in an audio signal sampled at 8192 samples per second. By default the signal is loaded into a vector "y" in Matlab, while the sampling rate is loaded into a variable "`Fs`." Execute the Matlab command `whos` to verify this (or look at the workspace window that shows the currently defined variables).
   (a) Play the sampled signal with `sound(y)`.
   (b) Plot the sampled signal. Based on the size of `y` and sample rate 8192 Hz, what is T, the length (in seconds) of the sampled sound?
   (c) The audio signal was sampled at frequency 8192. We can mimic a lower sampling rate $8192/m$ by taking every mth entry of the original vector `y`; this is called *downsampling*. In particular, let us try $m = 2$. Execute `y2 = y(1:2:10001);`. The downsampled vector `y2` is the same sound, but sampled at frequency $8192/2 = 4096$ Hz.
   Play the downsampled `y2` with `sound(y2,4096)`. The second argument to the `sound` command indicates the sound should be played back at the corresponding rate (in Hz).
   (d) Comment: why does the whistling sound in the original audio signal fall steadily in pitch, while the downsampled version seems to rise and fall?
4. You can clear from Matlab memory all variables defined to this point with the command `clear`. Do so now.
   (a) Find an image (JPEG will do) and store it in Matlab's current working directory. You can load the image into Matlab with

   ```
   z=imread('myimage.jpg');
   ```

   (change the name to whatever your image is called!) If the image is color, the "`imread`" command automatically converts the $m \times n$ pixel image (in whatever conventional format it exists) into three $m \times n$ arrays, one for each of red, blue, and green as in Section 1.3.5. Each array consists of unsigned 8-bit integers, that is, integers in the range 0–255. Thus, the variable z is now an $m \times n \times 3$ array. If the image is grayscale you will get only one array.

(b) The command

```
image(z);
```

displays the image in color, under the convention that z consists of unsigned integers and the colors are scaled in the 0–255 range. If we pass an $m \times n$ by 3 array of floating point numbers to the image command, it is assumed that each color array is scaled from 0.0 to 1.0. Consult the help page for the "image" command for more information.

(c) A simple way to construct an artificial grayscale image is by picking off one of the color components and using it as a grayscale intensity, for example, `zg = double(z(:,:,1));`. However, a slightly more natural result is obtained by taking a weighted average of the color components, as

```
zg = 0.2989*double(z(:,:,1))+0.5870*double(z(:,:,2))
+0.1140*double(z(:,:,3));
```

The `double` command indicates that the array should be converted from unsigned integers to double precision floating point numbers, for example, 13 becomes 13.0. It is not strictly necessary unless we want to do floating point arithmetic on `zg` (which we do). The weighting coefficients above stem from the NTSC (television) color scheme; see [14] for more information. Now we set up an appropriate grayscale color map with the commands

```
L = 255;
colormap([(0:L)/L; (0:L)/L; (0:L)/L]');
```

Type `help colormap` for more information on this command. Very briefly, the array that is passed to the colormap command should have three columns, any number of rows. The three entries in the kth row should be scaled in the 0.0–1.0 range, and indicate the intensity of RGB that should be displayed for a pixel that is assigned integer value k. In our `colormap` command above, the kth row consists of elements $((k-1)/255, (k-1)/255, (k-1)/255)$, which correspond to a shade of gray (equal amounts of red, blue, and green), with $k = 0$ as black and $k = 255$ as white.

Now display the grayscale image with

```
image(zg);
```

It is not necessary to execute the `colormap` command prior to displaying every image—once will do, unless you close the display window, in which case you must re-initialize the color map.

(d) The image is currently quantized at eight bit precision (each pixel's graylevel specified by one of $0, 1, 2, \ldots, 255$). We can mimic a cruder quantization level, say 6-bit, with

```
qz = 4*floor(zg/4);
```

followed by `image(qz);`. This command has the effect of rounding each entry of `zg` to the next lowest multiple of 4, so each pixel is now encoded as one of the 64 numbers $0, 4, 8, \ldots, 252$. Can you tell the difference in the image?
Compute the percent distortion introduced with

```
100*norm(zg-qz,'fro').^2/norm(zg,'fro').^2
```

The "`fro`" argument indicates that the Frobenius norm of Example 1.14 should be used for the matrices, that is, take the square root of the sum of the squares of the matrix entries.
Repeat the above computations for other b-bit quantizations with $b = 1, 2, 3, 4, 5$. Display the image in case, and compute the distortion. At what point does the quantization become objectionable?

(e) We can add noise to the image with

```
zn = zg + 50*(rand(size(zg))-0.5);
```

which should, of course, be followed by `image(zn);`. The `size` command returns the dimensions of the matrix `zg`, and the `rand` command generates a matrix of that size consisting of uniformly distributed random numbers (double precision floats) on the interval 0 to 1. Thus, `50*(rand(size(zg))-0.5)` yields an array of random numbers in the range −25 to 25 that is added to `zg`. Any values in `zn` that are out of range (< 0 or > 255) are "clipped" to 0 or 255, respectively.

5. This exercise illustrates aliasing for images. First, clear all variables with `clear`, and then execute the commands `L = 255;` and

```
colormap([(0:L)/L; (0:L)/L; (0:L)/L]');
```

as in the previous exercise.
Let us sample and display an analog image, say

$$f(x, y) = 128(1 + \sin(2\pi(20)x)\sin(2\pi(30)y)),$$

on the square $0 \le x, y \le 1$; we have chosen f to span the range 0–256. We will sample on an m by n grid for various values of m and n. This can be accomplished with

```
m = 50; X = [0:m-1]/m;
n = 50; Y = [0:n-1]/n;
f = 128*(1+sin(2*pi*30*Y)'*sin(2*pi*20*X));
```

The first portion `sin(2*pi*30*Y)'` are the y values of the grid points (as a column vector) and `sin(2*pi*20*X)` are the x values (as a row vector). Plot the image with `image(f)`.

Try various values of m and n from 10 to 500. Large values of m and n should produce a more "faithful" image, while small values should produce obvious visual artifacts; in particular, try $m = 20, n = 30$. The effect is highly dependent on screen resolution.

Exercises

1. Find at least five different color JPEG images on a computer (or with a web browser); they will have a ".jpg" suffix on the file name. Try to find a variety of images, for example, one that is "mostly dark" (astronomical images are a good bet). For each image determine the following:
- Its pixel-by-pixel size—how many pixels wide, how much tall. This can usually be determined by right mouse clicking on the image and selecting "Properties."
- The memory storage requirements for the image if it was stored byte-by-byte, with no compression.
- The actual memory storage requirement (the file size).
- Compute the ratio of the actual and "naive" storage requirements.

Summarize your work in a table. Note the range of compression ratios obtained. Can you detect any correlation between the nature or quality of the images and the compression ratios achieved?

2. Use Euler's identity (1.11) to prove that if x is real then
- $e^{-ix} = \overline{e^{ix}}$
- $e^{2\pi ix} = 1$ if and only if x is an integer.

3. The complex number z is an Nth root of unity if and only if $z^N = 1$. Draw the eighth roots of unity on the unit circle.

4. Prove that the entries of $\mathbf{E}_{N,k}$ are Nth roots of unity (see Exercise 3).

5. Suppose that

$$\begin{aligned} x(t) &= a\cos(\omega t) + b\sin(\omega t) \\ &= ce^{i\omega t} + de^{-i\omega t} \end{aligned}$$

for all real t. Show that

$$a = c + d, \quad b = ic - id,$$
$$c = \frac{a - ib}{2}, \quad d = \frac{a + ib}{2}.$$

6. Let $x(t) = 1.3t$ and $y(t) = \sin(\frac{\pi}{2}t)$ be analog signals on the interval $0 \le t \le 1$.

(a) Sample $x(t)$ at times $t = 0, 0.25, 0.5, 0.75$ to produce sampled vector $\mathbf{x} = (x(0), x(0.25), x(0.5), x(0.75)) \in \mathbb{R}^4$. Sample $y(t)$ at the same times to produce vector $\mathbf{y} \in \mathbb{R}^4$.
Verify that the sampled version (same times) of the analog signal $x(t) + y(t)$ is just $\mathbf{x} + \mathbf{y}$ (this should be painfully clear).

(b) Let q denote a function that takes any real number r and rounds it to the nearest integer, a simple form of quantization. Use q to quantize $\mathbf{x}$ from part (1) component by component, to produce a quantized vector $q(\mathbf{x}) = (q(x_0), q(x_1), q(x_2), q(x_3))$. Do the same for $\mathbf{y}$ and $\mathbf{x} + \mathbf{y}$. Show that $q(\mathbf{x}) + q(\mathbf{y}) \neq q(\mathbf{x} + \mathbf{y})$, and also that $q(2\mathbf{x}) \neq 2q(\mathbf{x})$. Quantization is a nonlinear operation!

7. Let $x(t) = \sin(\pi t/2)$ on $0 \le t \le 1$ (note $0 \le x(t) \le 1$ on this interval). Suppose we sample $x(t)$ at times $t = 0.2, 0.4, 0.6, 0.8$.

(a) Compute $x(t)$ at each time, to at least 10 significant figures.

(b) Suppose we quantize the sampled signal by rounding to the nearest multiple of 0.25. Compute the distortion so introduced, using the procedure of Example 1.26.

(c) Repeat part (b) if we round to the nearest multiple of 0.1 and 0.05.

(d) If we round to the nearest multiple of "h," what is the codebook we are using? What is the dequantization map?

8. Is the set of all quadratic polynomials in x with real-valued coefficients (with polynomial addition and scalar multiplication defined in the usual way) a vector space over $\mathbb{R}$? Why or why not? (Consider something like $a_0 + a_1x + 0x^2$ a quadratic polynomial.)

9. Is the set of all continuous real-valued functions $f(x)$ defined on $[0, 1]$ which satisfy

$$\int_0^1 f(x)\, dx = 3$$

a vector space over $\mathbb{R}$? Assume function addition and scalar multiplication are defined as usual.

10. Clearly $\mathbb{R}^n$ is a subset of $\mathbb{C}^n$. Is $\mathbb{R}^n$ a subspace of $\mathbb{C}^n$ (where $\mathbb{C}^n$ is considered a vector space over $\mathbb{C}$)?

11. Verify that the set in Example 1.3 with given operations is a vector space.

12. Verify that the set $L^2(\mathbb{N})$ in Example 1.5 with given operations is a vector space. Hint: This closely parallels Example 1.8, with summation in place of integrals.
Explain why $L^2(\mathbb{N})$ is a subspace of $L^\infty(\mathbb{N})$.

13. The point of this exercise is to prove the assertions in Proposition 1.1 for an abstract vector space.
(a) Show that the $\mathbf{0}$ vector is unique. To do this suppose there are two vectors, say $\mathbf{0}_1$ and $\mathbf{0}_2$, both of which play the role of the zero vector. Show $\mathbf{0}_1 = \mathbf{0}_2$. Hint: Consider $\mathbf{0}_1 + \mathbf{0}_2$.
(b) Below is a proof that $0\mathbf{u} = \mathbf{0}$ in any vector space. In this proof, $-\mathbf{u}$ denotes the additive inverse for $\mathbf{u}$, so $\mathbf{u} + (-\mathbf{u}) = \mathbf{0}$. What property or properties of the eight listed in Definition 1.1 justifies each step?

$$\begin{aligned}(1+0)\mathbf{u} &= 1\mathbf{u} + 0\mathbf{u},\\ 1\mathbf{u} &= \mathbf{u} + 0\mathbf{u},\\ \mathbf{u} &= \mathbf{u} + 0\mathbf{u},\\ \mathbf{u} + (-\mathbf{u}) &= (\mathbf{u} + (-\mathbf{u})) + 0\mathbf{u},\\ \mathbf{0} &= \mathbf{0} + 0\mathbf{u},\\ \mathbf{0} &= 0\mathbf{u}.\end{aligned}$$

(c) Show that if $\mathbf{u} + \mathbf{v} = \mathbf{0}$ then $\mathbf{v} = (-1)\mathbf{u}$ (this shows that the additive inverse $\mathbf{u}$ is $(-1)\mathbf{u}$).

14. Write out the basic waveforms $\mathbf{E}_{2,k}$ for $k = -2, -1, 0, 1, 2$, and verify that the resulting vectors are periodic with period 2 with respect to the index k. Repeat for $\mathbf{E}_{3,k}$ (same k range). Verify periodicity with period 3.

15. Let $a = re^{i\theta}$ be a complex number (where $r > 0$ and θ are real).
(a) Show that the function $f(t) = ae^{i\omega t}$ satisfies $|f(t)| = r$ for all t.
(b) Show that $f(t) = ae^{i\omega t}$ is shifted θ/ω units to the left, compared to $re^{i\omega t}$.

16. Show that each of the four types of waveforms in (1.19) can be expressed as a linear combination of waveforms $e^{\pm i\alpha x \pm i\beta y}$ of the form (1.17).

17. Let $\mathbf{C}_k$ be the vector obtained by sampling the function $\cos(2\pi kt)$ at the points $t = 0, \frac{1}{N}, \frac{2}{N}, \dots, \frac{N-1}{N}$, and let $\mathbf{S}_k$ be similarly defined with respect to the sine function. Prove the following vector analogs of equations (1.12)–(1.14) relating the exponential and trigonometric wave forms.

$$\mathbf{E}_k = \mathbf{C}_k + i\mathbf{S}_k, \quad \overline{\mathbf{E}_k} = \mathbf{C}_k - i\mathbf{S}_k,$$

$$\mathbf{C}_k = \frac{1}{2}(\mathbf{E}_k + \overline{\mathbf{E}_k}),\ \mathbf{S}_k = \frac{1}{2i}(\mathbf{E}_k - \overline{E_k}),$$
$$\mathbf{C}_k = \mathrm{Re}(E_k),\ S_k = \mathrm{Im}(E_k),$$

where $\mathbf{E}_k = \mathbf{E}_{N,k}$ is as defined in equation (1.22).

18. Show that we can factor the basic two-dimensional waveform $\mathcal{E}_{m,n,k,l}$ as

$$\mathcal{E}_{m,n,k,l} = \mathbf{E}_{m,k}\mathbf{E}^T_{n,l}$$

(recall superscript "T" denotes the matrix/vector transpose operation) where the vectors $\mathbf{E}_{m,k}$ and $\mathbf{E}_{m,k}$ are the discrete basic waveforms in one-dimension as defined in equation (1.22), as column vectors.

19. Consider an exponential waveform

$$f(x, y) = e^{2\pi i(px+qy)}$$

as was discussed in Section 1.5.2 (p and q need not be integers). Figure 1.7 in that section indicates that this waveform has a natural "direction" and "wavelength." The goal of this problem is to understand the sense in which this is true, and how these quantities depend on p and q.
Define $\mathbf{v} = (p, q)$, so $\mathbf{v}$ is a two-dimensional vector. Consider a line L through an arbitrary point (x_0, y_0) in the direction of a unit vector $\mathbf{u} = (u_1, u_2)$ (so $\|\mathbf{u}\| = 1$). The line L can be parameterized with respect to arc length as

$$x(t) = x_0 + tu_1,\quad y(t) = y_0 + tu_2.$$

(a) Show that the function $g(t) = f(x(t), y(t))$ with $x(t), y(t)$ as above (i.e., f evaluated along the line L) is given by

$$g(t) = Ae^{2\pi i\|\mathbf{v}\|\cos(\theta)t},$$

where A is some complex number that does not depend on t and θ is the angle between $\mathbf{v}$ and $\mathbf{u}$. Hint: Use equation (1.30).

(b) Show that if L is orthogonal to $\mathbf{v}$ then the function g (and so f) remains constant.

(c) Find the frequency (oscillations per unit distance moved) of g as a function of t, in terms of p, q, and θ.

(d) Find that value of θ which maximizes the frequency at which $g(t)$ oscillates. This θ dictates the direction one should move, relative to $\mathbf{v}$ so that f oscillates as rapidly as possible. How does this value of θ compare to the θ value in question (b)? What is this maximal frequency of oscillation, in terms of p and q?

(e) Find the wavelength (the shortest distance between the peaks) of the waveform $f(x, y)$, in terms of p and q.

20. Write out the vectors $\mathbf{E}_{6,0}, \mathbf{E}_{6,1}, \dots, \mathbf{E}_{6,5}$ as in Section 1.7.1. Determine all aliasing relations or redundancies (including conjugate aliasing) you can from the chart. (Remember to index the vector components from 0 to 5.)

21. For a pure 1D wave form of N samples prove the aliasing relation

$$\mathbf{E}_{N-k} = \overline{\mathbf{E}_k}.$$

22. Find all the aliasing relations you can (including conjugate aliasing) for $\mathcal{E}_{m,n,k,l}$. This can be done directly or you might use equation (1.26) and the aliasing relations for the $\mathbf{E}_{N,k}$.

23. Let $S = \{\mathbf{v}_1, \mathbf{v}_2, \mathbf{v}_3\}$ where $\mathbf{v}_1 = (1,1,0)$, $\mathbf{v}_2 = (-1,1,1)$, and $\mathbf{v}_3 = (1,-1,2)$ are vectors in $\mathbb{R}^3$.

(a) Verify that S is orthogonal with respect to the usual inner product. This shows S must be a basis for $\mathbb{R}^3$.

(b) Write the vector $\mathbf{w} = (3,4,5)$ as a linear combination of the basis vectors in S. Verify that the linear combination you obtain actually reproduces $\mathbf{w}$!

(c) Rescale the vectors in S as per Remark 1.11 to produce an equivalent set S' of orthonormal vectors.

(d) Write the vector $\mathbf{w} = (3,4,5)$ as a linear combination of the basis vectors in S'.

(e) Use the results of part (d) to check that Parseval's identity holds.

24. Let $S = \{\mathbf{E}_{4,0}, \mathbf{E}_{4,1}, \mathbf{E}_{4,2}, \mathbf{E}_{4,3}\}$ (these vectors are written out explicitly just prior to equation (1.24)). The set S is orthogonal and a basis for $\mathbb{R}^4$.

(a) Use Theorem 1.3 to write the vector $\mathbf{v} = (1,5,-2,3)$ as a linear combination of the basis vectors in S.

(b) Rescale the vectors in S as per Remark 1.11 to produce an equivalent set S' of orthonormal vectors.

(c) Write the vector $\mathbf{v} = (1,5,-2,3)$ as a linear combination of the basis vectors in S'.

(d) Use the results of part (c) to check that Parseval's identity holds.

25. There are infinitely many other inner products on $\mathbb{R}^n$ besides the standard dot product, and they can be quite useful too.
Let $\mathbf{d} = (d_1, d_2, \dots, d_n) \in \mathbb{R}^n$. Suppose that $d_k > 0$ for $1 \le k \le n$.

(a) Let $\mathbf{v} = (v_1, v_2, \dots, v_n)$ and $\mathbf{w} = (w_1, w_2, \dots, w_n)$ be vectors in $\mathbb{R}^n$. Show that the function

$$(\mathbf{v}, \mathbf{w})_d = \sum_{k=1}^{n} d_k v_k w_k$$

defines an inner product on $\mathbb{R}^n$. Write out the corresponding norm.

(b) Let $\mathbf{d} = (1, 5)$ in $\mathbb{R}^2$, and let $S = \{\mathbf{v}_1, \mathbf{v}_2\}$ with $\mathbf{v}_1 = (2, 1), \mathbf{v}_2 = (5, -2)$. Show that S is orthogonal with respect to the $(,)_d$ inner product.
(c) Find the length of each vector in S with respect to the norm induced by this inner product.
(d) Write the vector $\mathbf{w} = (-2, 5)$ as a linear combination of the basis vectors in S. Verify that the linear combination you obtain actually reproduces $\mathbf{w}$!

26. Let $\mathbf{v}$ and $\mathbf{w}$ be elements of a normed vector space. Prove the reverse triangle inequality,

$$|\|\mathbf{v}\| - \|\mathbf{w}\|| \leq \|\mathbf{v} - \mathbf{w}\|. \tag{1.54}$$

Hint: Start with $\mathbf{v} = (\mathbf{v} - \mathbf{w}) + \mathbf{w}$, take the norm of both sides, and use the usual triangle inequality.

27. Let $w(t)$ be a real-valued, positive, continuous function that is bounded away from 0 on an interval $[a, b]$, that is, $w(t) \geq \delta > 0$ for some δ and all $t \in [a, b]$. Verify that

$$(f, g)_w = \int_a^b w(t) f(t) \overline{g(t)}\, dt$$

defines an inner product on $C[a, b]$ (with complex-valued functions). Write out the corresponding norm.

28. Suppose V is an inner product space with inner product $(\mathbf{v}, \mathbf{w})$. Show that if we define

$$\|\mathbf{v}\| = \sqrt{(\mathbf{v}, \mathbf{v})}$$

then $\|\mathbf{v}\|$ satisfies the properties of a norm. Hints: All the properties are straightforward, except the triangle inequality. To show that, note that

$$\|\mathbf{v} + \mathbf{w}\|^2 = \|\mathbf{v}\|^2 + \|\mathbf{w}\|^2 + (\mathbf{v}, \mathbf{w}) + (\mathbf{w}, \mathbf{v}).$$

Apply the Cauchy–Schwarz inequality to both inner products on the right-hand side above, and note that for any $z \in \mathbb{C}$ we have $|Re(z)| \leq |z|$.

29. Show that in the real-valued case the choice $c = (\mathbf{v}, \mathbf{w})/(\mathbf{w}, \mathbf{w})$ as in the proof of Theorem 1.1 minimizes the value of $\|\mathbf{v} - c\mathbf{w}\|^2$ and makes $\mathbf{v} - c\mathbf{w}$ orthogonal to $\mathbf{w}$.

30. Suppose V is an inner product space, with a norm $\|\mathbf{v}\| = \sqrt{(\mathbf{v}, \mathbf{v})}$ that comes from the inner product.
(a) Show that this norm must satisfy the *parallelogram* identity

$$2\|\mathbf{u}\|^2 + 2\|\mathbf{v}\|^2 = \|\mathbf{u} + \mathbf{v}\|^2 + \|\mathbf{u} - \mathbf{v}\|^2.$$

(b) Let $f(x) = x$ and $g(x) = x(1 - x)$ be elements of the normed vector space $C[0, 1]$ with the supremum norm. Compute each of $\|f\|_\infty$, $\|g\|_\infty$, $\|f + g\|_\infty$, and $\|f - g\|_\infty$, and verify that the parallelogram identity from part (a) does not hold. Hence the supremum norm cannot come from an inner product.

31. Suppose that S is an orthogonal but not orthonormal basis for $\mathbb{R}^n$ consisting of vectors $\mathbf{v}_k$, $1 \le k \le n$. Show that Parseval's identity becomes

$$\|\mathbf{v}\|_2 = \sum_{k=1}^{n} |\alpha_k|^2 \|\mathbf{v}_k\|^2,$$

where $\mathbf{v} = \sum_{k=1}^{n} \alpha_k \mathbf{v}_k$. Hint: Just chase through the derivation of equation (1.36).

32. For the basic waveforms defined in equation (1.22) show that

$$(\mathbf{E}_{N,k}, \mathbf{E}_{N,k}) = N$$

with the inner product defined in Example 1.13.

33. For 2D wave forms which are $m \times n$ matrices defined in equation (1.25) prove that

$$(\mathcal{E}_{k,l}, \mathcal{E}_{p,q}) = 0$$

when $k \ne p$ or $l \ne q$, and

$$(\mathcal{E}_{k,l}, \mathcal{E}_{k,l}) = mn$$

with the inner product on $M_{m,n}(\mathbb{C})$ defined in Example 1.14. It may be helpful to look at Example 1.20.

34. Let $\mathbf{v}_k$, $1 \le k \le n$, be any set of vectors in $\mathbb{C}^n$, considered as column vectors. Define the n by n matrix

$$\mathbf{A} = \begin{bmatrix} \mathbf{v}_1 & \mathbf{v}_2 & \cdots & \mathbf{v}_n \end{bmatrix}.$$

Let $\mathbf{A}^* = (\overline{\mathbf{A}})^T$ (conjugate and transpose entries). What is the relationship between the entries $b_{m,k}$ of the n by n matrix $\mathbf{B} = (\mathbf{A}^*)(\mathbf{A})$ and the inner products $(\mathbf{v}_k, \mathbf{v}_m)$ (inner product defined as in Example 1.13)?

35. Use the result of the last exercise to show that if S is an orthonormal set of vectors $\mathbf{v}_k \in \mathbb{C}^n$, $1 \le k \le n$ and $\mathbf{A}$ is the matrix from the previous problem then $(\mathbf{A}^*)(\mathbf{A}) = \mathbf{I}$, where $\mathbf{I}$ denotes the n by n identity matrix.

36. Let ϕ_k, $1 \le k < \infty$ be an orthonormal set in an inner product space V (no assumption that ϕ_k is complete). Let $f \in V$ and define $c_k = (f, \phi_k)$. Prove Bessel's inequality,

$$\sum_{k=1}^{\infty} |c_k|^2 \le \|f\|^2. \tag{1.55}$$

Hint: Start with

$$0 \le \left(f - \sum_{k=1}^{n} c_k \phi_k, f - \sum_{k=1}^{n} c_k \phi_k \right)$$

(first explain why this inequality is true).

37. Suppose that V is a normed vector space and $\mathbf{v}_n$ a sequence of elements in V which converge to $\mathbf{v} \in V$, that is,

$$\lim_{n\to\infty} \|\mathbf{v}_n - \mathbf{v}\| = 0.$$

Show that

$$\lim_{n\to\infty} \|\mathbf{v}_n\| = \|\mathbf{v}\|.$$

Hint: Use equation (1.54) from Exercise 26.
Show the converse of this statement is false (provide a counterexample).

38. We can endow the vector space $L^2(\mathbb{N})$ in Example 1.5 (see also Exercise 12) with an inner product

$$(\mathbf{x}, \mathbf{y}) = \sum_{k=0}^{\infty} x_k y_k,$$

where $\mathbf{x} = (x_0, x_1, x_2, \ldots)$ and $\mathbf{y} = (y_0, y_1, y_2, \ldots)$ and all components are real-valued.

(a) Verify that this really does define an inner product. In particular, you should first show that the inner product of any two elements is actually defined, that is, the infinite sum converges. (Hint: Use $|x_k y_k| \le (x_k^2 + y_k^2)/2$ as in Example 1.8.)
What is the corresponding norm on $L^2(\mathbb{N})$?

(b) Let $\mathbf{e}_k$ denote that element of $L^2(\mathbb{N})$ which has $x_k = 1$ and all other $x_m = 0$. Show that $S = \cup_{k=0}^{\infty} \mathbf{e}_k$ is an orthonormal set.

(c) For an arbitrary $\mathbf{x} \in L^2(\mathbb{N})$, show that

$$\mathbf{x} = \sum_{k=0}^{\infty} \alpha_k \mathbf{e}_k$$

for a suitable choice of the α_k. Hint: This is very straightforward; just use Theorem 1.4.

39. Let V be an infinite-dimensional inner product space with orthonormal basis ϕ_k, $k \geq 1$. Suppose that $\mathbf{x}$ and $\mathbf{y}$ are elements of V and

$$\mathbf{x} = \sum_{k=1}^{\infty} a_k \phi_k \quad \mathbf{y} = \sum_{k=1}^{\infty} b_k \phi_k.$$

Of course $a_k = (\mathbf{x}, \phi_k)$ and $b_k = (\mathbf{y}, \phi_k)$.

(a) Define partial sums

$$\mathbf{x}_N = \sum_{k=1}^{N} a_k \phi_k \quad \mathbf{y}_N = \sum_{k=1}^{N} b_k \phi_k.$$

Show that

$$(\mathbf{x}_N, \mathbf{y}_N) = \sum_{k=1}^{N} a_k \overline{b}_k.$$

(b) Show that

$$(\mathbf{x}, \mathbf{y}) = \sum_{k=1}^{\infty} a_k \overline{b}_k.$$

Hint: Note that for any N

$$(\mathbf{x}, \mathbf{y}) - (\mathbf{x}_N, \mathbf{y}_N) = (\mathbf{x}, \mathbf{y} - \mathbf{y}_N) + (\mathbf{x} - \mathbf{x}_N, \mathbf{y}_N)$$

and of course $\mathbf{x}_N \to \mathbf{x}$ and $\mathbf{y}_N \to \mathbf{y}$. Use the Cauchy–Schwarz inequality to show the right-hand side above goes to zero, then invoke part (a). The result of Exercise 37 may also be helpful.

40. Show that if $\mathbf{v}$ and $\mathbf{w}$ are nonzero vectors then equality is obtained in the Cauchy–Schwarz inequality (i.e., $|(\mathbf{v}, \mathbf{w})| = \|\mathbf{v}\|\|\mathbf{w}\|$) if and only if $\mathbf{v} = c\mathbf{w}$ for some scalar c. (One direction is easy, the other is somewhat challenging.)

41. Let $\phi_k(t) = e^{i\pi kt}/\sqrt{2}$ for $k \in \mathbb{Z}$.

(a) Verify that the ϕ_k form an orthonormal set in $C[-1, 1]$ (complex-valued functions) with the inner product defined in Example 1.15.

(b) Find the Fourier coefficients α_k of the function $f(t) = t$ with respect to the ϕ_k explicitly in terms of k by using equation (1.48). Hint: It is a pretty easy integration by parts, and α_0 does not fit the pattern.

(c) Use the result of part (b) to prove the amusing result that

$$\sum_{k=1}^{\infty} \frac{1}{k^2} = \frac{\pi^2}{6}.$$

42. Let $f(t)$ be a real-valued continuous function defined on an interval $[-T, T]$.

(a) Show that if $f(t)$ is an even function ($f(-t) = f(t)$) then the Fourier series in (1.51) contains only cosine terms (and the constant term).

(b) Show that if $f(t)$ is an odd function ($f(-t) = -f(t)$) then the Fourier series in (1.51) contains only sine terms.

43. Let $\phi_k(t) = \cos(k\pi t)$ for $k \geq 0$.

(a) Verify that the ϕ_k are orthogonal on $[0, 1]$ with respect to the usual inner product (note $\phi_0(t) = 1$).

(b) Show that this set of functions is complete as follows (where we will make use of the given fact that the functions $e^{i\pi kt}$ are complete in $C[-1, 1]$). Let $f(t)$ be a function in $C[0, 1]$. We can extend f to an even function $\tilde{f}(t)$ in $C[-1, 1]$ as

$$\tilde{f}(t) = \begin{cases} f(t), & t \geq 0, \\ f(-t), & t < 0. \end{cases}$$

Now use the result of Exercise 42 to show that $\tilde{f}$ has a Fourier expansion in appropriate cosines on $[-1, 1]$. Why does this show $\cos(k\pi t)$ is complete on $[0, 1]$?

2

The Discrete Fourier Transform

2.1 Overview

In this chapter, we introduce the *discrete Fourier transform* (DFT) and *inverse discrete Fourier transform* (IDFT) for analyzing sampled signals and images. We have already developed the mathematics that underlies these tools: the formula $\alpha_k = (\mathbf{v}, \mathrm{E}_k)/N$ behind equation (1.34) in Chapter 1 is, up to a minor modification, the DFT, while equation (1.34) is itself the IDFT.

But rather than jumping straight to the definition of the DFT and IDFT, it may be useful to first discuss some basic facts about the notions of *time domain* and *frequency domain*. We will also pause to a look at a very simple computational example that can help build some important intuition about how the DFT and frequency domain are used. As usual, we will consider the one-dimensional case of signals first and then move to two dimensions.

Before proceeding further we should remark that the definitions of these transforms vary slightly from text to text. However, the definitions usually only differ by a scaling or "normalizing" constant. We selected our normalization to be algebraically convenient and consistent with Matlab. For a more comprehensive account of the DFT, see [1].

2.2 The Time Domain and Frequency Domain

In Chapter 1, we presented a few different mathematical models associated to one-dimensional signals. In the analog case, the signal is a function $x(t)$ of a real variable t, and we usually view t as time. In the discrete case, the signal is the vector $\mathbf{x}$ obtained by sampling the underlying function at regular times $t = a + k\Delta T$ for some range of the integer k. The indexing integer k thus acts as a kind of time variable. For this reason the original signal, either $x(t)$ or the sampled version $x_k = x(a + k\Delta T)$, is said to exist in the *time domain*. In the

Discrete Fourier Analysis and Wavelets: Applications to Signal and Image Processing, Second Edition.
S. Allen Broughton and Kurt Bryan.

Companion Website: www.wiley.com/go/Broughton/Discrete_Fourier_Analysis_and_Wavelets

two-dimensional image case, the phrase "time domain" does not make sense, but we will still use it. The original analog image $f(x, y)$ or sampled version might also be said to exist in the "image domain" or "spatial domain."

Every object that exists in the time domain has a representation in something called the *frequency domain*. We will make this more precise shortly. The DFT and IDFT allow us to move easily back and forth between the discrete time and frequency domains. If you have encountered the Laplace transform before, then you have already seen the general idea: every suitable function $f(t)$ has a counterpart $F(s)$ in the "Laplace s-domain" (a kind of frequency domain), and the Laplace and inverse Laplace transforms move us back and forth.

The motivation for moving to the frequency domain in the first place is that many operations are easier, both computationally and conceptually, in the frequency domain. We transform the time domain signal to the frequency domain, perform the operation of interest there, and then transform the altered signal back to the time domain.

In many cases, it turns out that removing noise from a time domain signal can be accomplished, at least approximately, by removing the signal's "high-frequency components" in the frequency domain. This is usually done according to the diagram in Figure 2.1. The precise mathematical structure of the frequency domain depends on the nature of the time domain signals we are trying to analyze and the transform used. In this chapter, we use the DFT for the analysis step to compute the frequency domain representation of the signal. In the smoothing step we eliminate or reduce the high frequencies, which is a simple algebraic operation. Finally, in the synthesis step we use the IDFT to resynthesize a less noisy version of the time domain signal from its altered frequency domain representation. This three-step process is typical of a lot of signal and image processing.

A specific example is presented in the following section. We will illustrate the transformation of a signal to the frequency domain via decomposition into basic waveforms, as well as how this frequency domain information can be presented graphically, altered to remove noise, and finally resynthesized into a denoised version of the original signal.

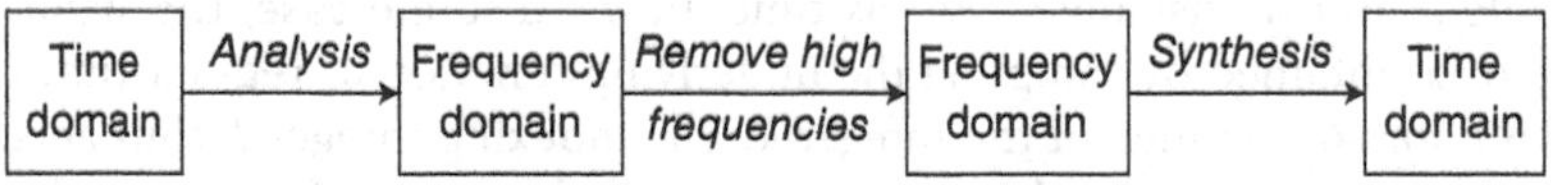

Figure 2.1 Processing a signal in the frequency domain.

2.3 A Motivational Example

2.3.1 A Simple Signal

Let us begin by considering the analog signal

$$x(t) = 2.0\cos(2\pi \cdot 5t) + 0.8\sin(2\pi \cdot 12t) + 0.3\cos(2\pi \cdot 47t) \tag{2.1}$$

on the time interval [0, 1]. As no other frequencies are present, $x(t)$ is built from a superposition of trigonometric waveforms with frequencies of 5, 12, and 47 Hz. A plot of $x(t)$ is shown in Figure 2.2.

Of course, in practice, only a sampled version of $x(t)$ is available (with quantization error too, but let us assume that is negligible for now). Suppose that we sample with period $\Delta T = 1/128$ and obtain a sample vector

$$\mathbf{x} = \left(x(0), x\left(\frac{1}{128}\right), x\left(\frac{2}{128}\right), \ldots, x\left(\frac{127}{128}\right)\right).$$

The goal for the moment is simple: Use the discretized signal $\mathbf{x}$ to determine, or at least approximate, the frequencies that make up the original analog signal $x(t)$. If sampling occurs at 128 Hz, then no aliasing will occur provided that we are willing to assume or assure that the analog signal contains no frequencies above 64 Hz.

The first 10 samples of $x(t)$ (the components of $\mathbf{x}$) are, to four significant figures,

$$2.300, 2.183, 2.474, 2.507, 1.383, 0.984, -0.023, -1.230, -1.289, -1.958.$$

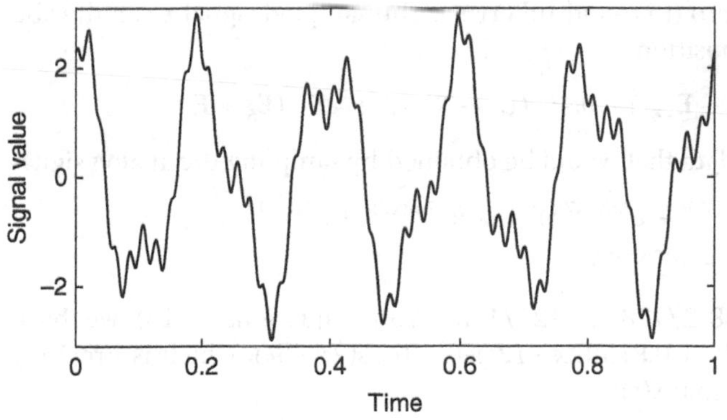

Figure 2.2 Time domain analog signal.

Of course, there are 118 more samples. Trying to ferret out the frequencies that make up $x(t)$ from these data looks like a formidable task, but we have already solved the problem in Chapter 1! Let us compute the expansion of $\mathbf{x}$ as a linear combination of the basic discrete waveforms $\mathbf{E}_{128,k}$ by using equation (1.34).

2.3.2 Decomposition into Basic Waveforms

First recall that because of the aliasing relation $\mathbf{E}_{128,k} = \mathbf{E}_{128,k+128m}$ for any integers k and m, we may as well restrict our attention to a range in the index k of length 128. For the moment we will use the range $-63 \leq k \leq 64$, though later it will be more convenient to use $0 \leq k \leq 127$. According to equation (1.34),

$$\mathbf{x} = \sum_{k=-63}^{64} c_k \mathbf{E}_k, \tag{2.2}$$

where we write just $\mathbf{E}_k$ instead of $\mathbf{E}_{128,k}$. The c_k for $-63 \leq k \leq 64$ are given by

$$c_k = \frac{(\mathbf{x}, \mathbf{E}_k)}{(\mathbf{E}_k, \mathbf{E}_k)} = \frac{1}{128} \sum_{m=0}^{127} x_m e^{-2\pi ikm/128},$$

where we have used equation (1.23) for the mth component of $\mathbf{E}_k$ and the fact that $(\mathbf{E}_k, \mathbf{E}_k) = 128$. Remember that the components of $\mathbf{E}_k$ must be conjugated because of the nature of the inner product on $\mathbb{C}^N$. We are also indexing the components of vectors in the time domain $\mathbb{R}^{128}$ from 0 to 127.

The c_k turn out to be

$$c_{-47} = 0.15, \quad c_{-12} = 0.4i, \quad c_{-5} = 1.0, \quad c_5 = 1.0,$$
$$c_{12} = -0.4i, \quad c_{47} = 0.15$$

and all other c_k are zero (to round-off error). The sampled signal $\mathbf{x}$ can thus be written as the superposition

$$\mathbf{x} = 0.15 \cdot (\mathbf{E}_{47} + \mathbf{E}_{-47}) - 0.4i \cdot (\mathbf{E}_{12} - \mathbf{E}_{-12}) + 1.0 \cdot (\mathbf{E}_5 + \mathbf{E}_{-5}).$$

This is precisely the data that would be obtained by sampling the analog signal

$$g(t) = 0.15(e^{2\pi i(47)t} - e^{2\pi i(-47)t}) - 0.4i(e^{2\pi i(12)t} + e^{2\pi i(-12)t})$$
$$+(e^{2\pi i(5)t} - e^{2\pi i(-5)t})$$

at times $t = 0, 1/128, 2/128, \ldots, 127/128$. From equations (1.13) we have $g(t) = 0.3\cos(2\pi \cdot 47t) + 0.8\sin(2\pi \cdot 12t) + 2.0\cos(2\pi \cdot 5t)$, which is precisely the original analog signal $x(t)$!

2.3.3 Energy at Each Frequency

It is worth examining this computation from the point of view of Remark 1.9, in which the quantity $||\mathbf{x}||^2$ is interpreted as the energy of a sampled signal $\mathbf{x}$.

With **x** expanded according to equation (2.2),

$$
\begin{aligned}
\|\mathbf{x}\|^2 &= (\mathbf{x}, \mathbf{x}) \\
&= \left(\sum_{k=-63}^{64} c_k \mathbf{E}_k, \sum_{m=-63}^{64} c_m \mathbf{E}_m \right) \\
&= \sum_{k=-63}^{64} \sum_{m=-63}^{64} c_k \overline{c_m} (\mathbf{E}_k, \mathbf{E}_m) \\
&= \sum_{k=-63}^{64} |c_k|^2 \|\mathbf{E}_k\|^2. \qquad (2.3)
\end{aligned}
$$

Note that we use different indexes of summation in the first and second argument to the inner product in line 2, to avoid confusion. Of course, we also use the fact that $(\mathbf{E}_k, \mathbf{E}_m) = 0$ for $k \neq m$.

The vector $c_k\mathbf{E}_k$ is that multiple of the basic waveform $\mathbf{E}_k$ needed to synthesize the sampled signal **x**. The quantity $|c_k|^2\|\mathbf{E}_k\|^2 = \|c_k\mathbf{E}_k\|^2$ is the energy of this constituent waveform. Equation (2.3) can thus be viewed as a decomposition of the total energy in the signal into the sum of the energies of the constituent orthogonal waveforms. In particular, for the example signal of interest we have (since $\|\mathbf{E}_k\|^2 = (\mathbf{E}_k, \mathbf{E}_k) = 128$ for all k)

$$
\begin{aligned}
\|\mathbf{x}\|^2 &= 128|c_{-5}|^2 + 128|c_5|^2 + 128|c_{-12}|^2 + 128|c_{12}|^2 \\
&\quad + 128|c_{-47}|^2 + 128|c_{47}|^2 \\
&= 128 + 128 + 20.48 + 20.48 + 2.88 + 2.88 \qquad (2.4) \\
&= 302.72.
\end{aligned}
$$

Now recall that the basic waveform $\mathbf{E}_k$ is the sampled version of $e^{2\pi ikt/T}$ and moreover that the frequency of $e^{2\pi ikt/T}$ (and $\mathbf{E}_k$) is $|k|/T$ Hz. In this case $T = 1$, so the frequency is merely $|k|$. Equation (2.4) states that the total energy contribution of the 5 Hz frequency components (index $k = \pm 5$) to **x** is $128|c_{-5}|^2 + 128|c_5|^2 = 256$. Of course, 128 units of this comes from the $k = -5$ waveform and 128 from the $k = 5$ waveform, but the quantities can be lumped together because we are interested in the energy at a given frequency. Similarly, the contribution of the 12 Hz frequency components is $128|c_{-12}|^2 + 128|c_{12}|^2 = 40.96$ and the 47 Hz components contribute energy $128|c_{-47}|^2 + 128|c_{47}|^2 = 5.76$. On a percentage basis the 5 Hz frequency component constitutes about 84.6% of the total signal energy, the 12 Hz contributes 13.5%, and the 47 Hz components constitute the remaining 1.9%.

2.3.4 Graphing the Results

One informative way to present frequency information is graphically as in Figure 2.3, where we plot the quantity $128|c_k|^2$ (the energy contributed by

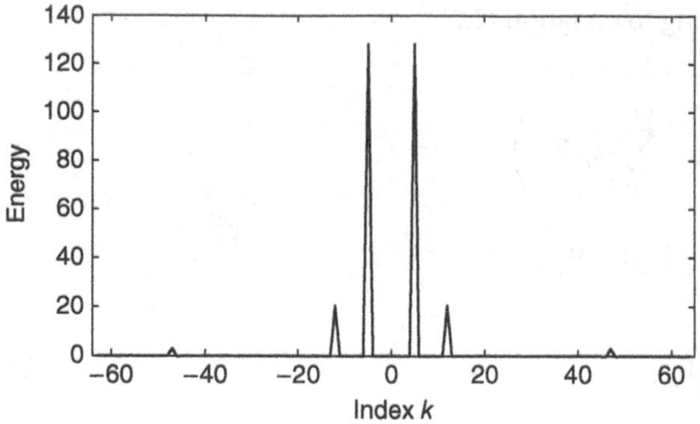

Figure 2.3 Spectrum of example signal.

the corresponding waveform) versus the index k for $-63 \leq k \leq 64$. The plot really consists of a discrete set of points $(k, 128|c_k|^2)$ for $-63 \leq k \leq 64$, but "connecting the dots" makes the plot more easily readable. Clearly, the plot is symmetric about $k = 0$, since in this example $|c_{-k}| = |c_k|$. As we will see, this is always the case when the signal $\mathbf{x}$ is real-valued, but not if the signal is complex-valued.

There is another, slightly different way to present such data. Recall that $\mathbf{E}_{N,k}$ is defined for any $k \in \mathbb{Z}$. As a consequence, the coefficient $c_k = (\mathbf{x}, \mathbf{E}_{N,k})/(\mathbf{E}_{N,k}, \mathbf{E}_{N,k})$ is also defined for any k. However, from the aliasing relation $\mathbf{E}_{N,k} = \mathbf{E}_{N,k+mN}$ it follows that $c_k = c_{k+mN}$; that is, the c_k are periodic with period N. In analyzing the frequency information embodied by the c_k, we can thus confine our attention to any range in k of length N. In Figure 2.3, we chose $-63 \leq k \leq 64$, but it will often be convenient to use the range $0 \leq k \leq N-1$ in the general case, or $0 \leq k \leq 127$ in this example.

If we adopt this convention for the present example, then the values of c_5, c_{12}, and c_{47} remain the same, and we obtain $c_{81} = c_{-47} = 0.15$, $c_{116} = c_{-12} = 0.4i$, and $c_{123} = c_{-5} = 1.0$. The resulting plot of the points $(k, 128|c_k|^2)$ for $0 \leq k \leq 127$ is shown in Figure 2.4. The previous symmetry at $k = 0$ is now symmetry about $k = 64$. In effect, the entire range $-63 \leq k \leq -1$ has been picked up and moved rigidly to the right 128 units, to the range $65 \leq k \leq 127$.

Finally, if our primary interest is the total energy contributed by the frequency $|k|$ waveforms, then we can lump together these energies and just plot the points $(|k|, 128(|c_k|^2 + |c_{-k}|^2))$ for $k = 1$ to 63. However, $k = 0$ and $k = 64$ are special. In the case $k = 0$ we should just plot $(0, 128|c_0|^2)$, for using $128(|c_k|^2 + |c_{-k}|^2)$ in this case double counts the energy. Similarly, for $k = 64$ we just plot $(64, 128|c_{64}|^2)$. Such a plot is shown in Figure 2.5.

The frequency information contained in the c_k over whatever k range we elect to work over is called the *spectrum* of the signal. How this spectral information is scaled or presented graphically varies widely.

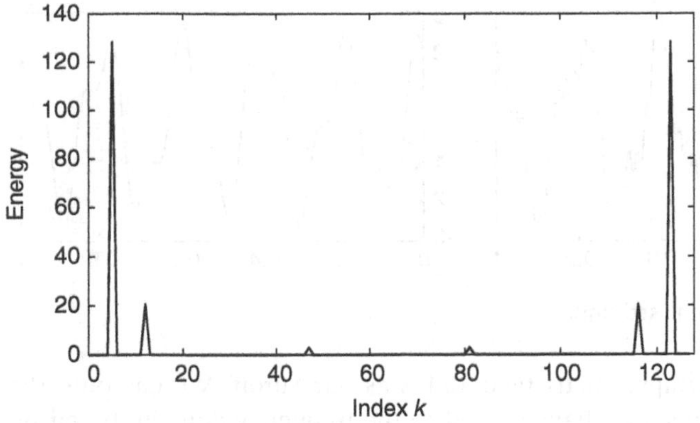

Figure 2.4 Spectrum of example signal.

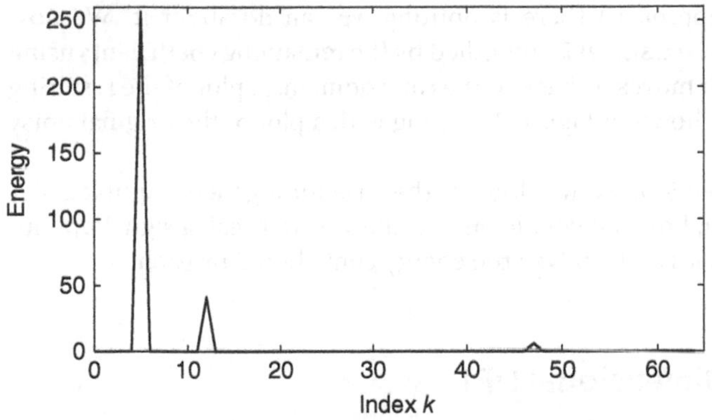

Figure 2.5 Spectrum of example signal, frequency energies lumped.

2.3.5 Removing Noise

Signals often contain noise, and one common task in signal and image processing is to "clean up" a signal or image by removing the noise. In the time domain, random noise can contribute to a jagged appearance of an otherwise smooth signal. In the frequency domain, random noise manifests itself across the entire spectrum, including high frequencies where an otherwise smooth signal should have little energy. Thus, one very simple approach to noise reduction is this: we decide (based on physical insight, preferably) on a "cutoff" frequency. Any spectral energy above this cutoff is considered to have come from noise, while anything below is signal. Of course, the delineation is rarely clean-cut, but it is sufficient for illustrative purposes at the moment.

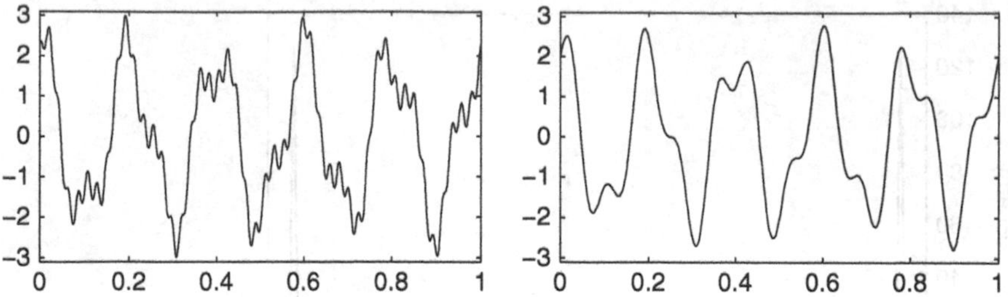

Figure 2.6 Noisy and denoised signal.

In the present example let us take 40 Hz as our cutoff. We compute the c_k as above; in doing so, we have moved to the frequency domain. Based on our cutoff frequency we decide that the c_{-47} and c_{47} coefficients are artifacts due to noise polluting the signal, and hence we zero them out, a very simple algebraic operation. The other coefficients may also be polluted by noise, but (at least with this approach) there is nothing we can do about it. We now reconstruct the denoised signal $\tilde{\mathbf{x}}$ embodied by the remaining coefficients using equation (2.2), which moves us back to the time domain. A plot of the resulting denoised signal $\tilde{\mathbf{x}}$ is shown in Figure 2.6, along with a plot of the original noisy signal.

In the following section, we will look at the case for a general sampled signal and, in particular, how the coefficients c_k allow us to break a signal up into constituent frequencies and analyze the energy contributed by each.

2.4 The One-Dimensional DFT

2.4.1 Definition of the DFT

In all that follows we assume that vectors in $\mathbb{R}^N$ or $\mathbb{C}^N$ are indexed from 0 to $N-1$, for example, $\mathbf{x} = (x_0, x_1, \ldots, x_{N-1})$. Whenever we use matrix methods, all vectors will be written in a column format. Whether a vector $\mathbf{x}$ is actually obtained by sampling an analog signal is irrelevant for now. Unless otherwise noted, we also use the range $0 \le k \le N-1$ for the basic vectors $\mathbf{E}_{N,k}$.

In the last section, we saw that the numbers

$$c_k = \frac{(\mathbf{x}, \mathbf{E}_{N,k})}{(\mathbf{E}_{N,k}, \mathbf{E}_{N,k})} \tag{2.5}$$

in the orthogonal decomposition of $\mathbf{x}$ carry a lot of information about a signal. We now define a transform $\mathbf{x} \to \mathbf{X}$ on signals, called the DFT, that collects all this information together into one vector.

Definition 2.1 Let $\mathbf{x} \in \mathbb{C}^N$ be a vector $(x_0, x_1, \ldots, x_{N-1})$. The DFT of $\mathbf{x}$ is the vector $\mathbf{X} \in \mathbb{C}^N$ with components

$$X_k = (\mathbf{x}, \mathbf{E}_{N,k}) = \sum_{m=0}^{N-1} x_m e^{-2\pi ikm/N} \tag{2.6}$$

for $0 \leq k \leq N-1$.

Comparison of equations (2.5) and (2.6) shows that $X_k = Nc_k$, since the $\mathbf{E}_{N,k}$ satisfy $(\mathbf{E}_{N,k}, \mathbf{E}_{N,k}) = N$. Traditionally, the c_k are themselves called the *Fourier coefficients* of $\mathbf{x}$, but some texts may use this terminology for X_k. When in doubt, we refer to the appropriate equation!

Remark 2.1 The coefficient $c_0 = X_0/N$ measures the contribution of the constant basic waveform $\mathbf{E}_0 = (1, 1, \ldots, 1)$ to $\mathbf{x}$. In fact,

$$c_0 = \frac{X_0}{N} = \frac{1}{N} \sum_{m=0}^{N-1} \mathbf{x}_m$$

is the average value of $\mathbf{x}$. The coefficient c_0 is usually referred to as the *dc coefficient* because it measures the strength of the constant or "direct current" component of the signal, in an electrical setting.

Remark 2.2 Equation (2.6) actually defines the DFT coefficients X_k for any index k, and the resulting X_k are periodic with period N in the index. We will thus sometimes refer to the X_k on other ranges of the index, for example, $-N/2 < k \leq N/2$ when N is even. Actually, even if N is odd, the range $-N/2 < k \leq N/2$ works because k is required to be an integer. Knowledge of X_k on such a range allows us to compute the DFT coefficients for all k.

The original signal $\mathbf{x}$ can be reconstructed from the c_k by using equation (1.33) in Theorem 1.3. Since $c_k = X_k/N$, this means we can reconstruct $\mathbf{x}$ from $\mathbf{X}$.

Definition 2.2 Let $\mathbf{X} \in \mathbb{C}^N$ be a vector $(X_0, X_1, \ldots, X_{N-1})$. The IDFT of $\mathbf{X}$ is the vector $\mathbf{x} = \frac{1}{N}\sum_{m=0}^{N-1} X_m \mathbf{E}_{N,m} \in \mathbb{C}^N$ with components

$$x_k = \frac{(\mathbf{X}, \overline{\mathbf{E}_{N,k}})}{N} = \frac{1}{N} \sum_{m=0}^{N-1} X_m e^{2\pi ikm/N}. \tag{2.7}$$

Note the $1/N$ term in front of the sum in (2.7) used to compensate for the fact that $X_k = Nc_k$. The IDFT shows how to synthesize $\mathbf{x}$ from the $\mathbf{E}_{N,k}$, which are the DFT basic waveforms. As mentioned in the first section of this chapter, various definitions of the DFT/IDFT pair exist. Some put a $1/\sqrt{N}$ factor in front of both sums in the DFT and IDFT. (The computer algebra system Maple scales transforms like this). In any case, the DFT and IDFT will "undo" each other.

We will occasionally write $\mathbf{X} = \text{DFT}(\mathbf{x})$ and $\mathbf{x} = \text{IDFT}(\mathbf{X})$ to indicate the application of the DFT or IDFT to a given vector.

We can now be a little more precise about the time and frequency domains for discretized one-dimensional signals. A signal vector $\mathbf{x} = (x_0, x_1, \ldots, x_{N-1})$ in $\mathbb{C}^N$ exists in the time domain. We can write out $\mathbf{x}$ with respect to the standard basis $\mathbf{e}_k$, $0 \leq k \leq N-1$, as

$$\mathbf{x} = \sum_{k=0}^{N-1} x_k \mathbf{e}_k.$$

The standard basis is most natural for working with signals in the time domain. The vector $\mathbf{X} = \text{DFT}(\mathbf{x}) \in \mathbb{C}^N$ is the frequency domain version of the time domain signal $\mathbf{x}$ and shows how to represent $\mathbf{x}$ as a linear combination of the basic waveforms $\mathbf{E}_k$,

$$\mathbf{x} = \frac{1}{N} \sum_{k=0}^{N-1} X_k \mathbf{E}_k.$$

Vectors in the time and frequency domain are paired one-to-one, via the transform relations $\mathbf{X} = \text{DFT}(\mathbf{x})$ and $\mathbf{x} = \text{IDFT}(\mathbf{X})$.

2.4.2 Sample Signal and DFT Pairs

In the following section, we will develop some of the mathematical properties of the DFT, but let us pause now to look at a few examples.

2.4.2.1 An Aliasing Example

Consider the analog signal $x(t)$ defined by equation (2.1) in Section 2.3, but this time let us sample $x(t)$ at only 64 Hz on the interval $0 \leq t \leq 1$. The Nyquist frequency drops to 32 Hz, below the frequency of the 47 Hz component of $x(t)$. The sampled signal is the vector $\mathbf{x} \in \mathbb{R}^{64}$ with components $x_k = x(k/64)$, $0 \leq k \leq 63$. In Figure 2.7, we plot $64(|c_{-k}|^2 + |c_k|^2)$ versus k, or equivalently, $(|X_{-k}|^2 + |X_k|^2)/64$ versus k. As in the previous case k corresponds to frequency in hertz. The 47 Hz energy now masquerades as energy at $64 - 47 = 17$ Hz, since

$$\begin{aligned}\cos\left(\frac{2\pi \cdot 47k}{64}\right) &= \cos\left(\frac{2\pi(-47)k}{64}\right)\\ &= \cos\left(\frac{2\pi(-47+64)k}{64}\right)\\ &= \cos\left(\frac{2\pi \cdot 17k}{64}\right)\end{aligned}$$

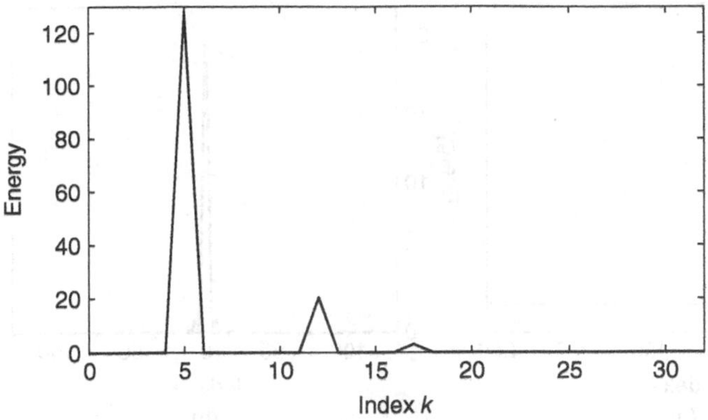

Figure 2.7 Spectrum of example signal, with aliasing.

for all k. The signal $x(t) = 2.0\cos(2\pi \cdot 5t) + 0.8\sin(2\pi \cdot 12t) + 0.3\cos(2\pi \cdot 47t)$ is thus identical to $\tilde{x}(t) = 2.0\cos(2\pi \cdot 5t) + 0.8\sin(2\pi \cdot 12t) + 0.3\cos(2\pi \cdot 17t)$ when sampled at 64 Hz.

2.4.2.2 Square Pulses

The previous examples have consisted of a superposition of relatively few basic waveforms. Let us examine some signals with energy at many frequencies.

Consider a discrete square waveform $\mathbf{x} \in \mathbb{R}^N$ with components $x_k = 1$ for $0 \le k \le R-1$ and $x_k = 0$ for $R \le k \le N-1$, where $0 < R < N$. In Exercise 14, you are asked to show that the DFT of this signal is given by

$$X_k = \frac{1 - e^{-2\pi i k R/N}}{1 - e^{-2\pi i k/N}}$$

for $k > 0$, while $X_0 = R$. The magnitude of X_k for $k > 0$ is

$$\begin{aligned}
|X_k| &= \frac{|1 - e^{-2\pi i k R/N}|}{|1 - e^{-2\pi i k/N}|} \\
&= \frac{((1 - \cos(2\pi k R/N))^2 + \sin^2(2\pi k R/N))^{1/2}}{((1 - \cos(2\pi k/N))^2 + \sin^2(2\pi k/N))^{1/2}} \\
&= \sqrt{\frac{1 - \cos(2\pi k R/N)}{1 - \cos(2\pi k/N)}},
\end{aligned}$$

where we make use of $|z_1 z_2| = |z_1||z_2|$ and $\sin^2(\omega) + \cos^2(\omega) = 1$.

In Figures 2.8 through 2.10, we plot $\mathbf{x}$ and the corresponding points $(k, |X_k|^2/N)$ for $N = 256$ and several values of R. In each case we use the range $-N/2 < k \le N/2$ for the DFT, rather than $0 \le k \le N-1$ (recall Remark 2.2; this allows us to show more clearly the relation between $\mathbf{x}$ and its DFT $\mathbf{X}$. In

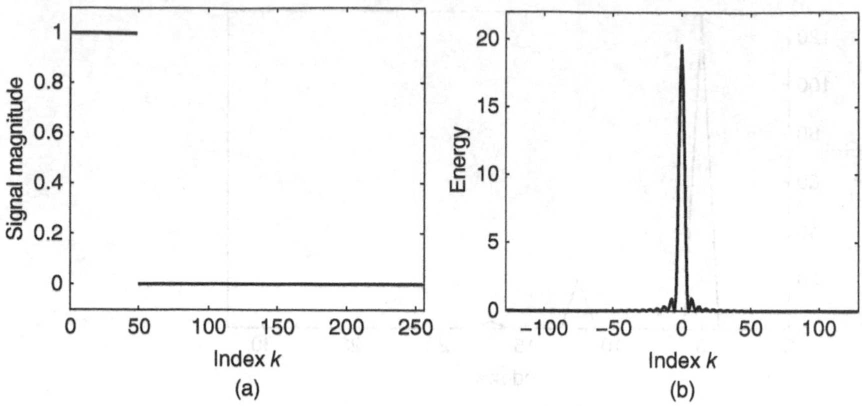

Figure 2.8 Time domain square wave and spectrum, $R = 50$.

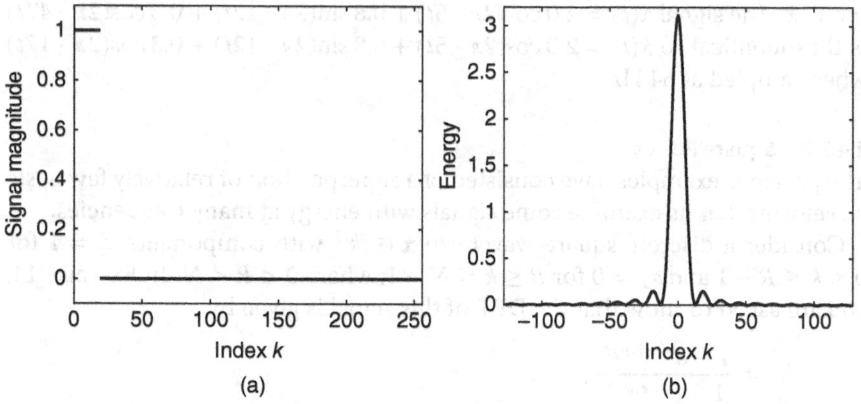

Figure 2.9 Time domain square wave and spectrum, $R = 20$.

Figure 2.8 we use $R = 50$, in Figure 2.9 we take $R = 20$, and in Figure 2.10 we take $R = 5$.

As the square wave pulse becomes shorter in the time domain, the DFT becomes broader in the frequency domain. This is generally true. Indeed in the extreme case that $R = 1$ (so $x_0 = 1$ and $x_k = 0$ for $k \geq 1$) we find that the DFT coefficients satisfy $|X_k| = 1$ for all k, so the spectral plot is flat; see Exercise 6.

2.4.2.3 Noise

Since we will occasionally have to deal with random noise, it will be helpful to say a bit about what noise looks like, in both the time and frequency domain. However, it is important to point out that almost all of the modeling and mathematics in this text is concerned with deterministic (nonrandom) signals and images. The study of signals that are predominantly random in nature is itself

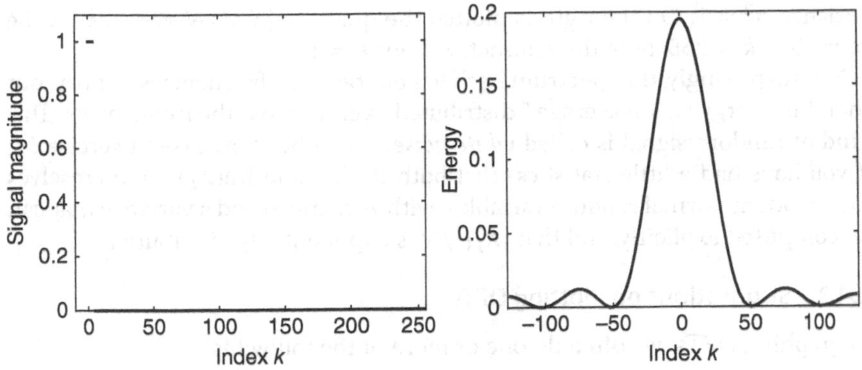

Figure 2.10 Time domain square wave and spectrum, $R = 5$.

a large area of applied mathematics and statistics, and usually goes under the name "time series analysis." Although this subject has many similarities and tools in common with the mathematics we have presented so far, it is based on somewhat different assumptions about the nature of the signals.

Nonetheless, here is a simple example of what noise often looks like. Consider a discrete signal $\mathbf{x} \in \mathbb{R}^N$ in which each of the samples x_k is a random variable. More precisely, let us assume that each x_k is a normally distributed random variable with zero mean and variance σ^2, and that the x_k are independent of each other. We could use any other distribution, but the normality assumption is a common one. (If you have not encountered the normal distribution, then the previous statements will not make sense. In that case, just think of x_k as a "random number" in the range $(-3\sigma, 3\sigma)$.) Equation (2.6) shows that the X_k will themselves be random variables. To illustrate, consider Figure 2.11. On the left is the time domain signal of the form described earlier with $N = 256$ and

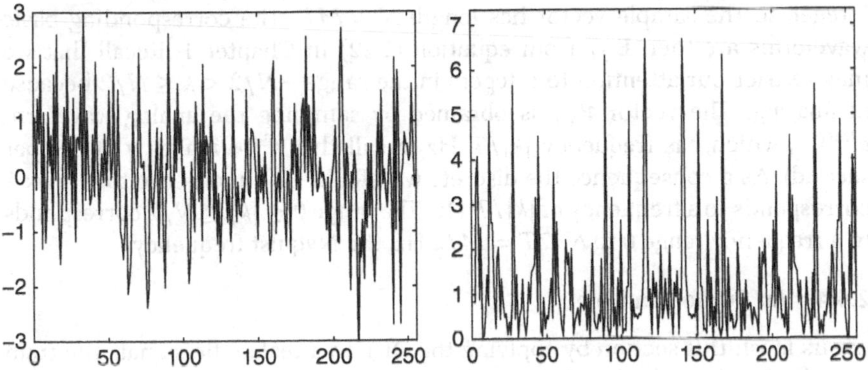

Figure 2.11 Random signal and its DFT magnitude.

variance $\sigma^2 = 1$. On the right is plotted the quantity $|X_k|^2/N$ versus k on the range $0 \le k \le 255$; note the symmetry about $k = 128$.

Not surprisingly, the spectrum itself has energy at all frequencies. It turns out that this energy is, "on average," distributed evenly across the frequencies. This kind of random signal is called *white noise.* It can be shown (see Exercise 15, if you have had a little statistics) that both $\text{Re}(X_k)$ and $\text{Im}(X_k)$ are themselves independent normal random variables, with zero mean and a variance that can be computed explicitly, and that $|X_k|^2/N$ is exponentially distributed.

2.4.3 Suggestions on Plotting DFTs

In graphing DFTs, we often do one or more of the following:

- Plot only the magnitude $|X_k|$, the squared magnitude $|X_k|^2$, or some rescaled versions thereof. In particular, we may not be concerned with the phase or argument of the complex number X_k.
- Make a change of variable by plotting $\log(|X_k|)$, $\log(1 + |X_k|)$, or something similar. This is very helpful when the $|X_k|$ vary widely in magnitude, which is often the case with images.
- Plot the spectral data $(k, |X_k|)$ on whatever k range best illustrates the point we want to make. For purely computational and analytical purposes we usually use $0 \le k \le N - 1$, but using a range in k that is symmetric about $k = 0$ is sometimes more informative when graphing. In fact, if all we care about is the energy at any given frequency, can we plot quantities $(|X_{-k}|^2 + |X_k^2|)/N$, $1 \le k \le N/2 - 1$, along with $|X_0|^2/N$ and $|X_{N/2}|^2/N$ as in Figure 2.5.

Remark 2.3 Up to this point, the horizontal axis on our spectral plots is always indexed by the integer k. It is worth remembering that if $\mathbf{x}$ comes from a sampled analog signal, then each k corresponds to a particular analog frequency, in hertz. Specifically, let the sampling rate be M samples per second over a time interval of length T seconds. For simplicity assume MT is an integer, so the sample vector has length $N = MT$. The corresponding basic waveforms are then $\mathbf{E}_{N,k}$ from equation (1.22) in Chapter 1. Recall that we may restrict our attention to integers in the range $-N/2 < k \le N/2$ because of aliasing. The vector $\mathbf{E}_{N,k}$ is obtained by sampling the analog waveform $e^{2\pi ikt/T}$, which has frequency $|k|/T$ Hz (recall that $e^{2\pi iqt}$ makes q cycles per second). As a consequence the discrete waveform $\mathbf{E}_{N,k}$ and hence the index k corresponds to a frequency of $|k|/T$ Hz. The range $0 \le |k| \le N/2$ corresponds to a frequency range 0 to $N/2T = M/2$ Hz, the Nyquist frequency.

2.4.4 An Audio Example

Let us finish this section by applying the DFT to a real audio signal. The train whistle signal graphed in Figure 2.12a comes as a part of the Matlab installation. It is sampled at $M = 8192$ Hz and is a vector of length $N = 12{,}880$

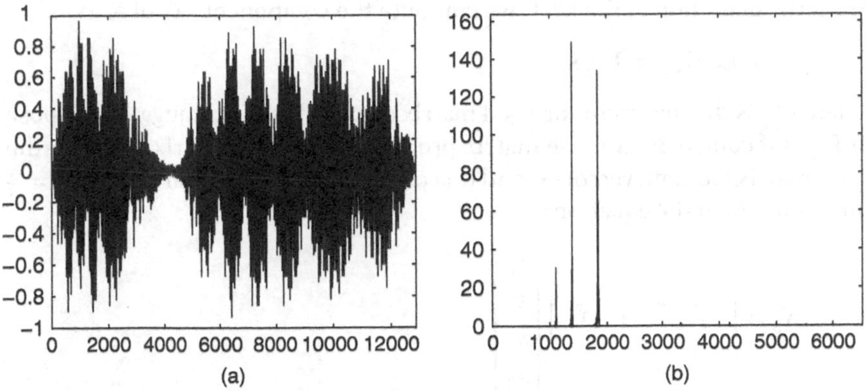

Figure 2.12 Train whistle signal and DFT magnitude.

samples. The time span of the sampled signal is thus $T = N/M \approx 1.5723$ s. In Figure 2.12b we graph the lumped frequency energies $(|X_k|^2 + |X_{N-k}|^2)/N$. In light of Remark 2.3, any given horizontal coordinate k in the spectral plot as follows corresponds to frequency $|k|/T \approx 0.636|k|$ Hz. The right end of the plot corresponds to the Nyquist frequency 4096 Hz. The DFT makes it clear that essentially three frequencies or closely grouped sets of frequencies dominate the signal. See the Matlab project later in this chapter, where you can play around with this signal yourself.

2.5 Properties of the DFT

In this section, we demonstrate some important algebraic properties of the DFT.

2.5.1 Matrix Formulation and Linearity

2.5.1.1 The DFT as a Matrix

The application of the DFT to a vector $\mathbf{x} \in \mathbb{C}^N$ can be implemented as a matrix multiplication $\mathbf{x} \to \mathbf{F}_N\mathbf{x}$ for an appropriate $N \times N$ matrix $\mathbf{F}_N$. Whenever we use matrix methods to analyze the DFT, it is important to write all vectors in a column format.

Let us start with a look at the specific case $N = 4$, and then proceed to the general case. Let $\mathbf{x}$ be a vector of samples

$$\mathbf{x} = \begin{bmatrix} x_0 \\ x_1 \\ x_2 \\ x_3 \end{bmatrix}.$$

From the definition of the DFT, we compute the components X_k of $\mathbf{X}$ as

$$X_k = (\mathbf{x}, \mathbf{E}_{4,k}) = \mathbf{E}^*_{4,k}\mathbf{x},$$

where $\mathbf{E}^*_{4,k}$ is the row vector or 1×4 matrix obtained as the conjugate transpose of $\mathbf{E}_{4,k}$. Of course, $\mathbf{E}^*_{4,k}\mathbf{x}$ is the matrix product of the 1×4 matrix $\mathbf{E}^*_{4,k}$ with the 4×1 matrix (column vector) $\mathbf{x}$, and this product is just the scalar X_k. For $N = 4$ this amounts to the equations:

$$X_0 = \begin{bmatrix} \overline{1} & \overline{1} & \overline{1} & \overline{1} \end{bmatrix} \begin{bmatrix} x_0 \\ x_1 \\ x_2 \\ x_3 \end{bmatrix},$$

$$X_1 = \begin{bmatrix} \overline{1} & \overline{i} & \overline{-1} & \overline{-i} \end{bmatrix} \begin{bmatrix} x_0 \\ x_1 \\ x_2 \\ x_3 \end{bmatrix},$$

$$X_2 = \begin{bmatrix} \overline{1} & \overline{-1} & \overline{1} & \overline{-1} \end{bmatrix} \begin{bmatrix} x_0 \\ x_1 \\ x_2 \\ x_3 \end{bmatrix},$$

$$X_3 = \begin{bmatrix} \overline{1} & \overline{-i} & \overline{-1} & \overline{i} \end{bmatrix} \begin{bmatrix} x_0 \\ x_1 \\ x_2 \\ x_3 \end{bmatrix}.$$

In more compact notation (and after we perform the complex conjugation), this is

$$\begin{bmatrix} X_0 \\ X_1 \\ X_2 \\ X_3 \end{bmatrix} = \begin{bmatrix} 1 & 1 & 1 & 1 \\ 1 & -i & -1 & i \\ 1 & -1 & 1 & -1 \\ 1 & i & -1 & -i \end{bmatrix} \begin{bmatrix} x_0 \\ x_1 \\ x_2 \\ x_3 \end{bmatrix}, \tag{2.8}$$

or $\mathbf{X} = \mathbf{F}_4\mathbf{x}$.

To analyze the general N-dimensional case, first recall from linear algebra that if $\mathbf{A}$ is an $N \times N$ matrix and $\mathbf{x}$ an N-dimensional column vector, then the product $\mathbf{Ax}$ is that N-dimensional column vector $\mathbf{v}$ with components

$$v_m = \sum_{k=0}^{N-1} A_{mk}x_k,$$

where we index from 0 to $N - 1$ as usual. Compare this to the definition (2.6) of the DFT—they are identical, with v_m above replaced by X_m and where $\mathbf{A}$ denotes the matrix with row m, column k entry $e^{-2\pi imk/N}$. We have shown the following theorem:

Theorem 2.1 *Let* $\mathbf{x} \in \mathbb{C}^N$ *and* $\mathbf{X} = \text{DFT}(\mathbf{x})$. *Then*

$$\mathbf{X} = \mathbf{F}_N\mathbf{x},$$

where $\mathbf{F}_N$ *is the* $N \times N$ *matrix (rows and columns indexed 0 to* $N - 1$*) with row* m*, column* k *entry* $e^{-2\pi imk/N}$.

If we let $z = e^{-2\pi i/N}$ (so $e^{2\pi imk/N} = z^{mk}$), then $\mathbf{F}_N$ can be written in the form

$$\mathbf{F}_N = \begin{bmatrix} 1 & 1 & 1 & 1 & \cdots & 1 \\ 1 & z & z^2 & z^3 & \cdots & z^{N-1} \\ 1 & z^2 & z^{2\cdot 2} & z^{2\cdot 3} & \cdots & z^{2\cdot(N-1)} \\ 1 & z^3 & z^{3\cdot 2} & z^{3\cdot 3} & \cdots & z^{3\cdot(N-1)} \\ \vdots & \vdots & \vdots & \vdots & \ddots & \vdots \\ 1 & z^{N-1} & z^{(N-1)\cdot 2} & z^{(N-1)\cdot 3} & \cdots & z^{(N-1)\cdot(N-1)} \end{bmatrix}. \tag{2.9}$$

The matrix $\mathbf{F}_N$ has a lot of structure. This structure can even be exploited to develop an algorithm called the "fast Fourier transform" (FFT) that provides a very efficient method for computing DFTs without actually doing the full matrix multiplication. We will look at this in Section 2.6.

2.5.1.2 The Inverse DFT as a Matrix

The IDFT defined by equation (2.7) also has a matrix formulation. Indeed essentially identical reasoning to that used for the DFT (with equation (2.7) in place of (2.6)) shows that $\mathbf{x} = \tilde{\mathbf{F}}_N\mathbf{X}$ where $\tilde{\mathbf{F}}_N$ is the $N \times N$ matrix with row m, column k entry $e^{2\pi imk/N}\ /N$. This matrix is symmetric, like $\mathbf{F}_N$.

In fact, $\tilde{\mathbf{F}}_N$ is easily obtained from $\mathbf{F}_N$. If we conjugate $\mathbf{F}_N$ entry by entry, we obtain the matrix $N\tilde{\mathbf{F}}_N$ so that $\overline{\mathbf{F}_N} = N\tilde{\mathbf{F}}_N$ or

$$\tilde{\mathbf{F}}_N = \frac{1}{N}\overline{\mathbf{F}_N}.$$

However, it will later be more useful to replace the entry-by-entry conjugation of $\mathbf{F}_N$ with the Hermitian transpose $\mathbf{F}_N^*$ of $\mathbf{F}_N$, a conjugation followed by transposition. Because $\mathbf{F}_N$ is symmetric, this transposition may seem pointless, but the Hermitian transpose operation is more standard in linear algebra and has some nice properties. In this case, we find that

$$\tilde{\mathbf{F}}_N = \frac{1}{N}\mathbf{F}_N^*.$$

We can summarize with the following theorem:

Theorem 2.2 *Let* $\mathbf{x} \in \mathbb{C}^N$ *and* $\mathbf{X} = DFT(\mathbf{x})$*. Let* $\mathbf{F}_N$ *denote the N-point DFT matrix. Then*

$$\mathbf{X} = \mathbf{F}_N \mathbf{x},$$
$$\mathbf{x} = \frac{1}{N}\mathbf{F}_N^* \mathbf{X}.$$

Of course, this also shows that $\mathbf{F}_N^{-1} = \frac{1}{N}\mathbf{F}_N^*$.

We have not stated the following explicitly, but Theorem 2.2 (or equations (2.6) and (2.7)) yields a subsequent important fact.

Theorem 2.3 *The discrete Fourier transform* $DFT : \mathbb{C}^N \to \mathbb{C}^N$ *is a linear mapping,*

$$\mathrm{DFT}(a\mathbf{x} + b\mathbf{y}) = a \cdot \mathrm{DFT}(\mathbf{x}) + b \cdot \mathrm{DFT}(\mathbf{y})$$

for $\mathbf{x}, \mathbf{y} \in \mathbb{C}^N$ *and* $a, b \in \mathbb{C}$*. The inverse DFT is also linear.*

2.5.2 Symmetries for Real Signals

In all the examples we have looked at so far the DFT has possessed obvious symmetries. The symmetries stemmed from the fact that the underlying signals were real-valued. This is quantified in the following proposition, in which we implicitly use the fact that X_k is defined for all k (recall Remark 2.2).

Proposition 2.1 *A vector* $\mathbf{x} \in \mathbb{C}^N$ *has all components* x_r *real if and only if* $\mathbf{X} = DFT(\mathbf{x})$ *has components that satisfy*

$$X_{-k} = \overline{X}_k,$$

or equivalently,

$$Re(X_{-k}) = Re(X_k) \quad and \quad Im(X_{-k}) = -Im(X_k)$$

for all $k \in \mathbb{Z}$*.*

Proof: Suppose that $\mathbf{x}$ has all components x_r real so that $\overline{x}_r = x_r$. Then

$$X_{-k} = \sum_{r=0}^{N-1} x_r e^{-2\pi i(-k)r/N}$$
$$= \sum_{r=0}^{N-1} x_r e^{2\pi ikr/N}$$

$$= \sum_{r=0}^{N-1} \overline{\overline{x}_r e^{-2\pi ikr/N}}$$

$$= \overline{\sum_{r=0}^{N-1} x_r e^{-2\pi ikr/N}}$$

$$= \overline{X_k}.$$

Here we have used $\overline{x}_r = x_r$, as well as elementary properties of complex conjugation. This shows that $X_{-k} = \overline{X_k}$.

To show the converse, suppose that $X_{-k} = \overline{X_k}$ for all k, so $X_k = \overline{X_{-k}}$. We have from the IDFT that

$$x_r = \frac{1}{N}\sum_{k=0}^{N-1} X_k e^{2\pi ikr/N}$$

$$= \frac{1}{N}\sum_{k=0}^{N-1} \overline{X_{-k}} e^{2\pi ikr/N}$$

$$= \overline{\frac{1}{N}\sum_{k=0}^{N-1} X_{-k} e^{-2\pi ikr/N}}$$

$$= \overline{\frac{1}{N}\sum_{n=0}^{N-1} X_n e^{2\pi inr/N}} \qquad \text{(substitute } n = -k\text{)}$$

$$= \overline{x}_r.$$

Since $x_r = \overline{x}_r$, we conclude that each x_r is real.

The symmetry and antisymmetry of the real and imaginary parts are a simple consequence of the definition of conjugation. Proposition 2.1 is really a special case of a more general symmetry for the DFT; see Exercise 10.

Remark 2.4 From Proposition 2.1, if $\mathbf{x}$ has real components, then $X_{-k} = \overline{X_k}$. Also, since X_k is periodic with period N in k (whether or not $\mathbf{x}$ is real-valued), we have $X_{-k} = X_{N-k}$. We conclude that

$$X_{N-k} = \overline{X_k}. \tag{2.10}$$

In particular, if $k = N/2 + j$ in equation (2.10), then

$$X_{N/2-j} = \overline{X_{N/2+j}}.$$

Thus, if $\mathbf{x}$ is real, X_k is conjugate-symmetric about $N/2$. (If N is odd, this still makes sense if we let $j = m + 1/2$ for $m \in \mathbb{Z}$.) We also conclude that $|X_{N/2-j}| =$

$|X_{N/2+j}|$, which is why for real-valued signals we need only plot $|X_k|$ in the range $0 \le k \le N/2$.

2.6 The Fast Fourier Transform

The FFT is an algorithm, or really a class of algorithms, for computing the DFT and IDFT efficiently. The FFT is one of the most influential algorithms of the twentieth century. The "modern" version of the algorithm was published by Cooley and Tukey [8] in 1965, but some of the ideas appeared earlier. Given the very regular structure of the matrix $\mathbf{F}_N$, it is not entirely surprising that shortcuts can be found for computing the matrix-vector product $\mathbf{F}_N\mathbf{x}$. All modern software uses the FFT algorithm for computing discrete Fourier transforms. The details are usually transparent to the user. Thus knowledge of the FFT algorithm is not generally essential for using or understanding DFTs.

There are several points of view on the FFT. A computer scientist would classify the FFT as a classic "divide and conquer" algorithm. A mathematician might view the FFT as a natural consequence of the structure of certain finite groups. Other practitioners might simply view it as a clever technique for organizing the sums involved in equation (2.6), or factoring the matrix $\mathbf{F}_N$ into simpler pieces. Our intention is merely to give a very brief introduction to the idea behind one form of the algorithm. The material in this section is not essential for anything that follows. For a more comprehensive treatment of the FFT, see [3].

2.6.1 DFT Operation Count

Let us begin by looking at the computational cost of evaluating the sums in equation (2.6), or equivalently, computing the matrix-vector product of Theorem 2.1. Let $\mathbf{x} \in \mathbb{C}^N$, and let $\mathbf{X}$ denote the DFT of $\mathbf{x}$. How many floating point operations does it take to compute $\mathbf{X}$ from $\mathbf{x}$ using equation (2.6)? The computation of any given X_k requires N complex multiplications (the product $x_m e^{-2\pi ikm/N}$ for $0 \le m \le N-1$) followed by $N-1$ additions, a total of $2N-1$ operations. Since we have to compute X_k for $0 \le k \le N-1$, we must perform $N(2N-1) = 2N^2 - N$ operations for this straightforward approach, where an "operation" means a complex addition or multiplication. Of course, this ignores the cost of computing the exponentials. However, if we let $z = e^{-2\pi i/N}$, then all exponentials are of the form z^m for m in the range $0 \le m \le N-1$, and these can be precomputed and tabled. We will thus ignore this computational cost, which, at any rate, is only on the order of N operations and insignificant with respect to N^2.

All in all, the straightforward approach to the DFT requires $2N^2 - N$ operations, or roughly just $2N^2$ if N is large. If the signal $\mathbf{x}$ is real, then there are some tricks that can be performed to lower the operation count, but the computational cost is still proportional to N^2.

2.6.2 The FFT

The FFT cuts the work for computing an N-sample DFT from $2N^2$ floating point operations down to $CN\log(N)$ operations for some constant C, a significant savings if N is large or many transforms have to be computed quickly. We will consider one of the most common varieties of the many FFT algorithms, and restrict our attention to the special but fairly typical case where $N = 2^n$ for some positive integer n.

Our first goal is to split the task of computing an N-point DFT into the computation of two transforms of size $N/2$. For the k index appearing in equation (2.6), we can write $k = (N/2)k_1 + k_0$, where $k_1 = 0$ or $k_1 = 1$ and $0 \leq k_0 < N/2$. Of course, as k_0 and k_1 range over these limits, k ranges from 0 to $N-1$. So we write the DFT in equation (2.6) as

$$\begin{aligned} X_{(N/2)k_1+k_0} &= \sum_{m=0}^{N-1} x_m e^{-2\pi i((N/2)k_1+k_0)m/N} \\ &= \sum_{m=0}^{N-1} x_m e^{-\pi i k_1 m} e^{-2\pi i k_0 m/N} \\ &= \sum_{m=0}^{N-1} x_m (-1)^{k_1 m} e^{-2\pi i k_0 m/N}, \end{aligned} \tag{2.11}$$

where we have split the exponential and used $e^{-\pi i k_1 m} = (-1)^{k_1 m}$. Next we split the sum over m in equation (2.11) into even and odd indexes to obtain

$$\begin{aligned} X_{(N/2)k_1+k_0} = &\sum_{m=0}^{N/2-1} x_{2m}(-1)^{2k_1 m} e^{-2\pi i k_0 (2m)/N} \quad \text{(even indexes)} \\ &+ \sum_{m=0}^{N/2-1} x_{2m+1}(-1)^{k_1(2m+1)} e^{-2\pi i k_0 (2m+1)/N} \quad \text{(odd indexes)} \end{aligned}$$

A bit of simplification yields

$$\begin{aligned} X_{(N/2)k_1+k_0} &= \sum_{m=0}^{N/2-1} x_{2m} e^{-2\pi i k_0 m/(N/2)} \\ &\quad + \sum_{m=0}^{N/2-1} x_{2m+1}(-1)^{k_1} e^{-2\pi i k_0/N} e^{-2\pi i k_0 m/(N/2)} \\ &= \sum_{m=0}^{N/2-1} x_{2m} e^{-2\pi i k_0 m/(N/2)} \\ &\quad + (-1)^{k_1} e^{-2\pi i k_0/N} \sum_{m=0}^{N/2-1} x_{2m+1} e^{-2\pi i k_0 m/(N/2)}. \end{aligned} \tag{2.12}$$

Let us use $F_1(k_0)$ and $F_2(k_0)$ to denote the first and second sums on the right in equation (2.12), that is,

$$F_1(k_0) = \sum_{m=0}^{N/2-1} x_{2m} e^{-2\pi i k_0 m/(N/2)} \text{ and } F_2(k_0) = \sum_{m=0}^{N/2-1} x_{2m+1} e^{-2\pi i k_0 m/(N/2)}.$$

Note that F_1 (as k_0 ranges from 0 to $N/2-1$) is exactly the DFT of the $N/2$-component vector $(x_0, x_2, x_4, \ldots, x_{N-2})$. Likewise, F_2 is exactly the DFT of the $N/2$-component vector $(x_1, x_3, x_5, \ldots, x_{N-1})$.

We can now write equation (2.12) as

$$X_{(N/2)k_1+k_0} = F_1(k_0) + (-1)^{k_1} e^{-2\pi i k_0/N} F_2(k_0). \tag{2.13}$$

Remember, $0 \le k_0 < N/2$ and $k_1 = 0$ or $k_1 = 1$. In the case where $k_1 = 0$, equation (2.13) yields

$$X_{k_0} = F_1(k_0) + e^{-(2\pi i/N)k_0} F_2(k_0), \tag{2.14}$$

while $k_1 = 1$ in (2.13) gives

$$X_{N/2+k_0} = F_1(k_0) - e^{-(2\pi i/N)k_0} F_2(k_0). \tag{2.15}$$

In short, if we compute the DFT of the even and odd indexed entries of $\mathbf{x}$, equations (2.14) and (2.15) tell us how to combine this information to get the full DFT of $\mathbf{x}$; equation (2.14) yields the first $N/2$ components of $\mathbf{X}$ and (2.15) yields the last $N/2$. An N-point transform is thus reduced to the computation and appropriate mixing of two $N/2$-point transforms.

2.6.3 The Operation Count

What is the operation count for the full N-point transform with this approach? Let $W_{N/2}$ denote the number of operations necessary to compute each transform F_1 and F_2, both of size $N/2$. The computation of F_1 and F_2 together requires a total of $2W_{N/2}$ operations. Also equations (2.14) and (2.15) require a total of N multiplications and N additions to weave together the full N-point transform. Thus, if W_N denotes the number of operations for the full N-point transform, we have

$$W_N = 2N + 2W_{N/2}.$$

But we can now apply the same approach recursively to compute each $N/2$-point transform, to conclude that $W_{N/2} = N + 2W_{N/4}$. Similarly, $W_{N/4} = N/2 + 2W_{N/8}$, and so on. We conclude that

$$\begin{aligned} W_N &= 2N + 2\left(\frac{N}{2} + 2\left(\frac{N}{4} + 2\left(\frac{N}{8} + \cdots + 2W_1\right)\right)\right) \\ &= 2(N + N + \cdots + N), \end{aligned}$$

where there are a total of $\log_2(N)$ terms in the sum. The net result is an algorithm that takes about $2N\log_2(N)$ complex operations.

The FFT is usually implemented without explicit recursion; see [3] for various practical implementations. The FFT is not limited to N that are powers of 2. In general, an efficient FFT algorithm can be derived most easily when N is "highly composite," that is, factors completely into small integers.

2.7 The Two-Dimensional DFT

Let $\mathbf{A} \in M_{m,n}(\mathbb{C})$; we might consider $\mathbf{A}$ as a sampled image. The general framework developed in Chapter 1 makes it easy to define the two-dimensional DFT of $\mathbf{A}$. We simply let the basic two-dimensional waveforms $\mathcal{E}_{m,n,k,l}$ take over for the one-dimensional waveforms $\mathbf{E}_{n,k}$. In what follows we write $\mathcal{E}_{k,l}$ instead of $\mathcal{E}_{m,n,k,l}$ for the two-dimensional waveforms, since m and n remain fixed in any given computation.

Recall that the $\mathcal{E}_{k,l}$ for $0 \le k \le m-1, 0 \le l \le n-1$ form an orthogonal basis for $M_{m,n}(\mathbb{C})$ with respect to the usual inner product. This allowed us to develop the expansion of equation (1.35). A slight variation on this yields the two-dimensional DFT.

Definition 2.3 Let $\mathbf{A} \in M_{m,n}(\mathbb{C})$ have components a_{rs}, $0 \le r \le m-1, 0 \le s \le n-1$. The two-dimensional discrete Fourier transform of $\mathbf{A}$ is the matrix $\hat{\mathbf{A}} \in M_{m,n}(\mathbb{C})$ with components

$$\hat{a}_{k,l} = (\mathbf{A}, \mathcal{E}_{k,l}) = \sum_{r=0}^{m-1}\sum_{s=0}^{n-1} a_{r,s} e^{-2\pi i(kr/m+ls/n)}, \tag{2.16}$$

where $0 \le k \le m-1, 0 \le l \le n-1$.

The definition of $\hat{a}_{k,l}$ above differs from the coefficients $(\mathbf{A}, \mathcal{E}_{k,l})/mn$ of equation (1.35) only in that $\hat{a}_{k,l}$ has no $1/mn$ factor in front. In general, we will use a "hat" symbol over a matrix $\mathbf{A}$ to denote the two-dimensional DFT of $\mathbf{A}$.

In light of equation (1.35) and the fact that the $\mathcal{E}_{k,l}$ form an orthogonal basis, we can reconstruct $\mathbf{A}$ using the two-dimensional IDFT defined as follows:

Definition 2.4 Let $\hat{\mathbf{A}} \in M_{m,n}(\mathbb{C})$ be a matrix with components $\hat{a}_{k,l}$. The two-dimensional IDFT of $\hat{\mathbf{A}}$ is the matrix

$$\mathbf{A} = \frac{1}{mn}\sum_{k=0}^{m-1}\sum_{l=0}^{n-1} \hat{a}_{k,l}\mathcal{E}_{k,l}$$

in $M_{m,n}(\mathbb{C})$ with components

$$a_{r,s} = \frac{1}{mn} \sum_{k=0}^{m-1} \sum_{l=0}^{n-1} \hat{a}_{k,l} e^{2\pi i(kr/m+ls/n)}. \tag{2.17}$$

The $1/mn$ factor is placed in the IDFT definition. Like the one-dimensional DFT/IDFT, definitions may vary slightly from text to text.

Remark 2.5 It is clear from equation (2.16) that the component $\hat{a}_{k,l}$ is defined for any integers k and l. Moreover, since $e^{-2\pi i(kr/m+ls/n)}$ is periodic in k with period m and in l with period n, the same is true for $\hat{a}_{k,l}$, so $\hat{a}_{k,l} = \hat{a}_{k+pm,l+qn}$ for any integers p and q. As a result we may, as in the case of the one-dimensional DFT, consider $\hat{a}_{k,l}$ on alternate ranges of k and l. One common variation is to use the range $-m/2 < k \le m/2, -n/2 < l \le n/2$.

Recall from Chapter 1 that each basic waveform $\mathcal{E}_{k,l}$ has a certain "frequency" and "direction" (recall the figures in Section 1.5.2). The two-dimensional DFT of an image indicates how much of each basic waveform is necessary to synthesize the image as a superposition of basic waveforms. The "low-frequency" Fourier coefficients (small k and l) stem from slowly varying features and patterns in the image, such as the blue sky in an outdoor picture. Rapidly varying features such as a striped shirt or other fine textures contribute to Fourier coefficients with larger indexes.

The two-dimensional DFT also has a matrix formulation.

Proposition 2.2 *The two-dimensional DFT in equation (2.16) may be computed as*

$$\hat{\mathbf{A}} = \mathbf{F}_m \mathbf{A} \mathbf{F}_n = \mathbf{F}_m \mathbf{A} \mathbf{F}_n^T,$$

where $\mathbf{F}_m$ *and* $\mathbf{F}_n$ *are the DFT matrices of Theorem 2.1.*

Proof: Consider the matrix product $\mathbf{F}_m\mathbf{A}$; note $\mathbf{F}_m$ is $m \times m$, while $\mathbf{A}$ is $m \times n$. The product is an $m \times n$ matrix with row k, column s entry $(\mathbf{F}_m\mathbf{A})_{k,s}$ given by

$$(\mathbf{F}_m\mathbf{A})_{k,s} = \sum_{r=0}^{m-1} a_{r,s} e^{-2\pi ikr/m}. \tag{2.18}$$

This follows from the definition of matrix multiplication, where note that $e^{-2\pi ikr/m}$ is the row k, column r entry of $\mathbf{F}_m$.

Now consider the product $(\mathbf{F}_m\mathbf{A})\mathbf{F}_n$. The matrix $(\mathbf{F}_m\mathbf{A})$ is $m \times n$, while $\mathbf{F}_n$ is $n \times n$. The product $(\mathbf{F}_m\mathbf{A})\mathbf{F}_n$ is thus $m \times n$. The row s, column l entry of $\mathbf{F}_n$ is

$e^{-2\pi i l s/n}$. If we make use of equation (2.18), we find that the row k, column l entry of the product $(\mathbf{F}_m\mathbf{A})\mathbf{F}_n$ is given by

$$
\begin{aligned}
(\mathbf{F}_m\mathbf{A}\mathbf{F}_n)_{k,l} &= \sum_{s=0}^{n-1} (\mathbf{F}_m\mathbf{A})_{k,s} e^{-2\pi i l s/n} \\
&= \sum_{s=0}^{n-1}\sum_{r=0}^{m-1} a_{r,s} e^{-2\pi i k r/m} e^{-2\pi i l s/n} \\
&= \sum_{r=0}^{m-1}\sum_{s=0}^{n-1} a_{r,s} e^{-2\pi i (kr/m + ls/n)}.
\end{aligned}
$$

This is precisely the row k, column l entry of the $m \times n$ matrix $\hat{\mathbf{A}} = \text{DFT}(\mathbf{A})$, so $\hat{\mathbf{A}} = \mathbf{F}_m\mathbf{A}\mathbf{F}_n$. The formula $\hat{\mathbf{A}} = \mathbf{F}_m\mathbf{A}\mathbf{F}_n^T$ follows from the fact that $\mathbf{F}_k^T = \mathbf{F}_k$.

Proposition 2.2 also has a nice algorithmic interpretation. First note these facts:

1. If $\mathbf{B}$ is an $m \times n$ matrix, then the operation $\mathbf{B} \rightarrow \mathbf{F}_m\mathbf{B}$ performs an m-point one-dimensional DFT on each column of $\mathbf{B}$.
2. The operation $\mathbf{B} \rightarrow \mathbf{B}\mathbf{F}_n^T$ performs an n-point one-dimensional DFT on each row of $\mathbf{B}$, and leaves each transformed row as a row vector. This is easy to see by noting that $\mathbf{B}^T$ swaps the rows and columns of $\mathbf{B}$ so that $\mathbf{F}_n\mathbf{B}^T$ performs an n-point transform on each column of $\mathbf{B}^T$ (i.e., rows of $\mathbf{B}$). If we now transpose back, the transformed columns are swapped back into rows, and the result is $(\mathbf{F}_n\mathbf{B}^T)^T = \mathbf{B}\mathbf{F}_n^T$.

According to the remarks above then, the operation

$$\mathbf{A} \rightarrow \mathbf{F}_m\mathbf{A}$$

performs a one-dimensional m-point DFT on each column of the matrix $\mathbf{A}$. The operation

$$(\mathbf{F}_m\mathbf{A}) \rightarrow (\mathbf{F}_m\mathbf{A})\mathbf{F}_n^T$$

then transforms the result row by row. From Proposition 2.2 the result is precisely the two-dimensional DFT of $\mathbf{A}$. Thus, the two-dimensional DFT can be computed by transforming $\mathbf{A}$ column by column, and then transforming the result row by row. We can also perform a two-dimensional DFT by first transforming the rows and then the columns. This corresponds to associating the product $\mathbf{F}_m\mathbf{A}\mathbf{F}_n^T$ as $\mathbf{F}_m(\mathbf{A}\mathbf{F}_n^T)$.

Like the one-dimensional DFT, the two-dimensional DFT is also linear and the DFT of a real-valued matrix possesses many symmetries; see Exercise 22. There are also two-dimensional versions of the FFT. Indeed, since the two-dimensional DFT can be computed as a sequence of one-dimensional

DFTs, the one-dimensional FFT has an easy extension to two dimensions (though a tailor-made two-dimensional FFT algorithm may be slightly more efficient).

2.7.1 Interpretation and Examples of the 2-D DFT

In Figure 2.13, we show a plotted image (336 × 448 pixels) and several renditions of the magnitude of the DFT for this image. We have done a logarithmic rescaling of the absolute value of the Fourier transform in Figure 2.13 (b) and (c), since there are a few relatively large coefficients and many that are quite small. For natural images, the dominant frequencies are almost always low (small k and l), since most of the energy is concentrated in the broad features of the image. By plotting $\log(1 + |\hat{a}_{k,l}|)$, we are able to see patterns over several orders of magnitude. We take $\log(1 + |\hat{a}_{k,l}|)$ rather than $\log(|\hat{a}_{k,l}|)$ to avoid taking the log of 0.

In Figure 2.13 (c) and (d), we have "centered" the DFT by using the range $-m/2 \le k \le m/2$, $-n/2 \le l \le n/2$. The centered transform puts low

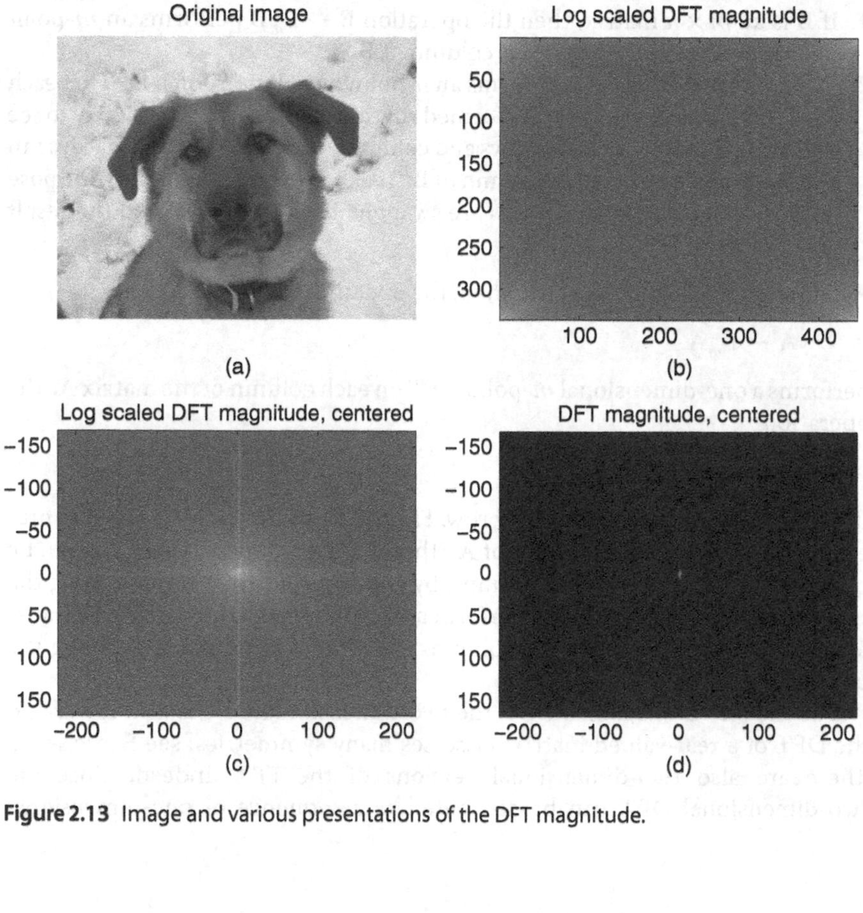

Figure 2.13 Image and various presentations of the DFT magnitude.

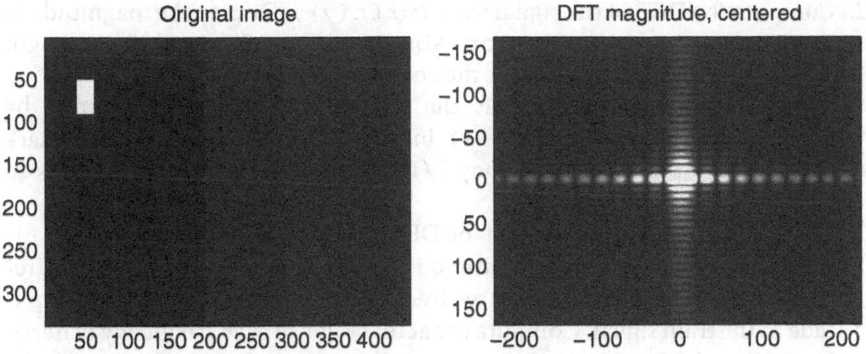

Figure 2.14 Artificial image and DFT magnitude $\ln(1 + |\hat{a}_{r,s}|)$.

frequencies with lots of energy in the middle and higher frequencies with lower energy at the boundary. This is especially clear from Figure 2.13 (d), in which we do not rescale with a logarithm. All the energy is obviously concentrated in the very lowest frequencies near the center of the plot, and the remaining frequencies are practically absent.

Our last example in Figure 2.14 is an artificially generated image that illustrates how the DFT reacts to discontinuities. The image is of a white rectangle on a black background. It is a product of rectangular pulses, namely $A_{r,s} = x_r y_s$ where $\mathbf{x}$ and $\mathbf{y}$ are two square wave pulse functions similar to the square wave example in Section 2.4.2. It can be shown (see Exercise 17) that the coefficients of $\hat{\mathbf{A}} = \text{DFT}(\mathbf{A})$ are the products $\hat{a}_{k,l} = X_k Y_l$ where $\mathbf{X}$ and $\mathbf{Y}$ are the one-dimensional DFTs of $\mathbf{x}$ and $\mathbf{y}$. Since each of $\mathbf{X}$ and $\mathbf{Y}$ looks like the DFTs in Figures 2.8 through 2.10, we see can why $\hat{\mathbf{A}}$ looks like it does. In general, discontinuities or other abrupt features generate a lot of energy at high frequencies.

2.8 Matlab Project

This section contains Matlab explorations that make use of the DFT for analyzing signals and images.

2.8.1 Audio Explorations

1. Start Matlab. Load in the "train" signal with the command `load ('train');`. Recall that the audio signal is loaded into a variable "`y`" and the sampling rate into "`Fs`." The sampling rate is 8192 Hz, and the signal contains 12,880 samples. If we consider this signal as sampled on an interval $[0, T]$, then $T = 12,880/8192 \approx 1.5723$ s.

2. Compute the DFT of the signal with `Y=fft(y);`. Display the magnitude of the Fourier transform with `plot(abs(Y))`. The DFT should have length 12,880 and be symmetric about the center.
   Since Matlab indexes from 1, the DFT coefficient Y_k as defined by equation (2.6) is actually `Y(k+1)` in Matlab. Also, according to Remark 2.3, Y_k corresponds to frequency $k/1.5723$, and so `Y(k)` corresponds to frequency $(k-1)/1.5723$ Hz.
3. You can plot only the first half of the DFT with `plot(abs(Y(1:6441)))`. Do so. Use the data cursor button on the plot window to pick out the frequency and amplitude of the three frequencies of (obviously) largest amplitude in the train signal. Compute the actual value of each frequency in hertz.
4. Let f_1, f_2, and f_3 denote the frequencies of largest amplitude, in hertz, and let A_1, A_2, A_3 denote the corresponding amplitudes from the plot above. Define these variables in Matlab. Synthesize a new signal using only these frequencies, sampled at 8192 Hz on the interval [0, 1.5], with

```
t = [0:1/8192:1.5];
ysynth = (A1*sin(2*pi*f1*t) + A2*sin(2*pi*f2*t)
+ A3*sin(2*pi*f3*t))/(A1+A2+A3);
```

   The division by $(A_1 + A_2 + A_3)$ guarantees that the synthesized signal `ysynth` lies in the range $[-1, 1]$, which is the range Matlab uses for audio signals.
5. Play the original train sound with `sound(y)`, and the synthesized version of only three frequencies with `sound(ysynth)`. Note that our computations do not take into account the phase information at these frequencies, merely the amplitude. Does the artificially generated signal capture the tone of the original?
6. Here is a simple approach to compressing an audio or other one-dimensional signal. The idea is to transform the audio signal to the frequency domain with the DFT. We then eliminate the insignificant frequencies by "thresholding," that is, zeroing out any Fourier coefficients below a given threshold. This becomes the compressed version of the signal. To recover an approximation to the signal, we use the IDFT to take the thresholded transform back to the time domain.
   For the train audio signal we can threshold as follows: First, we compute the maximum value of $|Y_k|$, with

```
M = max(abs(Y))
```

   Then choose a threshold parameter `thresh` between 0 and 1. Let us start with

```
thresh = 0.1
```

Finally, we zero out all frequencies in Y that fall below a value thresh*M in magnitude. This can be done with

```
Ythresh = (abs(Y)>thresh*M).*Y;
```

which installs the thresholded transform into "Ythresh." You should plot the thresholded transform with plot(abs(Ythresh)), just to make sure it worked. You can also see what fraction of the Fourier coefficients survived the cut with

```
sum(abs(Ythresh)>0)/12880
```

We will call this the "compression ratio."

To recover an approximation to the original signal inverse transform with

```
ythresh = real(ifft(Ythresh));
```

and play the "compressed" audio with sound(ythresh). The "real" command above truncates any vestigial imaginary round-off error in the ifft command.

You can compute the distortion of the compressed signal with

```
100*norm(y-ythresh).^2/norm(y).^2
```

Repeat the computations above for threshold values 0.001, 0.01, 0.1, and 0.5. In each case compute the compression ratio, the distortion, and of course, play the audio signal and rate its quality.

2.8.2 Images

1. Close any open graphics windows and clear all Matlab variables with clear. Then load an image into Matlab with

   ```
   z=imread('myimage.jpg');
   ```

   where "myimage.jpg" is any image you want (loaded from the current working directory). Construct a grayscale image with

   ```
   zg = 0.2989*double(z(:,:,1))+0.5870*double(z(:,:,2))
        +0.1140*double(z(:,:,3));
   ```

   and the colormap

   ```
   L = 255;
   colormap([(0:L)/L; (0:L)/L; (0:L)/L]');
   ```

   View the image with image(zg).

2. Compute the two-dimensional DFT of `zg` with

```
Z = fft2(zg);
```

You can view the DFT itself as an image as follows: First, as discussed in the text, we will look at $\log(1 + |\mathbf{Z}|)$ in order to better see components that vary over many orders of magnitude. Thus, let

```
Zlog = log(1+abs(Z));
```

In order to properly scale the image, it is helpful to compute the maximum value of `Zlog` with

```
M = max(max(Zlog))
```

Finally, we can view the transform (using the grayscale colormap set up previously) with

```
image(255*Zlog/M)
```

You should be able to see certain symmetries in the two-dimensional DFT, but otherwise, the data are difficult to interpret. The original image certainly is not evident.

3. As for audio signals, we can try a simple image compression technique based on thresholding the DFT (still stored in the variable `Z`).
To begin we compute the maximum value of $|Z_{k,l}|$, with

```
M = max(max(abs(Z)))
```

We also choose a threshold parameter `thresh` between 0 and 1. Let us start with

```
thresh = 0.0001
```

We zero out all frequencies in `Z` that fall below a value `thresh*M` in magnitude. This is done with

```
Zthresh = (abs(Z)>thresh*M).*Z;
```

which installs the thresholded transform into "`Zthresh`." You can see what fraction of the Fourier coefficients survived the cut with

```
sum(sum(abs(Zthresh)>0))/m/n
```

where m and n are the pixel counts in each direction. As in the audio case, we will call this the "compression ratio."
To uncompress and view the image, inverse transform with

```
zthresh = real(ifft2(Zthresh));
```

and view with `image(zthresh)`. As in the audio case, the `real` command above truncates any vestigial imaginary round-off error in the `ifft2` command.

You can compute the distortion of the compressed signal with

```
100*norm(zg-zthresh,'fro').^2/norm(zg,'fro').^2
```

Repeat the computation above for threshold values 0.001, 0.01, and 0.05, or even larger values. In each case, compute the compression ratio, the distortion, display the image, and rate its quality. Construct a plot of the distortion (vertical axis) versus compression ratio. It may be helpful to use a logarithmic scale on one or both axes.

Exercises

1. Prove the assertion in Remark 2.2, that the X_k from equation (2.6) are in fact periodic with period N.

2. Write out the matrix $\mathbf{F}_2$ that governs the 2-point DFT. Verify explicitly that $\frac{1}{2}\mathbf{F}_2^* = \mathbf{F}_2^{-1}$.

3. Use the matrix $\mathbf{F}_4$ to compute the DFT of the vector $\mathbf{x} = (1, 2, 0, -1)$.

4. Use the matrix $\mathbf{F}_4^*$ to compute the IDFT of the vector $\mathbf{X} = (3, 1 + i, 1, 1 - i)$.

5. Verify that $\mathbf{F}_4\mathbf{F}_4^* = 4\mathbf{I}$ directly, by performing the matrix multiplication.

6. Let $\mathbf{e}_k \in \mathbb{C}^N$ denote the standard basis vector with a "1" in the kth position, zeroes elsewhere (components indexed $0 \leq k \leq N - 1$). Find the DFT of $\mathbf{e}_k$, and also the magnitude of each DFT coefficient.

7. Let N be even. Compute the DFT of the vector $\mathbf{x} \in \mathbb{C}^N$ with components $x_k = (-1)^k$, $0 \leq k \leq N - 1$. Hint: $-1 = e^{i\pi}$.

8. Let $\mathbf{e}_{j,k}$ denote an $m \times n$ matrix that has a "1" in the row j, column k position, and zeroes elsewhere. Find the two-dimensional DFT of $\mathbf{e}_{j,k}$, and also the magnitude of each DFT coefficient.

9. In this exercise, we show that $\mathbf{F}_N^4 = N^2\mathbf{I}$, where $\mathbf{F}_N$ is the $N \times N$ DFT matrix and $\mathbf{I}$ is the $N \times N$ identity matrix. Thus, the DFT operation is, up to a scalar multiple, periodic with period 4.
   All following matrices should have columns and rows indexed from 0 to $N - 1$.
   (a) Use the definition of matrix multiplication to show that row j, column m entry of $\mathbf{F}_N^2$ is given by
   $$(\mathbf{F}_N)^2_{j,m} = 0$$

if $j+m \neq 0$ and $j+m \neq N$. Hint: Use $1+z+z^2+\cdots+z^{N-1} = (1-z^N)/(1-z)$. Show that if $j+m=0$ or $j+m=N$, then $(\mathbf{F}_N)^2_{j,m} = N$. You may find it helpful to explicitly write out $\mathbf{F}_4^2$, in order to see the pattern.

(b) Use part (a) to show that $\mathbf{F}_N^4 = N^2\mathbf{I}$. It may be helpful to note that $\mathbf{F}_N^2$ is symmetric, and each column or row of $\mathbf{F}_N^2$ has only one nonzero entry.

10. Let $\mathbf{x}$ and $\mathbf{y}$ be vectors in $\mathbf{C}^N$.

(a) Show that $y_r = \overline{x_r}$ for $0 \leq r \leq N-1$ if and only if the DFT coefficients satisfy $Y_m = \overline{X_{-m}}$.

(b) Deduce Proposition 2.1 as a special case of part (a).

11. Let $\mathbf{b} = \mathbf{a} - \text{mean}(\mathbf{a})$ denote the grayscale image $\mathbf{b}$ in which the average value of $\mathbf{a}$ is subtracted from each entry of $\mathbf{a}$. Show that $B_{0,0} = 0$ (where $\mathbf{B} = \text{DFT}(\mathbf{b})$) but argue that $\mathbf{b}$ and $\mathbf{a}$ encode the same image pattern.

12. Show that if $\mathbf{x} \in \mathbf{C}^N$ and $\mathbf{X} = \text{DFT}(\mathbf{x})$, then

$$N\|\mathbf{x}\|^2 = \|\mathbf{X}\|^2$$

where we use the usual norm on $\mathbf{C}^N$. Hint: It is a simple variation on Parseval's identity.

13. Let $\mathbf{A}$ be an $m \times n$ matrix with two-dimensional DFT $\hat{\mathbf{A}}$. Prove that $mn\|\mathbf{A}\|^2_{fro} = \|\hat{\mathbf{A}}\|^2_{fro}$, where

$$\|\mathbf{M}\|_{fro} = \left(\sum_{i=1}^{m}\sum_{j=1}^{n} |M_{ij}|^2\right)^{1/2}$$

denotes the "Frobenius" norm of $m \times n$ matrix $\mathbf{M}$. Hint: Use the result of Exercise 12 and Proposition 2.2.

14. Consider a discrete square waveform $\mathbf{x} \in \mathbb{R}^N$ with components $x_k = 1$ for $0 \leq k \leq R-1$ and $x_k = 0$ for $R \leq k \leq N-1$, where $0 < R < N$.

(a) Show that the DFT of this pulse is given by

$$X_m = \frac{1-e^{-2\pi imR/N}}{1-e^{-2\pi im/N}}$$

for $m > 0$, while $X_0 = R$. Hint: X_0 is easy. For X_m with $m > 0$ write out the DFT sum and use the geometric summation formula $1+z+z^2+\cdots+z^n = (1-z^{n+1})/(1-z)$.

(b) Plot the phase of X_m as a function of m for the case $N = 16$. What do you notice? How does the result depend on R?

15. Warning! This exercise requires a little background in statistics or probability, specifically, that if $x_0, x_1, \ldots, x_{N-1}$ are *independent* normal random variables (real-valued) with mean 0 and variance σ^2, then $x = \sum_k a_k x_k$ is also a normal random variable, with mean 0 and variance $\sigma^2 \sum_k a_k^2$ (where the a_k are fixed real numbers). Also, by these assumptions, the random variables $x = \sum_k a_k x_k$ and $y = \sum_k b_k x_k$ will be independent if $\sum_k a_k b_k = 0$. Consider a "noise signal" $\mathbf{x} \in \mathbb{R}^N$ as in Section 2.4.2, in which the x_k form a collection of independent normal random variables, each with mean 0 and variance σ^2. Let $\mathbf{X}$ denote the DFT of $\mathbf{x}$.

(a) Show that both $\mathrm{Re}(\mathbf{X}_m)$ and $\mathrm{Im}(\mathbf{X}_m)$ are normal random variables for $0 \le m \le N-1$, and that both have mean zero and variance $N\sigma^2/2$ for $m > 0$, while the mean and variance of X_0 are 0 and $N\sigma^2$, respectively. Hint: From $\cos^2(z) = \frac{1}{2} + \frac{1}{2}\cos(2z)$ we have

$$\sum_{k=0}^{N-1} \cos^2\left(\frac{2\pi km}{N}\right) = \frac{N}{2} + \frac{1}{2}\sum_{k=0}^{N-1} \cos\left(\frac{4\pi km}{N}\right).$$

Show the sum on the right above is zero. A similar identity $\sin^2(z) = \frac{1}{2} - \frac{1}{2}\cos(2z)$ holds for the sine.

(b) Show that $\mathrm{Re}(\mathbf{X}_m)$ and $\mathrm{Im}(\mathbf{X}_m)$ are independent.

(c) Find the probability distribution function for the random variable $(\mathrm{Re}(\mathbf{X}_m)^2 + \mathrm{Im}(\mathbf{X}_m)^2)/N$ for $m > 0$, and show that the expected value of $(\mathrm{Re}(\mathbf{X}_m)^2 + \mathrm{Im}(\mathbf{X}_m)^2)/N$ (the energy for waveform $\mathbf{E}_m$) is the same for all $m > 0$. Hence "on average" this signal has a flat spectrum. That is why such signals are called *white noise*—all frequencies are present in an equal amount.

16. Compute (by hand) the two-dimensional DFT of the matrix

$$\mathbf{A} = \begin{bmatrix} 1 & -1 \\ 2 & 0 \end{bmatrix}.$$

Then compute the inverse transform of the result.

17. Let $\mathbf{x}$ and $\mathbf{y}$ be two column vectors of length m and n, respectively, with one-dimensional DFTs $\mathbf{X}$ and $\mathbf{Y}$. Let $\mathbf{z}$ be the product matrix defined by

$$z_{r,s} = x_r y_s$$

for $0 \le r \le m-1, 0 \le s \le n-1$, and $\mathbf{Z}$ the two-dimensional DFT of $\mathbf{z}$.

(a) Show that $\mathbf{z}$ is an $m \times n$ matrix satisfying $\mathbf{z} = \mathbf{x}\mathbf{y}^T$ (where $\mathbf{y}^T$ is the transpose of $\mathbf{y}$).

(b) Show that

$$Z_{k,l} = X_k Y_l$$

or equivalently $\mathbf{Z} = \mathbf{X}\mathbf{Y}^T$, where $Z_{k,l}$ denotes the row k column l entry of $\mathbf{Z}$.

18. Let $\mathbf{z}$ denote an $m \times n$ matrix with $z_{j,k} = 1$ for $0 \leq j \leq R-1, 0 \leq k \leq S-1$, and $z_{j,k} = 0$ otherwise. Find the two-dimensional DFT of $\mathbf{z}$ as explicitly as possible. Hint: Make use of Exercises 14 and 17.

19. Let $\mathbf{x} \in \mathbb{C}^N$ have DFT $\mathbf{X}$. Let $\mathbf{y} \in \mathbb{C}^N$ be that vector obtained by circularly shifting $\mathbf{x}$ by m indexes as

$$y_k = x_{(k+m) \bmod N}.$$

Show that the DFT of $\mathbf{y}$ has components $Y_r = e^{2\pi irm/N}X_r$, and that $|X_r| = |Y_r|$ for all r.

20. Let $\mathbf{A}$ denote an $m \times n$ matrix with entries $a_{r,s}$ and DFT $\hat{\mathbf{A}}$ with components $\hat{a}_{k,l}$. Let $\mathbf{B}$ be the matrix obtained by circularly shifting $\mathbf{A}$ by p indexes and q indexes, as

$$b_{r,s} = a_{(r+p) \bmod m,(s+q) \bmod n}.$$

Show that the two-dimensional DFT of $\mathbf{B}$ has components

$$\hat{b}_{k,l} = e^{2\pi i(pk/m+ql/n)}\hat{a}_{k,l}$$

and that $|\hat{a}_{k,l}| = |\hat{b}_{k,l}|$ for all k and l.

21. This exercise involves Matlab:
(a) Load an image $\mathbf{A}$ into Matlab and let $\mathbf{B}$ be the shifted image as in Exercise 20. Shift by a reasonable amount, say 30–40 pixels in each direction. Print out both images.
(b) Compute and show both of $|\hat{\mathbf{A}}|$ and $|\hat{\mathbf{B}}|$. Use the centered view, logarithmically scaled. Are the results consistent with Exercise 20?

22. In Section 2.5.2, certain symmetry properties of the DFT were discussed, under the assumption that the time domain signal is real-valued. Show that if an $m \times n$ matrix $\mathbf{A}$ is real-valued then certain relationships (equality or "conjugate equality") exists among the 2-D Fourier coefficients $\hat{a}_{k,l}$, $\hat{a}_{m-k,l}$, $\hat{a}_{k,n-l}$, and $\hat{a}_{m-k,n-l}$. Take note of Remark 2.5.

23. This exercise involves Matlab. Try the audio compression technique of Section 2.8.1, part 6, on the "splat" signal in Matlab. Make a table showing the compression ratio obtained using various thresholds, the distortion introduced, and your subjective rating of the quality of the sound. It is probably more difficult to get compression comparable to that attainable with the "train" signal, while maintaining quality. Why?

3

The Discrete Cosine Transform

3.1 Motivation for the DCT—Compression

The goal of any compression algorithm is to make the data that represent the underlying signal or image as small as possible without undue loss of information. Other considerations include simplicity, speed of computation, and flexibility.

In this chapter, we present the discrete cosine transform (DCT) and introduce the notion of localizing frequency analysis by breaking a signal or image into smaller pieces. One motivation for this is to give the reader an understanding of the mathematical basis of classical JPEG compression. In order to keep the discussion focused on the essential mathematical principles, we will make a few simplifying assumptions. First, we will assume that the signal or image has already been digitized, and so there are no image capture or sampling issues involved. We will also assume that the quantization error when the signal is captured is negligible so that the vector space models of Chapter 1 continue to apply. The quantization that occurs when the image is first digitized is in the time domain or spatial domain. Quantization will also be an issue at a later point, but in the frequency domain. We will work only with monochrome images.

Before proceeding, it is helpful to define what we mean by "lossless" and "lossy" compression.

Definition 3.1 A compression algorithm or a specific step in a compression algorithm is called "*lossless*" if it is reversible, so that the input can be perfectly reconstructed from the output. A process that is not lossless is called "*lossy*."

Figure 3.1 illustrates the overall flow of a compression algorithm. The first step in the algorithms we will consider consists of applying a linear transform T like the discrete Fourier transform (DFT) to the object of interest, a vector, or matrix $\mathbf{x}$. Aside from round-off error, the transform is invertible, so this

Discrete Fourier Analysis and Wavelets: Applications to Signal and Image Processing, Second Edition.
S. Allen Broughton and Kurt Bryan.

Companion Website: www.wiley.com/go/Broughton/Discrete_Fourier_Analysis_and_Wavelets

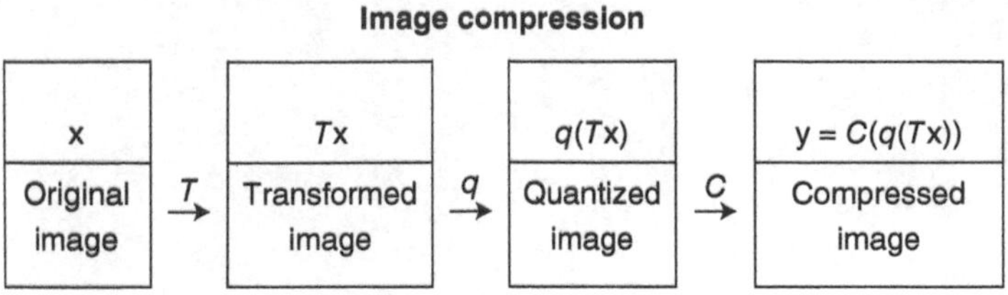

Figure 3.1 Overview of compression.

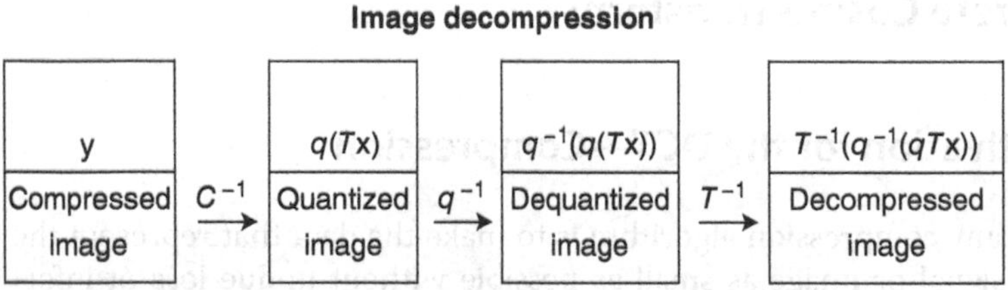

Figure 3.2 Overview of decompression.

step is essentially lossless. The next step is quantization of the frequency domain version of the signal using the techniques from Section 1.9; this is represented by the function q. Some care is required in the design of the transform and quantization scheme in order to achieve good compression. Ideally the quantized frequency domain representation of the original time domain object should have a large proportion of zeros, which will make the frequency domain data easy to compress. The quantization step is lossy.

The last step in the compression process is the application of some standard data compression technique, represented by C, not necessarily specific to audio or image compression. This often involves Huffman coding or simple run-length encoding. We will not discuss the details of this step in the process, since it really is quite separate from the mathematics we have developed so far. See chapter 8 of [14] for more information on this topic.

Decompressing the compressed image requires reversing the above steps and is illustrated in Figure 3.2. In most of this chapter, the transform T will be the DCT, described below. In later chapters, the transform T will become a *filter bank* or *wavelet* transform.

3.2 Other Compression Issues

We always face a compromise between good compression and fidelity to the original signal or image. There are several ways to quantify the efficiency of a compression method. The most obvious is to simply to compute the ratio of the original file size to the compressed file size. In the case of the full JPEG algorithm, however, this would involve the final compression step (the Huffman or run-length compression), which we do not want to get into here. A slightly cruder alternative is to look at the number of nonzero elements in the image after it has been transformed to the frequency domain and quantized. This is generally the approach we take.

In addition to measuring the compression achieved, we also need to measure image quality. Since image (or audio) quality is very subjective, it is impossible to find a perfect quantitative measure. However, the measure of distortion that we used in Chapter 1 is usually reasonable, since throwing away a lot of the energy in a signal or image generally means that the result has poor fidelity.

3.3 Initial Examples—Thresholding

Before proceeding, it will be very helpful to look at a simple compression scheme that makes use of the mathematics we already have at hand, in particular, the DFT. It will also serve to illustrate certain shortcomings in using the DFT for image compression. The examples below are closely related to the audio and image exercises from Section 2.8.

Consider a sampled signal $\mathbf{x}$ with components $x_k = f(k\Delta T)$, $0 \le k \le N-1$ for some function f defined on $[0, 1]$, where $\Delta T = 1/N$. The first simple compression scheme will be as follows: Choose a threshold parameter $0 \le c \le 1$. Then

1. compute $\mathbf{X} = \text{DFT}(\mathbf{x})$. Let $M = \max_{0 \le k \le N-1} (|X_k|)$;
2. define $\tilde{\mathbf{X}} \in \mathbb{C}^N$ with components

$$\tilde{X}_k = \begin{cases} X_k, & \text{if } |X_k| \ge cM, \\ 0, & \text{if } |X_k| < cM. \end{cases}$$

The vector $\tilde{\mathbf{X}} \in \mathbb{C}^N$ will become the compressed version of the signal. Of course, if most of the components $\tilde{X}_k = 0$, then $\tilde{\mathbf{X}}$ should be easy to compress. To transmit the image, we could just send the nonzero $\tilde{X}_k$. Decompression is done by inverse transforming, as $\tilde{\mathbf{x}} = \text{IDFT}(\tilde{\mathbf{X}})$. The vector $\tilde{\mathbf{x}}$ should approximate $\mathbf{x}$.

In practice, the vector $\mathbf{X}$ would not simply be thresholded but rather subjected to a more sophisticated quantization scheme like that in Section 1.9. Ideally this scheme should zero out many X_k, such as thresholding, and allow the remaining X_k to be stored economically. Nonetheless, we will proceed with a simple thresholding approach for now and apply more sophisticated methods later in this chapter.

We can crudely measure the efficiency of our compression scheme by computing the proportion $P(c)$ of the $|X_k|$ that exceed the threshold c, as

$$P(c) = \frac{\#\{k : |\tilde{X}_k| > 0\}}{N} = \frac{\#\{k : |X_k| > cM\}}{N},$$

where $\#\{A\}$ denotes the number of elements in the set A. Of course, $P(c) = 0$ indicates perfect compression, whereas $P(c) = 1$ indicates no compression at all. We will measure the distortion of the compressed image as in Chapter 1, as

$$D(c) = \frac{\|\mathbf{x} - \tilde{\mathbf{x}}\|^2}{\|\mathbf{x}\|^2}$$

or on a percentage basis as $100D(c)$. If $D(c) = 0$, then the reconstruction is perfect.

Ideally both $P(c)$ and $D(c)$ should be close to zero, but that cannot happen simultaneously. Let us look at a few specific examples and see how the numbers work out.

3.3.1 Compression Example 1: A Smooth Function

Let $f(t) = (t - t^2)^2$ in the scheme above, with $N = 256$. This function is plotted in Figure 3.3a. In Figure 3.3b, we plot the DFT $\mathbf{X}$, as pairs $(k, |X_k|^2/N)$ on the range $-127 \le k \le 128$. The plot makes it clear that most of the energy lies in a few low frequencies. Almost any threshold parameter $c > 0$ should leave these energetic frequencies unchanged while zeroing out the remainder. We thus expect good compression and little distortion. Table 3.1 shows various values of c and the corresponding values of $P(c)$ and $D(c)$. In this case, it is

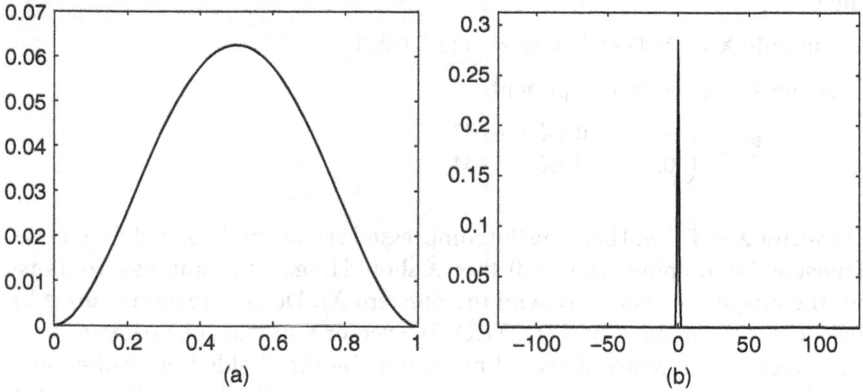

Figure 3.3 $f(t) = (t - t^2)^2$ and DFT.

Table 3.1 Compression data for $f(t) = (t - t^2)^2$.

c	0.5	0.1	0.01	0.001	10^{-4}	10^{-8}
$P(c)$	0.004	0.012	0.02	0.035	0.066	0.652
$D(c)$	0.300	1.2×10^{-3}	5.2×10^{-5}	1.0×10^{-6}	1.3×10^{-8}	$< 10^{-15}$

clear that we can obtain good compression and low distortion. Indeed we can compress the signal 10-fold and still obtain distortion less than 10^{-8}. At this level of compression, the original signal and reconstructed compressed signal are visually indistinguishable.

3.3.2 Compression Example 2: A Discontinuity

This time let

$$f(t) = \begin{cases} 1, & t \leq 0.2, \\ 0, & t > 0.2, \end{cases} \tag{3.1}$$

on $[0, 1]$, a discontinuous step function. We again take $N = 256$ and sample as in the previous example. In Figure 3.4a, we plot the DFT $\mathbf{X}$, as pairs $(k, |X_k|^2/N)$ on the range $-127 \leq k \leq 128$, while on the Figure 3.4b, we zoom in to the range $-9 \leq k \leq 9$. In this case, most of the energy still lies in the lower frequencies, but compared to the previous case the energy has spread out a bit. Table 3.2 shows various values of c and the corresponding values of $P(c)$ and $D(c)$. Compression here is a bit tougher than for the first example. In particular, a 10-fold compression ($P(c) = 0.1$) with this approach brings the distortion up to 2.9×10^{-2}. In Figure 3.5, we show the original signal and compressed/decompressed signal at this 10-fold compression level.

The essential difficulty in this example is the discontinuity: it takes a lot of different basic waveforms to synthesize a jump.

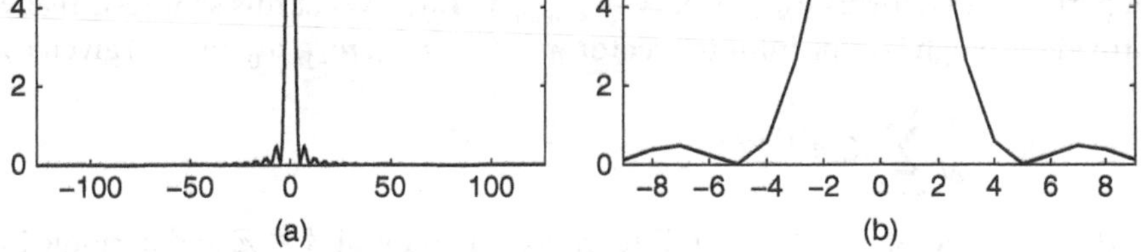

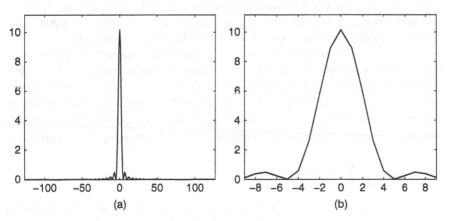

Figure 3.4 Two views of the DFT for step function $f(t)$.

Table 3.2 Compression data for discontinuous $f(t)$ of equation (3.1).

c	0.5	0.1	0.03	0.02	0.01	0.001
$P(c)$	0.027	0.082	0.262	0.543	0.793	0.988
$D(c)$	0.124	0.039	0.011	0.0035	3.64×10^{-4}	1.23×10^{-7}

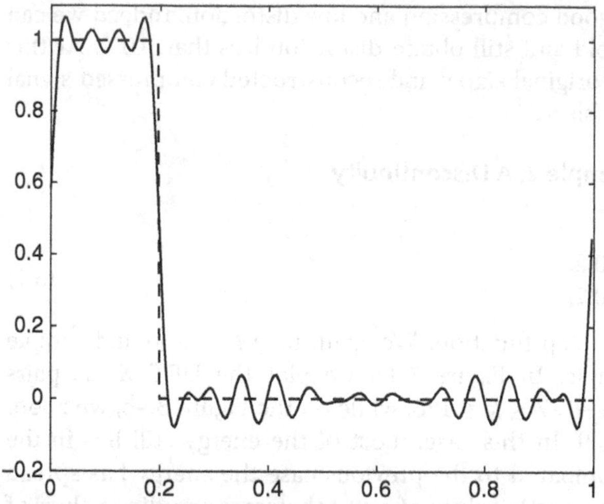

Figure 3.5 Original (dashed) and decompressed (solid) signal, 10-fold compression.

3.3.3 Compression Example 3

In this case, let $f(t) = t$ on the interval $[0, 1]$ with $N = 256$, sampled as above. The function f is continuous on the closed interval. If we apply the same procedure as above to this function, we obtain the data in Table 3.3. Compression for this signal also seems more difficult than for the first example. In particular, to achieve a 10-fold compression with this approach, we have to accept distortion 1.12×10^{-2}. In Figure 3.6, we show the original signal and compressed–decompressed signal at the 10-fold compression level.

Why is this nicely continuous function so hard to compress? The problem is that we need a slightly refined notion of continuity when we use the DFT. The difficulty in this case is that $f(0) \neq f(1)$. To see why this causes trouble, recall that the DFT/IDFT pair allows us to synthesize the vector $\mathbf{x}$ as a superposition of periodic waveforms $\mathbf{E}_{N,m}$, as $\mathbf{x} = \frac{1}{N}\sum_{m=0}^{N-1} X_m \mathbf{E}_{N,m}$. What this synthesis really provides though is a "bi-infinite" vector $\mathbf{w} = (\ldots, w_{-2}, w_{-1}, w_0, w_1, \ldots)$ given by

$$w_k = \frac{1}{N}\sum_{m=0}^{N-1} X_m e^{2\pi i k m/N}$$

with $x_k = w_k$ for $0 \le k \le N-1$. But w_k is defined for all $k \in \mathbb{Z}$ and periodic in k with period N. If $f(0) \neq f(1)$, then the DFT/IDFT must synthesize the jump

Table 3.3 Compression data for $f(t) = t$.

c	0.5	0.1	0.03	0.02	0.01	0.005
$P(c)$	0.004	0.027	0.082	0.129	0.254	0.574
$D(c)$	0.252	0.043	1.44×10^{-2}	9.2×10^{-3}	4.49×10^{-3}	1.5×10^{-3}

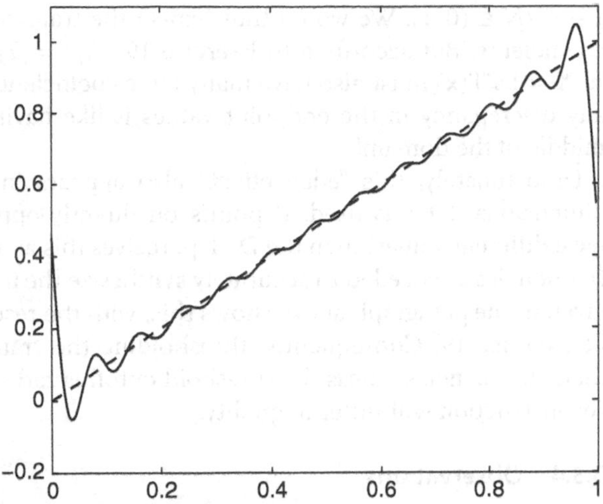

Figure 3.6 Original (dashed) and decompressed (solid) signal, 10-fold compression.

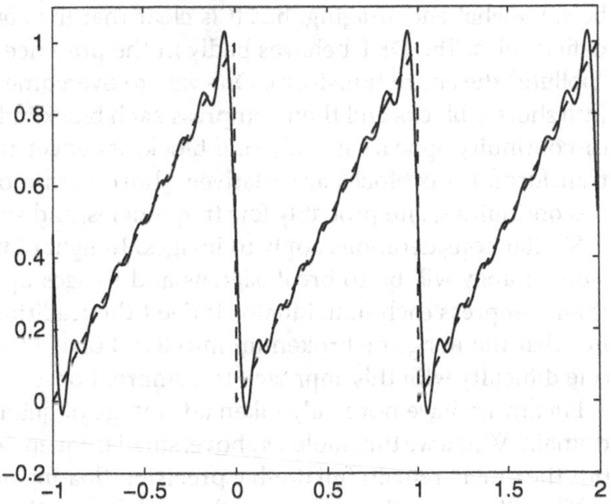

Figure 3.7 Partial Fourier synthesis (solid) and periodic extension w (dashed) of signal x from Figure 3.6.

in $\mathbf{w}$ that occurs whenever the index k is a multiple of N, just as if the jump occurred in the middle of the interval. This requires the superposition of many frequencies. The situation is illustrated in Figure 3.7 in which we plot $\mathbf{w}$, the periodic extension of the partial Fourier synthesis of $\mathbf{x}$ from Figure 3.6, superimposed over the periodic extension of the original signal.

Another way to look at the difficulty in compressing this signal is via Exercise 19, in which we consider a signal $\mathbf{y}$ obtained by shifting a signal $\mathbf{x}$ circularly to the left m units, as $y_k = x_{(k+m) \bmod N}$. If $\mathbf{x}$ stems from sampling a function $f(t)$ with $f(0) \neq f(1)$, then $\mathbf{y}$ can be obtained by sampling the function $g(t) = f((t + m/N) \bmod 1)$. If $1 \leq m \leq N-1$, then g has a discontinuity at

$1 - m/N \in (0, 1)$. We would thus expect the transform $\mathbf{Y}$ to have many large coefficients. But according to Exercise 19, $|X_k| = |Y_k|$ for $0 \le k \le N-1$, and so $\mathbf{X} = \text{DFT}(\mathbf{x})$ must also have many large coefficients. In the eyes of the DFT, any discrepancy in the endpoint values is like having a discontinuity in the middle of the domain!

Unfortunately, this "edge effect" also appears in images when the two-dimensional DFT is used. If points on directly opposite sides of the image have different values, then the DFT perceives this as a discontinuity, and many frequencies are needed to accurately synthesize the image. The same argument given in the paragraph above shows this, with the result of Exercise 20 in place of Exercise 19. Consequently, thresholding the transform will not zero out many frequencies unless the threshold cutoff is rather high, in which case the reconstruction will suffer in quality.

3.3.4 Observations

The examples above and those from the Matlab project of Section 2.8 should be somewhat encouraging, but it is clear that the current approach has some deficiencies. The DFT behaves badly in the presence of discontinuities, which "pollute" the entire transform. One way to overcome this is to break the signal into shorter blocks and then compress each block individually. Since any given discontinuity appears in only one block, its effect is confined to that block's transform. If the blocks are relatively short, then most blocks will contain no discontinuities, and probably few frequencies, and so compress more easily.

Similar considerations apply to images. In light of this, our overall compression strategy will be to break signals and images up into smaller blocks, and then compress each individually. Indeed the traditional JPEG standard specifies that the image be broken up into 8×8 pixel blocks. This partly alleviates one difficulty with this approach to compression.

Finally, we have not really taken advantage of quantization in the frequency domain. When we threshold as above, small frequencies are "quantized" to zero, but the others remain full double precision floating point.

We will address these issues in the next few sections.

3.4 The Discrete Cosine Transform

3.4.1 DFT Compression Drawbacks

The blocking strategy partially removes the DFT's difficulty with discontinuities, but as the example in the last section showed, even the edge discontinuities will be a problem. Moreover, breaking the signal up into blocks will make this even worse, since every block boundary is now a potential discontinuity! This can be overcome by replacing the DFT with the closely related DCT.

3.4.2 The Discrete Cosine Transform

Let us begin with the one-dimensional case. The overall approach to fixing the DFT's difficulty with edge discontinuities is to extend the signal to twice its original length by reflection about one end of the signal, compute the DFT of the double-length signal, and then "restrict back" to the appropriate length.

Here are the specifics. Let $\mathbf{x} \in \mathbb{C}^N$.

3.4.2.1 Symmetric Reflection

Compression difficulties arise when x_0 differs substantially from x_{N-1}. Let us thus define an extension $\tilde{\mathbf{x}} \in \mathbb{C}^{2N}$ of $\mathbf{x}$ as

$$\tilde{x}_k = \begin{cases} x_k, & 0 \le k \le N-1, \\ x_{2N-k-1}, & N \le k \le 2N-1. \end{cases} \tag{3.2}$$

We then have $\tilde{\mathbf{x}} = (x_0, x_1, \dots, x_{N-2}, x_{N-1}, x_{N-1}, x_{N-2}, \dots, x_1, x_0)$, so $\tilde{\mathbf{x}}$ is just $\mathbf{x}$ reflected about the right endpoint and $\tilde{x}_0 = \tilde{x}_{2N-1} = x_0$. When we take the $2N$-point DFT of $\tilde{\mathbf{x}}$, we will not encounter the kind of edge effects discussed above. Note that we duplicate x_{N-1} in our extension, called the *half-point symmetric extension*. There are other ways to extend $\mathbf{x}$; we will say more about this later.

3.4.2.2 DFT of the Extension

The DFT $\tilde{\mathbf{X}}$ of the vector $\tilde{\mathbf{x}} \in \mathbb{C}^{2N}$ has components

$$\tilde{X}_k = \sum_{m=0}^{2N-1} \tilde{x}_m e^{-2\pi ikm/(2N)} = \sum_{m=0}^{2N-1} \tilde{x}_m e^{-\pi ikm/N}. \tag{3.3}$$

Note that we use a $2N$-point DFT. We can split the sum on the right in equation (3.3) and obtain

$$\tilde{X}_k = \sum_{m=0}^{N-1} \left(\tilde{x}_m e^{-\pi ikm/N} + \tilde{x}_{2N-m-1} e^{-\pi ik(2N-m-1)/N} \right), \tag{3.4}$$

since as the index of summation m in (3.4) assumes values $m = 0, \dots, N-1$, the quantity $2N - m - 1$ in the second part of the summand assumes values $2N-1, \dots, N$ in that order. Thus the sums in (3.3) and (3.4) are in fact identical.

From equation (3.2), we have $\tilde{x}_{2N-m-1} = \tilde{x}_m = x_m$ for $0 \le m \le N-1$. Use this in (3.4), along with

$$e^{-\pi ik(2N-m-1)/N} = e^{i\pi k(m+1)/N} e^{2\pi ik} = e^{i\pi k(m+1)/N}$$

to obtain

$$\tilde{X}_k = \sum_{m=0}^{N-1} \left(x_m e^{-\pi ikm/N} + x_m e^{i\pi k(m+1)/N} \right)$$

$$= e^{\pi ik/2N} \sum_{m=0}^{N-1} \left(x_m e^{-\pi ik(m+1/2)/N} + x_m e^{\pi ik(m+1/2)/N}\right)$$

$$= 2e^{\pi ik/2N} \sum_{m=0}^{N-1} x_m \cos\left(\frac{\pi k(m+1/2)}{N}\right). \tag{3.5}$$

This defines the DFT coefficients $\tilde{X}_k$ on the range of $0 \leq k \leq 2N-1$. As usual however, the formula makes sense for any $k \in \mathbb{Z}$ and the resulting extension is periodic in k with period $2N$.

3.4.2.3 DCT/IDCT Derivation

Equation (3.5) is "almost" what we will call the DCT. Let us define

$$c_k = 2\sum_{m=0}^{N-1} x_m \cos\left(\frac{\pi k(m+1/2)}{N}\right) \tag{3.6}$$

so that $\tilde{X}_k = e^{\pi ik/2N} c_k$. Here are a few easy-to-check facts about c_k:

- $c_{-k} = c_k$ and $c_{k+2N} = -c_k$.
- If $\mathbf{x}$ is real-valued, then c_k is also real.

In particular, note that from the first bulleted item above $c_{2N-r} = c_{-2N+r} = -c_{-2N+r+2N} = -c_r$, so

$$c_{2N-r} = -c_r. \tag{3.7}$$

Taking $r = N$ shows that $c_N = 0$. If we consider the range $1 \leq r \leq N-1$ (so $2N - r$ ranges from $2N-1$ down to $N+1$), we see that knowledge of c_k on the range $0 \leq k \leq N-1$ allows us to compute c_k for $N \leq k \leq 2N-1$. All this stems from the fact that the underlying signal $\tilde{\mathbf{x}}$ possesses a symmetry obtained from reflection.

We can consider equation (3.6) as a transform that maps $\mathbf{x} \in \mathbb{C}^N$ to a vector $\mathbf{c} \in \mathbb{C}^N$, where $\mathbf{c} = (c_0, c_1, \ldots, c_{N-1})$. The transform must be invertible, for as remarked above, knowledge of c_k for $0 \leq k \leq N-1$ allows us to determine c_k for all $0 \leq k \leq 2N-1$. But we know that $\tilde{X}_k = e^{\pi ik/2N} c_k$ and so we can recover $\tilde{\mathbf{x}} = \text{IDFT}(\mathbf{X})$ and hence $\mathbf{x}$.

Let us write out this inverse transform explicitly. Begin with the IDFT ($2N$ points!) of $\mathbf{X}$

$$\tilde{x}_k = \frac{1}{2N} \sum_{r=0}^{2N-1} \tilde{X}_r e^{2\pi ikr/(2N)} = \frac{1}{2N} \sum_{r=0}^{2N-1} \tilde{X}_r e^{\pi ikr/N}.$$

Now fill in $\tilde{X}_r = e^{\pi ir/2N} c_r$ and split the sum as

$$\tilde{x}_k = \frac{1}{2N}\left(c_0 + \sum_{r=1}^{2N-1} c_r e^{\pi i(k+1/2)r/N}\right)$$

$$= \frac{1}{2N}\left(c_0 + \sum_{r=1}^{N-1} c_r e^{\pi i(k+1/2)r/N} + \sum_{r=N+1}^{2N-1} c_r e^{\pi i(k+1/2)r/N}\right)$$

$$= \frac{1}{2N}\left(c_0 + \sum_{r=1}^{N-1} c_r e^{\pi i(k+1/2)r/N} + \sum_{r=1}^{N-1} c_{2N-r} e^{\pi i(k+1/2)(2N-r)/N}\right)$$

$$= \frac{1}{2N}\left(c_0 + \sum_{r=1}^{N-1} \left(c_r e^{\pi i(k+1/2)r/N} - c_{2N-r} e^{-\pi i(k+1/2)r/N}\right)\right), \tag{3.8}$$

where we have explicitly broken out the $r = 0$ term and ignored $r = N$, since $c_N = 0$. In the third line, we made the substitution $r \to 2N - r$, and altered the range of summation accordingly. In the transition to (3.8), we also substituted

$$e^{\pi i(k+1/2)(2N-r)/N} = e^{2\pi i k} e^{i\pi} e^{-\pi i(k+1/2)r/N} = -e^{-\pi i(k+1/2)r/N}.$$

Finally, use equation (3.7) to write equation (3.8) in the form

$$\tilde{x}_k = \frac{1}{2N}\left(c_0 + \sum_{r=1}^{N-1} \left(c_r e^{\pi i(k+1/2)r/N} + c_r e^{-\pi i(k+1/2)r/N}\right)\right)$$

$$= \frac{c_0}{2N} + \frac{1}{N}\sum_{r=1}^{N-1} c_r \cos\left(\frac{\pi(k+1/2)r}{N}\right) \tag{3.9}$$

We can use equation (3.9) to recover $\tilde{x}_k$ for $0 \le k \le 2N - 1$. Then, of course, we can recover $x_k = \tilde{x}_k$ for $0 \le k \le N - 1$.

3.4.2.4 Definition of the DCT and IDCT

Equations (3.6) through (3.9) form a natural transform/inverse transform pair and are, up to a minor change of scale, the DCT. Let us modify the definition of the c_k slightly (we will of course compensate in equation (3.9)) and replace the c_k by C_k with

$$C_0 = \sqrt{\frac{1}{N}}\sum_{m=0}^{N-1} x_m \cos\left(\frac{\pi 0(m+1/2)}{N}\right) = \sqrt{\frac{1}{N}}\sum_{m=0}^{N-1} x_m, \tag{3.10}$$

$$C_k = \sqrt{\frac{2}{N}}\sum_{m=0}^{N-1} x_m \cos\left(\frac{\pi k(m+1/2)}{N}\right), \quad 1 \le k \le N-1. \tag{3.11}$$

Definition 3.2 Let $\mathbf{x} \in \mathbb{C}^N$. The discrete cosine transform (type II) of $\mathbf{x}$ is the vector $\mathbf{C} \in \mathbb{C}^N$ with components C_k defined by equations (3.10) and (3.11).

As with the DFT, the DCT is a linear transformation. The DCT maps $\mathbb{C}^N$ to $\mathbb{C}^N$, but if $\mathbf{x} \in \mathbb{R}^N$, then $\mathbf{C} \in \mathbb{R}^N$, unlike the DFT. The "type II" in the definition stems from the fact that there are many variations on the DCT; this version is the most commonly used.

If we compensate appropriately in equation (3.9), we obtain the inverse discrete cosine transform (IDCT).

Definition 3.3 Let $\mathbf{C} \in \mathbb{C}^N$. The IDCT of $\mathbf{C}$ is the vector $\mathbf{x} \in \mathbb{C}^N$ defined by

$$x_m = \frac{1}{\sqrt{N}} C_0 + \sqrt{\frac{2}{N}} \sum_{k=1}^{N-1} C_k \cos\left(\frac{\pi k(m+1/2)}{N}\right) \tag{3.12}$$

for $0 \leq m \leq N-1$.

3.4.3 Matrix Formulation of the DCT

As a linear mapping the DCT, like the DFT, can be represented by a matrix. From equations (3.10) and (3.11), it is easy to see that we can compute the DCT as $\mathbf{C} = C_N \mathbf{x}$, where

$$C_N = \begin{bmatrix} \frac{1}{\sqrt{N}} & \frac{1}{\sqrt{N}} & \cdots & \frac{1}{\sqrt{N}} \\ \sqrt{\frac{2}{N}} \cos\left(\frac{\pi}{N}\frac{1}{2}\right) & \sqrt{\frac{2}{N}} \cos\left(\frac{\pi}{N}\frac{3}{2}\right) & \cdots & \sqrt{\frac{2}{N}} \cos\left(\frac{\pi}{N}\frac{2N-1}{2}\right) \\ \vdots & \vdots & \ddots & \vdots \\ \sqrt{\frac{2}{N}} \cos\left(\frac{\pi k}{N}\frac{1}{2}\right) & \sqrt{\frac{2}{N}} \cos\left(\frac{\pi k}{N}\frac{3}{2}\right) & \cdots & \sqrt{\frac{2}{N}} \cos\left(\frac{\pi k}{N}\frac{2N-1}{2}\right) \end{bmatrix}. \tag{3.13}$$

The first row consists entirely of entries $1/\sqrt{N}$. For $k \geq 1$, the row k and column m entries are given by $\sqrt{2/N}\cos([\pi k(2m+1)]/2N)$.

A careful examination of the IDCT in equation (3.12) shows that $\mathbf{x} = C_N^T \mathbf{C}$ so that $C_N^{-1} = C_N^T$. The matrix C_N is thus orthogonal, that is,

$$C_N^T C_N = C_N C_N^T = \mathbf{I}_N. \tag{3.14}$$

This is the motivation for the $\sqrt{1/N}$ and $\sqrt{2/N}$ scaling factors in the DFT/IDFT definition—to make the relevant matrices orthogonal.

3.5 Properties of the DCT

3.5.1 Basic Waveforms for the DCT

The DFT allows us to write an arbitrary vector as a superposition of basic waveforms $\mathbf{E}_{N,k}$. The DCT does the same, even though we did not develop it in this manner. Specifically, consider the matrix form of the IDCT. Basic matrix algebra shows that we can write $\mathbf{x} = \mathcal{C}_N^T \mathbf{C}$ in the form

$$\mathbf{x} = \sum_{k=0}^{N-1} C_k \mathcal{C}_{N,k}, \tag{3.15}$$

where $\mathcal{C}_{N,k}$ denotes the kth column of $\mathcal{C}_N^T$ or the transpose of the kth the row of $\mathcal{C}_N$ (indexing rows and columns as usual from 0 to $N-1$). The vectors $\mathcal{C}_{N,k}$ play the role of the basic waveforms $\mathbf{E}_{N,k}$ for the DCT. Equation (3.14) implies that the vectors $\mathcal{C}_{N,k}$ are orthonormal in $\mathbb{R}^N$. The $\mathcal{C}_{N,k}$ are also orthonormal in $\mathbb{C}^N$ because the conjugation in the standard inner product makes no difference.

As with the DFT, we can compute the coefficients C_k by taking the inner product of both sides of (3.15) with the vector $\mathcal{C}_{N,k}$ and using the orthonormality of the $\mathcal{C}_{N,k}$ to find that

$$C_k = (\mathbf{x}, \mathcal{C}_{N,k}). \tag{3.16}$$

Of course, equation (3.16) is merely a restatement of the definition of $\mathbf{C} = \mathcal{C}_N \mathbf{x}$, the matrix form of the DCT.

3.5.2 The Frequency Domain for the DCT

Just as the waveforms, $\mathbf{E}_{N,k}$ can be obtained by sampling the function $e^{2\pi ikt/T}$ at times $t = 0, T/N, 2T/N, \ldots, (N-1)T/N$, we can obtain $\mathcal{C}_{N,k}$ by sampling the function

$$\phi_k(t) = \sqrt{\frac{2}{N}} \cos\left(\frac{\pi kt}{T}\right)$$

at times $t = T/2N, 3T/2N, 5T/2N, \ldots, [(2N-1)T]/2N$. These times are the midpoints of each interval of the form $[mT/N, ((m+1)T)/N]$, $0 \leq m \leq N-1$ within the interval $[0, T]$.

For $0 \leq k \leq N-1$, the function $\cos(\pi kt/T)$ ranges from frequency 0 Hz to $(N-1)/2T$ Hz. This is effectively the same frequency range, direct component (dc) to the Nyquist frequency, as in the complex exponential case or sine/cosine case (recall Remark 2.3).

Figure 3.8 shows a few of the waveforms $\mathcal{C}_{N,k}$ when $N = 64$ and $k = 0, 1, 2, 3, 4$.

If it does not cause confusion, we will write $\mathcal{C}_k$ for $\mathcal{C}_{N,k}$. Note that $\mathcal{C}_k \neq \mathrm{Re}(E_k)$. Observe that the cosine basis function $\mathcal{C}_k$ crosses the horizontal axis k times in the interval $0 \leq k \leq N-1$, in accordance with the frequency $k/2$ function $\cos(k\pi t/T)$ plotted on the range $0 \leq t \leq T$.

3.5.3 DCT and Compression Examples

In order to gain a better understanding of the DCT, let us go back to the thresholding compression examples from Section 3.3.

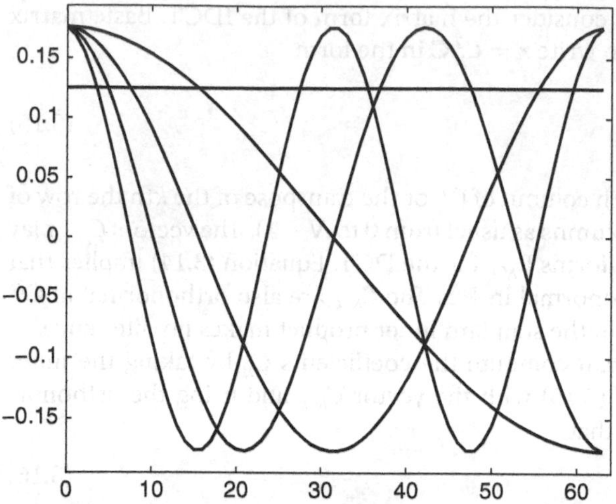

Figure 3.8 DCT basis functions $C_{64,k}$, $0 \leq k \leq 4$.

Before proceeding, we should point out that as the FFT provides a fast algorithm for computing the DFT, fast algorithms also exist for computing the DCT. Indeed, since we derived the DCT from an appropriate DFT computation, we can use the FFT to obtain an order $N\log(N)$ algorithm for the DCT. Specifically tailored algorithms for the DCT are bit more efficient, though.

Let us begin by reconsidering the first example of Section 3.3, but with the DCT taking the place of the DFT. In Figure 3.9, we plot the magnitude of the

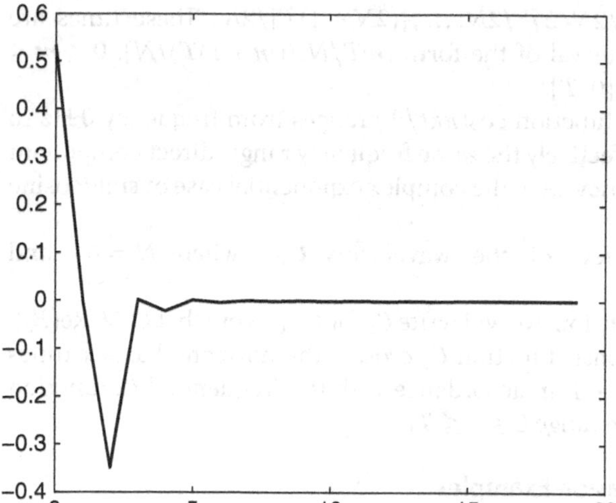

Figure 3.9 DCT of sampled function $f(t) = (t - t^2)^2$.

Table 3.4 Compression data for $f(t) = (t - t^2)^2$.

c	0.5	0.1	0.01	0.001	10^{-4}	10^{-8}
$P(c)$	0.008	0.008	0.012	0.035	0.070	0.805
$D(c)$	0.0013	0.0013	1.0×10^{-6}	8.1×10^{-7}	3.5×10^{-8}	$< 10^{-15}$

Table 3.5 Compression data for discontinuous f of equation (3.1).

c	0.5	0.1	0.03	0.02	0.01	0.001
$P(c)$	0.016	0.043	0.137	0.250	0.559	0.973
$D(c)$	0.124	0.039	0.012	0.0062	0.0015	5.8×10^{-7}

DCT for this signal, on the range $0 \leq k \leq 20$; the remaining coefficients C_k are all effectively zero. In Table 3.4, we show the compression and distortion for a number of threshold parameters. As in the DFT case, compression is efficient. We can get a 10-fold compression with distortion less than 10^{-8}.

For the second example involving a discontinuous f from Section 3.3, the data are shown in Table 3.5. These are pretty similar to the DFT data. In particular, at 10-fold compression, we obtain distortion 1.54×10^{-2}. This is expected because the discontinuity is in the interior of the interval. The reconstruction of the signal at this compression level looks quite similar to the DFT case.

For the third example from Section 3.3, the data are shown in Table 3.6. This is obviously a dramatic improvement over the DFT, which is not surprising since the DCT was designed to address this situation. In particular, with $c = 0.0002$ we obtain 10-fold compression but with distortion only 2.3×10^{-7}. The compressed and original signals are visually indistinguishable.

One way to summarize the data in this example is by plotting $\log(D(c))$ versus $P(c)$ for the data in Table 3.3. This is shown in the panel of Figure 3.10b, while the corresponding compression/distortion data for the DFT from Table 3.3 is shown in the panel of Figure 3.10a. In each case, the positive direction along the horizontal axis indicates poorer compression, while downward indicates lower distortion. This helps us visualize the compromise between low distortion and good compression. Note the rather different horizontal scales on each plot; the DCT achieves dramatically better compression on this signal while introducing much less distortion.

Table 3.6 Compression data for $f(t) = t$.

c	0.5	0.1	0.03	0.02	0.01	0.005
$P(c)$	0.008	0.008	0.012	0.016	0.020	0.023
$D(c)$	0.0036	0.0036	5.7×10^{-4}	2.0×10^{-4}	8.1×10^{-5}	3.6×10^{-5}

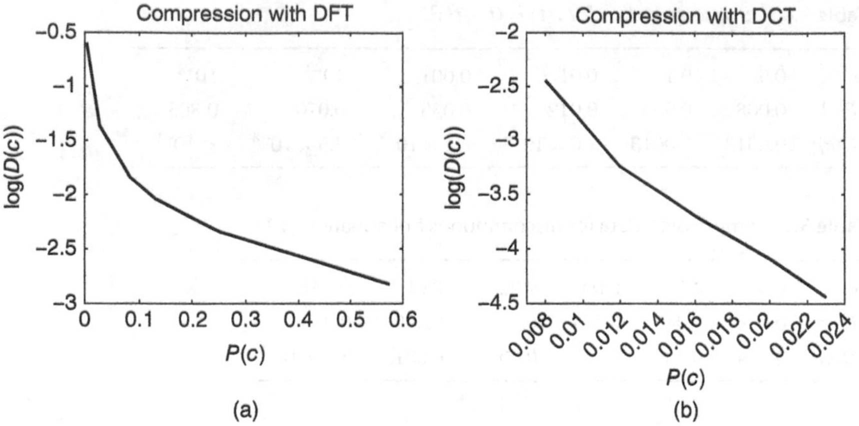

Figure 3.10 Data $\log(D(c))$ versus $P(c)$ for compression of $f(t) = t$ using DFT (a) and DCT (b).

3.6 The Two-Dimensional DCT

We can derive a two-dimensional version of the DCT in the same way we derived the one-dimensional DCT. An $m \times n$ image matrix $\mathbf{A}$ is extended to a $2m \times 2n$ matrix by appropriately reflecting about its bottom and right edge, and then reflecting one of these to fill in the fourth corner of the enlarged image. An example is shown in Figure 3.11 with the image of the dog from Chapter 2. DFT difficulties with mismatched edges are gone, since

Figure 3.11 Even reflections of dog image.

opposite edges now match. We can apply the two-dimensional DFT to the extended image and then restrict back to an appropriate set of $m \times n$ Fourier components, as in the one-dimensional case.

However, we are not going to work out the details of this approach, for it turns out to be equivalent to the following much simpler scheme. Specifically, we will define the two-dimensional DCT with inspiration from Proposition 2.2.

Definition 3.4 The two-dimensional DCT of an $m \times n$ matrix $\mathbf{A}$ is the $m \times n$ matrix

$$\hat{\mathbf{A}} = \mathcal{C}_m \mathbf{A} \mathcal{C}_n^T. \tag{3.17}$$

Exactly the same argument given for the DFT shows that this definition is equivalent to performing an m-point DCT on $\mathbf{A}$ column by column and then an n-point DCT on the result, row by row. Alternatively, we can transform rows first, then columns.

An explicit formula for the two-dimensional DCT easily falls out of equation (3.17) and is given by

$$\hat{a}_{k,l} = u_k v_l \sum_{r=0}^{m-1} \sum_{s=0}^{n-1} a_{r,s} \cos\left(\frac{\pi}{m} k \left(r + \frac{1}{2}\right)\right) \cos\left(\frac{\pi}{n} l \left(s + \frac{1}{2}\right)\right), \tag{3.18}$$

where

$$\begin{aligned} u_0 &= \sqrt{\frac{1}{m}}, \quad & u_k &= \sqrt{\frac{2}{m}}, \quad k > 0, \\ v_0 &= \sqrt{\frac{1}{n}}, \quad & v_l &= \sqrt{\frac{2}{n}}, \quad l > 0. \end{aligned} \tag{3.19}$$

The basic waveforms are the $m \times n$ matrices $\mathcal{C}_{m,n,k,l}$ (or $\mathcal{C}_{k,l}$ when m, n are fixed) where $0 \le k \le m-1, 0 \le l \le n-1$. The row r and column s entries of $\mathcal{C}_{k,l}$ are given by

$$\mathcal{C}_{k,l}(r, s) = u_k v_l \cos\left(\frac{\pi}{m} k \left(r + \frac{1}{2}\right)\right) \cos\left(\frac{\pi}{n} l \left(s + \frac{1}{2}\right)\right).$$

The two-dimensional IDCT is easily derived from equations (3.14) and (3.17), and is given by (see Exercise 14)

$$\mathbf{A} = \mathcal{C}_m^T \hat{\mathbf{A}} \mathcal{C}_n. \tag{3.20}$$

3.7 Block Transforms

As was discussed in Section 3.3, it can aid compression if we cut up the signal or image into smaller blocks and transform each piece. For a variety of reasons, the block size for traditional JPEG compression has been standardized to 8×8 pixel blocks, but in theory any block size could be used. To get some feeling for the process, let us apply these ideas to the image of the dog from previous chapters.

In Figure 3.12, we show both the DCT (a) and "block DCT" (b) magnitude of the 336×448 pixel dog image, logarithmically scaled. The standard DCT on the left is computed with equation (3.18), with $0 \leq k \leq 335, 0 \leq l \leq 447$. It does not really have any discernible information, at least to the naked eye.

For the block DCT, the image is divided up into disjoint 8×8 pixel submatrices or blocks. Each 8×8 block of the original grayscale image is then replaced by the two-dimensional DCT of that block, itself an 8×8 (real-valued) matrix. The row k and column l elements of each such submatrix correspond to a frequency vector pair (k, l) in equation (3.18), where (k, l) is one of the pairs from Table 3.7. It is interesting that we can now perceive the original image in some fashion in this block DCT (think about why!)

(a)

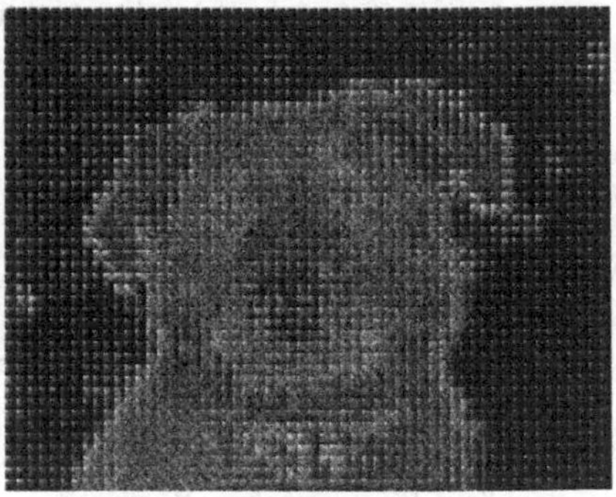

(b)

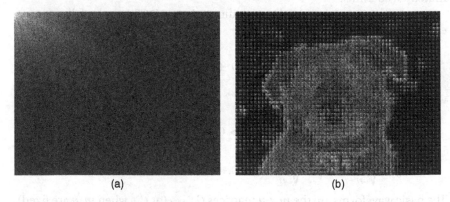

Figure 3.12 DCT and block DCT for dog image.

Table 3.7 Frequency vectors for the block DCT.

$(0,0)$	$(0,1)$	$(0,2)$	$(0,3)$	$(0,4)$	$(0,5)$	$(0,6)$	$(0,7)$
$(1,0)$	$(1,1)$	$(1,2)$	$(1,3)$	$(1,4)$	$(1,5)$	$(1,6)$	$(1,7)$
$(2,0)$	$(2,1)$	$(2,2)$	$(2,3)$	$(2,4)$	$(2,5)$	$(2,6)$	$(2,7)$
$(3,0)$	$(3,1)$	$(3,2)$	$(3,3)$	$(3,4)$	$(3,5)$	$(3,6)$	$(3,7)$
$(4,0)$	$(4,1)$	$(4,2)$	$(4,3)$	$(4,4)$	$(4,5)$	$(4,6)$	$(4,7)$
$(5,0)$	$(5,1)$	$(5,2)$	$(5,3)$	$(5,4)$	$(5,5)$	$(5,6)$	$(5,7)$
$(6,0)$	$(6,1)$	$(6,2)$	$(6,3)$	$(6,4)$	$(6,5)$	$(6,6)$	$(6,7)$
$(7,0)$	$(7,1)$	$(7,2)$	$(7,3)$	$(7,4)$	$(7,5)$	$(7,6)$	$(7,7)$

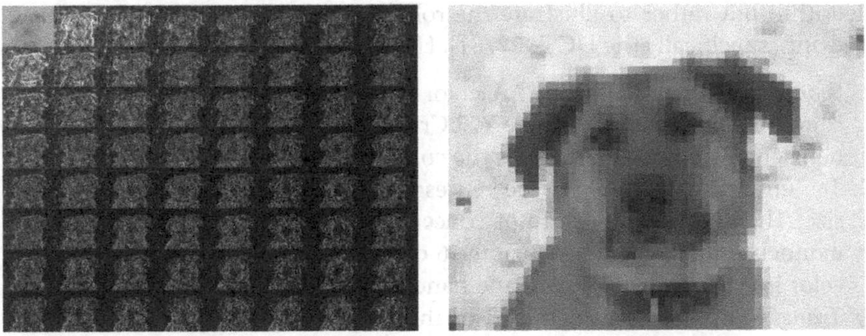

Figure 3.13 Regrouped block DCT (a) and dc components only (b).

The block transform can have more visual impact if we regroup the block DCT coefficients by frequency vector rather than block location. We show this "frequency grouping" presentation in Figure 3.13a. Each of the 64 distinct subblocks encodes the information from one of the (k, l) frequency pairs. For example, in each 8×8 subblock in Figure 3.12b, the (0, 0) or dc coefficient measures the energy of the constant function in the DCT expansion of the corresponding block. When the dc coefficients are all grouped together in the upper left block of size $(336/8) \times (448/8) = 42 \times 56$ in the image of Figure 3.13a, we get a "thumbnail" of the original image. The other 63 blocks are obtained by gathering together coefficients of like frequency vectors from the other blocks. The image in Figure 3.13b is a close-up of the dc coefficient 42×56 subblock from the image in Figure 3.13a.

Remark 3.1 For an 8×8 DCT calculation, it is actually faster in Matlab to use matrix multiplication than to use the two-dimensional DCT command "`dct2`". The `dct2` command is much faster than matrix multiplication when the images are somewhat larger. The matrix multiplication computation of the DCT for an 8×8 matrix $\mathbf{A}$ is done using equation (3.17), as $\mathbf{A} \to C_8 \mathbf{A} C_8^T$.

3.8 JPEG Compression

3.8.1 Overall Outline of Compression

We are now ready to discuss the fundamentals of how traditional JPEG compression works. There are many options for JPEG compression and many parameters that may be specified. Our goal is not a "bit-by-bit" account of the algorithm but rather to illustrate the role of the mathematics we have been studying, specifically the DCT [22, 31]. Here is the basic procedure:

1. *Separate color*, if applicable: A color image is decomposed into its color components, usually using the YCbCr color scheme. This color representation scheme, like RGB, dictates the color of each pixel with three numbers. The entire image can thus be represented by three arrays of appropriate size. The details here need not concern us, since we will be working with monochrome images. See chapter 6 of [14] for more information on how color images are represented and handled.
2. *Transform*: Perform a block DCT on the image using 8×8 pixel blocks. If the image pixel count in either dimension is not a multiple of 8, the image is padded in some way to a higher multiple of 8. Again, the details are unimportant at the moment.
3. *Quantize*: Each 8×8 pixel block has an 8×8 DCT consisting of real numbers. Each of the 64 components or frequencies in this DCT is quantized in the manner described in Section 1.9. This is the main lossy step in the compression.
4. *Compress*: The image is compressed by using run-length encoding on each block, and then Huffman coding the result. The dc coefficient is often treated separately. This is the step where actual file compression occurs.

Decoding reverses the steps:

4′. *Decompress*: Recover the quantized DCT blocks.
3′. *Dequantize*: Again, this is done using a scheme as in Section 1.9.
2′. *Inverse transform*: Apply the IDCT to each block.
1′. *Mix colors*: If applicable, and display block.

3.8.2 DCT and Quantization Details

Since we are considering only monochrome images, we need not worry about step 1 above, but let us take a closer look at steps 2 and 3. This is where all the mathematics we have developed comes into play. Suppose that $\mathbf{B}$ is an 8×8 pixel block from the original image with each pixel encoded by a number $-127 \le b_{k,m} \le 128$ (if the grayscale image was originally in the 0–255 range, we just subtract 127 from each pixel). We apply the DCT to $\mathbf{B}$ to obtain $\hat{\mathbf{B}} = \text{DCT}(\mathbf{B})$. If the original 8×8 pixel matrix $\mathbf{B}$ is in the range $-127 \le b_{k,m} \le 128$, then the DCT $\hat{\mathbf{B}}$ has all entries in the range $(-2048, 2048)$ (see Exercise 15), although the entries will generally be much closer to zero.

The next step is quantization. We quantize $\hat{\mathbf{B}}$ component by component as

$$q(\hat{b}_{k,m}) = \text{round}\left(\frac{\hat{b}_{k,m}}{e_{k,m}}\right), \tag{3.21}$$

where the "round" function means round to the nearest integer. Here $e_{k,m}$ is a scale factor that depends on the particular frequency (k, m) being quantized. The 8×8 matrix $\mathbf{e}$ with components $e_{k,m}$ is called the quantization matrix. A typical quantization matrix is (see [34])

$$\mathbf{e} = \begin{bmatrix} 16 & 11 & 10 & 16 & 24 & 40 & 51 & 61 \\ 12 & 12 & 14 & 19 & 26 & 58 & 60 & 55 \\ 14 & 13 & 16 & 24 & 40 & 57 & 69 & 56 \\ 14 & 17 & 22 & 29 & 51 & 87 & 80 & 62 \\ 18 & 22 & 37 & 56 & 68 & 109 & 103 & 77 \\ 24 & 35 & 55 & 64 & 81 & 104 & 113 & 92 \\ 49 & 64 & 78 & 87 & 103 & 121 & 120 & 101 \\ 72 & 92 & 95 & 98 & 112 & 100 & 103 & 99 \end{bmatrix}. \tag{3.22}$$

The quantized matrix $q(\hat{\mathbf{B}})$ will consist mostly of small integers and, we hope, many zeros. As a result $q(\hat{\mathbf{B}})$ should be easy to compress.

The next step is the actual compression. Here a combination of run-length encoding followed by Huffman coding is employed. In run-length encoding, the length of sequences of zeros in $q(\hat{\mathbf{B}})$ are recorded in a list, along with the nonzero elements that interrupt the sequences. The placement of the zeros may be exploited by using the so-called zigzag encoding (see [34, p. 406] for further details).

The reconstruction consists of decoding the compressed data stream to obtain each quantized DCT $q(\hat{\mathbf{B}})$. For any given 8×8 DCT, we then dequantize each entry as

$$\tilde{b}_{k,m} = q(\hat{b}_{k,m})e_{k,m}, \quad 0 \le k, m \le 7, \tag{3.23}$$

where $e_{k,m}$ is the corresponding element of the quantization matrix. The IDCT is then applied to the 8×8 array $\tilde{\mathbf{B}}$ to obtain the relevant 8×8 portion of the reconstructed image. The resulting inverse transform consists of floating point numbers, but it would be rounded or quantized to integers.

Example 3.1 *Here is an example that shows how a single* 8×8 *block might be handled. Consider the matrix* $\mathbf{B}$ *of grayscale intensities below:*

$$\mathbf{B} = \begin{bmatrix} 47 & 32 & 75 & 148 & 192 & 215 & 216 & 207 \\ 36 & 82 & 161 & 196 & 205 & 207 & 190 & 140 \\ 86 & 154 & 200 & 203 & 213 & 181 & 143 & 82 \\ 154 & 202 & 209 & 203 & 159 & 145 & 147 & 127 \\ 184 & 207 & 199 & 147 & 134 & 127 & 137 & 138 \\ 205 & 203 & 125 & 72 & 123 & 129 & 150 & 115 \\ 209 & 167 & 126 & 107 & 111 & 94 & 105 & 107 \\ 191 & 129 & 126 & 136 & 106 & 54 & 99 & 165 \end{bmatrix}.$$

This image is scaled from 0 to 255. Viewed as a 8-bit grayscale image, the matrix **B** *is shown in Figure 3.14a. If we subtract 127 from each entry and perform a DCT, we obtain (rounded to one figure after the decimal)*

$$\hat{\mathbf{B}} = \begin{bmatrix} 157.0 & 9.5 & -43.5 & -4.0 & -11.0 & 2.8 & -16.1 & 17.7 \\ 66.2 & -216.5 & -191.1 & 5.0 & -38.6 & 13.6 & 6.8 & -10.0 \\ -88.1 & -148.3 & 11.4 & 22.7 & 72.1 & 22.3 & 12.6 & 0.6 \\ -19.5 & -55.2 & 90.3 & 60.9 & -16.4 & -12.6 & -2.2 & 4.0 \\ 16.3 & -22.8 & 42.1 & -44.0 & 40.8 & 12.6 & 8.3 & -11.1 \\ -5.2 & 15.7 & 11.1 & 23.4 & 6.0 & -0.2 & -6.3 & -9.0 \\ -9.5 & -9.4 & 11.8 & 24.7 & -1.7 & -14.8 & -9.1 & 8.1 \\ -6.5 & 16.6 & -17.8 & -2.3 & 4.2 & -4.7 & 10.2 & 4.7 \end{bmatrix}.$$

Quantization of $\hat{\mathbf{B}}$ *with equation (3.21) and the quantization matrix in equation (3.22) yields*

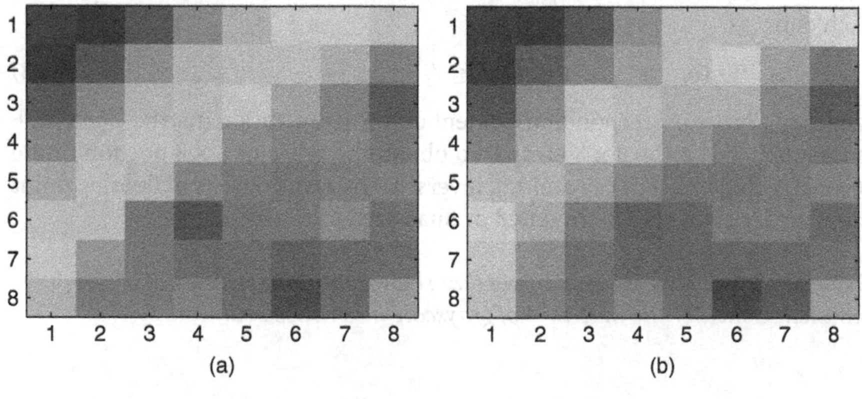

Figure 3.14 Grayscale view of **B** and compressed/decompressed $\tilde{\mathbf{B}}$.

$$q(\hat{\mathbf{B}}) = \begin{bmatrix} 10 & 1 & -4 & 0 & 0 & 0 & 0 & 0 \\ 6 & -18 & -14 & 0 & -1 & 0 & 0 & 0 \\ -6 & -11 & 1 & 1 & 2 & 0 & 0 & 0 \\ -1 & -3 & 4 & 2 & 0 & 0 & 0 & 0 \\ 1 & -1 & 1 & -1 & 1 & 0 & 0 & 0 \\ 0 & 0 & 0 & 0 & 0 & 0 & 0 & 0 \\ 0 & 0 & 0 & 0 & 0 & 0 & 0 & 0 \\ 0 & 0 & 0 & 0 & 0 & 0 & 0 & 0 \end{bmatrix}.$$

The matrix above would become part of the compressed image. As desired, $q(\hat{\mathbf{B}})$ contains a large fraction of zeros and many other integer entries that are close to zero.

Let us look at what happens when we decompress the image. From equation (3.23), the dequantized version of the DCT for this block is given by

$$\tilde{\mathbf{B}} = \begin{bmatrix} 160 & 11 & -40 & 0 & 0 & 0 & 0 & 0 \\ 72 & -216 & -196 & 0 & -26 & 0 & 0 & 0 \\ -84 & -143 & 16 & 24 & 80 & 0 & 0 & 0 \\ -14 & -51 & 88 & 58 & 0 & 0 & 0 & 0 \\ 18 & -22 & 37 & -56 & 68 & 0 & 0 & 0 \\ 0 & 0 & 0 & 0 & 0 & 0 & 0 & 0 \\ 0 & 0 & 0 & 0 & 0 & 0 & 0 & 0 \\ 0 & 0 & 0 & 0 & 0 & 0 & 0 & 0 \end{bmatrix}.$$

Compare this to $\hat{\mathbf{B}}$ above. If we apply the IDCT to $\tilde{\mathbf{B}}$ above, add 127 to all entries and round to the nearest integer, we obtain

$$\begin{bmatrix} 48 & 35 & 77 & 164 & 211 & 201 & 202 & 229 \\ 49 & 92 & 142 & 181 & 211 & 216 & 179 & 132 \\ 81 & 155 & 208 & 207 & 198 & 193 & 149 & 88 \\ 145 & 190 & 221 & 204 & 167 & 144 & 135 & 130 \\ 199 & 200 & 182 & 150 & 128 & 130 & 139 & 144 \\ 215 & 192 & 139 & 94 & 103 & 140 & 139 & 106 \\ 206 & 166 & 119 & 97 & 100 & 110 & 114 & 113 \\ 196 & 139 & 118 & 135 & 106 & 54 & 85 & 167 \end{bmatrix}.$$

The matrix above is shown as a grayscale image in Figure 3.14b. This distortion is about 6.8%. Below we show for comparison the difference of that matrix minus the original matrix **B***:*

$$\begin{bmatrix} 1 & 3 & 2 & 16 & 19 & -14 & -14 & 22 \\ 13 & 10 & -19 & -15 & 6 & 9 & -11 & -8 \\ -5 & 1 & 8 & 4 & -15 & 12 & 6 & 6 \\ -9 & -12 & 12 & 1 & 8 & -1 & -12 & 3 \\ 15 & -7 & -17 & 3 & -6 & 3 & 2 & 6 \\ 10 & -11 & 14 & 22 & -20 & 11 & -11 & -9 \\ -3 & -1 & -7 & -10 & -11 & 16 & 9 & 6 \\ 5 & 10 & -8 & -1 & 0 & 0 & -14 & 2 \end{bmatrix}.$$

At an additional level of sophistication, the entire quantization matrix **e** *could be multiplied by a number r greater or less than 1 depending on the characteristics of the block. The choice $r > 1$ then reduces the number of bits required to store each entry by producing a "coarser" quantization. This, of course, decreases the fidelity of the reconstruction. Taking $r < 1$ has the opposite effect. The choice of r could vary on a block-by-block basis.*

Figure 3.15 shows the original image and reconstructions for r=0.1,0.5, and 2, along with the original. The distortions are 2.4%, 5.0%, and 9%, respectively.

3.8.3 The JPEG Dog

In Figure 3.16a, we see the original dog image and in panel (b) the dog image processed with the quantization matrix **e** of equation (3.22) and then reconstructed. The resulting quantized block DCT is 90% zeros, which should compress well. The distortion is 2.4%. Upon close examination some blocking artifacts can be seen around the muzzle, and some whiskers have disappeared.

3.8.4 Sequential versus Progressive Encoding

After the block DCT is computed and quantized, the data can be organized for compression in a couple different ways. One way is *sequentially*, block by block, so that as the image is decoded and displayed each 8×8 block is displayed with full resolution, typically starting from the upper left and sweeping left to right and down until the bottom right corner of the image is displayed.

An alternative is progressive encoding/decoding, in which the lowest frequency DCT coefficients (starting with the dc coefficients) for each 8×8 block are transmitted, followed by the higher frequency components. As the information arrives and is decoded and displayed, the visual effect is that of a blurry image that gets progressively sharper and sharper. This is especially helpful

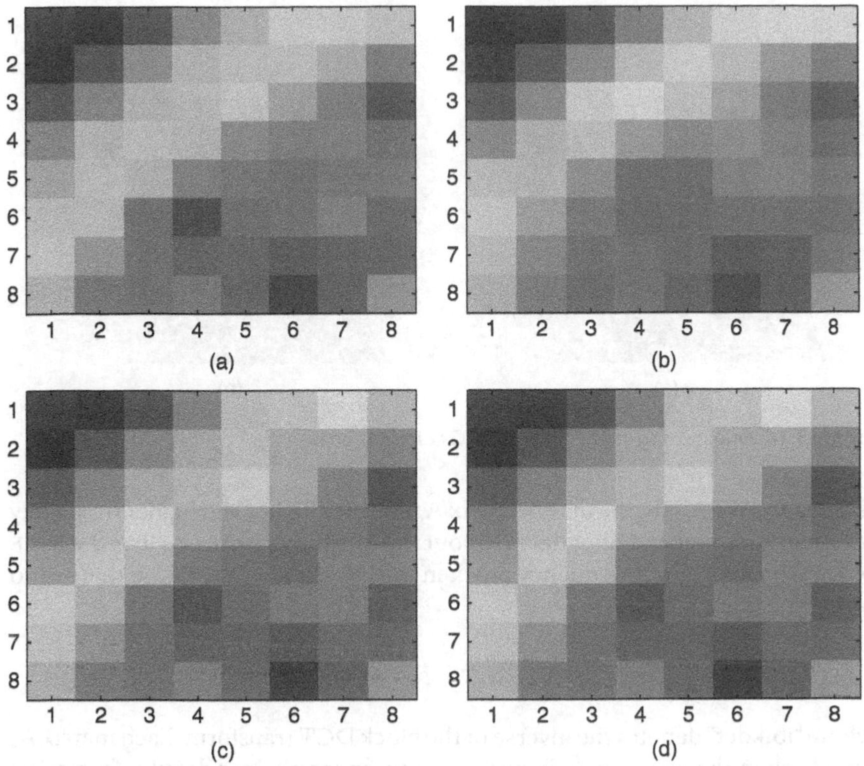

Figure 3.15 Original image (a) and reconstructions with $r = 0.1$ (b), $r = 0.5$ (c), and $r = 2.0$ (d).

when loading large images from a gallery of many images, or the transmission of high resolution medical images. Below we outline one simple approach to this problem.

To be more specific, let $\widehat{\mathbf{A}}$ be the full $m \times n$ block DCT of the $m \times n$ image $\mathbf{A}$. Let $\widehat{\mathbf{A}}_{k,l}$ be the $m \times n$ matrix obtained by zeroing out all DCT coefficients in $\widehat{\mathbf{A}}$ except the one in each 8×8 block corresponding to the frequency vector (k, l) given in Table 3.7. For example, $\widehat{\mathbf{A}}_{0,0}$ would be obtained by zeroing all elements in each 8×8 block *except* the dc coefficient.

Define $\widehat{\mathbf{A}}_s$ as

$$\widehat{\mathbf{A}}_s = \sum_{k+l=s} \widehat{\mathbf{A}}_{k,l},$$

for example, $\widehat{\mathbf{A}}_0 = \widehat{\mathbf{A}}_{0,0}$ $\widehat{\mathbf{A}}_1 = \widehat{\mathbf{A}}_{1,0} + \widehat{\mathbf{A}}_{0,1}$, and $\widehat{\mathbf{A}}_2 = \widehat{\mathbf{A}}_{2,0} + \widehat{\mathbf{A}}_{1,1} + \widehat{\mathbf{A}}_{0,2}$. If $k + l$ is small, then $\widehat{\mathbf{A}}_{k,l}$ carries low frequency information about the image, which corresponds to low resolution features in the image. If $k + l$ is large, then $\widehat{\mathbf{A}}_{k,l}$ carries higher frequency information, which corresponds to detailed features in

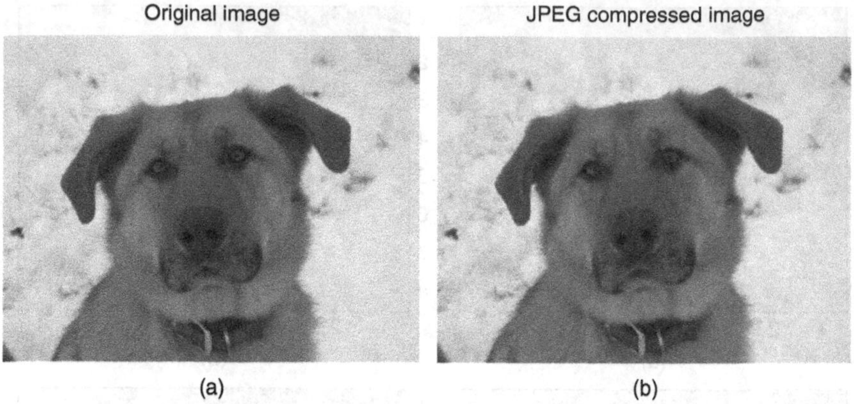

Figure 3.16 Original dog image (a) and JPEG compressed dog (b).

the image. As a result, the $m \times n$ matrix $\hat{\mathbf{A}}_s$ encodes higher and higher frequency information (finer and finer detail) about the image as s increases. For the 8×8 block size, the highest frequency block in the DCT is $k = 7, l = 7$, so there is no point in using $s > 14$.

Let us also define

$$\mathbf{A}_s = \text{iblkdct}(\hat{\mathbf{A}}_s)$$

where "iblkdct" denotes the inverse of the block DCT transform. Each matrix $\mathbf{A}_s$ lives back in the image domain and contains finer and finer detail information about the original image as s increases. The quantity

$$\sum_{s=0}^{p} \mathbf{A}_s \tag{3.24}$$

amalgamates the information in the $\mathbf{A}_s$ and provides a sharper and sharper version of the original image as p increases. If $p = 14$, the sum in (3.24) reconstructs the full image $\mathbf{A}$, since

$$\hat{\mathbf{A}} = \sum_{k=0}^{7} \sum_{l=0}^{7} \hat{\mathbf{A}}_{k,l} = \sum_{s=0}^{14} \sum_{k+l=s} \hat{\mathbf{A}}_{k,l} = \sum_{s=0}^{14} \hat{\mathbf{A}}_s$$

so applying the linear operator "iblkdct" to both sides above yields

$$\mathbf{A} = \sum_{s=0}^{14} \mathbf{A}_s.$$

If the data are suitably arranged (transmit the dc coefficient for all blocks first, followed by the (0, 1) frequency for each block, then the (1, 1), the (2, 0), (1, 1), (0, 2) frequencies, etc.), we can implement progressive transmission as follows:

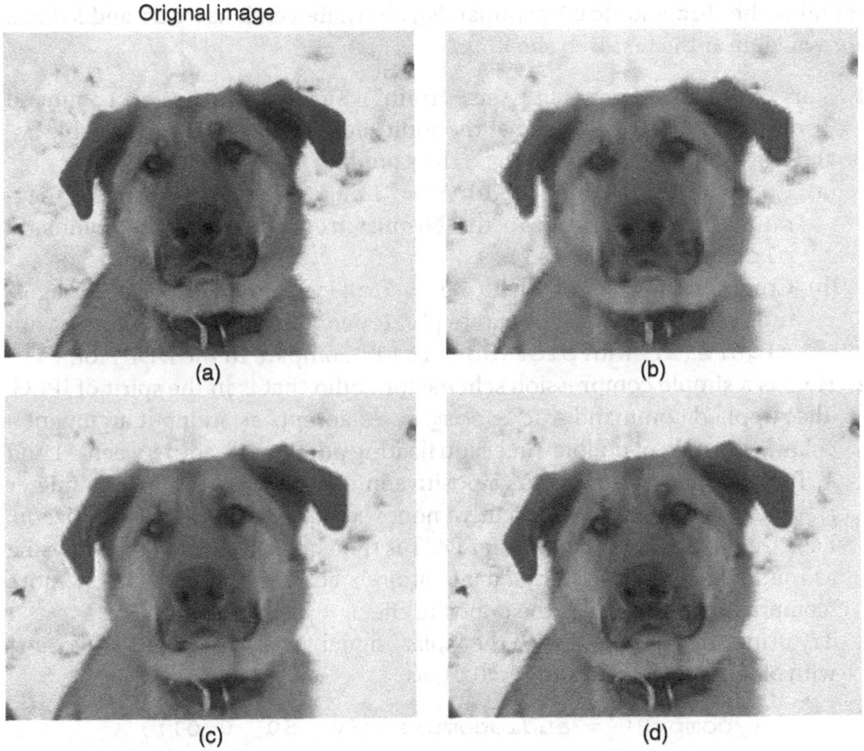

Figure 3.17 Original image (a), $\mathbf{A}_1$ (b), $\mathbf{A}_2$ (c), and $\mathbf{A}_3$ (d).

1. Compute each $\hat{\mathbf{A}}_s$ and $\mathbf{A}_s$ as the data become available.
2. For each $0 \leq p \leq 14$, once all $\mathbf{A}_s$ for $0 \leq s \leq p$ have been constructed, display the sum in equation (3.24) while waiting for further data to arrive.

Example 3.2 *The original image* $\mathbf{A}$ *is shown in Figure 3.17a. The image embodied by* $\mathbf{A}_0$ *is that obtained by reconstructing only from the dc coefficients and was shown in Figure 3.13b. The image in Figure 3.17b is* $\mathbf{A}_1$ *in the scheme's notation given above. The images shown in panels (c) and (d) are* $\mathbf{A}_2$ *and* $\mathbf{A}_3$*, respectively. Obvious blocking artifacts can be seen in the image* $\mathbf{A}_1$*, around the ears and muzzle of the dog.*

3.9 Matlab Project

In this section, we make use of Matlab to explore the DCT and its use in compression. If you do not have the Matlab Signal Processing Toolbox (which contains the `dct` and `dct2` commands), alternate versions `kdct` and `kdct2` are available at the text web site [32].

1. Start Matlab. Load in the "train" signal with the command `load('train');` Recall that the audio signal is loaded into a variable "y" and the sampling rate into "`Fs`." The sampling rate is 8192 Hz.
   (a) Compute the DFT of `y` with `Y = fft(y);`, then display the magnitude of the DFT up to the Nyquist frequency with the command `plot(abs(Y(1:6441)))`.
   (b) Compute the DCT of `y` with `Yc = dct(y);`, then display the magnitude of the DCT on a separate plot (open a new plotting window with `figure(2)`) with `plot(abs(Yc))`. Compare to the DFT plot.
2. Here is a simple compression scheme for audio that is in the spirit of JPEG. The supplied command `audiocompress` accepts as an input argument a column vector **y** of double-precision floating point numbers between −1 and 1. The command also accepts a positive integer m and a scaling parameter r. A DCT is applied to consecutive nonoverlapping sections of **y** of size m. Each frequency Y_k in the m-point DCT is then quantized to an integer value as round(Y_k/r). If $r > 1$, the quantization is coarser and presumably more "compressible"; $r < 1$ has the opposite effect.
   Try using the command with the "splat" signal (`load('splat');`). Start with block size $m = 50$ and $r = 0.01$, as

   ```
   [ycomp,P] = audiocompress(y, 50, 0.01);
   ```

   The output variable `ycomp` is the compressed/decompressed audio signal, and `P` is the fraction of frequencies in the block DCT that are not zeroed out by the quantization. In the case above, this should be about one-third.
   Try varying the block size m by choosing a few other values in the range $m = 10 - -5000$. In each case, adjust r so that the fraction of nonzero quantized coefficients is about one-third, listen to the compressed/decompressed signal, and compute the distortion. Make a table showing the value of m, the distortion, and your subjective rating of the sound quality. Is there an optimal choice for m?
3. Alter the `audiocompress` command to use the DFT instead of the DCT (change the `dct` to `fft`, and `idct` to `ifft`; you will also need to enclose the `ifft` command inside `real`, to eliminate any imaginary round-off errors). Repeat the last problem for each block size and compare the performance of the DFT to the DCT for this application. Is the DFT consistently better or worse than the DCT? Why?
   How high can the compression go before you can detect it by listening to the sound?
4. The supplied command `jpegdemo` accepts as an input argument an array **A** of integers or double-precision floats, assumed to encode a grayscale image

in the range 0–255. It also accepts a scaling parameter r. The image array is block transformed and quantized as described in Section 3.8. The output of the command is the "compressed/decompressed" version of the image as an array of double-precision floating numbers in the interval [0,255], and the fraction of nonzero DCT coefficients remaining. In reality, we would return the output image as unsigned integers, but with the floating point numbers we can do arithmetic.

Load in an image file, preferably a losslessly encoded image—for example, something in a TIFF format (but JPEG will do) with the command `y = imread('myimage.tif');`. Use this to simulate a grayscale image; for example, if the image is loaded into `z`, then use

```
zg = 0.2989*double(z(:,:,1))+0.5870*double(z(:,:,2))
+0.1140*double(z(:,:,3));
```

Subject the image to simulated JPEG compression with

```
[z2,P] = jpegdemo(zg, 1.0);
```

Take note of the compression factor P. And of course, display the image and compute the distortion as in equation (1.39). Recall that the command `colormap([(0:n)/n; (0:n)/n; (0:n)/n]');` sets up a grayscale color map on the range 0 to n.

Vary the scaling parameter r over a wide range (e.g., 0.1–100). Make a table showing the value of r, the compression ratio, distortion, and visual appearance of the image.

5. Does repeatedly subjecting an image to JPEG compression continue to degrade the image? Execute `[z2,P] = jpegdemo(zg, 1.0);` and then run the command

```
[z2,P] = jpegdemo(z2, 1.0);
```

10 times or more, to repeatedly subject the image to compression and decompression. Compute the distortion when done, and compare to the distortion after only one iteration. Can you see any degradation in the image? Where in this process would any degradation (after the first application of `jpegdemo`) occur?

6. The supplied command `jpegprogressive` is identical to the previously introduced command `jpegdemo`, but accepts a third integer argument "p" as in equation (3.24). As described in Section 3.8.1, with the choice $p = 0$ the image will be reconstructed from the dc coefficients only, while $p = 14$ will yield the full resolution reconstruction.

Use the your image from the last problem and execute

```
[z2,P] = jpegprogressive(zg, 1.0, p);
```

for $p = 0, 1, 2, 5, 10, 14$. Why does the $p = 0$ image look the way it does?

Exercises

1. Write out the matrix C_3, and use it to compute the DCT of the vector $\mathbf{x} = (1, 2, -1)$.

2. Use the matrix C_3^T to compute the IDCT of the vector $\mathbf{X} = (3, 0, 1)$.

3. Write out the matrix C_3 and verify that it is orthogonal ($C_3 C_3^T = \mathbf{I}$).

4. Let $\mathbf{e}_m$, $0 \leq m \leq N - 1$, denote the mth standard basis vector in $\mathbb{R}^N$. Write out the DCT of $\mathbf{e}_m$ explicitly.

5. Let $f(t) = t^2$ on the interval $0 \leq t \leq 1$. Let $\mathbf{x}$ be the vector in $\mathbb{R}^{1024}$ obtained by sampling $f(t)$ at $t = k/1024$ for $0 \leq k \leq 1023$. Compute the DCT of $\mathbf{x}$ in Matlab. Threshold the DCT $\mathbf{X}$ at levels suitable to obtain compression to (roughly) 0.1%, 0.5%, 1.0%, and 5.0% of the original size of $\mathbf{X}$ (i.e., eliminate 99.9%, 99.5%, 99.0%, and 95.0% of the DCT coefficients). For each case, compute the distortion in the reconstructed signal.
Repeat with the DFT in place of the DCT. Which transform yields better results, and why?

6. Use Matlab to make up a random 3×3 matrix, then compute `dct(dct(A)')'` (or use the supplied command `kdct`). Verify this gives the same result as `dct2(A)` (or `kdct2d(A)`).

7. In Matlab, the command `dct(A)` computes the DCT of the matrix `A`, column by column (as opposed to `dct2`, which is the full two-dimensional DCT). Explain why the kth row of the matrix obtained from `dct(eye(N))` is the vector $C_{N,k}$.

8. Use Matlab to plot the cosine basis vectors or waveforms $C_{N,k}$ for $N = 100$ and several values of k. Verify that the graph of $C_{N,k}$ crosses the horizontal axis k times. Exercise 7 may be useful.

9. Let

$$\mathbf{A} = \begin{bmatrix} 0 & 1 & 5 & 7 & 9 \\ 1 & 1 & 1 & 1 & 1 \\ 0 & 5 & 3 & 1 & 4 \\ 5 & 3 & 3 & 0 & 0 \end{bmatrix}.$$

Use Matlab to compute the two-dimensional DCT of $\mathbf{A}$ (command `dct2`). Quantize/dequantize the DCT $\hat{\mathbf{A}}$ by applying the function

$x \to d\,\text{round}(x/d)$ to $\hat{\mathbf{A}}$ component by component, using $d = 1.0$ (thus rounding each $\hat{a}_{k,l}$ to the nearest integer), and then inverse transform. Compute the percentage error $100\|\mathbf{A} - \tilde{\mathbf{A}}\|^2/\|\mathbf{A}\|^2$, where $\tilde{\mathbf{A}}$ is the inverse transformed quantized matrix. Here $\|\mathbf{A}\|$ is the "Frobenius" norm of Example 1.14. In Matlab, execute `norm(A,'fro')`.
Repeat with $d = 0.01, 0.1, 10.0$, and $d = 100.0$, and report the percentage error in each case.

10. Show that the two-dimensional DCT of the $N \times N$ identity matrix is again the $N \times N$ identity matrix.

11. Compute (by hand) the two-dimensional DCT of the matrix

$$\mathbf{A} = \begin{bmatrix} 1 & -1 \\ 2 & 0 \end{bmatrix}.$$

Then inverse transform the result.

12. Show that the DCT is an orthogonal transform, that is,

$$\|\text{DCT}(\mathbf{x})\|^2 = \|\mathbf{x}\|^2$$

using the usual Euclidian norm for vectors in $\mathbb{C}^N$. This is the DCT version of Parseval's identity. Hint: For any vector $\mathbf{v} \in \mathbb{C}^N$, we have $\|\mathbf{v}\|^2 = \mathbf{v}^*\mathbf{v}$, where $\mathbf{v}^*$ is the row vector obtained as the conjugate transpose of $\mathbf{v}$.

13. Let $\mathbf{x}$ be a vector in $\mathbb{R}^N$ with DCT $\mathbf{X}$. Show that

$$X_0 = \sqrt{N}\,\text{ave}(x),$$

where $\text{ave}(x)$ denotes the mean value of the components of $\mathbf{x}$.
Show also that if $\mathbf{A} \in M_{m,n}(\mathbb{R})$ has two-dimensional DCT $\hat{\mathbf{A}}$ then

$$\hat{a}_{0,0} = \sqrt{mn}\,\text{ave}(A).$$

14. Prove equation (3.20). Use this to find the corresponding explicit inversion formula for the two-dimensional DCT (analogous to the explicit formula (3.18) for the two-dimensional DCT).

15. Suppose that $\mathbf{x} \in \mathbb{R}^N$ has components that satisfy $|x_k| \le M$. Show that if $\mathbf{X} = \text{DCT}(\mathbf{x})$, then $|X_k| \le M\sqrt{2N}$.
Derive an analogous bound for the two-dimensional DCT.

16. Suppose that $\mathbf{x} \in \mathbb{R}^N$ has DCT $\mathbf{X}$. Let $\tilde{\mathbf{x}}$ denote the vector with components $\tilde{x}_k = x_{N-k-1}$, that is,

$$\tilde{\mathbf{x}} = (x_{N-1}, x_{N-2}, \ldots, x_1, x_0).$$

(a) Show that the DCT $\tilde{\mathbf{X}}$ of $\tilde{\mathbf{x}}$ has components $\tilde{X}_k = (-1)^k X_k$.
(b) Suppose that "R" denotes the reversal operation that takes $\mathbf{x}$ to $\tilde{\mathbf{x}}$ and "C" denotes compression via the DCT by thresholding. Explain why the result of part (a) shows that $R(C(\mathbf{x})) = C(R(\mathbf{x}))$. That is, our approach to compression is invariant with respect to "time-reversal." Hint: Show that $R(C(\mathbf{x}))$ and $C(R(\mathbf{x}))$ have the same DCT (and since the DCT is invertible, must be the same vector).

17. Use Matlab to compute $\mathcal{C}_3^k$ for $k = 1, 2, 3, \ldots$. Does it appear that $\mathcal{C}_3^k$ is a multiple of $\mathbf{I}$ for any choice of k? (Compare to Exercise 9 in Chapter 2.) Is there any evidence that $\mathcal{C}_N^k$ is a multiple of $\mathbf{I}$ for some $N > 2$ and some k?

18. This is a simple variation on Exercise 17, in which the DCT replaces the DFT.
Let $\mathbf{x}$ and $\mathbf{y}$ be two column vectors of length m and n, respectively, with one-dimensional DCT's $\mathbf{X}$ and $\mathbf{Y}$. Let $\mathbf{z}$ be the product image defined by

$$z_{r,s} = x_r y_s$$

for $0 \le r \le m-1, 0 \le s \le n-1$, and $\mathbf{Z}$ the two-dimensional DCT of $\mathbf{z}$.
(a) Show that $\mathbf{z}$ is an $m \times n$ matrix satisfying $\mathbf{z} = \mathbf{x}\mathbf{y}^T$.
(b) Show that

$$Z_{k,l} = X_k Y_l$$

or equivalently $\mathbf{Z} = \mathbf{X}\mathbf{Y}^T$.

19. This is a 2D version of Exercise 16. Let $\mathbf{A}$ denote be an $m \times n$ image matrix and let $\mathbf{B}$ denote the "90° clockwise rotation" of $\mathbf{A}$. (This is not quite the same as taking the transpose!) Note that $\mathbf{B}$ is $n \times m$, and in fact

$$b_{r,s} = a_{m-s-1,r}$$

with all rows and columns indexed from 0. Make sure you believe this, for example, $b_{0,0} = a_{m-1,0}$, $b_{0,m-1} = a_{0,0}$, $b_{n-1,m-1} = a_{0,n-1}$, and $b_{n-1,0} = a_{m-1,n-1}$.
(a) Find the coefficients $\hat{b}_{r,s}$ of the two-dimensional DCT $\hat{\mathbf{B}}$ in terms of the coefficients $\hat{a}_{r,s}$ of $\hat{\mathbf{A}}$, the DCT of $\mathbf{A}$.
(b) Suppose that "R" denotes the clockwise rotation operation that takes $\mathbf{A}$ to $\mathbf{B}$ (i.e., $\mathbf{B} = R(\mathbf{A})$) and "$C$" denotes compression via the

DCT and thresholding. Explain why the result of part (b) shows that $C(R(\mathbf{x})) = R(C(\mathbf{x}))$ (rotating and compressing yields the same result as compressing and then rotating, or equivalently, $R^{-1}(C(R(\mathbf{x}))) = C(\mathbf{x})$). Our approach to compression is thus invariant with respect to rotation.
Hint: Work in the frequency domain. Show that $R(C(\mathbf{A}))$ and $C(R(\mathbf{A}))$ have the same 2D DCT (and since the DCT is invertible, must be the same vector).

20. Suppose that $\mathbf{A}$ is a grayscale image in which each pixel consists of an integer in the range 0–255. We compress and then decompress $\mathbf{A}$, using the JPEG scheme outlined in the text, to produce image $\tilde{\mathbf{A}}$, consisting of integers in the range 0–255. Ignore any floating point round-off error; that is, assume all computations involving the block DCT are done "exactly." At which point(s) in this process is information lost?

21. The quantization used in JPEG in the frequency domain is of the form

$$q(x) = \text{round}\left(\frac{x}{d}\right)$$

for some real number d (see equation (3.21)). What are the quantization intervals in this scheme, and how do they depend on d? (There are in principle infinitely many quantization intervals here, so this quantization is not perfectly in accordance with Section 1.9.)
If we dequantize as $\tilde{q}(q(x)) = dq(x)$ (equation (3.23)), what are the corresponding codewords?

22. Is there any choice for the quantization matrix $\mathbf{e}$ for which the quantization process of equations (3.21) through (3.23) will yield lossless compression? Why or why not?

4
Convolution and Filtering

4.1 Overview

The term *filtering* refers to the systematic alteration of the frequency content of a signal or image. In particular, we sometimes want to "filter out" certain frequencies. The operation is often linear and usually (but not always) performed on the time domain version of the signal.

One important tool for filtering in the time domain is *convolution*. The form taken by convolution depends on the vector space in which the signals reside. We will begin by examining convolution for finite-length one-dimensional signals, then look at convolution in two-dimensions (images), and finally we will examine convolution for infinite and bi-infinite signals. We also introduce the *z-transform*, a powerful tool in signal and image processing.

4.2 One-Dimensional Convolution

4.2.1 Example: Low-Pass Filtering and Noise Removal

Recall the example of Section 2.3, in which the function $x(t) = 2.0\cos(2\pi \cdot 5t) + 0.8\sin(2\pi \cdot 12t) + 0.3\cos(2\pi \cdot 47t)$ was sampled on the time interval $[0, 1]$ at times $t = k/128$ for integers $0 \leq k \leq 127$. The goal in that example was the removal of the 47 Hz component in the sampled signal, perhaps because it represented noise. In that case, the noise was deterministic, not random, but that does not matter for the moment. We removed the high-frequency portion of the signal with frequency domain methods, by performing a discrete Fourier transform (DFT) (though we had not actually defined the DFT at that point), zeroing out the 47 Hz component, and then resynthesizing the signal without the noise by using an inverse DFT.

Let us consider how such a goal might be achieved, at least approximately, with strictly time domain methods. Let $\mathbf{x} \in \mathbb{R}^N$ be obtained by sampling some function $x(t)$ on time interval $[0, 1]$, at times $t = k/N$ for $0 \leq k \leq N-1$.

Discrete Fourier Analysis and Wavelets: Applications to Signal and Image Processing, Second Edition.
S. Allen Broughton and Kurt Bryan.

Companion Website: www.wiley.com/go/Broughton/Discrete_Fourier_Analysis_and_Wavelets

As usual, we index vector components from 0 to $N-1$. For simplicity, we assume that $x(t)$ is periodic in time with period 1, and that $x(t)$ consists of a superposition of low frequencies that represent the "real" signal and high frequencies that represent unwanted noise, random or deterministic. The frequency content of the analog signal $x(t)$ is assumed to lie in the range 0–N/2 Hz, so aliasing is not an issue.

A simple approach to partially removing the high frequencies from the signal while leaving low frequencies is to use a *moving average*, by replacing each sample x_k with the average value of itself and nearby samples. For example, let us take $\mathbf{w} \in \mathbb{R}^N$ as the vector defined by

$$w_k = \frac{1}{2}x_{k-1} + \frac{1}{2}x_k \tag{4.1}$$

for $0 \leq k \leq N-1$. Equation (4.1) presents a problem in the case where $k = 0$, since x_{-1} is undefined. We will thus interpret all indexes "modulo N," so, for example, $x_{-1} = x_{N-1}$. This is in keeping with the assumption that the function $x(t)$ is periodic with period 1, so $x_{-1} = x(-1/N) = x((N-1)/N) = x_{N-1}$. We can thus consider x_k as defined for all k via $x_k = x_{k \bmod N}$, if necessary.

The vector $\mathbf{w}$ defined by equation (4.1) will be a smoothed version of $\mathbf{x}$, with much less high-frequency content. The is best seen by considering this two-point averaging scheme's effect in the case that $x(t)$ consists of a basic waveform, complex exponential, or trigonometric. For example, if $x(t) = \sin(2\pi qt)$, then

$$\begin{aligned} w_k &= \frac{1}{2}x_{k-1} + \frac{1}{2}x_k \\ &= \frac{1}{2}\sin\left(\frac{2\pi q(k-1)}{N}\right) + \frac{1}{2}\sin\left(\frac{2\pi qk}{N}\right) \\ &= \left(\frac{1+\cos(2\pi q/N)}{2}\right)\sin\left(\frac{2\pi qk}{N}\right) - \frac{1}{2}\sin\left(\frac{2\pi q}{N}\right)\cos\left(\frac{2\pi qk}{N}\right) \\ &= A\sin\left(\frac{2\pi qk}{N}\right) - B\cos\left(\frac{2\pi qk}{N}\right), \end{aligned}$$

where $A = (1+\cos(2\pi q/N))/2$, $B = \frac{1}{2}\sin(2\pi q/N)$, and we make use of $\sin(a-b) = \sin(a)\cos(b) - \cos(a)\sin(b)$ with $a = 2\pi qk/N$, $b = 2\pi q/N$. If q is close to zero (more accurately, if q/N is close to zero), then $A \approx 1$ and $B \approx 0$. As a consequence, $w_k \approx \sin(2\pi qk/N) = x_k$. In short, a low-frequency waveform passes through the two-point averaging process largely unchanged. On the other hand, if $q \approx N/2$ (the Nyquist frequency), then $A \approx 0$ and $B \approx 0$, so $w_k \approx 0$. The highest frequencies that we can represent will be nearly zeroed out by this process.

Example 4.1 *In Figure 4.1, we show the original signal vector* $\mathbf{x}$ *from Section 2.3 in panel (a) and the vector* $\mathbf{w}$ *produced by application of equation (4.1)*

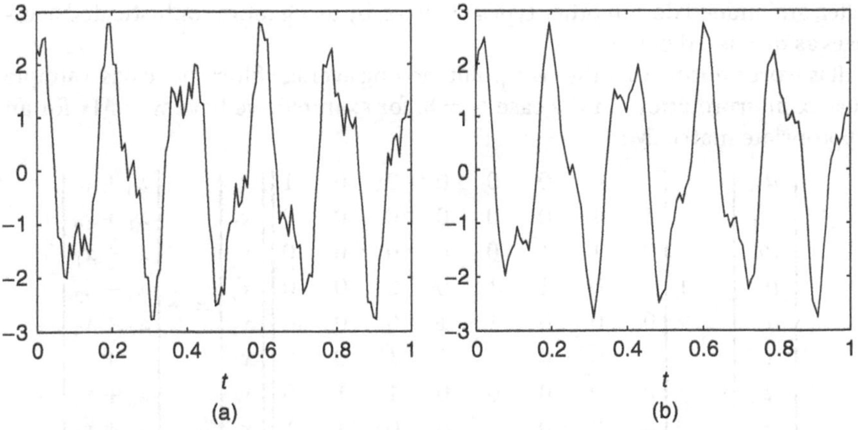

Figure 4.1 Original and smoothed signal.

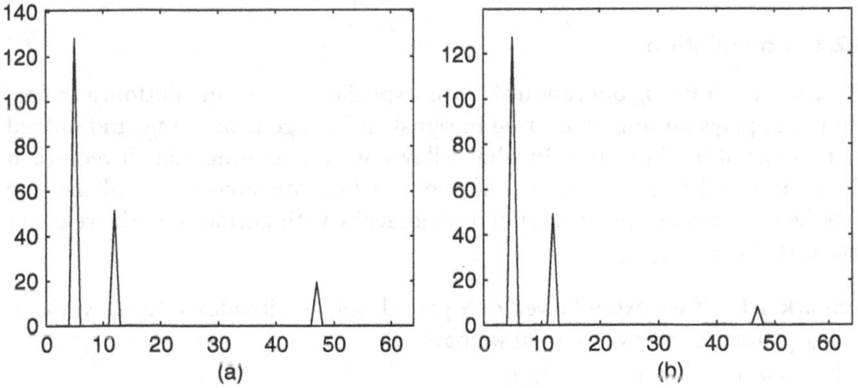

Figure 4.2 Original and smoothed signal DFT.

in panel (b), both plotted as functions on the interval [0, 1]. *In Figure 4.2, we plot the DFT magnitude for both the original and smoothed signals, on the frequency range 0–64 Hz. The magnitude of the 5 Hz spike has changed from* 128 *to* 126.1, *while the 12 Hz spike has changed from* 51.2 *to* 49.0. *The 47 Hz portion has diminished in magnitude from* 19.2 *to* 7.8, *considerably more (on a percentage basis) than the lower two frequencies.*

Example 4.1 illustrates *low-pass filtering*, in which one attempts to remove high frequencies from a signal while leaving lower frequencies unchanged. The two-point moving average procedure is obviously imperfect in this application, for the 5 and 12 Hz frequencies were slightly altered and the 47 Hz component was not completely removed. We can improve the performance of this low-pass filter, and indeed design other types of filters, by using more sophisticated averages as discussed below.

It is worth noting that the two-point moving average filter above has a simple matrix interpretation. In the case $N = 8$, for example, we have $\mathbf{w} = \mathbf{Mx}$ for an appropriate matrix $\mathbf{M}$:

$$\begin{bmatrix} w_0 \\ w_1 \\ w_2 \\ w_3 \\ w_4 \\ w_5 \\ w_6 \\ w_7 \end{bmatrix} = \frac{1}{2}\begin{bmatrix} 1 & 0 & 0 & 0 & 0 & 0 & 0 & 1 \\ 1 & 1 & 0 & 0 & 0 & 0 & 0 & 0 \\ 0 & 1 & 1 & 0 & 0 & 0 & 0 & 0 \\ 0 & 0 & 1 & 1 & 0 & 0 & 0 & 0 \\ 0 & 0 & 0 & 1 & 1 & 0 & 0 & 0 \\ 0 & 0 & 0 & 0 & 1 & 1 & 0 & 0 \\ 0 & 0 & 0 & 0 & 0 & 1 & 1 & 0 \\ 0 & 0 & 0 & 0 & 0 & 0 & 1 & 1 \end{bmatrix}\begin{bmatrix} x_0 \\ x_1 \\ x_2 \\ x_3 \\ x_4 \\ x_5 \\ x_6 \\ x_7 \end{bmatrix} = \frac{1}{2}\begin{bmatrix} x_0 + x_7 \\ x_1 + x_0 \\ x_2 + x_1 \\ x_3 + x_2 \\ x_4 + x_3 \\ x_5 + x_4 \\ x_6 + x_5 \\ x_7 + x_6 \end{bmatrix}.$$

A two-point average for larger sample vectors has an analogous structure. In particular, it is clear that this process is linear.

4.2.2 Convolution

The low-pass filtering operation above is a special case of convolution, an operation that plays an important role in signal and image processing, and indeed many areas of mathematics. In what follows, we will assume that all vectors in $\mathbb{C}^N$ are indexed from 0 to $N - 1$. Moreover, when convenient, we will assume that the vectors have been extended periodically with period N in the relevant index, via $x_k = x_{k \bmod N}$.

Remark 4.1 If we extend a vector $\mathbf{x}$ periodically to all index values k via $x_k = x_{k \bmod N}$, then for any value of m we have

$$\sum_{k=m}^{m+N-1} x_k = \sum_{k=0}^{N-1} x_{k+m} = \sum_{k=0}^{N-1} x_k.$$

In short, the sum of any N consecutive components of $\mathbf{x}$ is the same, a fact we will use over and over again (see Exercise 4).

4.2.2.1 Convolution Definition

Let us recast the filtering operation above in a more general format. We begin with a definition.

Definition 4.1 Let $\mathbf{x}$ and $\mathbf{y}$ be vectors in $\mathbb{C}^N$. The circular convolution of $\mathbf{x}$ and $\mathbf{y}$ is the vector $\mathbf{w} \in \mathbb{C}^N$ with components

$$w_r = \sum_{k=0}^{N-1} x_k y_{(r-k) \bmod N} \tag{4.2}$$

for $0 \leq r \leq N - 1$. The circular convolution is denoted $\mathbf{w} = \mathbf{x} * \mathbf{y}$.

For now, we will drop the "circular" prefix and refer to this operation as just "convolution." Of course, since we assume that all vectors are extended periodically, we can ignore the modulo N arithmetic on indexes, but we put it explicitly in equation (4.2) for the sake of clarity.

Convolution looks a bit confusing, but here is a simple way to visualize it. We compute the quantity w_0 by taking the vector $\mathbf{x}$ and the vector $\mathbf{y}$ indexed in reverse order starting at $k = 0$, and lining them up:

$$\begin{array}{ccccc} x_0 & x_1 & x_2 & \cdots & x_{N-1} \\ y_0 & y_{N-1} & y_{N-2} & \cdots & y_1. \end{array}$$

Then w_0 is simply the dot product of these two rows, $w_0 = x_0y_0 + x_1y_{N-1} + \cdots + x_{N-1}y_1$. To compute w_1, take the y_1 at the end of the second row and move it to the front, pushing other components to the right, as

$$\begin{array}{ccccc} x_0 & x_1 & x_2 & \cdots & x_{N-1} \\ y_1 & y_0 & y_{N-1} & \cdots & y_2. \end{array}$$

The dot product of these two row vectors is w_1. To compute w_2, move y_2 to the front, push all other components to the right, and compute the dot product. Do this a total of N times, at which point $\mathbf{y}$ will be back in its original position. The overall computation can be summarized as

$$\begin{array}{c|ccccc} & x_0 & x_1 & x_2 & \cdots & x_{N-1} \\ \hline w_0 & y_0 & y_{N-1} & y_{N-2} & \cdots & y_1 \\ w_1 & y_1 & y_0 & y_{N-1} & \cdots & y_2 \\ w_2 & y_2 & y_1 & y_0 & \cdots & y_3 \\ & & & & \vdots & \\ w_{N-1} & y_{N-1} & y_{N-2} & y_{N-3} & \cdots & y_0. \end{array}$$

Each w_r is obtained as the dot product of the corresponding row with the vector $(x_0, x_1, \ldots, x_{N-1})$.

Example 4.2 *The two-point low-pass filtering operation from Section 4.2.1 can be cast as a convolution $\mathbf{w} = \mathbf{x} * \ell$ where $\ell \in \mathbb{C}^{128}$ has components $\ell_0 = \frac{1}{2}$, $\ell_1 = \frac{1}{2}$, and all other $\ell_k = 0$. To see this note that $\ell_{(r-k) \bmod N} = \frac{1}{2}$ when $k = r$ or $k = r - 1$ and 0 otherwise. Equation (4.2) then becomes $w_r = x_r/2 + x_{r-1}/2$, which is equation (4.1). Indeed all the filtering operations with which we will be concerned can be implemented via an appropriate convolution.*

4.2.2.2 Convolution Properties

The following theorem summarizes some important algebraic properties of convolution.

Theorem 4.1 *Let* $\mathbf{x}$, $\mathbf{y}$, *and* $\mathbf{w}$ *be vectors in* $\mathbb{C}^N$. *The following hold:*

1. *Linearity:* $\mathbf{x} * (a\mathbf{y} + b\mathbf{w}) = a(\mathbf{x} * \mathbf{y}) + b(\mathbf{x} * \mathbf{w})$ *for any scalars* a, b.
2. *Commutativity:* $\mathbf{x} * \mathbf{y} = \mathbf{y} * \mathbf{x}$.
3. *Matrix formulation: If* $\mathbf{w} = \mathbf{x} * \mathbf{y}$, *then* $\mathbf{w} = \mathbf{M_y x}$, *where* $\mathbf{M_y}$ *is the* $N \times N$ *matrix*

$$\mathbf{M_y} = \begin{bmatrix} y_0 & y_{N-1} & y_{N-2} & \cdots & y_1 \\ y_1 & y_0 & y_{N-1} & \cdots & y_2 \\ y_2 & y_1 & y_0 & \cdots & y_3 \\ & & & \vdots & \\ y_{N-1} & y_{N-2} & y_{N-3} & \cdots & y_0 \end{bmatrix}.$$

   In particular, the row k *and column* m *entries of* $\mathbf{M_y}$ *are* $y_{(k-m) \bmod N}$, *rows and columns indexed from 0. Moreover,* $\mathbf{M_x M_y} = \mathbf{M_{x*y}}$.
   The matrix $\mathbf{M_y}$ *is called the* circulant matrix *for* $\mathbf{y}$. *Note that the rows of* $\mathbf{M_y}$, *or the columns, can be obtained by the circular shifting procedure described after equation (4.2).*
4. *Associativity:* $\mathbf{x} * (\mathbf{y} * \mathbf{w}) = (\mathbf{x} * \mathbf{y}) * \mathbf{w}$.
5. *Periodicity: If* x_k *and* y_k *are extended to be defined for all* k *with period* N, *then the quantity* w_r *defined by equation (4.2) is defined for all* r *and satisfies* $w_r = w_{r \bmod N}$.

Proof: The linearity relation has a very straightforward proof (see Exercise 5). To prove commutativity (list item 2), let $\mathbf{w} = \mathbf{x} * \mathbf{y}$ so that from equation (4.2), we have

$$w_r = \sum_{k=0}^{N-1} x_k y_{r-k}$$

in which we have suppressed the modulo N arithmetic on indexes. The sum is defined for any r, since y_m is defined for any index m. The change of index $k = r - j$ (so $j = r - k$) yields

$$w_r = \sum_{j=r}^{r-N+1} x_{r-j} y_j.$$

The sequence $x_{r-j}y_j$ is periodic in j with period N and the sum for w_r above involves N consecutive terms of this sequence. By Remark 4.1, we know that we can instead sum over the N consecutive indexes $j = 0$ to $j = N - 1$ to find

$$w_r = \sum_{j=0}^{N-1} x_{r-j} y_j,$$

which is precisely the definition of the rth component of the vector $\mathbf{y} * \mathbf{x}$.

To prove the matrix formulation (list item 3), start with the formal summation definition of the matrix–vector product $\mathbf{w} = \mathbf{M}_\mathbf{y}\mathbf{x}$, which is

$$w_r = \sum_{k=0}^{N-1} (\mathbf{M}_\mathbf{y})_{rk} x_k,$$

where $(\mathbf{M}_\mathbf{y})_{rk}$ denotes the row r and column k entries of $\mathbf{M}_\mathbf{y}$, defined to be $y_{(r-k) \bmod N}$. We then have

$$w_r = \sum_{k=0}^{N-1} x_k y_{(r-k) \bmod N}.$$

This is precisely the definition of $\mathbf{w} = \mathbf{x} * \mathbf{y}$ from equation (4.2).

To show that $\mathbf{M}_\mathbf{x}\mathbf{M}_\mathbf{y} = \mathbf{M}_{\mathbf{x}*\mathbf{y}}$, we start from the definition of matrix multiplication

$$\begin{aligned}(\mathbf{M}_\mathbf{x}\mathbf{M}_\mathbf{y})_{k,m} &= \sum_{r=0}^{N-1} (\mathbf{M}_\mathbf{x})_{k,r}(\mathbf{M}_\mathbf{y})_{r,m} \\ &= \sum_{r=0}^{N-1} x_{k-r} y_{r-m}. \end{aligned} \tag{4.3}$$

For fixed k and m, the sequence $x_{k-r}y_{r-m}$ (considered as a sequence in r) is periodic with period N. In view of Remark 4.1, we can shift the index of summation r in equation (4.3) (as $r \to r + m$) to find

$$(\mathbf{M}_\mathbf{x}\mathbf{M}_\mathbf{y})_{k,m} = \sum_{r=0}^{N-1} x_{k-m-r} y_r.$$

The quantity on the right is by definition the $k - m$ component of the vector $\mathbf{x} * \mathbf{y}$, and by definition, this is precisely the row k and column m elements of $\mathbf{M}_{\mathbf{x}*\mathbf{y}}$.

Associativity (list item 4) is easy to prove from the matrix formulation of convolution and the associativity of matrix multiplication, for

$$\begin{aligned}\mathbf{x} * (\mathbf{y} * \mathbf{w}) &= \mathbf{M}_\mathbf{x}(\mathbf{M}_\mathbf{y}\mathbf{w}) \\ &= (\mathbf{M}_\mathbf{x}\mathbf{M}_\mathbf{y})\mathbf{w} \\ &= \mathbf{M}_{\mathbf{x}*\mathbf{y}}\mathbf{w} \\ &= (\mathbf{x} * \mathbf{y}) * \mathbf{w}.\end{aligned}$$

Finally, property (list item 5) is left as Exercise 6.

4.3 Convolution Theorem and Filtering

4.3.1 The Convolution Theorem

Computing the convolution of two vectors in the time domain may look a bit complicated, but the frequency domain manifestation of convolution is very simple.

Theorem 4.2 ***The Convolution Theorem*** *Let* $\mathbf{x}$ *and* $\mathbf{y}$ *be vectors in* $\mathbb{C}^N$ *with DFT's* $\mathbf{X}$ *and* $\mathbf{Y}$*, respectively. Let* $\mathbf{w} = \mathbf{x} * \mathbf{y}$ *have DFT* $\mathbf{W}$*. Then*

$$W_k = X_k Y_k \tag{4.4}$$

for $0 \le k \le N-1$.

Proof: This is a simple and direct computation. From the definition of the DFT, we have

$$W_k = \sum_{m=0}^{N-1} e^{-2\pi ikm/N} w_m.$$

Since $\mathbf{w} = \mathbf{x} * \mathbf{y}$, we have $w_m = \sum_{r=0}^{N-1} x_r y_{m-r}$. Substitute this into the formula for W_k above and interchange the summation order to find

$$W_k = \sum_{r=0}^{N-1} \sum_{m=0}^{N-1} e^{-2\pi ikm/N} x_r y_{m-r}.$$

Make a change of index in the m sum by substituting $n = m - r$ (so $m = n + r$). With the appropriate change in the summation limits and a bit of algebra, we obtain

$$\begin{aligned} W_k &= \sum_{r=0}^{N-1} \sum_{n=-r}^{N-1-r} e^{-2\pi ik(n+r)/N} x_r y_n \\ &= \left(\sum_{r=0}^{N-1} e^{-2\pi ikr/N} x_r \right) \left(\sum_{n=-r}^{N-1-r} e^{-2\pi ikn/N} y_n \right), \end{aligned}$$

where in the last line we have broken apart the exponential and grouped disjoint sums together. Now note that for any fixed r the sum in n above represents N consecutive terms of the sequence $e^{-2\pi ikn/N} y_n$, which is periodic in n. By Remark 4.1, we can shift the summation limits to $n = 0$ to $n = N - 1$ (also N consecutive terms) and obtain

$$\begin{aligned} W_k &= \left(\sum_{r=0}^{N-1} e^{-2\pi ikr/N} x_r \right) \left(\sum_{n=0}^{N-1} e^{-2\pi ikn/N} y_n \right) \\ &= X_k Y_k. \end{aligned}$$

This proves the theorem.

Theorem 4.2 will provide essential insight into the operation of filtering signals.

4.3.2 Filtering and Frequency Response

Let $\mathbf{x} \in \mathbb{C}^N$ have DFT $\mathbf{X}$. We will think of $\mathbf{x}$ as a sampled signal, in which case it will have real components, but that does not matter at the moment. Let $\mathbf{h} \in \mathbb{C}^N$ denote a "filter" vector whose components we choose, with DFT $\mathbf{H}$. Define $\mathbf{w} = \mathbf{x} * \mathbf{h}$. The operation of convolving $\mathbf{x}$ with $\mathbf{h}$ is a type of *filtering*, with $\mathbf{x}$ as the input to the filter and $\mathbf{w}$ as the output. Theorem 4.2 quantifies exactly how the input $\mathbf{x}$ and output $\mathbf{w}$ are related in the frequency domain.

Below we look a little more closely at the effect of filtering in the time domain, especially how filtering affects the basic waveforms that make up the input signal. We also show how Theorem 4.2 can be used to design filters that predictably and systematically alter the frequency content of the input signal, entirely eliminating some while accentuating others.

4.3.2.1 Filtering Effect on Basic Waveforms

It is important to note that filtering is a linear process, for it is a convolution. As such we can analyze the effect of filtering a signal by considering the effect of the filter on each of the basic waveforms $\mathbf{E}_{N,m}$ that make up the input signal, one waveform at a time. Filtering has a very simple effect on a basic waveform—it merely multiplies it by a constant!

To see this, suppose that a basic waveform $\mathbf{E}_m \in \mathbb{C}^N$ is used as the input to the filter $\mathbf{h}$ and let $\mathbf{w} = \mathbf{E}_m * \mathbf{h}$ be the output; we suppress the dependence of $\mathbf{E}_m$ on N for now. The DFT of the input in this case is the vector $\mathbf{X} \in \mathbb{C}^N$ with a single nonzero component, namely $X_m = N$ and $X_k = 0$ if $k \neq m$. This follows immediately from the definition of the DFT, equation (2.6). According to Theorem 4.2, $W_k = H_k X_k$, so the output $\mathbf{w}$ also has a DFT $\mathbf{W}$ with a single nonzero component $W_m = NH_m$ while $W_k = 0$ for $k \neq m$. If we apply the inverse DFT to $\mathbf{W}$, we find that $\mathbf{w} = H_m \mathbf{E}_m$, which follows easily from the definition of the inverse DFT (equation (2.7)). Thus

$$\mathbf{E}_m * \mathbf{h} = H_m \mathbf{E}_m. \tag{4.5}$$

In summary, the effect of filtering a basic waveform $\mathbf{E}_m$ is to multiply the waveform by the (probably complex) scalar H_m.

The equation $\mathbf{E}_m * \mathbf{h} = H_m \mathbf{E}_m$ can be cast in matrix form as $\mathbf{M_h E}_m = H_m \mathbf{E}_m$, where $\mathbf{M_h}$ is the circulant matrix for $\mathbf{h}$ and $\mathbf{E}_m$ is written as a column vector. Equation (4.5) immediately yields the following theorem.

Theorem 4.3 *Let $\mathbf{h} \in \mathbb{C}^N$ have DFT $\mathbf{H}$. The eigenvectors of the circulant matrix $\mathbf{M_h}$ are the basic waveform vectors $\mathbf{E}_{N,m}$, with corresponding eigenvalues H_m.*

Remark 4.2 The basic waveforms are complex exponential, with real and imaginary parts that are sines and cosines. It is worth looking at how filtering affects the amplitude and phase of $\mathbf{E}_{N,m}$. Recall that the kth component of $\mathbf{E}_{N,m}$ is $e^{2\pi ikm/N}$. In particular, we may consider the vector $\mathbf{E}_{N,m}$ as a sampled version of the function

$$\phi_m(t) = e^{2\pi imt}$$

at sample points $t = k/N$, $0 \le k \le N-1$. Note that $|\phi_m(t)| = 1$ for all t and $\phi_m(t)$ oscillates with period $1/m$. The filtered version of the waveform is $H_m\mathbf{E}_{N,m}$, and this may be considered as the sampled version of the function

$$\begin{aligned} H_m\phi_m(t) &= H_m e^{2\pi imt} \\ &= |H_m| e^{i\,\arg(H_m)}\, e^{2\pi imt} \\ &= |H_m| e^{2\pi im[t+(\arg(H_m)/2\pi m)]} \end{aligned} \tag{4.6}$$

at the same times $t = k/N$, where we have used the fact that $A = |A|e^{i\,\arg(A)}$ for any complex A. The function $H_m\phi_m(t)$ satisfies $|H_m\phi_m(t)| = |H_m|$ for all t and is advanced in time (graphically, shifted to the left) by an amount $\arg(H_m)/2\pi m$, which is a fraction $\arg(H_m)/2\pi$ of the period of $\phi_m(t)$.

In summary, when filtered, the basic waveform $\mathbf{E}_{N,m}$ is amplified in magnitude by $|H_m|$ and phase-shifted in time by a fraction $\arg(H_m)/2\pi$ of one oscillation period, as summarized in Table 4.1

A general input $\mathbf{x}$ is a superposition of such basic waveforms. When passed through the filter, this amplification and phase shifting occurs frequency by frequency.

One can similarly analyze the effect of filtering on the basic trigonometric waveforms obtained by sampling $\cos(2\pi mt)$ and $\sin(2\pi mt)$ (see Exercise 20).

Example 4.3 *To illustrate the ideas above, let us revisit the two-point moving average filter from Section 4.2.1. We can filter any vector* $\mathbf{x} \in \mathbb{R}^{128}$ *with a two-point moving average as* $\mathbf{w} = \mathbf{x} * \ell$*, where* $\ell \in \mathbb{R}^{128}$ *with* $\ell_0 = \frac{1}{2}, \ell_1 = \frac{1}{2}$*, and all other* $\ell_k = 0$*. In Figure 4.3, we plot the magnitude and complex argument of* $\mathbf{L} = DFT(\ell)$ *on the range* $-63 \le k \le 64$*, so lower frequencies are near the center and high frequencies at the edges. The magnitude is shown in panel (a), argument is shown in panel (b).*

Table 4.1 Effect of filtering on basic waveforms.

Input		Output
basic waveform	Filter $\mathbf{E}_m \to \mathbf{h} * \mathbf{E}_m$ $\longrightarrow$	magnified, shifted waveform
$\mathbf{E}_m$		$\mathbf{h} * \mathbf{E}_m = H_m\mathbf{E}_m$

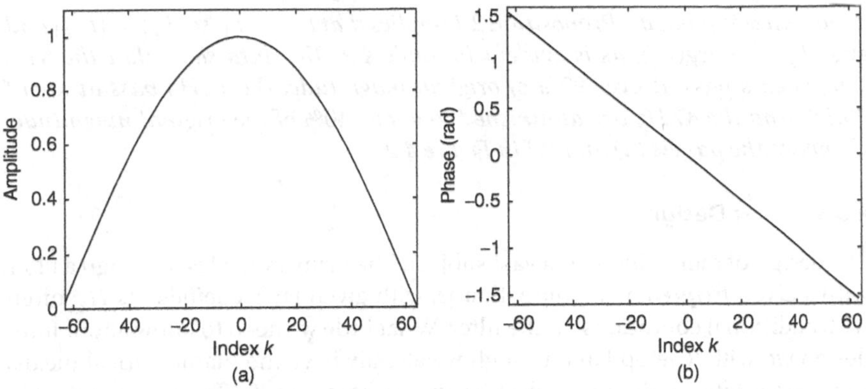

Figure 4.3 Amplitude and argument plot of **L**.

From the plot on the left, we can see that frequencies near zero in any input vector **x** *have their magnitude largely unchanged by the filtering operation, for* $L_k \approx 1$ *in this region, hence* $W_k = L_k X_k \approx X_k$*. On the other hand, frequencies near the edges* $k = \pm 64$ *(corresponding to the Nyquist frequency) have their magnitude attenuated to nearly zero, for in this region* $L_k \approx 0$ *and so* $W_k = L_k X_k \approx 0$*. Thus* ℓ *acts as a crude low-pass filter, with no sharp delineation between frequencies that are allowed to pass and those that are blocked.*

The plot of the complex argument of the components of **L** *on the right shows that input waveforms with frequency near zero have their phase shifted very little, for here* $\arg(L_k) \approx 0$*. On the other hand, frequencies near* $k = 64$ *will have their phase shifted by a fraction of about* $-1.6/2\pi$ *(really* $(-\pi/2)/2\pi = -\frac{1}{4}$*) period (graphically, to the right), while those near* $k = -64$ *will be shifted about* $\frac{1}{4}$ *period to the left.*

Let us look at how the basic waveforms that comprise the specific input **x** *from the example in Section 4.2.1 are altered by the filter. Table 4.2 shows the value of* L_k*, as well as* $|L_k|$ *and* $\arg(L_k)$ *for* $k = -5, 5, -12, 12, -47,$ *and 47. Of course, since*

Table 4.2 Moving average effect on basic waveforms.

k	L_k	$\lvert L_k \rvert$	$\text{Arg}(L_k)$
−5	0.9850 + 0.1215i	0.9925	0.1227
5	0.9850 − 0.1215i	0.9925	−0.1227
−12	0.9157 + 0.2778i	0.9569	0.2945
12	0.9157 − 0.2778i	0.9569	−0.2945
−47	0.1642 + 0.3705i	0.4052	1.1536
47	0.1642 − 0.3705i	0.4052	−1.1536

ℓ has real components, Proposition 2.1 implies that $L_{-k} = \overline{L_k}$, so $|L_k| = |L_{-k}|$ and $\arg(-L_k) = -\arg(L_k)$, as is evident in Table 4.2. The data show that the 5 Hz components pass at over 99% of original magnitude, the 12 Hz pass at about 95.7%, and the 47 Hz are diminished to about 40% of the original magnitude. Compare the panels (a) and (b) in Figure 4.2.

4.3.3 Filter Design

The design of digital filters is a vast subject. The primary goal is to design a filter with a given *frequency response*, that is, with given DFT coefficients H_k, often with additional conditions on the filter. What little we need to know about filter design we will develop later, though we already have the mathematical means to design a filter with any desired frequency response if efficiency is not a big concern.

Specifically, suppose that we want a filter $\mathbf{h}$ so that if $\mathbf{w} = \mathbf{h} * \mathbf{x}$, then $W_k = H_k X_k$, where we choose each H_k (possibly complex). By choosing the H_k suitably, we can make the filter do whatever we want to each frequency in the input. By Theorem 4.2, we can take the filter vector $\mathbf{h}$ as the inverse DFT of the vector $\mathbf{H}$. The primary drawback to this approach is that all filter coefficients h_k will almost certainly be nonzero. This is not a big deal if we are working with vectors of length 10, but it is if the vectors are of length 10^6. The convolution $\mathbf{x} * \mathbf{h}$ would be pretty slow, of order N^2 for vectors of length N (see Exercise 23). Indeed it would probably be faster to perform an FFT on $\mathbf{x}$, multiply each X_k in $\mathbf{X}$ by H_k, and then perform an inverse FFT. If we really want to filter strictly in the time domain, then it would be helpful if we could come close to the desired frequency response with a filter that has few nonzero coefficients.

Definition 4.2 The nonzero coefficients h_k in a filter $\mathbf{h}$ are called the "taps" for the filter.

There are a variety of strategies for designing filters with a given frequency response and few taps. One simple (not always ideal) approach is to specify the filter frequency response $\mathbf{H}$, take $\mathbf{h}$ as the inverse DFT of $\mathbf{H}$, and then zero out coefficients in $\mathbf{h}$ that lie below some threshold.

Example 4.4 *We will again work with input signals $\mathbf{x} \in \mathbb{C}^{128}$, but this time let us design a high-pass filter that will pass unchanged all basic waveforms with frequency in the range 30–64 and block all waveforms with frequency 29 and below. If we index from 0 to 127, then these lower frequencies correspond to waveforms $\mathbf{E}_{128,k}$ with $0 \le k \le 29$ or $99 \le k \le 127$ (recall that $\mathbf{E}_{128,N-k} = \mathbf{E}_{128,-k}$, so the range $99 \le k \le 127$ corresponds to frequencies 1–29).*

The DFT of the desired filter is the vector $\mathbf{H} \in \mathbb{C}^{128}$ with components $H_k = 0$ for $0 \le k \le 29$ and $99 \le k \le 127$, and $H_k = 1$ otherwise. The inverse DFT of $\mathbf{H}$

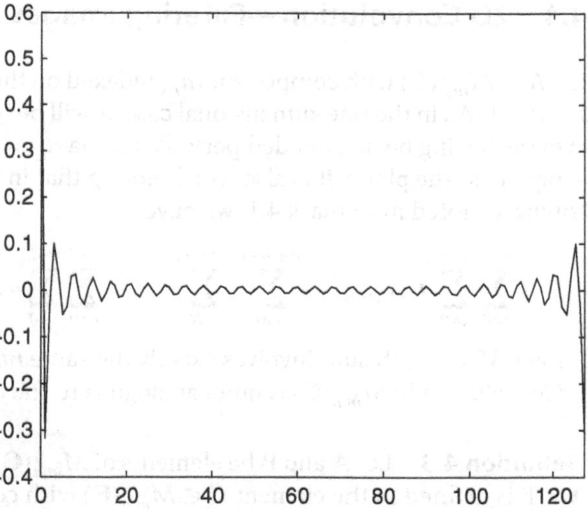

Figure 4.4 High-pass filter coefficients.

yields a filter vector $\mathbf{h}$, *which is real-valued due to the* $H_{N-k} = H_k$ *symmetry, with components* h_k *plotted for* $0 \leq k \leq 127$ *in Figure 4.4.*

All the h_k *turn out to be nonzero. We can eliminate some filter taps by thresholding. Define the vector* $\mathbf{h}_\epsilon$ *for* $\epsilon > 0$ *by zeroing out all components of* $\mathbf{h}$ *that have magnitude less than* ϵ. *We will use* $\epsilon = 0.01, 0.05$, *and 0.1, which yields filters with 43, 7, and 3 taps, respectively. Increasing* ϵ *decreases the number of taps, but the resulting DFT of* $\mathbf{h}_\epsilon$ *deviates more and more from the ideal high-pass frequency response. In Figure 4.5, we show the magnitude of the DFT coefficients for each filter* $\mathbf{h}_\epsilon$, *in each case superimposed over the ideal high-pass filter's DFT. Whether this degradation in the frequency response is acceptable depends on the situation. Phase distortion (not shown here) may also be a concern.*

We will return to filter design in Chapter 7, in the context of "filter banks."

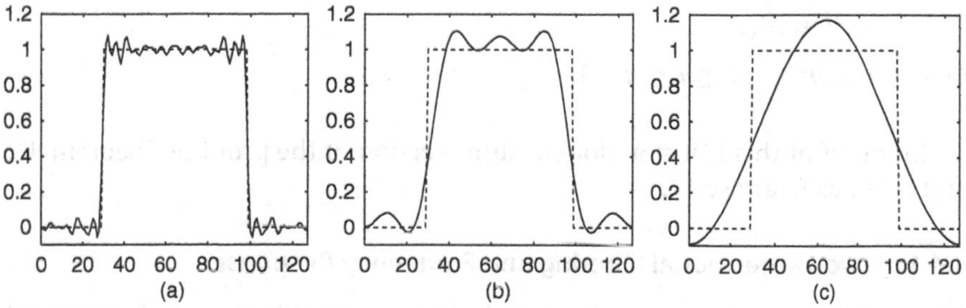

Figure 4.5 DFT's for $\mathbf{h}_\epsilon$: (a) $\epsilon = 0.01$, (b) $\epsilon = 0.05$, (c) $\epsilon = 0.1$.

4.4 2D Convolution—Filtering Images

Let $\mathbf{A} \in M_{m,n}(\mathbb{C})$ with components $a_{r,s}$ indexed on the range $0 \leq r \leq m-1, 0 \leq s \leq n-1$. As in the one-dimensional case, it will be helpful to think of any such array as having been extended periodically via $a_{r,s} = a_{r \bmod m, s \bmod n}$ so that the array "tiles" the plane. It is also worth noting that in this case, by the same reasoning as noted in Remark 4.1, we have

$$\sum_{r=0}^{m-1}\sum_{s=0}^{n-1} a_{r+M,s+N} = \sum_{r=M}^{M+m-1}\sum_{s=N}^{N+n-1} a_{r,s} = \sum_{r=0}^{m-1}\sum_{s=0}^{n-1} a_{r,s}$$

for any M, N. Each sum involves exactly the same mn summands.

Convolution in $M_{m,n}(\mathbb{C})$ is quite analogous to the one-dimensional case.

Definition 4.3 Let **A** and **B** be elements of $M_{m,n}(\mathbb{C})$. The circular convolution $\mathbf{A} * \mathbf{B}$ is defined as the element $\mathbf{C} \in M_{m,n}(\mathbb{C})$ with components

$$c_{p,q} = \sum_{r=0}^{m-1}\sum_{s=0}^{n-1} a_{r,s} b_{p-r,q-s}. \tag{4.7}$$

Like its one-dimensional counterpart, two-dimensional circular convolution has various useful algebraic properties: it is linear, commutative, associative, and if $\mathbf{C} = \mathbf{A} * \mathbf{B}$, then $c_{r,s} = c_{r \bmod m, s \bmod n}$. There is no convenient matrix formulation for two-dimensional convolution, however, except in certain special cases (see Exercise 24).

We also have a two-dimensional version of the convolution theorem:

Theorem 4.4 ***The Convolution Theorem*** *Let* **A** *and* **B** *be elements of* $M_{m,n}(\mathbb{C})$ *with (two-dimensional) DFT's* $\hat{\mathbf{A}}$ *and* $\hat{\mathbf{B}}$, *respectively (components* $\hat{a}_{k,l}, \hat{b}_{k,l}$*). Let* $\mathbf{C} = \mathbf{A} * \mathbf{B}$ *have two-dimensional DFT* $\hat{\mathbf{C}}$, *components* $\hat{c}_{k,l}$. *Then*

$$\hat{c}_{k,l} = \hat{a}_{k,l}\hat{b}_{k,l}$$

for $0 \leq k \leq m-1, 0 \leq l \leq n-1$.

The proof of this is just a "double-sum" version of the proof of Theorem 4.2 and is left as Exercise 17.

4.4.1 Two-Dimensional Filtering and Frequency Response

We can filter two-dimensional images in a manner similar to one-dimensional signals. Specifically, given an image array $\mathbf{A} \in M_{m,n}(\mathbb{C})$ and element

$\mathbf{D} \in M_{m,n}(\mathbb{C})$, we filter by convolving $\mathbf{A}$ with $\mathbf{D}$. If $\mathbf{B} = \mathbf{A} * \mathbf{D}$, then, by Theorem 4.4, the DFT components satisfy

$$\hat{b}_{k,l} = \hat{a}_{k,l}\hat{d}_{k,l}.$$

In the context of image filtering, the array $\mathbf{D}$ is often called a mask.

It is easy to see the effect of filtering a two-dimensional basic waveform with a mask $\mathbf{D}$. Let $\mathcal{E}_{m,n,k,l}$ be a two-dimensional basic waveform (recall the definition of these waveforms from equation (1.25); note that k and l here indicate which basic waveform we are considering and are fixed. Let $\mathcal{G} \in M_{m,n}(\mathbb{C})$ denote the two-dimensional DFT of $\mathcal{E}_{m,n,k,l}$. The definition of the two-dimensional DFT shows that the $m \times n$ array $\mathcal{G}$ has a single nonzero component:

$$\mathcal{G}_{r,s} = \begin{cases} mn, & \text{if } r = k \text{ and } s = l, \\ 0, & \text{else.} \end{cases}$$

By Theorem 4.4, if we filter $\mathcal{E}_{m,n,k,l}$ as $\mathbf{B} = \mathcal{E}_{m,n,k,l} * \mathbf{D}$, then $\hat{b}_{r,s} = \hat{d}_{r,s}\mathcal{G}_{r,s}$. The DFT $\hat{\mathbf{B}}$ will have the same form as $\mathcal{G}$, with a single nonzero component $\hat{b}_{r,s} = mn\hat{d}_{r,s}$ when $r = k$ and $s = l$, while $\hat{b}_{r,s} = 0$ for all other indexes r, s. An inverse DFT applied to $\hat{\mathbf{B}}$ then shows that

$$\mathcal{E}_{m,n,k,l} * \mathbf{D} = \hat{d}_{k,l}\mathcal{E}_{m,n,k,l}, \tag{4.8}$$

the two-dimensional counterpart to equation (4.5). As in the one-dimensional case, when a basic 2D waveform is convolved with a mask, the effect is to multiply the waveform by a (complex) scalar.

Although two-dimensional convolution has no convenient matrix formulation, equation (4.8) shows that each of the mn basic waveforms $\mathcal{E}_{m,n,k,l}$ is a kind of "eigenvector" or "eigenfunction" for the mapping $\phi : M_{m,n}(\mathbb{C}) \to M_{m,n}(\mathbb{C})$ defined by $\phi(\mathbf{A}) = \mathbf{A} * \mathbf{D}$. The corresponding eigenvalues are the constants $\hat{d}_{k,l}$.

4.4.2 Applications of 2D Convolution and Filtering

4.4.2.1 Noise Removal and Blurring

Define a mask $\mathbf{D} \in M_{m,n}(\mathbb{C})$ with components $d_{r,s}$ given by

$$d_{r,s} = \begin{cases} d_{r,s} = \dfrac{1}{9}, & r, s = -1, 0, 1, \\ d_{r,s} = 0, & \text{otherwise.} \end{cases}$$

Recall we extend $\mathbf{D}$ periodically so, for example, $d_{m-1,n-1} = d_{-1,-1}$. The effect of convolving an image array $\mathbf{A}$ with $\mathbf{D}$ is to average each pixel in $\mathbf{A}$ with its eight nearest neighbors, analogous to the two-point moving average used to denoise one-dimensional signals. Specifically, the (k, l) entry of $\mathbf{A} * \mathbf{D}$ is

$$\begin{aligned}(\mathbf{A} * \mathbf{D})_{k,l} = \frac{1}{9}(&a_{k-1,l-1} + a_{k-1,l} + a_{k-1,l+1} \\ &+ a_{k,l-1} + a_{k,l} + a_{k,l+1} \\ &+ a_{k+1,l-1} + a_{k+1,l} + a_{k+1,l+1}).\end{aligned}$$

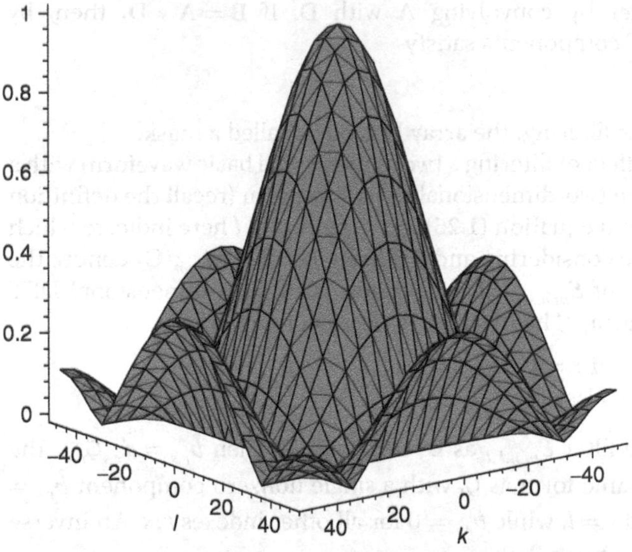

Figure 4.6 Magnitude of nine-point averaging mask DFT.

The magnitude of the DFT of $\mathbf{D}$ is plotted in Figure 4.6, in the case where $m = n = 100$, for the frequency range $-m/2 < k \leq m/2, -n/2 < l \leq n/2$. Low frequencies are near the center, higher frequencies near the edges. In this case, the direct component (dc) is passed with magnitude unchanged. In general, lower frequencies pass with slightly diminished magnitude, while higher frequencies are strongly attenuated.

This mask can be used in a simple scheme for denoising images. The rationale is that noise (especially random noise) contains a lot of high-frequency energy, so the filter should substantially remove it while allowing more slowly varying image features to pass. The images in Figure 4.7 provide an example. Panel (a) is a grayscale image without noise. Panel (b) is the same image with random noise added. In Figure 4.8a, we show the result of applying the nine-point averaging mask to the noisy image. It is clear that the noise is somewhat reduced. In panel (b), we show the result of applying the mask five times. The noise is much reduced, but at the expense of substantial blurring of the image. This illustrates an essential difficulty in the use of averaging for noise reduction. Sharp edges in a photograph require high frequencies to synthesize, so the same averaging or low-pass filter that "smears out" the noise also smears the edges.

4.4.2.2 Edge Detection

Let us consider the problem of detecting sharp edges in a grayscale image, an important task in image processing. If we think of the analog case where the image intensity is given by a function $f(x, y)$, such an edge corresponds to an

(a) (b)

Figure 4.7 Original and noisy image.

(a) (b)

Figure 4.8 Noisy image filtered once and five times.

abrupt change in the value of f in some direction. In this case, $\partial f/\partial x$ and/or $\partial f/\partial y$ will likely be large in magnitude; more generally, $\|\nabla f\|$ should be large, where ∇f denotes the gradient of f and $\|\nabla f\|$ the magnitude of this vector.

To detect such edges in the discrete case, let $\mathbf{D} \in M_{m,n}(\mathbb{C})$ be the mask defined by $d_{0,0} = 1$, $d_{1,0} = -1$, and all other $d_{r,s} = 0$. For a sampled image $\mathbf{A} \in M_{m,n}(\mathbb{R})$, the row r and column s entries of $(\mathbf{A} * \mathbf{D})$ are given by

$$(\mathbf{A} * \mathbf{D})_{r,s} = a_{r,s} - a_{r-1,s}. \tag{4.9}$$

If we think of $\mathbf{A}$ as a sampled analog image $f(x, y)$ via $a_{r,s} = f(x_s, y_r)$ on a rectangular grid $x_s = s(\Delta x), y_r = r(\Delta y)$, then

$$a_{r,s} - a_{r-1,s} = f(x_s, y_r) - f(x_s, y_{r-1}) \approx (\Delta y)\frac{\partial f}{\partial y}(x_s, y_r),$$

where $\Delta y = y_r - y_{r-1}$. From the equation above, we conclude that $(\mathbf{A} * \mathbf{D})_{r,s}$ is approximately proportional to $\partial f(x_s, y_r)/\partial y$, and so $|(\mathbf{A} * \mathbf{D})_{r,s}|$ should be large if the image has an abrupt change in the y direction at this location. This corresponds to a horizontal edge in the image.

The mask $\mathbf{V}$ with components $v_{0,0} = 1, v_{0,1} = -1$, and all other $v_{r,s} = 0$ does the same for vertical edges (abrupt changes in $\partial f/\partial x$).

Figure 4.9a and b shows a grayscale image $\mathbf{A}$ and the filtered version $|\mathbf{A} * \mathbf{D}|$, respectively, with the absolute value taken component by component, to highlight horizontal edges. The image $|\mathbf{A} * \mathbf{D}|$ is displayed in an 8-bit "reverse" grayscale scheme, with 0 as white and 255 as black, which makes it a bit easier to see the edges (since most of the image does not consist of edges). In Figure 4.10, we show the similarly scaled image $|\mathbf{A} * \mathbf{V}|$ (top) as well as the quantity

$$\sqrt{|\mathbf{A} * \mathbf{V}|^2 + |\mathbf{A} * \mathbf{D}|^2}$$

with squaring and square roots performed component by component. The latter quantity effectively amalgamates the vertical and horizontal edge images into a single image that picks out all of the edges, and is a discrete approximation to $\|\nabla f\|$, at least if $\Delta x = \Delta y$ (see also Exercise 25).

4.5 Infinite and Bi-Infinite Signal Models

The one-dimensional discrete signals considered have so far consisted of vectors in $\mathbb{R}^N$ or $\mathbb{C}^N$, extended by periodicity when convenient. We may think of these vectors as having been obtained by sampling a function on some finite time interval $[0, T]$. In the same spirit, our two-dimensional discrete images are $m \times n$ matrices obtained by sampling a function of two variables on a rectangle.

But it will be useful, especially in later chapters, to consider signals that are infinite or bi-infinite in extent. For example, in Example 1.3, we considered an

Figure 4.9 Original and horizontal edge filtered image.

analog function $f(t)$ sampled at times $t_k = a + k(\Delta T)$ for $k = 0, 1, 2, \ldots$, to yield a sampled signal $\mathbf{x} = (x_0, x_1, x_2, \ldots)$. We might also consider sampling a function from the indefinite past to indefinite future, at $t_k = a + k(\Delta T)$ for all $k \in \mathbb{Z}$, to yield a sample vector $\mathbf{x} = (\ldots, x_{-1}, x_0, x_1, x_2, \ldots)$.

In this section, we discuss how the mathematics of transforms, convolution, and filtering extend to these settings. It very closely parallels the finite case.

(a)

(b)

Figure 4.10 Vertical edge-filtered (a) and composite edge image (b).

4.5.1 $L^2(\mathbb{N})$ and $L^2(\mathbb{Z})$

4.5.1.1 The Inner Product Space $L^2(\mathbb{N})$

The vector space $L^2(\mathbb{N})$ consists of elements $\mathbf{x} = (x_0, x_1, x_2, \dots)$ of complex numbers x_k such that

$$\sum_{k=0}^{\infty} |x_k|^2 < \infty$$

and was introduced in Example 1.5. Let us now verify that

$$(\mathbf{x},\mathbf{y}) = \sum_{k=0}^{\infty} x_k \overline{y_k} \tag{4.10}$$

defines an inner product on this vector space; the algebraic properties 1 through 3 of Definition 1.3 are straightforward verifications (see Exercise 27). The harder part is the verification that the sum on the right in equation (4.10) actually converges for any $\mathbf{x}$ and $\mathbf{y}$ in $L^2(\mathbb{N})$, which we now demonstrate.

The sum in equation (4.10) will converge if the real and imaginary parts on the right, $\sum_k \mathrm{Re}(x_k\overline{y_k})$ and $\sum_k \mathrm{Im}(x_k\overline{y_k})$, each converge. To show this, we use the fact that for any two complex numbers z and w, we have the inequalities $|w||z| \leq \frac{1}{2}(|w|^2 + |z|^2)$, as well as $|\mathrm{Re}(zw)| \leq |z||w|$ and $|\mathrm{Im}(zw)| \leq |z||w|$ (see Exercise 1). To show that the sum for the real part in equation (4.10) converges, note that for any $n > 0$,

$$\begin{aligned}\sum_{k=0}^{n} |\mathrm{Re}(x_k\overline{y_k})| &\leq \sum_{k=0}^{n} |x_k||\overline{y_k}| && (\text{use } |\mathrm{Re}(zw)| \leq |z||w|),\\ &\leq \frac{1}{2}\sum_{k=0}^{n}(|x_k|^2 + |y_k|^2) && \left(\text{use } |z||w| \leq \frac{(|w|^2+|z|^2)}{2}\right).\end{aligned}$$

If we take the limit of both sides above as $n \to \infty$, we find that

$$\lim_{n\to\infty}\sum_{k=0}^{n} |\mathrm{Re}(x_k\overline{y_k})| \leq \frac{1}{2}\lim_{n\to\infty}\sum_{k=0}^{n}(|x_k|^2 + |y_k|^2) < \infty, \tag{4.11}$$

since $\sum_{k=0}^{\infty}|x_k|^2$ and $\sum_{k=0}^{\infty}|y_k|^2$ both converge by assumption. From equation (4.11), we conclude that the real series $\sum_{k=0}^{\infty}\mathrm{Re}(x_k\overline{y_k})$ converges absolutely, and hence converges. A similar argument works for $\sum_{k=0}^{\infty}\mathrm{Im}(x_k\overline{y_k})$. The series defining the inner product $(\mathbf{x},\mathbf{y})$ in equation (4.10) converges to some complex number.

The space $L^2(\mathbb{N})$ is thus an inner product space with inner product defined by (4.10). The norm on $L^2(\mathbb{N})$ is defined by

$$\|\mathbf{x}\| = \sqrt{(\mathbf{x},\mathbf{x})} = \left(\sum_{k=0}^{\infty}|x_k|^2\right)^{1/2}.$$

4.5.1.2 The Inner Product Space $L^2(\mathbb{Z})$

The vector space $L^2(\mathbb{Z})$ consists of elements $\mathbf{x} = (\ldots, x_{-2}, x_{-1}, x_0, x_1, x_2, \ldots)$, where each $x_k \in \mathbb{C}$ and

$$\sum_{k=-\infty}^{\infty}|x_k|^2 < \infty.$$

This is an inner product space with the inner product

$$(\mathbf{x}, \mathbf{y}) = \sum_{k=-\infty}^{\infty} x_k \overline{y_k}.$$

The verification is almost identical to that for $L^2(\mathbb{N})$.

We can impose the additional requirement that the components x_k of vectors in $L^2(\mathbb{N})$ or $L^2(\mathbb{Z})$ be real numbers. The resulting spaces are still inner product spaces with the inner products as above, but of course no conjugation is needed.

In what follows, we will work for the most part in the space $L^2(\mathbb{Z})$, bi-infinite sequences. Results for $L^2(\mathbb{N})$ can be obtained by considering elements $\mathbf{x} \in L^2(\mathbb{Z})$ with $x_k = 0$ for $k < 0$.

4.5.2 Fourier Analysis in $L^2(\mathbb{Z})$ and $L^2(\mathbb{N})$

The DFT/IDFT pair allows us to decompose a vector in $\mathbb{C}^N$ into basic complex exponential waveforms. As we have seen, this facilitates a lot of useful analysis. We will now look at the analogue of the DFT in $L^2(\mathbb{Z})$.

4.5.2.1 The Discrete Time Fourier Transform in $L^2(\mathbb{Z})$

What follows in this paragraph is intended as "inspiration" for a reasonable definition for the Fourier transform in $L^2(\mathbb{Z})$, to be followed by the precise definition. In the finite-dimensional case we were able to write each component of a vector $\mathbf{x} \in \mathbb{C}^N$ as a sum

$$x_k = \frac{1}{N} \sum_{m=0}^{N-1} X_m e^{2\pi ikm/N}$$

for $0 \le k \le N-1$ and where $\mathbf{X}$ is the DFT of $\mathbf{x}$. The sum above is just the inverse DFT. Since $e^{2\pi ikm/N}$ is periodic in both k and m with period N, we may "shift" the indexes k and m and write

$$x_k = \frac{1}{N} \sum_{m=-N/2+1}^{N/2} X_m e^{2\pi ikm/N} \tag{4.12}$$

for $-N/2 + 1 \le k \le N/2$, where we have effectively used the periodic extensions $x_k = x_{k \bmod N}$ and $X_m = X_{m \bmod N}$. Now the sum in equation (4.12) is precisely a right Riemann sum appropriate to integrating a function $X(f)e^{2\pi ifk}$ on the interval $-\frac{1}{2} \le f \le \frac{1}{2}$ where $X_m = X(m/N)$; we use "f" for the integrand variable since X should have frequency as its independent variable. Taking the limit as $N \to \infty$ in (4.12) strongly suggests that we should try to write

$$x_k = \int_{-1/2}^{1/2} X(f) e^{2\pi ikf}\, df \tag{4.13}$$

for an appropriate function $X(f)$. Presumably the function $X(f)$ would be computed linearly from the samples x_k with a summation in k.

We have already done this computation in Section 1.10.4! In particular, equations (1.47) and (1.48), which we reproduce here for convenience in the case $T = \frac{1}{2}$, state that for a continuous function h defined on the interval $[-\frac{1}{2}, \frac{1}{2}]$, we have

$$h(f) = \sum_{k=-\infty}^{\infty} x_k e^{2\pi ikf}, \tag{4.14}$$

where

$$x_k = \int_{-1/2}^{1/2} h(f)e^{-2\pi ikf}\, df, \tag{4.15}$$

and we have used f for the independent variable instead of t. Compare equations (4.13) and (4.15)—they are quite similar. If we define $X(f) = h(-f)$ (so $h(f) = X(-f)$), then equation (4.14) defines the transform we want, and equation (4.15) (after a compensating change of variable $f \to -f$) defines a corresponding inverse transform.

Definition 4.4 For an element $\mathbf{x} \in L^2(\mathbb{Z})$, the "discrete time Fourier transform" (DTFT) is the function $X(f)$ on the interval $-\frac{1}{2} \leq f \leq \frac{1}{2}$ defined by

$$X(f) = \sum_{k=-\infty}^{\infty} x_k e^{-2\pi ikf}. \tag{4.16}$$

The "inverse discrete time Fourier transform" (IDTFT) is defined by

$$x_k = \int_{-1/2}^{1/2} X(f)e^{2\pi ikf}\, df. \tag{4.17}$$

Equation (4.17) shows how to synthesize the sequence $\mathbf{x} \in L^2(\mathbb{Z})$ as an integral superposition of basic waveforms $e^{2\pi ikf}$. Equation (4.16) indicates how much $X(f)$ of the waveform $e^{2\pi ikf}$ is required. Here f plays the role of a frequency variable while the index k is essentially a time variable. The time domain in this case is $\mathbb{Z}$ and the frequency domain is the real interval $[-\frac{1}{2}, \frac{1}{2}]$. In essence, the DTFT/IDTFT pair is just Fourier series turned around, with the roles of the time and frequency domains swapped, and the forward and inverse transforms swapped.

4.5.2.2 Aliasing and the Nyquist Frequency in $L^2(\mathbb{Z})$

We might think of the vector $\mathbf{x}$ in Definition 4.4 as having been obtained by sampling an analog signal $x(t)$ as $x_k = x(k)$, a sampling rate of 1 Hz. As such the Nyquist frequency is $\frac{1}{2}$ Hz, which is why the domain of the DTFT $X(f)$ is $-\frac{1}{2} \leq f \leq \frac{1}{2}$. However, just as in the finite-dimensional case, frequencies outside this range will be aliased when sampled.

The forward/inverse transforms of Definition 4.4 are valid for any $\mathbf{x} \in L^2(\mathbb{Z})$, of course, even if $\mathbf{x}$ was obtained by sampling at a different rate, say $x_k = x(kT)$, (sampling interval T, sampling frequency $F = 1/T$,Hz, Nyquist rate $1/2T$). However, in this case, it is sometimes helpful to rescale the transforms in (4.16) and (4.17) to reflect the different sampling rate. Specifically, if we sample at $F = 1/T$ Hz then we would expect the DTFT to be defined on the range $-1/2T \leq f \leq 1/2T$. The formulas (4.16) and (4.17) can be adapted to reflect this via a simple change of variable. Let $\tilde{f} = f/T$ (so $f = T\tilde{f}$ and $df = T\,d\tilde{f}$) in the integral on the right in (4.17) to obtain

$$x_k = \int_{-1/2T}^{1/2T} X(\tilde{f}T)e^{2\pi ik\tilde{f}T}\,T\,d\tilde{f}, \tag{4.18}$$

where the function X is still defined by (4.16). From (4.16), we have

$$X(\tilde{f}T) = \sum_{k=-\infty}^{\infty} x_k e^{-2\pi ik\tilde{f}T}. \tag{4.19}$$

Define a rescaled DTFT $\tilde{X}(\tilde{f}) := X(\tilde{f}T)$. Then from (4.18), we have

$$x_k = \int_{-1/2T}^{1/2T} \tilde{X}(\tilde{f})e^{2\pi ik\tilde{f}T}\,T\,d\tilde{f}, \tag{4.20}$$

where from (4.19), $\tilde{X}(\tilde{f})$ is defined as

$$\tilde{X}(\tilde{f}) = \sum_{k=-\infty}^{\infty} x_k e^{-2\pi ik\tilde{f}T}. \tag{4.21}$$

Equations (4.20) and (4.21) comprise a rescaled version of the forward/inverse transforms (4.16) and (4.17). Equation (4.20) shows how to synthesize the signal $x(t)$ at $t = kT$ from the waveforms $e^{2\pi i\tilde{f}t}$ sampled at $t = kT$, using frequencies $\tilde{f} \in [-1/2T, 1/2T]$ (the Nyquist range).

In general, however, we will not worry about how $\mathbf{x}$ was sampled, or whether it was even obtained via sampling. We will simply use formulas (4.16) and (4.17).

Remark 4.3 If the number of nonzero components in $\mathbf{x}$ is finite, then the sum in (4.16) is finite and defines a continuous (indeed, infinitely differentiable) function $X(f)$. However, if $\mathbf{x}$ is a general vector in $L^2(\mathbb{Z})$, then the convergence of the sum (4.16) is slightly more problematic, and precisely the same analytic issues are raised as those in Section 1.10.5 with regard to Fourier series. It can be shown nevertheless that for any $\mathbf{x} \in L^2(\mathbb{Z})$ the sum (4.16) converges to a function in the space $L^2(-\frac{1}{2}, \frac{1}{2})$ as defined in Example 1.8, that is, to a function that satisfies $\int_{-1/2}^{1/2} |X(f)|^2\,df < \infty$. The function $X(f)$ does not need to be continuous in this case, and indeed not even Riemann integrable.

4.5.2.3 The Fourier Transform on $L^2(\mathbb{N})$

We can consider an element $\mathbf{x} \in L^2(\mathbb{N})$ as an element of $L^2(\mathbb{Z})$ by extending the vector $\mathbf{x}$ as $x_k = 0$ for all $k < 0$. In this case, we obtain a transform/inverse transform pair from equations (4.16) and (4.17), which become

$$X(f) = \sum_{k=0}^{\infty} x_k e^{-2\pi i k f}$$

and

$$x_k = \int_{-1/2}^{1/2} X(f) e^{2\pi i k f}\, df,$$

for $k \geq 0$.

4.5.3 Convolution and Filtering in $L^2(\mathbb{Z})$ and $L^2(\mathbb{N})$

We now discuss convolution and filtering for infinite and bi-infinite sequences, a common conceptual operation in digital signal processing. There are two-dimensional analogs as well, which we discuss briefly at the end of this section.

The convolution of two elements $\mathbf{x}$ and $\mathbf{y}$ in $L^2(\mathbb{Z})$ is defined by

$$(\mathbf{x} * \mathbf{y})_m = \sum_{k=-\infty}^{\infty} x_k y_{m-k}, \tag{4.22}$$

where $m \in \mathbb{Z}$ and $(\mathbf{x} * \mathbf{y})_m$ denotes mth component of the convolution. As with circular convolution in $\mathbb{C}^N$, a diagram is helpful in understanding equation (4.22). When $m = 0$, the sum may be computed as the "dot product" of the infinite row vectors

$$\begin{array}{ccccccc} \cdots & x_{-2} & x_{-1} & x_0 & x_1 & x_2 & \cdots \\ \cdots & y_2 & y_1 & y_0 & y_{-1} & y_{-2} & \cdots \end{array}$$

For $m \neq 0$, the quantity $(\mathbf{x} * \mathbf{y})_m$ is computed as the dot product of

$$\begin{array}{ccccccccc} \cdots & x_{-2} & x_{-1} & x_0 & x_1 & x_2 & \cdots & x_m & \cdots \\ \cdots & y_{m+2} & y_{m+1} & y_m & y_{m-1} & y_{m-2} & \cdots & y_0 & \cdots \end{array}$$

in which the bottom row above has been shifted m steps to the right (a left shift if $m < 0$).

Convergence of the infinite sum in (4.22) is not obvious, but in fact the sum converges absolutely (and hence converges), since

$$\sum_{k=-\infty}^{\infty} |x_k||y_{m-k}| \leq \frac{1}{2} \sum_{k=-\infty}^{\infty} (|x_k|^2 + |y_{m-k}|^2) = \frac{1}{2}(\|\mathbf{x}\|^2 + \|\mathbf{y}\|^2) < \infty$$

and $\mathbf{x}$ and $\mathbf{y}$ are in $L^2(\mathbb{Z})$. Note that $\sum_{k=-\infty}^{\infty} |y_{m-k}|^2 = \sum_{k=-\infty}^{\infty} |y_k|^2$ for any m.

Remark 4.4 The argument just given demonstrates that if $\mathbf{w} = \mathbf{x} * \mathbf{y}$ then $|w_m| \leq \frac{1}{2}(\|\mathbf{x}\|^2 + \|\mathbf{y}\|^2)$ for all $m \in \mathbb{Z}$. We conclude that the components w_m all satisfy a common bound, so $\mathbf{w}$ is in the vector space $L^\infty(\mathbb{Z})$ defined in Example 1.4. However, it need not be the case that $\mathbf{w}$ itself lies in $L^2(\mathbb{Z})$, so convolution is not well-defined as an operation from $L^2(\mathbb{Z}) \times L^2(\mathbb{Z})$ to $L^2(\mathbb{Z})$ (see Exercise 31). Under some circumstances, however, we can conclude that $\mathbf{w} \in L^2(\mathbb{Z})$, for example, if either $\mathbf{x}$ or $\mathbf{y}$ has only finitely many nonzero components (see Exercise 30).

To convolve vectors $\mathbf{x}$ and $\mathbf{y}$ in $L^2(\mathbb{N})$, just treat them as vectors in $L^2(\mathbb{Z})$ with all components equal to zero for negative indexes. In this case, convolution can be written

$$(\mathbf{x} * \mathbf{y})_m = \sum_{k=0}^{m} x_k y_{m-k},$$

since $x_k = 0$ for $k < 0$ and $y_{m-k} = 0$ for $k > m$.

Convolution in $L^2(\mathbb{N})$ has the following properties.

Proposition 4.1 *Let* $\mathbf{x}, \mathbf{y}$, *and* $\mathbf{w}$ *all be elements in either* $L^2(\mathbb{Z})$ *or* $L^2(\mathbb{N})$. *Then*

- *Commutativity:* $\mathbf{x} * \mathbf{y} = \mathbf{y} * \mathbf{x}$.
- *Associativity:* $\mathbf{x} * (\mathbf{y} * \mathbf{w}) = (\mathbf{x} * \mathbf{y}) * \mathbf{w}$.
- *Linearity:* $\mathbf{x} * (a\mathbf{y} + b\mathbf{w}) = a\mathbf{x} * \mathbf{y} + b\mathbf{x} * \mathbf{w}$ *for any scalars* a *and* b.

The proof of each point is a straightforward manipulation of the relevant summations.

4.5.3.1 The Convolution Theorem

Convolution in $\mathbb{C}^N$ became point-by-point multiplication in the frequency domain. The same thing happens for $L^2(\mathbb{Z})$.

Theorem 4.5 *Let* $\mathbf{x}$ *and* $\mathbf{y}$ *be elements of* $L^2(\mathbb{Z})$ *with DTFT* $X(f)$ *and* $Y(f)$, *respectively. If* $\mathbf{w} = \mathbf{x} * \mathbf{y}$, *then the DTFT* $W(f)$ *of* $\mathbf{w}$ *is*

$$W(f) = X(f)Y(f).$$

The equation $W(f) = X(f)Y(f)$ is just the frequency-by-frequency product of the transforms, analogous to $W_k = X_k Y_k$ for the DFT.

Proof: The proof of Theorem 4.5 in the most general context requires a bit of analysis. We will content ourselves with a proof for the case of primary interest, in which $\mathbf{x}$ and $\mathbf{y}$ contain only finitely many nonzero components. We thus avoid some delicate issues concerning the convergence of the relevant sums and integrals.

Let us write the DTFT of the vectors **x** and **y** in the form

$$X(f) = \sum_{k=-\infty}^{\infty} x_k z^{-k},$$

$$Y(f) = \sum_{m=-\infty}^{\infty} y_m z^{-m},$$

for $-\frac{1}{2} \leq f \leq \frac{1}{2}$, where $z = e^{2\pi i f}$ (note z depends on the frequency f, but we will not explicitly indicate this). Since **x** and **y** are assumed to have only finitely many nonzero components, the sums above involve only finitely many terms and the convergence of any sum is not in question. The product $X(f)Y(f)$ is given by

$$X(f)Y(f) = \sum_{k=-\infty}^{\infty} \sum_{m=-\infty}^{\infty} x_k y_m z^{-k-m},$$

which also involves only finitely many nonzero terms. We collect together like powers of z by setting $s = k + m$ (so $m = s - k$) and performing a change of summation index in the inner sum, to obtain

$$X(f)Y(f) = \sum_{k=-\infty}^{\infty} \sum_{s=-\infty}^{\infty} x_k y_{s-k} z^{-s} = \sum_{s=-\infty}^{\infty} \left(\sum_{k=-\infty}^{\infty} x_k y_{s-k} \right) z^{-s}, \tag{4.23}$$

where we also interchange the summation order. Since $\mathbf{w} = \mathbf{x} * \mathbf{y}$, the parenthesized sum on the right in equation (4.23) is exactly w_s (recall equation (4.22)). Equation (4.23) yields

$$X(f)Y(f) = \sum_{k=-\infty}^{\infty} w_s z^{-s},$$

and the right side is precisely $W(f)$.

Example 4.5 *Consider the vector $\ell \in L^2(\mathbb{Z})$ with components*

$$\ell_0 = \frac{1}{2}, \quad \ell_1 = \frac{1}{2}, \quad \ell_k = 0, \textit{otherwise}.$$

In Example 4.3, it was shown that the finite length ($\mathbb{C}^N$) version of ℓ is a crude low-pass filter. The same holds true in $L^2(\mathbb{Z})$.

To see this, consider the convolution $\mathbf{y} = \ell * \mathbf{x}$ for a general $\mathbf{x} \in L^2(\mathbb{Z})$. In the frequency domain, we have

$$Y(f) = L(f)X(f).$$

By Definition 4.4, $L(f)$ is given by

$$L(f) = \sum_{k=-\infty}^{\infty} \ell_k e^{-2\pi i k f} = \frac{1}{2}(1 + e^{-2\pi i f}).$$

The magnitude of $L(f)$ is $|L(f)| = \cos(\pi f)$ on the Nyquist range $-\frac{1}{2} \leq f \leq \frac{1}{2}$ (this follows from $L(f) = e^{-\pi if}\frac{1}{2}(e^{\pi if} + e^{-\pi if}) = e^{-\pi if}\cos(\pi f)$). We have $L(0) = 1$. Therefore, the dc component of $\mathbf{x}$ *passes unchanged, $L(f)$ tapers to zero at $f = \pm\frac{1}{2}$, and frequencies in* $\mathbf{x}$ *near the Nyquist rate are attenuated to zero.*

4.5.4 The *z*-Transform

We can remove the restriction that $z = e^{2\pi if}$ in the proof of Theorem 4.5 and consider $X(z)$ as a function of a "general" variable z. The resulting expression, a type of infinite series

$$X(z) = \sum_{k=-\infty}^{\infty} x_k z^{-k}, \tag{4.24}$$

is called the *bilateral z-transform* of the sequence $\mathbf{x}$, though we will usually leave off the adjective "bilateral." The *unilateral* z-transform is applied to sequences $\mathbf{x} = (x_0, x_1, \ldots)$ and will not concern us. The "z" in "z-transform" may or may not be capitalized, depending on the author/text.

4.5.4.1 Two Points of View

But this definition of the z-transform is a bit ambiguous—what can z be? There are two distinct ways we can approach the z-transform:

1. We can treat the z-transform *formally*, by considering z as an indeterminate "placeholder." We do not think about plugging in real or complex numbers for z and then summing the resulting series. In this approach, we can write down the z-transform of any bi-infinite sequence.
2. We can treat the z-transform *analytically*, by considering $X(z)$ as a function of a complex variable z. In this case, we must concern ourselves with the convergence of the infinite series (e.g., see Exercise 34). On the bright side, it means we can apply powerful tools from complex analysis to z-transforms.

Example 4.6 *Let* $\mathbf{x}$ *be the sequence in $L^2(\mathbb{Z})$ with components $x_{-1} = 2, x_0 = 1, x_1 = 3, x_2 = -2$, and all other components equal to zero. The z-transform of* $\mathbf{x}$ *is*

$$X(z) = 2z + 1 + 3z^{-1} - 2z^{-2}.$$

We can consider $X(z)$ formally, as a Laurent polynomial *in z (a finite sum of positive and negative powers of z). Or, we can treat $X(z)$ analytically, by considering it as a function defined on the complex plane. In this case, $X(z)$ is well-defined for any $z \neq 0$.*

Example 4.7 *Let* $\mathbf{x}$ *be the sequence in $L^2(\mathbb{Z})$ with components $x_0 = 0$ and $x_k = 1/|k|$ for $k \neq 0$. In this case, the z-transform is*

$$X(z) = \cdots + \frac{1}{2}z^2 + z + 0 + z^{-1} + \frac{1}{2}z^{-2} + \cdots .$$

Formally, $X(z)$ is merely another way to write $\mathbf{x}$*. Analytically $X(z)$ is problematic. A simple ratio test shows that the sum of the positive powers of z diverges for $|z| > 1$, while the series consisting of the negative powers diverges for $|z| < 1$. As a function of a complex variable $X(z)$ is defined—at best—only on the unit circle $|z| = 1$. This greatly limits what we can do with complex analysis.*

Example 4.8 *Let* $\mathbf{x}$ *be the sequence in $L^2(\mathbb{Z})$ with $x_k = 1/2^k$ for $k \geq 0$, $x_k = 0$ for $k < 0$. The z-transform of* $\mathbf{x}$ *is, formally,*

$$X(z) = \sum_{k=0}^{\infty} \frac{1}{2^k} z^{-k} = 1 + \frac{1}{2}z^{-1} + \frac{1}{4}z^{-2} + \cdots .$$

If we consider $X(z)$ as a function of a complex variable z, then a ratio test shows that the series defined by $X(z)$ converges for $|z| > \frac{1}{2}$ and diverges for $|z| < \frac{1}{2}$. In the region of convergence, we find $X(z) = 1/(1 - (2z)^{-1})$ (it is a simple geometric sum).

Now consider the sequence $\mathbf{y}$ *with components $y_k = 0$ for $k \geq 0$ and $y_k = -2^{-k}$ for $k < 0$ (the components y_k grow as k decreases, so this vector is not in $L^2(\mathbb{Z})$, but let us press on). Then*

$$Y(z) = -\sum_{k=-\infty}^{-1} \frac{1}{2^k} z^{-k} = \cdots - 8z^3 - 4z^2 - 2z.$$

From the formal perspective, $X(z)$ and $Y(z)$ are different. Considered as a function of z, however, it is not hard to show (see Exercise 34) that $Y(z) = 1/(1 - z^{-1}/2)$ for $|z| < \frac{1}{2}$ and diverges for $|z| > \frac{1}{2}$. Superficially, it appears that from the analytical point of view we have $X(z) = Y(z)$, but this is not correct! A function always comes with a domain of definition. Since $X(z)$ and $Y(z)$ do not have the same domain of definition, they are distinct functions. Without a domain of definition, $1/(1 - z^{-1}/2)$ is just an expression, not a function.

In this text, we will take the formal point of view with regard to z-transforms, for the simple reason that this is sufficient for all of our needs. In particular, most of the signals that we z-transform, though ostensibly in $L^2(\mathbb{Z})$, have only finitely many nonzero components. An analytic treatment will not be necessary.

4.5.4.2 Algebra of z-Transforms; Convolution

The z-transform is a useful tool in signal processing. As with the other transforms that we have studied, many operations in the time domain have simple parallels in the z-frequency domain. In particular, it is easy to see that the z-transform is linear: if $\mathbf{x}$ and $\mathbf{y}$ are any bi-infinite sequences with z-transforms $X(z)$ and $Y(z)$, respectively, and $\mathbf{w} = c\mathbf{x} + d\mathbf{y}$, then

$$W(z) = cX(z) + dY(z).$$

Note also that the proof of Theorem 4.5 works perfectly well for "any" z, so we have already shown.

Theorem 4.6 *Let* $\mathbf{x}$ *and* $\mathbf{y}$ *be elements of* $L^2(\mathbb{Z})$ *with z-transforms* $X(z)$ *and* $Y(z)$, *respectively. Let* $\mathbf{w} = \mathbf{x} * \mathbf{y}$ *(note* $\mathbf{w} \in L^\infty(\mathbb{Z})$*). The z-transform* $W(z)$ *of* $\mathbf{w}$ *is*

$$W(z) = X(z)Y(z).$$

Theorem 4.5 becomes a special case of Theorem 4.6 where $z = e^{2\pi i f}$.

From the formal perspective, the z-transform thus provides a convenient way to do algebra—especially addition and convolution—for signals in $L^2(\mathbb{Z})$.

4.5.5 Convolution in $\mathbb{C}^N$ versus $L^2(\mathbb{Z})$

If $\mathbf{x}$ and $\mathbf{y}$ are elements of $\mathbb{C}^N$, then we can also consider each as an element of $L^2(\mathbb{Z})$ by extending each with zeros. However, it is not generally the case that the convolution of $\mathbf{x}$ and $\mathbf{y}$ in $L^2(\mathbb{Z})$ will equal the circular convolution of $\mathbf{x}$ and $\mathbf{y}$ in $\mathbb{C}^N$, even if we interpret indexes modulo N.

Example 4.9 *Let* $\mathbf{x} = (x_0, x_1, x_2)$ *and* $\mathbf{y} = (y_0, y_1, y_2)$ *be vectors in* $\mathbb{C}^3$, *where we also consider*

$$\mathbf{x} = (\dots, 0, x_0, x_1, x_2, 0, \dots) \quad and \quad \mathbf{y} = (\dots, 0, y_0, y_1, y_2, 0, \dots)$$

as elements of $L^2(\mathbb{Z})$. *The circular convolution* $\mathbf{w} = \mathbf{x} * \mathbf{y}$ *in* $\mathbb{C}^3$ *has components*

$$w_0 = x_0y_0 + x_1y_2 + x_2y_1, \quad w_1 = x_0y_1 + x_1y_0 + x_2y_2,$$
$$w_2 = x_0y_2 + x_1y_1 + x_2y_0.$$

But the vector $\tilde{\mathbf{w}} = \mathbf{x} * \mathbf{y} \in L^2(\mathbb{Z})$ *has five nonzero components given by*

$$\tilde{w}_0 = x_0y_0, \quad \tilde{w}_1 = x_0y_1 + x_1y_0, \quad \tilde{w}_2 = x_0y_2 + x_1y_1 + x_2y_0,$$
$$\tilde{w}_3 = x_1y_2 + x_2y_1, \quad \tilde{w}_4 = x_2y_2.$$

There is a simple relation between $\mathbf{w}$ and $\tilde{\mathbf{w}}$, however, and this relation is most elegantly understood in terms of the z-transform. The following ideas and notation will also be very useful in Chapter 7.

4.5.5.1 Some Notation

Consider a Laurent polynomial $p(z)$ of the form

$$p(z) = \sum_{k=-m}^{n} a_k z^k$$

consisting of (possibly) negative and positive powers of z. We will use the notation "$p(z)$ mod z^{-N}" with $N > 0$ to denote that unique polynomial in the variable z^{-1} obtained by identifying in $p(z)$ all powers of z that are congruent modulo N. More formally,

$$p(z) \bmod z^{-N} := \sum_{j=0}^{N-1} \left(\sum_{k \equiv j \bmod N} a_k \right) z^{-j}.$$

Thus $p(z)$ mod z^{-N} is a sum of powers $z^{-(N-1)}, \dots, z^{-2}, z^{-1}, z^0$.

Example 4.10 *Let*

$$p(z) = 2z^{-2} - z^{-1} + 3 + \frac{1}{2}z + z^2 + 5z^3.$$

Then

$$p(z) \bmod z^{-3} = \left(2 + \frac{1}{2}\right) z^{-2} + (-1+1)z^{-1} + (3+5)z^0 = \frac{5}{2}z^{-2} + 8,$$

since we identify the powers $z^{-2} \equiv z^1, z^{-1} \equiv z^2$, *and* $z^0 \equiv z^3$.

4.5.5.2 Circular Convolution and z-Transforms

To relate convolution in $\mathbb{C}^N$ and $L^2(\mathbb{Z})$, let $\mathbf{x}$ and $\mathbf{y}$ be vectors in $\mathbb{C}^N$, and define $X(z)$ and $Y(z)$ as

$$X(z) = \sum_{k=0}^{N-1} x_k z^{-k}, \quad Y(z) = \sum_{k=0}^{N-1} y_k z^{-k},$$

polynomials in z^{-1}. These are merely the z-transforms of $\mathbf{x}$ and $\mathbf{y}$, respectively, if we consider $\mathbf{x}$ and $\mathbf{y}$ as elements of $L^2(\mathbb{Z})$ via extension by zero. Let $\mathbf{w} = \mathbf{x} * \mathbf{y}$ (circular convolution in $\mathbb{C}^N$), and let

$$W(z) = \sum_{k=0}^{N-1} w_k z^{-k}.$$

Theorem 4.7 *If* $\mathbf{x}, \mathbf{y} \in \mathbb{C}^N$ *and* $\mathbf{w} = \mathbf{x} * \mathbf{y}$ *(circular convolution in* $\mathbb{C}^N$*), then*

$$W(z) = X(z)Y(z) \bmod z^{-N}.$$

Proof: The proof is easy enough to write out directly and is similar to the proof of Theorem 4.2 or 4.5. However, a slightly shorter proof is obtained by making use of Theorem 4.2.

For each integer k in the range $0 \le k \le N-1$ let $\omega_k = e^{2\pi ik/N}$, an Nth *root of unity*; we will suppress the dependence of ω_k on N. Let $\mathbf{X}$, $\mathbf{Y}$, and $\mathbf{W}$ denote the DFT's of $\mathbf{x}, \mathbf{y}, \mathbf{w}$, respectively. From the definition of the DFT, we have $X_k =$

$X(\omega_k)$, $Y_k = Y(\omega_k)$, and $W_k = W(\omega_k)$. From Theorem 4.2, we have $W_k = X_k Y_k$, or

$$W(\omega_k) = X(\omega_k)Y(\omega_k) \tag{4.25}$$

for $0 \le k \le N-1$. Now define the function $P(z)$ as

$$P(z) := X(z)Y(z) \bmod z^{-N} \tag{4.26}$$

(so $P(z)$ consists of powers $z^{-(N-1)}, \dots, z^{-2}, z^{-1}, z^0$, while the full product $X(z)Y(z)$ consists of powers $z^{-2(N-1)}, \dots, z^{-2}, z^{-1}, z^0$). Then

$$X(\omega_k)Y(\omega_k) = P(\omega_k) \tag{4.27}$$

for $0 \le k \le N-1$, since $\omega_k^m = \omega_k^n$ whenever $m \equiv n \bmod N$. We conclude from (4.25) and (4.27) that

$$W(\omega_k) = P(\omega_k) \tag{4.28}$$

for $0 \le k \le N-1$.

From equation (4.28), the polynomials $z^{N-1}W(z)$ and $z^{N-1}P(z)$ of degree $N-1$ agree at N distinct points $z = \omega_k$, $0 \le k \le N-1$. Equivalently, the polynomial $z^{N-1}(W(z) - P(z))$ of degree $N-1$ has N distinct roots. Since a nonzero nth degree polynomial has at most n complex roots, this forces $z^{N-1}(W(z) - P(z))$ to be the zero polynomial, or $W(z) = P(z)$. The theorem follows from equation (4.26).

4.5.5.3 Convolution in $\mathbb{C}^N$ from Convolution in $L^2(\mathbb{Z})$

Theorem 4.7 makes it easy to see how convolution in $\mathbb{C}^N$ and $L^2(\mathbb{Z})$ are related. Let $\mathbf{x}$ and $\mathbf{y}$ be vectors in $\mathbb{C}^N$ (indexed from 0) and $\mathbf{w} = \mathbf{x} * \mathbf{y} \in \mathbb{C}^N$, circular convolution. Let $\tilde{\mathbf{w}} = \mathbf{x} * \mathbf{y}$, the convolution in $L^2(\mathbb{Z})$ (so $\tilde{\mathbf{w}} \in L^\infty(\mathbb{Z})$). We already know by Theorem 4.5 that $\tilde{w}_m$ is the coefficient of z^{-m} in the product $X(z)Y(z)$, consisting of powers $z^{-2(N-1)}, \dots, z^{-1}, z^0$. By Theorem 4.7, we know that w_m is the coefficient of z^{-m} in $X(z)Y(z) \bmod z^{-N}, 0 \le m \le N-1$. Now the coefficient of z^{-m} in $X(z)Y(z) \bmod z^{-N}$ comes from adding the coefficients of the powers z^{-m} and z^{-m-N} in the product $X(z)Y(z)$. These coefficients are just $\tilde{w}_m$ and $\tilde{w}_{m+N}$, and we conclude that

$$w_m = \tilde{w}_m + \tilde{w}_{m+N} \tag{4.29}$$

for $0 \le m \le N-1$. Equation (4.29) shows how to convert convolution in $L^2(\mathbb{Z})$ into circular convolution in $\mathbb{C}^N$.

Example 4.11 *Let $\mathbf{x}$ and $\mathbf{y}$ be as in Example 4.9. From equation (4.29), we expect $w_0 = \tilde{w}_0 + \tilde{w}_3$, $w_1 = \tilde{w}_1 + \tilde{w}_4$, and $w_2 = \tilde{w}_2 + \tilde{w}_5$. These are easily seen to be true (note $\tilde{w}_5 = 0$).*

Remark 4.5 The convolution command "conv" in Matlab is not circular convolution, although it acts on finite vectors. The conv command produces the convolution of vectors $\mathbf{x}$ and $\mathbf{y}$ in $\mathbb{C}^N$ by considering them as elements of $L^2(\mathbb{Z})$ via zero-extension. We thus need to construct a circular convolution command "manually." The simplest approach is to use the built-in conv command and then to employ equation (4.29). The Signal Processing Toolbox in Matlab also contains a command cconv that performs circular convolution.

We may also circularly convolve vectors $\mathbf{x} \in \mathbb{C}^M$ and $\mathbf{y} \in \mathbb{C}^N$ for $M \neq N$ by "padding" the shorter vector with zeros, and then performing a circular convolution in $\mathbb{C}^{\max(M,N)}$.

4.5.6 Some Filter Terminology

Let $\mathbf{x} = (\dots, x_{-1}, x_0, x_1, \dots)$ be a bi-infinite signal and $\mathbf{h}$ another bi-infinite vector such that

$$h_k = 0 \quad \text{for } k < L$$

and

$$h_k = 0 \quad \text{for } k > M$$

for some integers L, M. We will consider $\mathbf{h}$ to be a filter. The vector $\mathbf{x} * \mathbf{h}$ (which equals $\mathbf{h} * \mathbf{x}$) has components given by

$$(\mathbf{x} * \mathbf{h})_m = \sum_{k=L}^{M} h_k x_{m-k},$$

so the value of $(\mathbf{x} * \mathbf{h})_m$ only depends on x_k for $m - M \leq k \leq m - L$. Only finitely many values of x_k are needed to compute $(\mathbf{x} * \mathbf{h})_m$, and since the sum is finite, there are no convergence issues.

We say that $\mathbf{h}$ as above is a *finite impulse response* (FIR) filter. If $L \geq 0$, then the computation of $(\mathbf{x} * \mathbf{h})_m$ involves only values of x_k for $k \leq m$ (no "future" values, $k > m$). We then say that the filter $\mathbf{h}$ is causal. If, on the other hand, we have $M \leq 0$, then

$$h_k = 0 \quad \text{for } k > 0$$

and $\mathbf{h}$ is an *acausal* filter; the computation of $(\mathbf{x} * \mathbf{h})_m$ depends on future times. If

$$h_k = h_{-k},$$

then $\mathbf{h}$ is a *symmetric filter*, and it is neither causal nor acausal (unless h_0 is the only nonzero entry). In the real-time filtering of data, the use of causal filters is obviously preferred because future samples are not available. For images, all the data are known at one time and the concept of causality is generally irrelevant. In this setting, many filters are in fact symmetric.

4.5.7 The Space $L^2(\mathbb{Z} \times \mathbb{Z})$

Although we will not actually need them, the results of Section 4.5 extend in a straightforward way to the vector space $L^2(\mathbb{Z} \times \mathbb{Z})$ consisting of "arrays" **A** with real or complex components $a_{r,s}$ for $-\infty < r, s < \infty$ such that

$$\sum_{r,s=-\infty}^{\infty} |a_{r,s}|^2 < \infty.$$

This set has an obvious vector space structure and inner product

$$(\mathbf{A}, \mathbf{B}) = \sum_{r,s=-\infty}^{\infty} a_{r,s}\overline{b_{r,s}}.$$

The Fourier transform of **A** is the function $\hat{A}(f_1, f_2)$ defined by

$$\hat{A}(f_1, f_2) = \sum_{r,s=-\infty}^{\infty} a_{r,s} e^{-2\pi i(rf_1 + sf_2)}$$

for $-\frac{1}{2} \leq f_1, f_2 \leq \frac{1}{2}$. The inverse transform is

$$a_{r,s} = \int_{-1/2}^{1/2} \int_{-1/2}^{1/2} \hat{A}(f_1, f_2) e^{2\pi i(rf_1 + sf_2)}\, df_1\, df_2.$$

We define convolution (which is commutative) as

$$(\mathbf{A} * \mathbf{B})_{r,s} = \sum_{k=-\infty}^{\infty} \sum_{m=-\infty}^{\infty} a_{k,m} b_{r-k,s-m}. \tag{4.30}$$

If $\mathbf{C} = \mathbf{A} * \mathbf{B}$, then

$$\hat{C}(f_1, f_2) = \hat{A}(f_1, f_2)\hat{B}(f_1, f_2), \tag{4.31}$$

where $\hat{C}$ denotes the transform of **C**.

The proof of these facts is analogous to the one-dimensional case.

4.6 Matlab Project

4.6.1 Basic Convolution and Filtering

1. Use Matlab to construct a one-second audio signal of the form

$$y(t) = 0.5\sin(2\pi(200t)) + 0.2\sin(2\pi(455t)) - 0.3\cos(2\pi(672t))$$

sampled at 8192 Hz with the commands

```
t= [0:8191]/8192;
y=0.5*sin(2*pi*200*t) + 0.2*sin(2*pi*455*t)
  -0.3*cos(2*pi*672*t);
```

Play the signal with `sound(y)`. Because of the choice of 8192 for the length of y, the coefficient Y_k (for $0 \le k \le 4096$) in the DFT **Y** corresponds to exactly k hertz. However, since Matlab indexes vectors starting with index 1, Y_k corresponds to `Y(k+1)` in Matlab.

2. Design a high-pass filter **h** of length 8192 that, when convolved with another vector $\mathbf{y} \in \mathbb{C}^{8192}$, filters out all DFT components Y_k with $0 \le k \le 500$ and the corresponding conjugate coefficients, while leaving all other frequencies unchanged. To do this, recall that if $\mathbf{w} = \mathbf{y} * \mathbf{h}$, then $W_k = Y_k H_k$, where **H** is the DFT of **h**. Thus we want $H_k = 0$ for $0 \le k \le 500$ and $7692 \le k \le 8191$ (the conjugate frequencies for $1 \le k \le 500$), as well as $H_k = 1.0$ for $501 \le k \le 7691$. Accounting for the fact that Matlab indexes from 1, we can construct **H** with

```
H = zeros(1,8192);
  H(502:7692)=1.0;
```

The (real-valued) filter itself is obtained as

```
h = real(ifft(H));
```

We need to take the real part, since `ifft` will likely leave some vestigial imaginary part due to round-off.
Plot the filter vector **h** with `plot(h)`. Most of the coefficients should be very close to zero. It might be more illuminating to plot only `h(1:100)`. Count the number of nonzero coefficients with `sum(abs(h)>0)`; almost all coefficients should be nonzero.

3. Filter **y** as $\mathbf{w} = \mathbf{y} * \mathbf{h}$. You can use the supplied routine `circconv` as

```
w = circconv(y,h);
```

(The same effect can be achieved in the frequency domain by zeroing out all appropriate Y_k.) Play the audio signal w. It should consist of a pure 672 Hz tone.
Note that since **h** has essentially all components nonzero, the straightforward circular convolution of $\mathbf{y} * \mathbf{h}$ in the time domain requires on the order of N^2 operations where $N = 8192$. It would actually be more efficient to implement the convolution via an FFT/IFFT.

4. Determine the smallest threshold δ so that the number of coefficients in **h** that exceed δ is 10 or fewer. For example, the command

```
sum(abs(h)>0.01)
```

will count how many coefficients in **h** exceed 0.01 in magnitude.

Once you have an appropriate δ, construct a filter **g** consisting only of those coefficients of **h** that exceed the threshold, as

```
g = h.*(abs(h)>delta);
```

where `delta` is the appropriate threshold. The filter **g** should be an "economized" version of **h** that provides approximately the same frequency response with many fewer coefficients.

Plot the frequency response of the filter **g** by computing `G = fft(g);` and executing

```
plot(abs(G(1:4097)))
```

(dc to the Nyquist frequency). Compare to the ideal frequency response given by the filter **h**.

5. Construct the filtered vector $\mathbf{w} = \mathbf{y} * \mathbf{g}$ with the `circconv` command. Play **w** with the `sound` command and compare to the perfectly high-pass filtered version from question 2. In addition, plot the DFT of **w** on the range from dc to the Nyquist frequency. Are the frequencies below 500 Hz still present? Experiment with various values for the threshold δ, and note the compromise between filter length and high-pass performance.

4.6.2 Audio Signals and Noise Removal

1. Load in the "train whistle" signal with `load train;` (which loads the signal into a vector `y`). Recall that the audio signal is a vector of double-precision floating point numbers in the range -1 to 1. Add random noise to the signal with

```
yn = y + 0.2*(rand(size(y))-0.5);
```

The noise added is uniformly distributed in the range -0.1 to 0.1, though we could just as well use Gaussian noise. Play the noisy signal with `sound(yn)`. You may wish to multiply `yn` by a suitable constant (0.9 will work) to make sure that the noisy signal is still in the range -1 to 1. This avoids "clipping," whereby a signal that exceeds its assigned range is clipped to the maximum or minimum permissible value. The audio signal should have a noticeable background hiss.

2. Use the techniques from Section 4.6.1 to design a low-pass filter with cutoff frequency 2500 Hz and no more than 40 filter coefficients. It may be helpful to refer back to Section 2.8.1 in the Matlab portion of Chapter 2, to recall the relation between frequency in hertz and the Fourier coefficients Y_k. Plot the magnitude of the filter vector's DFT. Use `circconv` to filter the noisy train whistle signal, and play the resulting audio signal.

Experiment with the cutoff frequency and/or number of filter coefficients. Is there any good compromise between fidelity to the original audio signal and effective noise removal?

4.6.3 Filtering Images

1. Load your favorite image into Matlab with `z = imread('myimage.jpg');`. Let us suppose that the image is $m \times n$ pixels. Construct a grayscale version `zg` of the image as in previous chapters, and then construct a noisy version of `zg` with

   ```
   zn = double(zg) + 50*(rand(size(zg))-0.5);
   ```

   The image `zn` is just `zg` with each pixel corrupted by a uniformly distributed random variable in the −25 to 25 range; vary the 50 multiplier to get more or less noise. Set up an appropriate color map for viewing grayscale images (refer back to Section 1.11) with

   ```
   L = 255;
   colormap([(0:L)/L; (0:L)/L; (0:L)/L]');
   ```

   and view the image with `image(zn)`. Some pixels in `zn` will likely lie outside the range 0–255, but will be clipped to black or white when displayed.
2. Try low-pass filtering the corrupted image `zn` to get rid of the noise, by circularly convolving `zn` with the mask **h** that has (indexing starting at zero) $h_{0,0} = 0.2, h_{0,1} = 0.2, h_{1,0} = 0.2, h_{m-1,0} = 0.2$, and $h_{0,n-1} = 0.2$. This mask has the effect of averaging each pixel with its four nearest neighbors. Set up the mask with `h = zeros(m,n)`, and then manually set `h(1,1)=0.2`, and so forth (remember, you now index from 1!).
   Performing the convolution in the "image" domain can be rather slow. Instead, apply the Fourier transform to both `h` and `zn` with the `fft2` command to obtain transforms `H` and `Zn`, multiply "point by point" as `Zdn = Zn.*H;` in the frequency domain, then apply the inverse transform as `zdn=real(ifft2(Zdn));`. Display the denoised image `zdn`. Is this approach to noise removal effective?
3. To see that the mask **h** is indeed a low-pass filter, image the magnitude of the DFT **H** (use a 0–255 grayscale colormap) with `image(c*abs(H))` where c is a suitably chosen scaling constant. You should find that the components of **H** that correspond to high frequencies are small and so display as darker shades of gray or black, while the components that correspond to low frequencies are large and display as lighter shades or white. Recall also that in the indexing of $H_{j,k}$ with $0 \leq j \leq m-1, 0 \leq k \leq n-1$, high frequencies are near the center, and low frequencies near the corners.

4. Try filtering the noisy image zn repeatedly. This can be done painlessly with the command

```
Zdn = Zn.*(H.^N);
```

which applies the filter to zn N times in the frequency domain (think about why). Then set zdn=real(ifft2(Zdn));. What happens when you filter repeatedly?

5. Let us perform some edge detection as in Section 4.4.2. Use the same steps as in the last exercise (operate on the noiseless image z1, though) to construct and apply an $m \times n$ mask $\mathbf{h}$ with components $h_{0,0} = 1, h_{1,0} = -1$ and all other components zero to the image array z1. As remarked prior to equation (4.9), the convolution $\mathbf{z} * \mathbf{h}$ should detect horizontal edges. Display $|\mathbf{z} * \mathbf{h}|$ as a grayscale image; it may be helpful to rescale first.
Similarly construct an $m \times n$ mask $\mathbf{v}$ with $v_{0,0} = 1, v_{0,1} = -1$ and all other components zero. The quantity $|\mathbf{z1} * \mathbf{v}|$ should detect vertical edges. Display this quantity as a grayscale image.
Display also the quantity $\sqrt{|\mathbf{z1} * \mathbf{v}|^2 + |\mathbf{z1} * \mathbf{h}|^2}$, a discrete version of the magnitude of the image function gradient. As above, it may be helpful to rescale before displaying, or altering the colormap so that 255 corresponds to black and 0 to white.

6. Redo list item 3 above to compute and display as images the magnitude of the DFT's $\mathbf{H}$ and $\mathbf{V}$ of the masks from list item 5. Correlate the appearance of these DFT's with the job that each mask performs.

Exercises

1. Prove that for complex numbers w, z we have the inequalities $|w||z| \leq \frac{1}{2}(|w|^2 + |z|^2)$, $|\text{Re}(zw)| \leq |z||w|$, and $|\text{Im}(zw)| \leq |z||w|$.

2. Write out the circulant matrix for $\mathbf{g} = (1, 5, 7, 2)^T$, and use it to compute the circular convolution of $\mathbf{g}$ with the vector $\mathbf{x} = (1, -1, 1, 2)^T$.

3. Let $\mathbf{w} = (1, 0, \ldots, 0) \in \mathbb{C}^N$ (indexed from 0, so $w_0 = 1$), and let $\mathbf{x}$ be any vector in $\mathbb{C}^N$. Show that $\mathbf{x} * \mathbf{w} = \mathbf{x}$ by
(a) Using Definition 4.1 directly.
(b) Computing the DFT $\mathbf{W}$ of $\mathbf{w}$ and using the Convolution Theorem 4.2.

4. Prove the assertion of Remark 4.1.

5. Prove list item (1) in Theorem 4.1 (the linearity of convolution).

6. Prove list item (5) of Theorem 4.1, that if x_k and y_k are defined for all $k \in \mathbb{Z}$ via $x_k = x_{k \bmod N}$ and $y_k = y_{k \bmod N}$, then w_r as defined in equation (4.2) is defined for all r and satisfies $w_r = w_{r \bmod N}$.

7. Suppose that circulant matrices satisfy $\mathbf{M_g} = \mathbf{M_h}$ for vectors $\mathbf{g}$ and $\mathbf{h}$ in $\mathbb{C}^N$. Show $\mathbf{g} = \mathbf{h}$.

8. Let $\mathbf{h} \in \mathbb{C}^N$ and $\mathbf{M}_h$ the corresponding circulant matrix. Show that the DFT matrix $\mathbf{F}_N$ of Theorem 2.1 diagonalizes $\mathbf{M}_h$, that is,

$$\mathbf{M}_h = \mathbf{F}_N \mathbf{D}_H \mathbf{F}_N^{-1},$$

where $\mathbf{D}_H$ denotes the diagonal matrix with diagonal entries H_m, $0 \leq m \leq N-1$, the components of the DFT $\mathbf{H}$ of the vector $\mathbf{h}$. Assume that the diagonal entries are ordered so that H_m corresponds to the basic waveform $\mathbf{E}_{N,m}$. Hint: Use Theorem 4.3.

9. (a) Define the *time reversal* (also called the *adjoint*) of a filter $\mathbf{h} \in \mathbb{R}^N$ as the vector $\mathbf{h}' \in \mathbb{R}^N$ with components

$$h'_k = h_{-k \bmod N}$$

for $0 \leq k \leq N-1$ (though we may extend $\mathbf{h}'$ periodically in the index k as well). Show that the circulant matrices for $\mathbf{h}$ and $\mathbf{h}'$ are related as $\mathbf{M}_{\mathbf{h}'} = \mathbf{M}_{\mathbf{h}}^T$.

(b) We say that a filter $\mathbf{h}$ is symmetric if $\mathbf{h} = \mathbf{h}'$. Show that $\mathbf{h}$ is symmetric if and only if the circulant matrix $\mathbf{M_h}$ is a symmetric matrix.

(c) For a filter $\mathbf{h} \in \mathbb{C}^N$ let us define by adjoint filter $\mathbf{h}' \in \mathbb{C}^N$ via

$$h'_k = \overline{h_{-k \bmod N}}$$

(the overline denotes complex conjugation). Show that the circulant matrices for $\mathbf{h}$ and $\mathbf{h}'$ are related as $\mathbf{M}_{\mathbf{h}'} = \mathbf{M}_{\mathbf{h}}^*$, where the superscript "$*$" denotes the conjugate transpose.

10. (Refer to Exercise 9 for notation.) What is the relationship among $\mathbf{g}$, $\mathbf{h}$, and $(\mathbf{g}' * \mathbf{h}')'$? Assume that the filters are real-valued. Hint: Use circulant matrices; list item (3) of Theorem 4.1 may be useful, and also Exercise 7.

11. *Matlab experiment*: Use Matlab to construct a sampled version $\mathbf{f}$ of the function

$$f(t) = \begin{cases} \sin(10\pi t), & 0 \leq t < \frac{1}{2}, \\ \frac{1}{2}, & \frac{1}{2} \leq t < 1, \end{cases}$$

for $t \in [0, 1)$, say at $N = 256$ or more points. Let $\mathbf{v}$ be the vector in $\mathbb{R}^N$ with components

$$v_k = \begin{cases} \frac{1}{M}, & 0 \leq k \leq M-1, \\ 0, & M \leq N-1, \end{cases}$$

where M is to be chosen. Compute and plot the circular convolution $\mathbf{v} * \mathbf{f}$ for $M = N, 50, 20, 10, 5, 1$. Explain what you see.

12. *Matlab experiment:* Find out what a "Toeplitz" matrix is, and then read the help page for and experiment with Matlab's "`toeplitz`" command to find a simple way to construct the circulant matrix for a filter $\mathbf{h} \in \mathbb{C}^N$. It may be helpful to note that the first column of $\mathbf{M_h}$ is $\mathbf{h}$, whereas the first row of $\mathbf{M_h}$ is the time reversal $\mathbf{h}' = (h_0, h_{N-1}, h_{N-2}, \ldots, h_1)$ as in Exercise 9. Note also that if `h` is an N-dimensional vector in Matlab (indexed from 1 to N), then the time reversal of `h` is the vector `[h(1) h(N:-1:2)]`. Show your code and verify that it works on a vector in $\mathbb{R}^4$.

13. *Matlab experiment:* Let $\mathbf{M_h}$ be the circulant matrix for $\mathbf{h} \in \mathbb{C}^N$ and let $\mathbf{E}_{N,k}$ denote the basic exponential waveform defined by equation (1.23). Theorem 4.3 says that $\mathbf{E}_{N,k}$ an eigenvector of $\mathbf{M_h}$ with eigenvalue H_k, where $\mathbf{H}$ denotes the DFT of $\mathbf{h}$.
Choose a nontrivial vector $\mathbf{h} \in \mathbb{C}^4$, and compute its DFT. In addition, construct the circulant matrix for $\mathbf{h}$ (see Exercise 12), and compute the eigenvalues of this matrix. Verify that the eigenvalues are the H_k.

14. For symmetric filters (real coefficients) show that the eigenvalues discussed in Exercise 13 are real-valued. The result of Exercise 9 and Theorem 4.3 may help.

15. Consider a filter $\mathbf{h}$ with coefficients

$$h_0 = \frac{1}{2}, \; h_1 = -\frac{1}{2}, \quad h_r = 0 \quad \text{otherwise.}$$

This is called a "two-point differencing" filter. Compute the DFT of $\mathbf{h}$ and plot the magnitude and phase of the DFT, as was done for the two-point averaging filter in Figure 4.3. Why is the two-point differencing filter an example of a "high-pass" filter?

16. Consider the convolutional equation $\mathbf{f} = \mathbf{x} * \mathbf{g}$ in which $\mathbf{f}$ and $\mathbf{g}$ are given vectors in $\mathbb{C}^N$ and we are to solve for $\mathbf{x} \in \mathbb{C}^N$.

(a) Let $\mathbf{g} = (2, 1, 0, 3)$ and $\mathbf{f} = (4, 8, 20, 16)$. Find a vector $\mathbf{x} \in \mathbb{C}^4$ that satisfies $\mathbf{f} = \mathbf{x} * \mathbf{g}$. Hint: Use the DFT and the convolution theorem. You can use Matlab or do the DFT with the matrix $\mathbf{F}_4$ from equation (2.8).

(b) For a general vector $\mathbf{g} \in \mathbb{C}^N$, find a simple condition on $\mathbf{g}$ that guarantees that $\mathbf{x}$ can be solved for uniquely for any $\mathbf{f} \in \mathbb{C}^N$.

(c) If your condition on $\mathbf{g}$ in part (b) is not met, might we still be able to find some $\mathbf{x}$ that satisfies $\mathbf{f} = \mathbf{x} * \mathbf{g}$? Would $\mathbf{x}$ be unique?

17. Prove Theorem 4.4. Hint: Mimic the proof of the one-dimensional case, Theorem 4.2.

18. Consider the two-point averaging process of equation (4.1) applied to a basic complex exponential vector $\mathbf{x} = \mathbf{E}_{N,m}$ to produce output $\mathbf{w}$. Show that $\mathbf{w} = A_m\mathbf{x}$ for some complex constant A_m. Compute $|A_m|$ and in particular, $|A_0|$ and $|A_{N/2}|$. Explain how this demonstrates that the two-point averaging filter is a low-pass filter.
In addition, compute $\arg(A_m)$, and use this to explain the linear appearance of the graph on the right in Figure 4.3.

19. In each case, take vectors of length $N = 8$; use Matlab liberally.

(a) Use the inverse Fourier transform approach of Section 4.3.3 to design a filter $\mathbf{h}$ that leaves any input $\mathbf{x}$ unchanged. Write out $\mathbf{h}$ explicitly, and note why it makes perfect sense.

(b) Use the inverse Fourier transform approach to design a filter $\mathbf{h}$ that leaves the dc component of the input signal unchanged, but zeros out all other frequencies. Again, write out $\mathbf{h}$ explicitly and note why it makes perfect sense.

(c) Design a filter $\mathbf{h}$ that leaves the Fourier coefficients X_2 and X_6 of the input $\mathbf{x}$ unchanged, flips the sign of X_1 and X_7, and zeros out all other frequencies. The filter $\mathbf{h}$ should have real coefficients—why?

20. Let $\mathbf{x} \in \mathbb{R}^N$ be obtained by sampling the function $\phi(t) = \cos(2\pi mt)$ at times $t = k/N$ for $0 \leq k \leq N - 1$. Let $\mathbf{h} \in \mathbb{R}^N$ be a filter vector with DFT $\mathbf{H}$, and let $\mathbf{w} = \mathbf{x} * \mathbf{h}$. All vectors have real-valued components.
Show that $\mathbf{w}$ is a sampled version of the function $|H_m|\phi(t + d)$ where $d = arg(H_m)/(2\pi m)$. (This is just an amplified and phase-shifted version of $\phi(t)$.) Hint: Use $\phi(t) = (e^{2\pi imt} + e^{-2\pi imt})/2$, and employ the reasoning that led to equation (4.6). Note also that $H_{N-k} = H_{-k} = \overline{H_k}$.
Show that the same conclusion holds for $\phi(t) = \sin(2\pi mt)$.

21. Let $\mathbf{h} \in \mathbb{C}^N$ be a filter vector and $\mathbf{w} = \mathbf{x} * \mathbf{h}$ (circular convolution in $\mathbb{C}^N$). Show that

$$\|\mathbf{w}\| \leq \|\mathbf{x}\|$$

for all inputs $\mathbf{x}$ if and only if $\mathbf{H}$ satisfies $|H_k| \leq 1$ for all k. Thus filtering by $\mathbf{h}$ conserves or attenuates the energy of all input signals if and only if $\mathbf{h}$ does not increase the energy at any given frequency. *Suggestion*: Use the convolution theorem and the result of Exercise 12.

22. Write out explicitly and plot (analogous to Figure 4.6) the magnitude of the 2D DFT for the vertical and horizontal edge detection masks used to produce Figures 4.9 and 4.10, for the case of 100×100 pixel images.

23. A filter or mask $\mathbf{h}$ is called "sparse" if the number of nonzero elements in $\mathbf{h}$ is small in comparison to the size of the vectors or arrays with which $\mathbf{h}$ is convolved. Here the term "small" is somewhat subjective.

(a) Let $\mathbf{h} \in \mathbb{C}^N$ be a filter vector with only r nonzero components and $\mathbf{x} \in \mathbb{C}^N$ a vector with possibly all components nonzero. Determine the maximum number of operations required to compute the circular convolution $\mathbf{h} * \mathbf{x}$ in the time domain. Count each complex add, subtract, multiply, or divide as a single operation.

(b) Suppose that we have an FFT algorithm that can compute each of $\mathbf{H}$ and $\mathbf{X}$ with $2N\log_2(N)$ operations, and we have a correspondingly efficient IFFT algorithm. How many operations will be needed to compute $\mathbf{h} * \mathbf{x}$ using these FFT/IFFT algorithms and the convolution theorem?

(c) What value for r will make a direct circular convolution more efficient than using the FFT? (The answer will depend on N.)

(d) Repeat the analysis above for the problem of convolving an $m \times n$ mask $\mathbf{D}$ with an $m \times n$ array $\mathbf{A}$. Assume that $\mathbf{D}$ has r nonzero elements and that the FFT/IFFT algorithms have $2mn\log_2(mn)$ efficiency.

24. A mask $\mathbf{D} \in M_{m,n}(\mathbb{C})$ with components $d_{r,s}$ is called "factorizable" if

$$d_{r,s} = f_r g_s$$

for vectors $\mathbf{f} \in \mathbb{C}^m$, $\mathbf{g} \in \mathbb{C}^n$. Equivalently $\mathbf{D} = \mathbf{f}\mathbf{g}^T$ where $\mathbf{f}$ and $\mathbf{g}$ are treated as column vectors.

Show that if $\mathbf{A} \in M_{m,n}(\mathbb{C})$, then

$$\mathbf{A} * \mathbf{D} = \mathbf{M}_\mathbf{f} \mathbf{A} \mathbf{M}_\mathbf{g}{}^T,$$

where $\mathbf{M}_\mathbf{f}$ and $\mathbf{M}_\mathbf{g}$ are the circulant matrices for $\mathbf{f}$ and $\mathbf{g}$.

25. To detect all edges in an image as in Section 4.4.2 and Figure 4.10, we could simply filter for both simultaneously as

$$\mathbf{A} \rightarrow (\mathbf{A} * \mathbf{V}) * \mathbf{D},$$

where **V** and **D** are the vertical and horizontal edge detection masks from Section 4.4.2. Why will not this work well?

26. Let $\mathbf{x} \in L^2(\mathbb{Z})$ have components

$$x_k = \begin{cases} 0, & k < 0, \\ e^{-k}, & k \geq 0. \end{cases}$$

(a) Verify that **x** is a member of $L^2(\mathbb{Z})$.
(b) Compute the discrete time Fourier transform $X(f)$ of **x**. Hint: Use $1 + z + z^2 + \cdots = 1/(1 - z)$ for $|z| < 1$.

27. Show that the expression defined in equation (4.10) satisfies properties 1–3 of Definition 1.3, and so defines an inner product on $L^2(\mathbb{N})$.

28. Let $\mathbf{x} \in L^2(\mathbb{Z})$, and define $\mathbf{y} \in L^2(\mathbb{Z})$ with components $y_k = x_{k-m}$ for some fixed integer m. How are the discrete time Fourier transforms $X(f)$ and $Y(f)$ related? If we define $y_k = x_{-k}$, how are $X(f)$ and $Y(f)$ related?

29. Let $\mathbf{x} = (x_0, x_1, \ldots, x_{N-1}) \in \mathbb{C}^N$, and let $\tilde{\mathbf{x}}$ denote that element of $L^2(\mathbb{Z})$ defined by $\tilde{x}_k = x_k$ for $0 \leq k \leq N-1$, $\tilde{x}_k = 0$ otherwise (so $\tilde{\mathbf{x}}$ is just **x** extended by zeros).
(a) Show that the DFT coefficients X_m of **x** can be computed from $X(f)$, the discrete time Fourier transform of $\tilde{\mathbf{x}}$.
(b) Show that $X(f)$ can be expressed in terms of the X_m as

$$X(f) = \frac{1}{N} \sum_{m=0}^{N-1} \sum_{k=0}^{N-1} X_m e^{2\pi i k(m/N - f)}.$$

Hint: First use the IDFT to express x_k in terms of the X_m.
(c) Show that if $f \neq m/N$ for any integer m in the range $0 \leq m \leq N-1$, then

$$X(f) = \frac{1}{N} \sum_{m=0}^{N-1} \left(\frac{1 - e^{2\pi i(m - fN)}}{1 - e^{2\pi i(m/N - f)}} \right) X_m.$$

Hint: Use part (b) and the identity (1.31).

30. Suppose that **x** and **y** are elements of $L^2(\mathbb{Z})$ and that **x** has only finitely many nonzero components. Let $\mathbf{w} = \mathbf{x} * \mathbf{y}$. Show that $\mathbf{w} \in L^2(\mathbb{Z})$. Hint: Start by supposing that **x** has only one nonzero component x_r. Show in this case that $\|\mathbf{x} * \mathbf{y}\| = |x_r|\|\mathbf{y}\|$. Then suppose that $\mathbf{x}$ has finitely many nonzero components and make use of linearity and the triangle inequality.

31. To see that the convolution of two vectors in $L^2(\mathbb{N})$ (or $L^2(\mathbb{Z})$) need not be in $L^2(\mathbb{N})$, let $\mathbf{x}$ be the vector with components $x_k = 1/k^\alpha$ for $k \geq 1$, where $\alpha \in (1/2, 3/4)$. We will show $\mathbf{x} * \mathbf{x}$ is not in $L^2(\mathbb{N})$.

(a) Show that $\mathbf{x} \in L^2(\mathbb{N})$. Hint: Recall that a "$p$-series"

$$\sum_{k=1}^{\infty} \frac{1}{k^p}$$

converges if and only if $p > 1$.

(b) Let $\mathbf{w} = \mathbf{x} * \mathbf{x}$, so

$$w_r = \sum_{k=1}^{r-1} \frac{1}{k^\alpha} \frac{1}{(r-k)^\alpha}.$$

All terms are positive, as is w_r. Show that

$$w_r \geq \frac{1}{(r-1)^\alpha} \sum_{k=1}^{r-1} \frac{1}{k^\alpha}. \tag{4.32}$$

(c) Show that

$$\sum_{k=1}^{r-1} \frac{1}{k^\alpha} \geq \int_1^{r-1} \frac{dx}{x^\alpha}. \tag{4.33}$$

Hint: Show that the left side above is a left Riemann sum for the integral on the right (with rectangle base width 1, and one "extra" rectangle); note also that $1/x^\alpha$ is decreasing in x. A sketch may be helpful.

(d) Evaluate the integral in (4.33), and combine this with (4.32) to show that

$$w_r \geq \frac{1}{(1-\alpha)(r-1)^\alpha}((r-1)^{1-\alpha} - 1). \tag{4.34}$$

(e) Show that if $r \geq r_0$ where r_0 is chosen as $r_0 \geq 2^{1/(1-\alpha)} + 1$, then $(r-1)^{1-\alpha} - 1 \geq \frac{1}{2}(r-1)^{1-\alpha}$. Conclude that for $r \geq r_0$,

$$w_r \geq \frac{1}{2(1-\alpha)(r-1)^\alpha}(r-1)^{1-\alpha} = \frac{1}{2(1-\alpha)} \frac{1}{(r-1)^{2\alpha-1}}.$$

Why is $2^{1/(1-\alpha)} + 1$ necessarily finite?

(f) Based on part (e), explain why if $\frac{1}{2} < \alpha < \frac{3}{4}$ the sum

$$\sum_{k=r_0}^{\infty} w_r^2$$

cannot converge (and hence the sum starting at $r = 1$ cannot converge either). Hint: Use the p-series test again.

32. Prove the convolution theorem analogous to Theorem 4.5 for the space $L^2(\mathbb{Z} \times \mathbb{Z})$ (refer to equation (4.30) and Section 4.5.7) for the case in which the Fourier transforms of **A** and **B** have only finitely many nonzero coefficients (it holds more generally).

33. Let $\mathbf{x} = (-1, 2, 0, 4)$ and $\mathbf{y} = (1, 1, 2, -1)$ be vectors in $\mathbb{C}^4$. We will also consider $\mathbf{x}$ and $\mathbf{y}$ as elements of $L^2(\mathbb{Z})$, via extension by zeros.

(a) Compute the z-transforms $X(z)$ and $Y(z)$, and the product $X(z)Y(z)$.

(b) Use the result of part (a) to write out the convolution of $\mathbf{x}$ and $\mathbf{y}$ in $L^2(\mathbb{Z})$.

(c) Use part (a) to compute the circular convolution of $\mathbf{x}$ and $\mathbf{y}$ in $\mathbb{R}^4$ by using Theorem 4.7.

(d) Use part (b) to compute the circular convolution of $\mathbf{x}$ and $\mathbf{y}$ in $\mathbb{R}^4$ by using equation (4.29).

34. Show that the z-transforms $X(z)$ and $Y(z)$ in Example 4.8 are equal to $1/(1 - z^{-1}/2)$ on the domains asserted.

35. Adapt Example 4.5 to show that the vector $\mathbf{h} \in L^2(\mathbb{Z})$ with components $h_0 = \frac{1}{2}, h_1 = -\frac{1}{2}$, and all other $h_k = 0$ acts as a high-pass filter under convolution.

36. Find a vector $\mathbf{x} \in L^2(\mathbb{Z})$ such that $\mathbf{x} * \mathbf{x} = \mathbf{y}$, where

$$y_0 = 1, \quad y_1 = 0, \quad y_2 = -4, \quad y_3 = 0, \quad y_4 = 4$$

and all other $y_k = 0$. Hint: Use the z-transform and Theorem 4.5.

37. Suppose that $\mathbf{x} \in L^2(\mathbb{Z})$ with z-transform $X(z)$. Let $\mathbf{y}$ be defined by $y_k = x_{k+r}$ for some integer r. Show that $Y(z) = z^r X(z)$.

5

Windowing and Localization

5.1 Overview: Nonlocality of the DFT

Let us begin with an example.

Example 5.1 *Consider two different audio signals $f(t)$ and $g(t)$ defined on $0 \leq t \leq 1$, given by*

$$f(t) = 0.5\sin(2\pi(96)t) + 0.5\sin(2\pi(235)t)$$

and

$$g(t) = \begin{cases} \sin(2\pi(96)t), & 0 \leq t < \dfrac{1}{2}, \\ \sin(2\pi(235)t), & \dfrac{1}{2} \leq t \leq 1. \end{cases}$$

Both are composed of the same two basic waveforms, $\sin(2\pi(96)t)$ and $\sin(2\pi(235)t)$. While in f the waveforms are present throughout, in g each waveform is present for exactly half of the time interval $[0, 1]$.

Let us sample each at 1000 Hz to produce sample vectors $\mathbf{f}$ and $\mathbf{g}$, and then compute the DFT of each sampled signal. The magnitude of each is plotted in Figure 5.1, DFT($\mathbf{f}$) on the left and DFT($\mathbf{g}$) on the right, where in each graph the horizontal index k corresponds to k Hertz. It is obvious from the plots that each signal contains dominant frequencies near 96 and 235 Hz, and the magnitude of the DFTs are otherwise fairly similar. But the signals $f(t)$ and $g(t)$ are quite different in the time domain, a fact that is difficult to glean from the DFT graphs.

This example illustrates one of the shortcomings of traditional Fourier transforms: the coefficient X_k in $\mathbf{X} = \text{DFT}(\mathbf{x})$ quantifies the amount of the corresponding waveform in $\mathbf{x}$ as if that waveform is present throughout the entire signal at constant amplitude. We can see this clearly in the inverse DFT

Discrete Fourier Analysis and Wavelets: Applications to Signal and Image Processing, Second Edition.
S. Allen Broughton and Kurt Bryan.

Companion Website: www.wiley.com/go/Broughton/Discrete_Fourier_Analysis_and_Wavelets

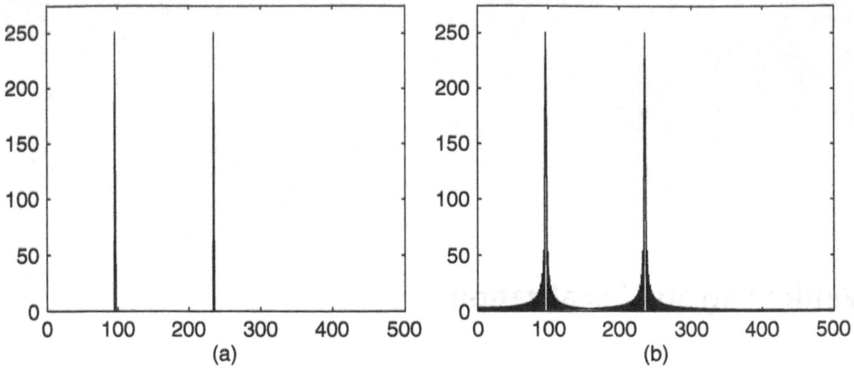

Figure 5.1 Magnitude of DFT for **f** (a) and **g** (b).

formula

$$\mathbf{x} = \frac{1}{N}\sum_{k=0}^{N-1} X_k \mathbf{E}_{N,k}$$

in which the basic waveform $\mathbf{E}_{N,k}$ appears multiplied by X_k throughout. However, it is quite common that a frequency appears in one part of a signal and not another, as in $g(t)$ above. The DFT does not make it easy to recognize or quantify this. The essence of the problem is the "nonlocality" or global nature of the basis vectors $\mathbf{E}_{N,k}$ or underlying analog waveforms $e^{2\pi ikt/T}$: they are nonzero at almost all points.

Discontinuities are particularly troublesome. For example, consider a function that is identically 1 for $t \in [0, \frac{1}{2})$ and 0 for $t \in [\frac{1}{2}, 1]$. In any discretization of the signal the discontinuity excites "most" of the coefficients X_k in the DFT (see Exercise 2), but this is in some sense a misrepresentation of the signal. The signal consists of pure dc for both halves of the time interval, with a single highly localized discontinuity. However, the excitation of many X_k gives the impression that the entire signal is highly oscillatory.

What we would like is the ability to *localize* our frequency analysis to smaller portions of the signal. We have already employed this strategy for both audio signals and images in the Matlab project of Section 3.9. Indeed we saw that certain frequencies may be present in some parts of a signal and not others, which is what led to the "JPEG idea" of breaking the image up into blocks and compressing each independently. The notion of localizing the analysis of a signal by performing block-by-block transforms dates back to (at least) 1946 [13]. In conjunction with Fourier methods this leads to "*windowed Fourier transforms,*" which we will take a look at in this chapter, as well as "*spectrograms,*" a way to present such data.

We will focus on windowing discrete signals of finite length, but analogous results hold for signals in $L^2(\mathbb{Z})$ and analog signals (functions) (see Exercise 10).

5.2 Localization via Windowing

5.2.1 Windowing

Consider a sampled signal $\mathbf{x} \in \mathbb{C}^N$, indexed from 0 to $N-1$ as usual. We wish to analyze the frequencies present in $\mathbf{x}$, but only within a certain index (time) range. To accomplish this, we choose integers $m, M \geq 0$ such that $m + M \leq N$ and define a vector $\mathbf{w} \in \mathbb{C}^N$ as

$$w_k = \begin{cases} 1, & m \leq k \leq m+M-1, \\ 0, & \text{otherwise}. \end{cases} \tag{5.1}$$

Define a vector $\mathbf{y}$ with components

$$y_k = w_k x_k \tag{5.2}$$

for $0 \leq k \leq N-1$. We use the notation $\mathbf{y} = \mathbf{w} \circ \mathbf{x}$ to denote the component-by-component product defined in equation (5.2) and refer to the vector $\mathbf{w}$ as the *window*. The vector $\mathbf{w}$ has M nonzero components, so we say that "$\mathbf{w}$ is a window of length M."

The reason for this terminology is simple: the operation $\mathbf{x} \to \mathbf{w} \circ \mathbf{x}$ zeros out all components x_k with $k < m$ and $k > m + M - 1$ while leaving the other M components unchanged, and thus it provides a "window" on $\mathbf{x}$ in the range $m \leq k \leq m + M - 1$. We localize our frequency analysis of $\mathbf{x}$ by analyzing $\mathbf{y}$. By changing m and/or M, we can systematically localize the analysis to different portions of the signal $\mathbf{x}$. The window vector $\mathbf{w}$ defined in equation (5.1) is called a *rectangular window*. We will discuss others later.

The relation between the DFTs of $\mathbf{x}$ and $\mathbf{y}$ requires a bit of analysis, however. The operation of windowing induces a specific type of distortion into the process, even in the case where $\mathbf{x}$ consists of only a few basic waveforms.

Example 5.2 *Let us consider the analog signal* $f(t) = 0.5\sin(2\pi(96)t) + 0.5\sin(2\pi(235)t)$ *from Example 5.1, sampled at times* $t = k/1000$ *for* $0 \leq k \leq 999$ *to produce sample vector* $\mathbf{f} \in \mathbb{C}^{1000}$. *The DFT of* $\mathbf{f}$ *was shown in Figure 5.1a. Now consider a windowed version* $\tilde{\mathbf{f}} \in \mathbb{C}^{1000}$ *obtained as* $\tilde{\mathbf{f}} = \mathbf{w} \circ \mathbf{f}$ *where* $\mathbf{w} \in \mathbb{C}^{1000}$ *has components* $w_k = 1$ *for* $100 \leq k \leq 149$, *and* $w_k = 0$ *otherwise. The magnitude of the DFT of* $\tilde{\mathbf{f}}$ *is shown in Figure 5.2a, in the frequency range 0–500 Hz (the Nyquist frequency).*

The windowed signal $\tilde{\mathbf{f}}$ *is 95% zeros, however; it seems inefficient to perform a 1000-point DFT on a vector with only 50 nonzero components. In Figure 5.2b we show the 50-point DFT of the nonzero portion* $(f_{100}, \ldots, f_{149})$ *of the windowed vector. The frequency range remains dc to 500 Hz, but now the DFT index* k *corresponds to frequency* $20k$ *Hz, hence the index range is 0–25.*

It is clear that both DFTs in Figure 5.2 are closely related to that in Figure 5.1a. We want to understand this relation quantitatively.

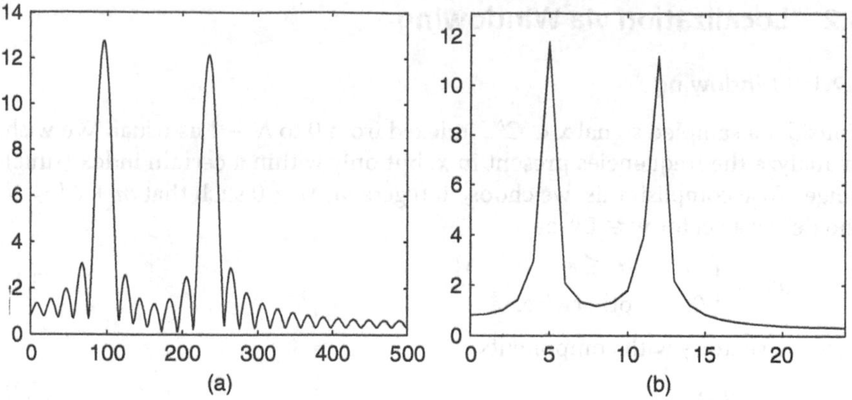

Figure 5.2 DFT magnitude for two windowed versions of **f**.

5.2.2 Analysis of Windowing

Suppose that we have a sampled signal $\mathbf{x} \in \mathbb{C}^N$ and produce a windowed version $\mathbf{y} \in \mathbb{C}^N$ as

$$\mathbf{y} = \mathbf{x} \circ \mathbf{w}, \tag{5.3}$$

where $\mathbf{w} \in \mathbb{C}^N$. We previously chose $\mathbf{w}$ according to equation (5.1), but more general choices are possible. For the moment let us assume that as in (5.1), $\mathbf{w}$ has components $w_k = 0$ for $k < m$ and $k \geq m + M$, but leave w_k unspecified in the range $m \leq k \leq m + M - 1$. This will allow greater flexibility in windowing later. Windowing according to equation (5.3) will still zero out every component of $\mathbf{x}$ that is not in the range $m \leq k \leq m + M - 1$. After windowing, we perform an M-point DFT on just the nonzero portion $\tilde{\mathbf{x}} = (w_m x_m, w_{m+1} x_{m+1}, \ldots, w_{m+M-1} x_{m+M-1})$ of the windowed signal $\mathbf{y}$.

Our goal is to determine the relationship between the M-point DFT of $\tilde{\mathbf{x}}$ and the original N-point DFT of the full signal $\mathbf{x}$. In what follows we will assume that M divides N so that $N = qM$ for some integer q.

The following two propositions will be helpful in answering this question. The first is a counterpart to the convolution theorem in which the roles of the time and frequency domains are reversed, as well as the roles of the DFT and IDFT.

Proposition 5.1 *Let* $\mathbf{x}$ *and* $\mathbf{w}$ *be vectors in* $\mathbb{C}^N$ *with DFTs* $\mathbf{X}$ *and* $\mathbf{W}$, *respectively. Let* $\mathbf{y} = \mathbf{x} \circ \mathbf{w}$ *as defined in equation (5.2) have DFT* $\mathbf{Y}$. *Then*

$$\mathbf{Y} = \frac{1}{N} \mathbf{X} * \mathbf{W}, \tag{5.4}$$

where $*$ *is circular convolution in* $\mathbb{C}^N$.

In short, a component-by-component product in the time domain becomes a convolution in the frequency domain. The proof is almost identical to that of Theorem 4.2 and is left to Exercise 3.

From Exercise 19 in Chapter 2 we also have the following proposition.

Proposition 5.2 *Let* $\mathbf{y} \in \mathbb{C}^N$ *have DFT* $\mathbf{Y}$. *Let* $\tilde{\mathbf{y}} \in \mathbb{C}^N$ *be the vector obtained by circularly shifting* $\mathbf{y}$ *as*

$$\tilde{y}_k = y_{(k+m) \bmod N}.$$

Then the DFT of $\tilde{\mathbf{y}}$ *has components* $\tilde{Y}_r = e^{2\pi irm/N} Y_r$.

We will determine the relationship between the original N-point DFT $\mathbf{F}$ and the M-point DFT of the windowed signal in three steps:

1. Quantify the relation between the N-point DFT of $\mathbf{x}$ and N-point DFT of the windowed version $\mathbf{y}$ of equation (5.3).
2. Determine the frequency domain effect of shifting the windowed signal $\mathbf{y}$ circularly as in Proposition 5.2, to the vector

$$\tilde{\mathbf{y}} = (w_m x_m, w_{m+1} x_{m+1}, \dots, w_{m+M-1} x_{m+M-1}, 0, \dots, 0) \in \mathbb{C}^N. \qquad (5.5)$$

3. Determine the relation between the N-point DFT of $\tilde{\mathbf{y}}$ in (5.5) and the M-point DFT of $\tilde{\mathbf{x}} = (w_m x_m, w_{m+1} x_{m+1}, \dots, w_{m+M-1} x_{m+M-1}) \in \mathbb{C}^M$.

5.2.2.1 Step 1: Relation of X and Y

Proposition 5.1 provides the answer: if we perform an N-point DFT on $\mathbf{y}$ as defined in equation (5.3), then equation (5.4) shows

$$\mathbf{Y} = \frac{1}{N}(\mathbf{X} * \mathbf{W}). \qquad (5.6)$$

Equation (5.6) makes it obvious that the N-point DFT $\mathbf{Y}$ is not identical to $\mathbf{X}$ but is distorted by convolution with $\mathbf{W}$, the Fourier transform of the window vector.

Example 5.3 *In the case where* $\mathbf{w}$ *is the rectangular window defined in equation (5.1), then* $\mathbf{W}$ *can be worked out in closed form. It is easy to check that* $W_0 = M$ *while*

$$\begin{aligned} W_k &= \sum_{j=m}^{m+M-1} e^{-2\pi ijk/N} \\ &= e^{-2\pi imk/N} \sum_{j=0}^{M-1} e^{-2\pi ijk/N} \\ &= \frac{e^{-2\pi imk/N}(1 - e^{-2\pi iMk/N})}{1 - e^{-2\pi ik/N}} \end{aligned}$$

for $k \neq 0$, where we yet again make use of the identity $1 + z + z^2 + \cdots + z^{M-1} = (1 - z^M)/(1 - z)$. One can also check that

$$|W_k| = \left(\frac{1 - \cos(2\pi M\, k/N)}{1 - \cos(2\pi k/N)}\right)^{1/2}$$

for $k \neq 0$, while $|W_0| = M$. Note the value of m is irrelevant.

Figure 5.3 shows the magnitude of $\mathbf{W}$ for the window vector $\mathbf{w}$ used in Example 5.2. The many ripples away from the central peak are called the "side lobes" for the window. The operation of windowing $\mathbf{f} \in \mathbb{C}^{1000}$ in that example with $\mathbf{w}$ in the time domain effectively convolves the DFT $\mathbf{F}$ (Figure 5.1a) with $\mathbf{W}$. This gives some indication of why the plot in Figure 5.2a looks the way it does.

5.2.2.2 Step 2: Effect of Index Shift

The shifted vector $\widetilde{\mathbf{y}} \in \mathbb{C}^N$ in equation (5.5) has components $\widetilde{y}_k = y_{(k+m) \bmod N}$ with $\mathbf{y}$ defined in (5.3). According to Proposition 5.2,

$$\begin{aligned}\widetilde{Y}_r &= e^{2\pi imr/N} Y_r \\ &= \frac{e^{2\pi imr/N}}{N} (\mathbf{X} * \mathbf{W})_r,\end{aligned} \tag{5.7}$$

where $(\mathbf{X} * \mathbf{W})_r$ is the rth component of $\mathbf{X} * \mathbf{W}$ and the last line follows from equation (5.6).

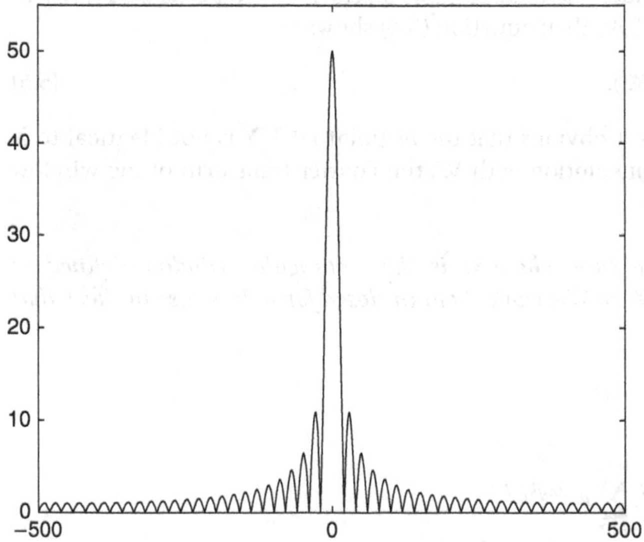

Figure 5.3 Magnitude for DFT **W** of window vector **w**.

5.2.2.3 Step 3: *N*-Point versus *M*-Point DFT

The N-point DFT of the vector $\tilde{\mathbf{y}}$ is given by

$$\begin{aligned}\tilde{Y}_r &= \sum_{k=0}^{N-1} \tilde{y}_k e^{-2\pi ikr/N}\\ &= \sum_{k=0}^{M-1} \tilde{y}_k e^{-2\pi ikr/N}\end{aligned} \tag{5.8}$$

for $0 \leq r \leq N-1$, where the last line follows since $\tilde{y}_k = 0$ for $k > M-1$. However, the M-point DFT of the vector $\tilde{\mathbf{x}} = (\tilde{y}_0, \tilde{y}_1, \ldots, \tilde{y}_{M-1}) \in \mathbb{C}^M$ is given by

$$\tilde{X}_s = \sum_{k=0}^{M-1} \tilde{y}_k e^{-2\pi iks/M} \tag{5.9}$$

for $0 \leq s \leq M-1$, where we use $\tilde{\mathbf{X}} \in \mathbb{C}^M$ to denote this DFT.

Now recall that we are assuming that M divides N, so $N = qM$ for some integer q. If we substitute $M = N/q$ into the exponential on the right in equation (5.9), we obtain

$$\tilde{X}_s = \sum_{k=0}^{M-1} \tilde{y}_k e^{-2\pi ik(qs)/N}.$$

But from equation (5.8) this is exactly $\tilde{Y}_r$ in the case of $r = qs$. In short, the relation between the N-point DFT $\tilde{\mathbf{Y}}$ and M-point DFT $\tilde{\mathbf{X}}$ is

$$\tilde{X}_s = \tilde{Y}_{qs}. \tag{5.10}$$

If we combine equations (5.7) and (5.10), we can summarize our results in the following theorem:

Theorem 5.1 *Let $\mathbf{x} \in \mathbb{C}^N$ be windowed according to equation (5.3), where $\mathbf{w} \in \mathbb{C}^N$ with $w_k = 0$ for $k < m$ and $k \geq m + M$. Assume that $N = qM$. The relation between the N-point DFT $\mathbf{X}$ of $\mathbf{x}$ and the M-point DFT $\tilde{\mathbf{X}}$ of $\tilde{\mathbf{x}} = (w_m x_m, w_{m+1}x_{m+1}, \ldots, w_{m+M-1}x_{m+M-1})$ is given by*

$$\tilde{X}_s = \frac{e^{2\pi imqs/N}}{N}(\mathbf{X} * \mathbf{W})_{qs} = \frac{e^{2\pi ims/M}}{N}(\mathbf{X} * \mathbf{W})_{qs} \tag{5.11}$$

for $0 \leq s \leq M-1$.

The last equality in (5.11) follows from $q/N = 1/M$.

Example 5.4 *Theorem 5.1 quantifies the distortion induced by windowing a signal. It is worth looking at this effect in some specific cases. In Figure 5.4 are*

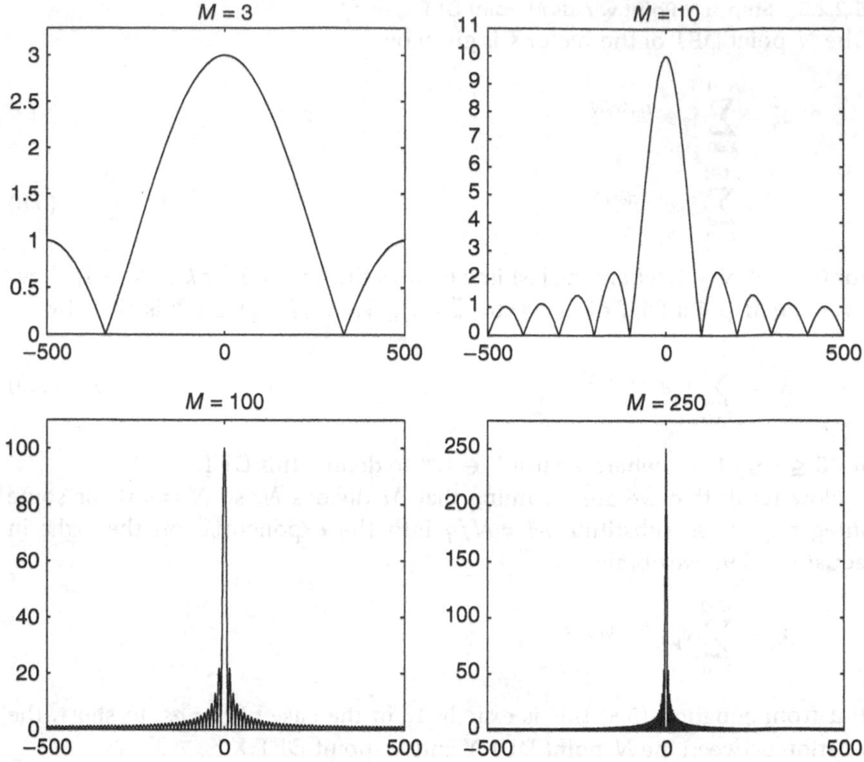

Figure 5.4 DFT magnitude $|\mathbf{W}|$ for rectangular windows of width $M = 3, 10, 100$, and 250.

plots showing the magnitude of the DFT $\mathbf{W}$ *for rectangular windows of the form (5.1) for window widths* $M = 3, 10, 100, 250$ *(all with* $m = 0$*, but* m *does not influence the plot), in the case of* $N = 1000$*. It is apparent that as the window* $\mathbf{w}$ *gets narrower (and so encompasses a more and more localized portion of the signal), the DFT* $\mathbf{W}$ *gets wider. In the frequency domain the windowing process* $\mathbf{x} \to \mathbf{w} \circ \mathbf{x}$ *is* $\mathbf{X} \to \mathbf{X} * \mathbf{W}$*, and this frequency domain convolution tends to "smear out" the DFT of* $\mathbf{X}$*. As we will see in the following examples, this can hamper our ability to distinguish closely spaced frequencies in a signal, but this is the price we pay for localizing the frequency analysis.*

In Exercise 4, you are asked to analyze the extreme cases where $M = 1$ and $M = N$.

5.2.3 Spectrograms

A common approach to analyzing a signal with changing frequency content is to choose some window of length ΔT in the time domain, short enough so that the frequency content is relatively stable over the window length, and then to perform a frequency analysis on many windowed sub-blocks of length ΔT of the original signal. A typical windowed sub-block would be $(x_m, x_{m+1}, \dots, x_{m+M-1})$, where M samples corresponds to a time interval of length ΔT; this also assumes that we use the rectangular window of (5.1). By changing the starting position m, we obtain a moving window on the signal. Analysis of the data in this window yields local frequency information on the signal. The sub-blocks into which we break the original signal may or may not overlap. This process of isolating different time slices of a signal and performing a Fourier transform on each is called a *windowed Fourier transform* or *short-time Fourier transform.*

Dividing the signal into many sub-blocks and computing the DFT of each produces a lot of data. The data can be summarized efficiently with a plot called a *spectrogram*. Suppose that the block size is M samples, with the kth block of data (starting with $k = 0$) given by

$$(x_{kn}, x_{kn+1}, \dots, x_{kn+M-1}), \tag{5.12}$$

where $n \geq 1$. The integer n controls the distance from the start of one block to the start of the next block. If $n = M$, the kth block is $(x_{kM}, \dots, x_{(k+1)M-1})$ and the kth and $k + 1$ sub-blocks do not overlap. By taking $n < M$, we obtain overlapping blocks, which can be useful. The magnitude of the M-point DFT of the kth block is stored as the kth column of a matrix, and this matrix is then displayed as an image, perhaps grayscale or color-coded. Choosing n so that adjacent blocks overlap 50–80% is common.

This is best illustrated with a specific example.

Example 5.5 *Consider the signal $g(t)$ from Example 5.1, sampled at 1000 Hz. The DFT of the full sampled signal vector $\mathbf{g}$ was shown in Figure 5.1b, and it gives the mistaken impression that energy at frequencies 96 and 235 Hz is present throughout the signal. Let us break the signal vector $\mathbf{g}$ into sub-blocks of length $\Delta T = 0.05$ s (size $M = 50$ samples) and take $n = 20$ in (5.12), so the k and $k + 1$ blocks will overlap by 30 samples, or 60%. Each block is transformed and becomes a "column" in the grayscale spectrogram of Figure 5.5. The DFTs are displayed on a logarithmic scale, with lighter shades indicating greater magnitude.*

Figure 5.5 makes it easy to see that the 96-Hz energy is present for only the first 0.5s, and then the 235-Hz sinewave kicks in. Of course, there is some untidiness at the transition.

In Figure 5.6, we illustrate the effect of using narrower or wider windows. In panel (a), we use a rectangular window of width $M = 20$ ($\Delta T = 0.02$) and panel (b), we use a window of length $M = 200$ ($\Delta T = 0.2$), both with 60% overlap between successive transforms. The narrower window more clearly picks up the abrupt frequency transition at $t = 0.5$, but it has less overall resolution in the

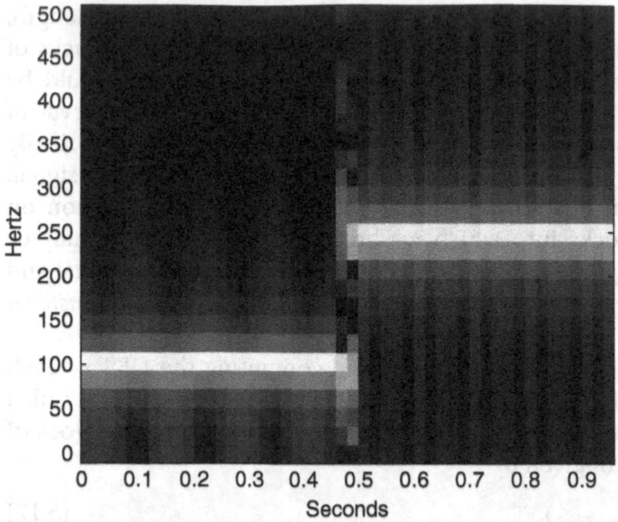

Figure 5.5 Spectrogram for **g**, window length $\Delta T = 0.05$ s.

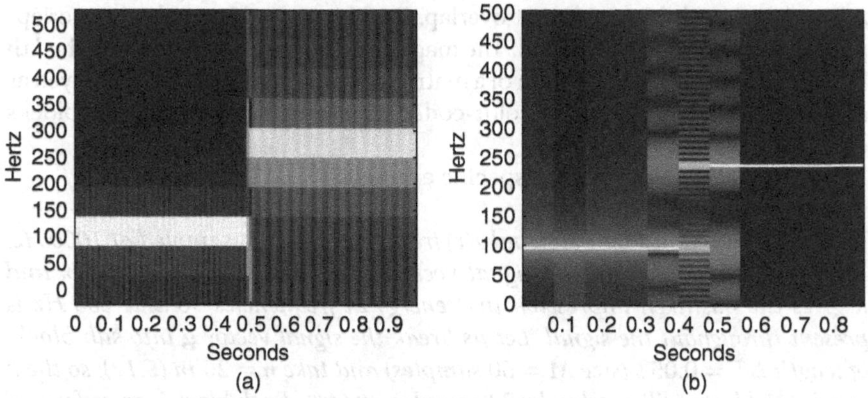

Figure 5.6 Two different window widths in the spectrogram for **g**, $\Delta T = 0.02$ (a) and $\Delta T = 0.2$ (b).

frequency (vertical) domain. Conversely, the wider window resolves the change at $t = 0.5$ *less crisply but gives better frequency resolution, at least when the frequency is stable. This is illustrated by the plots of Figure 5.4: the narrower the window, the broader the window's DFT. Convolution of this broader window DFT with the original signal's DFT tends to smear adjacent frequencies together.*

Example 5.6 *Consider a signal consisting of two closely spaced frequencies and a third component with varying frequency, namely*

$$f(t) = 1.0\sin(2\pi \cdot 111t) + 0.5\sin(2\pi \cdot 123t) + 0.5\sin(2\pi\omega(t)t)$$

for $0 \le t \le 1$, *where* $\omega(t) = 150 + 50\cos(2\pi t)$. *We sample* $f(t)$ *at 1000 Hz to produce sample vector* $\mathbf{f}$ *with components* $f_k = f(k/1000)$, $0 \le k \le 999$. *The DFT* $\mathbf{F}$ *of* $\mathbf{f}$ *is shown in Figure 5.7, scaled as* $\log(1 + |F_k|)$ *over the Nyquist range 0–500 Hz: The stable frequencies at* 111 *and* 123 *Hz are obvious, but the plot makes it clear that the signal also contains energy at many other frequencies. Unfortunately, the plot does not show how this energy is distributed in time. Figure 5.8 is a spectrogram with sub-blocks of length* $M = 100$ ($\Delta T = 0.1$), *with 80% overlap between blocks* ($n = 20$ *in (5.12)). Now it is easy to see the portion* $0.5\sin(2\pi \cdot \omega(t)t)$ *of the signal. However, the localization to sub-blocks of length* $M = 100$ *has blurred together the stable* 111 *and* 123 *Hz components, though the presence of these stable frequencies is clear.*

Unfortunately, there is no perfect compromise that allows us to maintain the highest frequency resolution (to distinguish the closely spaced 111 *and* 123 *Hz components) while tracking the changing frequency. In Figure 5.9a is the spectrogram obtained by taking* $M = 500$, *which provides better resolution of frequencies but much less localization in time. In Figure 5.9b is the*

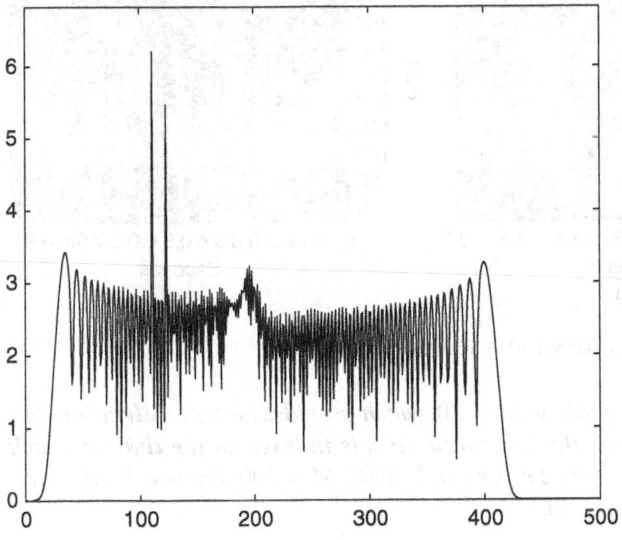

Figure 5.7 Plot of $\log(1 + |\mathbf{F}|)$ for signal $\mathbf{f}$.

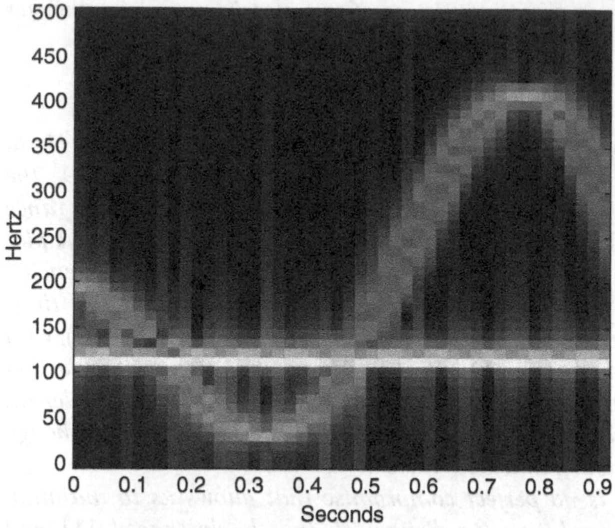

Figure 5.8 Spectrogram of signal **f**, $M = 100$.

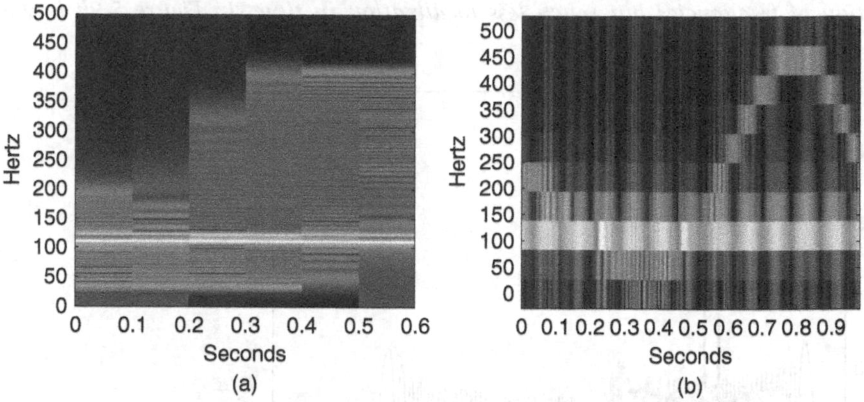

Figure 5.9 Spectrograms of signal **f**, $M = 500$ (a) and $M = 20$ (b).

spectrogram obtained by taking $M = 20$*, which provides better localization, but less frequency resolution. The horizontal axis is indexed to the time at which the relevant sub-block starts. For example, with* $M = 500$*, the last block starts at* $t = 0.5$ *(see also Exercise 7).*

5.2.4 Other Types of Windows

As noted, the rectangular window defined in equation (5.1) has a DFT with magnitude that is rather oscillatory. The DFT has prominent side lobes as shown in Figure 5.3, since the rectangular window is to good approximation a sampled version of a discontinuous function. When we use a rectangular window for localization, we end up convolving the "true" spectrum of the signal with the DFT of the window. This introduces the type of distortion shown in Figure 5.2.

But the rectangular window is not the only option. We may choose something other than 1 for the nonzero components of **w**. Since the prominent side lobes in the DFT of a rectangular window stem from the underlying discontinuity, it may be helpful to choose **w** so that the nonzero components taper gradually to zero at the ends of the window. This typically helps suppress the magnitude of the side lobes.

For example, we might take **w** to be defined as

$$w_j = \begin{cases} \dfrac{2j}{M}, & m \le j \le m + M/2, \\[2ex] \dfrac{M+m-1-j}{M/2+m-1}, & m + M/2 < j < m + M, \\[2ex] 0, & \text{else}. \end{cases}$$

the so-called *triangular window* (or "Bartlett" window). The case $N = 1000$, $m = 100, M = 49$ (comparable to the rectangular window with the DFT shown in Figure 5.3) is shown in Figure 5.10, along with the full 1000-point DFT of the window. Compare the DFT on the right with that of Figure 5.3. It is clear that the side lobes are diminished. If we use this window for a signal **x**, then the energy at each frequency is not dissipated into oscillatory side lobes quite so much. However, this benefit comes at the price of "fattening" the central lobe, an issue explored in the Matlab project of Section 5.3.

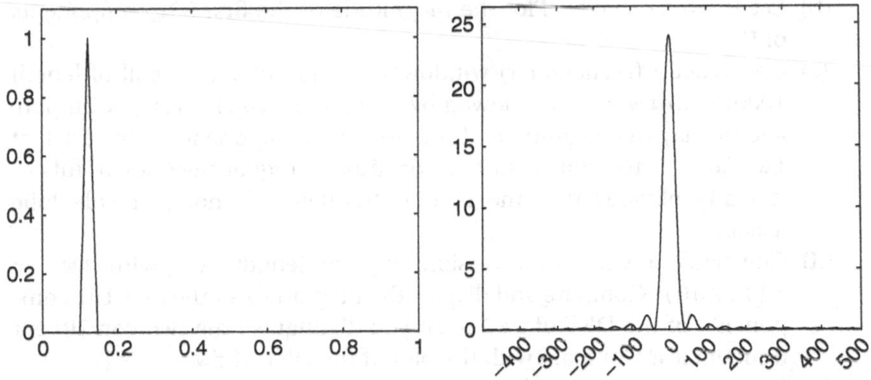

Figure 5.10 Triangular window and DFT.

Many other types of windows have been developed in an attempt to balance various competing needs: suppress side lobe energy, keep the central peak narrow, obtain good localization in time, and so forth. Some examples are the Gaussian window

$$w_j = \phi\left(\frac{j-(m+(M-1)/2)}{M/2}\right), \tag{5.13}$$

where $\phi(t) = Ce^{-\alpha t^2}$ and C and α are constants that can be adjusted to alter the shape of and/or normalize the window. As defined, w_j peaks at $j = m + M/2$ and is symmetric about this point; very small values of w_j might be truncated to zero. Another option is the *Hamming window*, defined by

$$w_j = \begin{cases} \phi((j-m)/M), & m \le j \le m+M, \\ 0, & \text{else}, \end{cases} \tag{5.14}$$

with $\phi(t) = 0.54 - 0.46\cos(2\pi t)$.

Many other windows—Hanning, Blackwell, Chebyshev, Kaiser, and the like—also exist in many variations. Any can be used to perform localized frequency analysis and construct spectrograms. Some of the trade-offs are illustrated in the Matlab exercises given in the following section.

5.3 Matlab Project

5.3.1 Windows

Let $f(t) = \sin(2\pi(137)t) + 0.4\sin(2\pi(147)t)$ for $0 \le t \le 1$.

1. (a) Use Matlab to sample $f(t)$ at 1000 points $t_k = k/1000$, $0 \le k \le 999$, as

```
t = [0:999]/1000;
f = sin(2*pi*137*t) + 0.4*sin(2*pi*147*t);
```

   (b) Let `F = fft(f);`. Plot the magnitude of the first 501 components of `F`.

   (c) Construct a (rectangular) windowed version of f (but still of length 1000) with `fw = f;` followed by `fw(201:1000)=0.0;`. Compute and display the magnitude of the first 501 components of the DFT of `fw`. Can you distinguish the two constituent frequencies? Be careful—is it really obvious that the second frequency is not just side lobe leakage?

   (d) Construct a windowed version of f of length 200, with `fw2 = f(1:200)`. Compute and display the magnitude of the first 101 components of the DFT of `fw2`. Can you distinguish the two constituent frequencies? Compare with the plot of the DFT of `fw`.

   (e) Repeat parts (c) and (d) using shorter windows, such as 100 or 50, (and longer windows too). How short can the time window be and still allow resolution of the two separate frequencies? Does it seem to matter whether we treat the windowed signal as a vector of length 1000 as in part (c) or as a shorter vector as in part (d)? Does the side lobe energy confuse the issue?

2. Repeat problem 1, but use a triangular window. You can, for example, construct a window vector w of length 201 as

```
w = zeros(1,201);
w(1:101) = [0:100]/100;
w(102:201) = [99:-1:0]/100;
```

Then construct a windowed signal of length 1000 as `fw = zeros(size(f));` and `fw(1:201) = f(1:201).*w;`.
Try varying the window length. What is the shortest window that allows you to clearly distinguish the two frequencies?

3. Repeat problems 1 and 2 for the Hamming window.

5.3.2 Spectrograms

1. Construct a signal `f` with varying frequency content on the time interval $0 \le t \le 1$ sampled at 1 kHz as

```
t = [0:999]/1000;
p = 100+10*cos(2*pi*t);
f = sin(2*pi*p.*t);
```

Compute the DFT of `f` and display it is magnitude up to the Nyquist frequency 500 Hz.

2. Use the supplied "spectrogram" routine `spectro` to compute a spectrogram of the signal `f`, using a rectangular window. You might start with

```
spectro(f,1.0,0.1,50,'rect');
```

(signal of length 1.0 s, rectangular windows of length 0.1 s, 50% overlap between windows).
Experiment with the window length and overlap.

3. Repeat problem 2 using a triangular window (string argument "tri" to the spectrogram command).
4. Repeat problem 2 using a Gaussian window (string argument "gauss" to the spectrogram command).
5. Load in the "splat" sound with `load splat` (recall the sampling rate is 8192 Hz). Create an informative spectrogram, and comment on its relation to what you hear when the sound is played.

Exercises

1. Let $\mathbf{x} = (x_0, x_1, x_2, x_3)$ be a signal and $\mathbf{X} = (X_0, X_1, X_2, X_3)$ the DFT of $\mathbf{x}$. Let $\mathbf{w} = (0, 1, 1, 0)$ be the window vector and $\mathbf{y} = \mathbf{w} \circ \mathbf{x}$.

a. Compute the four-point DFT $\mathbf{X}$ explicitly/symbolically (the matrix $\mathbf{F}_4$ in equation (2.8) might be helpful).

b. Compute the four-point DFTs $\mathbf{W}$ and $\mathbf{Y}$, and verify that equation (5.6) holds.

c. Compute the two-point DFT $\widetilde{\mathbf{X}}$ of just the nonzero portion $\tilde{\mathbf{x}} = (x_1, x_2)$ of the windowed vector, and verify that Theorem 5.1 holds.

2. Let $x(t)$ be defined on the interval $[0, 1]$ as

$$x(t) = \begin{cases} 1, & t < \dfrac{1}{2}, \\ 0, & t \geq \dfrac{1}{2}. \end{cases}$$

Let $\mathbf{x} \in \mathbb{R}^N$ with components $x_m = x(m/N)$ for $0 \leq m \leq N - 1$ be a sampled version of $x(t)$ and assume, for simplicity, that N is even. Show that the DFT coefficient X_k of $\mathbf{x}$ is given by $X_0 = N/2$, $X_k = 0$ if k is even and positive, and

$$X_k = \frac{2}{1 - e^{-2\pi i k/N}}$$

if k is odd. Hence, over half of the DFT coefficients are nonzero, even though $x(t)$ consists of two "dc pieces." Hint: Write out X_k explicitly from the definition of the DFT and use the geometric summation formula $1 + z + \cdots + z^{r-1} = (1 - z^r)/(1 - z)$ for $z \neq 1$.

3. Prove Proposition 5.1.

4. Consider analyzing the local frequency content of a signal $\mathbf{x} \in \mathbb{R}^N$ by taking (rectangular) windows of length M, of the form

$$\tilde{\mathbf{x}} = (x_m, x_{m+1}, \ldots, x_{m+M-1})$$

for some value of m, then performing an M-point DFT of $\tilde{\mathbf{x}}$.

(a) Examine the extreme case where $M = 1$. In particular, compute the DFT of $\tilde{\mathbf{x}}$. Argue that $M = 1$ provides perfect localization in time, but essentially no frequency information except "local dc."

(b) Examine the extreme case $M = N$ (with $m = 0$). In particular, compute the DFT of $\tilde{\mathbf{x}}$. Argue that $M = N$ provides no localization in time but undistorted frequency information.

5. (a) Use Matlab to plot the Gaussian window defined in equation (5.13), with $N = 1000$, $m = 100$, $M = 49$, $C = 1$, and $\alpha = 5$. Compute and plot the magnitude of the DFT of this window.
(b) Change α in part (a). Verify that as α gets larger, the window $\mathbf{w}$ narrows and the DFT $\mathbf{W}$ widens, and vice versa.
(c) Use Matlab to plot the Hamming window defined in equation (5.14), with $N = 1000$, and $m = 100$, and $M = 49$. Vary M and observe the effect on the window and its DFT.

6. Investigate one of the other windows listed at the end of Section 5.2.4 (or any other windowing function). Plot the window function, and modify the supplied `spectro.m` file to have this window as an option.

7. In Example 5.6 we considered a signal that contains a time-varying frequency component $\sin(2\pi\omega(t)t)$ with $\omega(t) = 150 + 50\cos(2\pi t)$. One might expect that the "local" frequency at any time t_0 is just $\omega(t_0)$, so in Example 5.6 we had expect a low frequency of 100 Hz and a high of 200 Hz. However, Figure 5.8 indicates this is not quite the case—we see local frequencies in the spectrogram as low as (roughly) 30 Hz at time $t = 0.32$ s and as high as 400 Hz, around $t = 0.75$.
To understand this, consider a signal $f(t) = \sin(\omega(t)t)$ in a time interval $t_0 \le t \le t_0 + \Delta t$. Assume that $\omega(t)$ is differentiable.
(a) Justify the approximation

$$\omega(t)t \approx \omega(t_0)t_0 + (\omega(t_0) + \omega'(t_0)t_0)(t - t_0)$$

for $t_0 \le t \le t_0 + \Delta t$ if Δt is small. Hint: Apply the tangent line approximation to the function $\omega(t)t$.
(b) Show that to a good approximation we have

$$f(t) = \sin(a + \omega_0 t),$$

where $\omega_0 = \omega(t_0) + \omega'(t_0)t_0$ and a is some constant. Thus, ω_0 is the "local" frequency of f near $t = t_0$.
(c) Use part (b) to explain the minimum and maximum frequencies seen in the spectrogram of Figure 5.8. In particular, plot ω_0 as a function of t_0 for $0 \le t_0 \le 1$.
(d) Show that the instantaneous frequency ω_0 at time t_0 is given by

$$\omega_0 = \frac{d}{dt}(\omega(t)t)|_{t=t_0}.$$

The function $\omega(t)t$ is called the *phase function* for $\sin(\omega(t)t)$.

8. Consider the following windowed or short-time Fourier transform. Given a signal vector $\mathbf{x} \in \mathbb{C}^N$, where N factors as $N = mn$, we compute the transform $\mathbf{X} \in \mathbb{C}^N$ as follows: for each integer k in the range $0 \le k \le m - 1$, let

$\tilde{\mathbf{x}}_k$ denote the vector

$$\tilde{\mathbf{x}}_k = (x_{kn}, x_{kn+1}, \ldots, x_{kn+n-1})$$

in $\mathbb{C}^n$. Let $\tilde{\mathbf{X}}_k$ denote the n-point DFT of $\tilde{\mathbf{x}}_k$, and let the full transform $\mathbf{X} \in \mathbb{C}^N$ of $\mathbf{x}$ be obtained as the concatenation of the $\tilde{\mathbf{X}}_k$, that is,

$$\mathbf{X} = (\tilde{\mathbf{X}}_1, \tilde{\mathbf{X}}_2, \ldots, \tilde{\mathbf{X}}_m)^T$$

if we treat $\mathbf{X}$ as a column vector. In short, we transform $\mathbf{x}$ by breaking the vector into nonoverlapping pieces of length n, take the DFT of each piece, and then concatenate the results. No windowing is performed prior to transforming each piece.

(a) What does the matrix representation of the transform $\mathbf{x} \to \mathbf{X}$ look like, in relation to $\mathbf{F}_m$ (the m-point DFT matrix)? Show that $\mathbf{x} \to \mathbf{X}$ is invertible, and explain how to compute the inverse.

(b) Show that each component X_j, $0 \le j \le N-1$, of $\mathbf{X}$ can be computed as an inner product

$$X_j = (\mathbf{x}, \mathbf{v}_j)$$

for a suitable vector $\mathbf{v}_j \in \mathbb{C}^N$. (Note that X_j refers to a single component of $\mathbf{X}$, not the block $\mathbf{X}_j$.) Show that the set $\mathbf{v}_j$, $0 \le j \le N-1$ forms an orthogonal basis for $\mathbb{C}^N$.

(c) Suppose we window each piece prior to transforming; that is, take

$$\tilde{\mathbf{x}}_k = (w_0 x_{kn}, w_1 x_{kn+1}, \ldots, w_{n-1} x_{kn+n-1})$$

for some vector $\mathbf{w} \in \mathbb{R}^n$. What conditions on $\mathbf{w}$ are necessary if the transform $\mathbf{x} \to \mathbf{X}$ is to be invertible? Would you expect the corresponding vectors $\mathbf{v}_j$ from part (b) to still be orthogonal?

9. Let $\mathbf{x}$ and $\mathbf{w}$ be vectors in $L^2(\mathbb{Z})$ with discrete time Fourier transforms $X(f)$ and $W(f)$ for $-\frac{1}{2} \le f \le \frac{1}{2}$, respectively (recall Definition 4.4 and formula 4.17). For simplicity, we will assume that $\mathbf{w}$ has only finitely many nonzero components. Define $\mathbf{z}$ to have components $z_k = x_k w_k$. In this case $\mathbf{z} \in L^2(\mathbb{Z})$. Define the (circular) convolution of X and W as

$$(X * W)(f) = \int_{-1/2}^{1/2} X(g)W(f-g)\,dg,$$

where as in Chapter 4 we extend X and W periodically with period 1. Show that $Z = X * W$, where Z is, of course, the discrete time Fourier transform of $\mathbf{z}$. Assume that the $X(f)$ and $W(f)$ are "well-behaved" functions, for example, double integrals in which these functions appear can be integrated in any order. Thus, just as in the finite case, convolution in the frequency domain corresponds to a pointwise product in the time domain.

10. Let $\mathbf{x} = (\dots, x_{-1}, x_0, x_1, \dots) \in L^2(\mathbb{Z})$ have discrete time Fourier transform

$$X(f) = \sum_{k=-\infty}^{\infty} x_k e^{-2\pi ikf},$$

where $-\frac{1}{2} \le f \le \frac{1}{2}$ (recall Definition 4.4). Suppose that we window $\mathbf{x}$ symmetrically about index 0 to construct a vector

$$\tilde{\mathbf{x}} = (\dots, 0, x_{-M}, x_{-M+1}, \dots, x_{M-1}, x_M, 0, \dots)$$

in $L^2(\mathbb{Z})$. Let $\tilde{X}(f)$ denote the discrete time Fourier transform of $\tilde{\mathbf{x}}$.

(a) Show that $\tilde{X} = X * W$ where

$$W(f) = \frac{\sin((2M+1)\pi f)}{\sin(\pi f)}.$$

Hint: Use Exercise 9 with an appropriate choice for $\mathbf{w}$. Thus, windowing a signal in $L^2(\mathbf{Z})$ distorts its spectrum, but in a quantifiable way.

(b) Use part (a) and the fact that

$$\lim_{R\to\infty} \int_{-1/2}^{1/2} \frac{\sin(R\pi x)}{\sin(\pi x)} \phi(x)\, dx = \phi(0)$$

(if ϕ is continuous at $x = 0$) to show that for any fixed f we have that $\tilde{X}(f)$ converges to $X(f)$ (if X is continuous at f). Thus, the distortion disappears in the limit that the window width goes to infinity (as expected).

6
Frames

6.1 Introduction

In this chapter, we extend the notion of basic waveforms to create redundant transform methods, namely there are more waveforms than are minimally needed to analyze and synthesize a signal without loss. These expanded sets of waveforms will be called frames. The redundancy of frames can be used in at least two ways. First, we can reconstruct a signal if we have lost some the analyzed signal. Second, we can gain more information about a signal using the additional information. As an example, we will revisit the short-time Fourier transform of Chapter 5 in the guise of Gabor frames.

In this chapter, we first look at packet loss as a motivation for frames. Next, we develop the notion of frames and a perfect reconstruction formula for frames. We follow up with a look at several examples of frames, especially those that can be built up through the use of matrix group techniques. We end with brief descriptions of some applications and Matlab explorations of some of the concepts and examples of the chapter.

Frames have garnered much more interest now that vast quantities of cheap computation and communication are available. We can only cover basic ideas and simple examples since we are only devoting a chapter to the topic in this text. The goal of the chapter is to suggest an additional direction for analysis/synthesis of signals, though still based on the linear algebra techniques of the rest of the book. An excellent reference for frames from an advanced undergraduate perspective is given in [16]. Some more advanced references are [7, 21].

6.2 Packet Loss

Suppose that we have sampled, transformed, compressed, and transmitted a band limited signal as a series of packets, but one or more of the packets is

Discrete Fourier Analysis and Wavelets: Applications to Signal and Image Processing, Second Edition.
S. Allen Broughton and Kurt Bryan.

Companion Website: www.wiley.com/go/Broughton/Discrete_Fourier_Analysis_and_Wavelets

lost or corrupted. If we encoded the signal using *discrete Fourier transform* (*DFT*) or *discrete cosine transform* (*DCT*) methods then it is unlikely that we can reconstruct the signal without significant error. In this chapter, we study how to remedy this problem by using oversampling via frames. First, we start by looking a very simplified scenario for data loss, followed by a numerical example.

To make the calculations simple, we are going to use exponential signal representation and the *DFT*. Since we are assuming a band limited signal, we can assume for the purpose of our example that our signal, before sampling, has the form

$$x(t) = \sum_{k=-n}^{n} c_k \exp(2\pi ikt), \quad 0 \le t \le 1. \tag{6.1}$$

These signals form a $(d = 2n + 1)$-dimensional vector space with basis $\exp(2\pi ikt)$, $-n \le k \le n$. We know that by sampling the signal at the points $t_j = \frac{j}{2n+1}$, $0 \le j \le 2n$, we can form the sample column vector

$$\mathbf{x} = \left[f(t_0) \cdots f(t_{2n})\right]^t,$$

which completely determines the signal $x(t)$. Indeed, we can go back and forth between the permuted coefficient vector

$$\mathbf{C} = \begin{bmatrix} c_0 & \cdots & c_n & c_{-n} & \cdots & c_{-1} \end{bmatrix}^t$$

and sample vector $\mathbf{x}$ by the *DFT* formulas

$$\mathbf{x} = F\mathbf{C},$$
$$\mathbf{C} = \frac{1}{d}F^*\mathbf{x},$$

where F^* is the *DFT* matrix. So, the sampling map $x(t) \to \mathbf{x}$ is an isomorphism and all the entries of $\mathbf{x}$ are needed to reconstruct the original $x(t)$.

Given our signal sampled at the prescribed points, we transmit $\mathbf{X} = F^*\mathbf{x} = d\mathbf{C}$ in some compressed fashion. The transmission is lossy but of high fidelity. The compressed signal will be sent as a series of packets of suitable transmission size. If one or more of the packets is lost or corrupted then, even if we can partially reconstruct the signal, there will be some loss. In order to show how things can go wrong assume that we just send the vector $\mathbf{X}$, without compression, as d/s packets each of length s. That is,

$$\mathbf{X} = \left[\mathbf{X}_1^t \cdots \mathbf{X}_{d/s}^t\right]^t$$

where $\mathbf{X}_j \in \mathbb{C}^s$ is column vector. We assume that at least one of the $\mathbf{X}_j$ is lost or corrupted. In the following numerical example, we assume that the packets are of size 1 and that we lose 2 packets $\mathbf{X}_j$ and $\mathbf{X}_{j'}$ whose loss keeps the signal real.

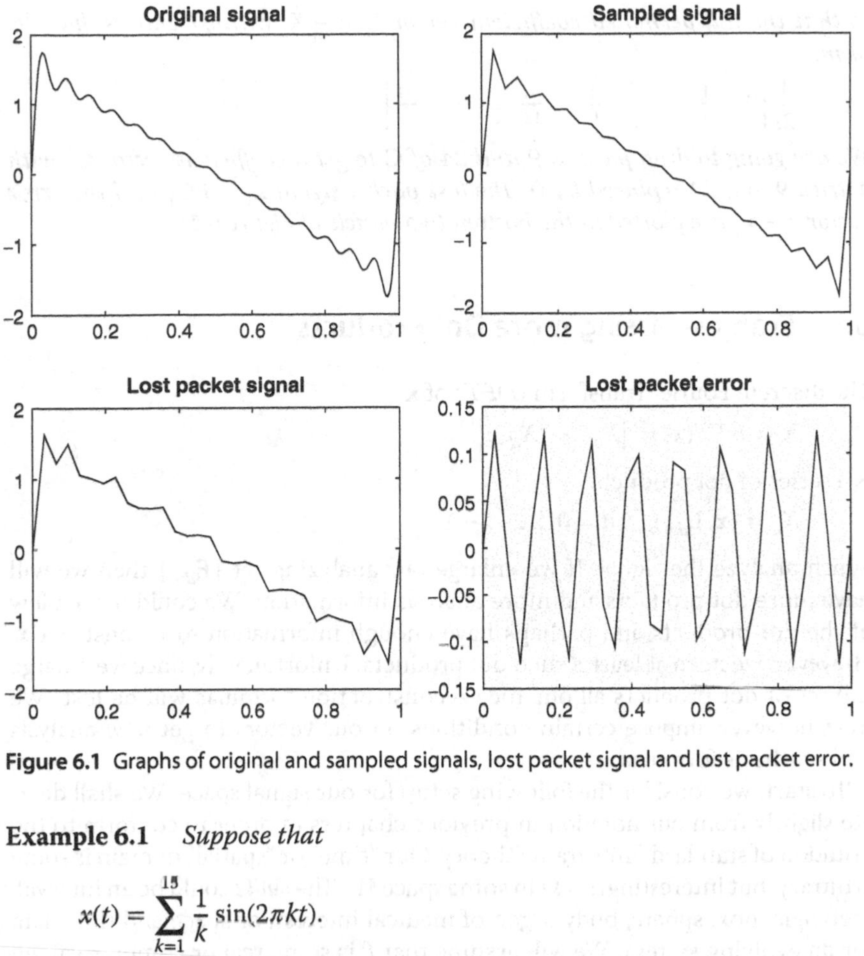

Figure 6.1 Graphs of original and sampled signals, lost packet signal and lost packet error.

Example 6.1 *Suppose that*

$$x(t) = \sum_{k=1}^{15} \frac{1}{k} \sin(2\pi kt).$$

The graph of the function and the samples at 31 points are given in the top two panels of Figure 6.1. To put $x(t)$ in the form given in equation (6.1) we rewrite:

$$\begin{aligned}
x(t) &= \sum_{k=1}^{15} \frac{1}{k} \sin(2\pi kt) \\
&= \sum_{k=1}^{15} \frac{1}{k} \frac{\exp(2\pi ikt) - \exp(-2\pi ikt)}{2i} \\
&= \frac{1}{2i} \left(\sum_{k=-15}^{-1} \frac{1}{k} \exp(2\pi ikt) + \sum_{k=1}^{15} \frac{1}{k} \exp(2\pi ikt) \right),
\end{aligned}$$

so that the full permuted coefficient vector $\mathbf{C} = \frac{1}{31}\mathbf{X}$, *with 31 entries, has the form:*

$$\frac{1}{2i}\begin{bmatrix} 0 & \frac{1}{1} & \cdots & \frac{1}{15} & \frac{-1}{15} & \cdots & \frac{-1}{1} \end{bmatrix}^t.$$

We are going to drop packets 9 and 24 of $\mathbf{C}$ *to get a coefficient vector* $\mathbf{C}_1$ *with entries 9 and 24 replaced by 0. The lost packet signal* $\mathbf{x}_1 = F\mathbf{C}_1$ *and the error vector* $\mathbf{x} - \mathbf{x}_1$ *are plotted in the bottom two panels of Figure 6.1.*

6.3 Frames—Using more Dot Products

The discrete Fourier transform (*DFT*) of $\mathbf{x}$

$$\mathbf{X} = DFT(\mathbf{x}) = \begin{bmatrix} X_0 & \cdots & X_{d-1} \end{bmatrix}^t$$

is a series of dot products

$$X_n = (\mathbf{x}, \mathbf{E}_{d,n}), \quad n = 0, \ldots, d-1,$$

which analyze the signal. If we enlarge our analyzing set $\{\mathbf{E}_{d,n}\}$ then we will have more dot products and more analysis information. We could drop a few of the dot products and perhaps have enough information to reconstruct $\mathbf{x}$. However, we need at least d such dot products. Unfortunately, once we enlarge the set of dot products all our nice reconstruction formulas will be lost. We may, however, impose certain conditions on our vectors to get new analysis and synthesis formulas.

To start, we consider the following setup for our signal space. We shall deviate slightly from our notation in previous chapters in order to conform to the notation of standard finite frame theory. Our "time" or "spatial" domain is some arbitrary, but interesting, set Ω in some space $\mathbb{R}^r$. The set Ω could be an interval, rectangle, box, sphere, body organ of medical interest, or space and time data for an evolving system. We will assume that f is some real or complex valued function defined at all $\omega \in \Omega$. Pick an appropriate, finite set $\{\omega_1, \ldots, \omega_d\} \subset \Omega$ and form the sample vector or sample signal

$$\mathbf{x}_f = (f(\omega_1), \ldots, f(\omega_d)) \in \mathbb{C}^d \text{or } \mathbb{R}^d,$$

to get a finite approximation vector for f. Uniform 1D linear, 2D rectangular, and 3D box grids are obvious choices for $\{\omega_1, \ldots, \omega_d\} \subset \Omega$. We shall consider vectors $\mathbf{x}$ obtained in this way, but not worry too much at the moment about how they were obtained. Now let $\Phi = \boldsymbol{\phi}_1, \ldots, \boldsymbol{\phi}_N$ be any sequence of vectors in $\mathbb{C}^d$. We think of the sequence $\boldsymbol{\phi}_1, \ldots, \boldsymbol{\phi}_N$ as basic discrete waveforms obtained by sampling suitable waveforms on Ω. The structure of the space Ω may guide us on how we select waveforms in Φ. We will do analysis and synthesis operations using the vectors in Φ and a (non-unique) dual set of vectors Ψ.

We are finally ready to define a *frame*.

Definition 6.1 Let $\Phi = \phi_1, \dots, \phi_N$ be a finite sequence of vectors in $\mathbb{C}^d$ (possibly with repeats) and suppose that there are constants $0 < A \leq B$ such that for all $\mathbf{x} \in \mathbb{C}^d$

$$A\|\mathbf{x}\|^2 \leq \sum_{n=1}^{N} |(\mathbf{x}, \phi_n)|^2 \leq B\|\mathbf{x}\|^2. \tag{6.2}$$

Then Φ is called a (finite) frame with *frame bounds A* and *B*.

If the inequalities are sharp then A and B are called *optimal bounds.* For optimal bounds, the ratio $\rho = \rho(\Phi) = B/A$ is called the *frame bounds ratio.*

Some specific types of frames are:

1. If $A = B$ then Φ is called a *tight frame.*
2. If all vectors satisfy $\|\phi_n\| = L > 0$ then Φ is called a *uniform frame* and if $\|\phi_n\| = 1$ for all n then Φ is called a *unit norm frame.*
3. If $(\phi_m, \phi_n) = c$ for all $n \neq m$ and all vectors have the same length then Φ is called an *equiangular frame.*

Remark 6.1 The frame Φ is a tight frame if and only if there is a constant $A > 0$ satisfying:

$$A\|\mathbf{x}\|^2 = \sum_{n=1}^{N} |(\mathbf{x}, \phi_n)|^2. \tag{6.3}$$

If the frame Φ is an orthogonal basis then by Parseval's equation we have a tight frame with $A = 1$. So we may consider equation (6.3) as a generalization of Parseval's equation. By scaling $\phi_n \to \phi_n/\sqrt{A}$, we can make the constant $A = 1$, though we no longer have a unit norm frame. Such a frame is called a Parseval frame, with Parseval equation

$$\|\mathbf{x}\|^2 = \sum_{n=1}^{N} |(\mathbf{x}, \phi_n)|^2. \tag{6.4}$$

Remark 6.2 For any finite set $\Phi \subset \mathbb{C}^d$ equation (6.2) is satisfied for

$$A = \min_{\|\mathbf{x}\|\neq 0} \frac{1}{\|\mathbf{x}\|^2} \sum_{n=1}^{N} |(\mathbf{x}, \phi_n)|^2, \quad B = \max_{\|\mathbf{x}\|\neq 0} \frac{1}{\|\mathbf{x}\|^2} \sum_{n=1}^{N} |(\mathbf{x}, \phi_n)|^2, \tag{6.5}$$

and these bounds will be sharp. We automatically have $B > 0$ if Φ contains a nonzero vector. If $A = 0$, then there is a unit vector $\mathbf{x}$ orthogonal to all ϕ_n. Thus, $A > 0$ if and only if Φ spans $\mathbb{C}^d$.

Remark 6.3 Suppose Φ is a unit norm frame. If A is very small then there will be vectors that are nearly orthogonal to every vector in Φ. This is not desirable since we cannot capture the nature of a signal nearly perpendicular to the vectors in Φ. If B is very large then the vectors of Φ bunch up more in some directions. This is also not desirable. So, having the frame bound ratio B/A close to 1 says that the vectors in Φ are "nicely distributed" in $\mathbb{C}^d$. These ideas are illustrated in Figures 6.2 and 6.6. Later, in Section 6.4.5, we shall see that a small frame bounds ratio is important for numerical stability in analysis and synthesis of signals.

Example 6.2 *The Mercedes-Benz frame is the simplest interesting frame. It is a tight, equiangular, unit norm frame, pictured in Figure 6.2. The vectors are:*

$$\boldsymbol{\phi}_1 = \begin{bmatrix} 1 \\ 0 \end{bmatrix}, \quad \boldsymbol{\phi}_2 = \begin{bmatrix} -\frac{1}{2} \\ \frac{\sqrt{3}}{2} \end{bmatrix}, \quad \boldsymbol{\phi}_3 = \begin{bmatrix} -\frac{1}{2} \\ \frac{-\sqrt{3}}{2} \end{bmatrix}.$$

If $\mathbf{x} = \begin{bmatrix} x_1 \\ x_2 \end{bmatrix}$ *then*

$$\sum_{n=1}^{3} (\mathbf{x}, \boldsymbol{\phi}_n)^2 = x_1^2 + \left(-\frac{1}{2}x_1 + \frac{\sqrt{3}}{2}x_2\right)^2 + \left(-\frac{1}{2}x_1 - \frac{\sqrt{3}}{2}x_2\right)^2$$
$$= \frac{3}{2}x_1^2 + \frac{3}{2}x_2^2 = \frac{3}{2}\|\mathbf{x}\|^2$$

and the optimal frame bounds both equal $\frac{3}{2}$.

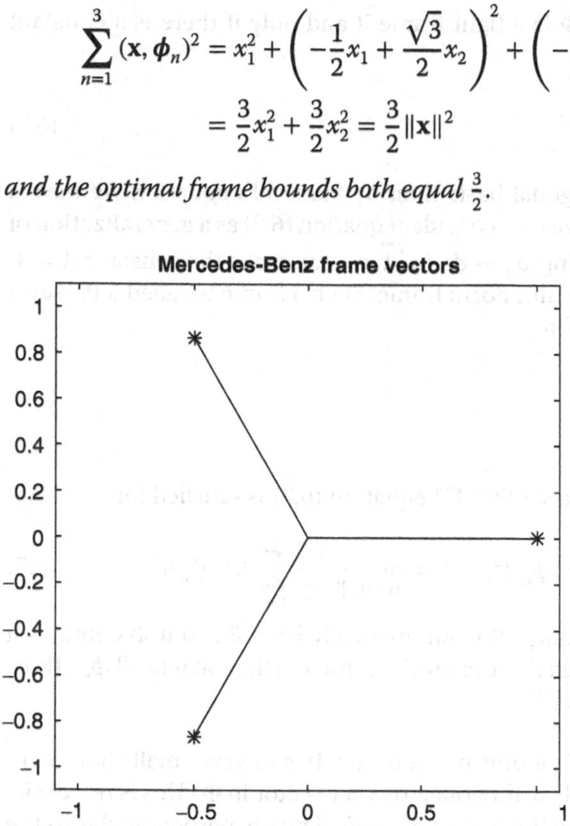

Figure 6.2 Mercedes-Benz frame.

Example 6.3 *Frames are ubiquitous. Indeed, by Remark 6.2 every finite subset of $\mathbb{C}^d$ containing a basis of $\mathbb{C}^d$ is a frame (see Exercise 1).*

Example 6.4 *Tight frames are also ubiquitous. We can construct tight frames from other tight frames, and construct tight frames from generic frames. See Proposition 6.1 and the discussion in Section 6.5 on canonical tight frames.*

Proposition 6.1 *If Φ and Φ' are tight frames with N and N' vectors, respectively, and frame constants A and A', respectively, then the concatenation*

$$\Phi \frown \Phi' = \phi_1, \ldots, \phi_N, \phi'_1, \ldots, \phi'_{N'}$$

of the two frames is a tight frame with constant $A + A'$.

Proof: Since Φ and Φ' are tight frames then $A\|\mathbf{x}\|^2 = \sum_{n=1}^{N} |(\mathbf{x}, \phi_n)|^2$ and $A'\|\mathbf{x}\|^2 = \sum_{n=1}^{N'} |(\mathbf{x}, \phi'_n)|^2$. Therefore,

$$(A + A')\|\mathbf{x}\|^2 = \sum_{n=1}^{N} |(\mathbf{x}, \phi_n)|^2 + \sum_{n=1}^{N'} |(\mathbf{x}, \phi'_n)|^2.$$

It follows that the concatenation of Φ and Φ' is a tight frame.

6.4 Analysis and Synthesis with Frames

In this section, we consider the analogues of Fourier analysis and synthesis operations using frames. In particular we are interested in perfect reconstruction formulas and partial reconstruction. For these operations tight frames or frames with small frame bounds ratio B/A are important.

In this section and subsequent sections, we shall use freely some facts from linear algebra on the *adjoint, Hermitian transpose,* or *conjugate transpose* $A^* = \overline{A}^t$ of a matrix A. In particular the scalar product $(\mathbf{x}, \mathbf{y})$ of two column vectors $\mathbf{x}$ and $\mathbf{y}$ is given by

$$(\mathbf{x}, \mathbf{y}) = \mathbf{y}^*\mathbf{x}$$

See Exercise 6 for more detail on the Hermitian transpose and scalar products.

6.4.1 Analysis and Synthesis

The goal of analysis and synthesis is to find to *frame coefficients* c_n so that for a given $\mathbf{x}$

$$\mathbf{x} = \sum_{n=1}^{N} c_n \phi_n.$$

The c_n will be found with the aid of a dual frame Ψ, which we introduce shortly. To perform analysis and synthesis with frames, we will use matrix methods as in the case of the *DFT* and *DCT*.

We define the *frame matrix F* in $M_{d,N}(\mathbb{C})$ by

$$F = \left[\boldsymbol{\phi}_1\ \boldsymbol{\phi}_2\ \cdots\ \boldsymbol{\phi}_N\right]. \tag{6.6}$$

The *analysis coefficients* of **x** with respect to Φ are given by

$$(\mathbf{x}, \boldsymbol{\phi}_n) = \boldsymbol{\phi}_n^* \mathbf{x}.$$

The vector consisting of the $(\mathbf{x}, \boldsymbol{\phi}_n)$ is denoted by $\mathbf{X}_\Phi \in \mathbb{C}^N$, which we shall call the *(frame) Fourier transform* of **x** with respect to Φ. In matrix form, the transform is given by

$$\mathbf{X}_\Phi = \begin{bmatrix} (\mathbf{x}, \boldsymbol{\phi}_1) \\ (\mathbf{x}, \boldsymbol{\phi}_2) \\ \vdots \\ (\mathbf{x}, \boldsymbol{\phi}_N) \end{bmatrix} = \begin{bmatrix} \boldsymbol{\phi}_1^* \mathbf{x} \\ \boldsymbol{\phi}_2^* \mathbf{x} \\ \vdots \\ \boldsymbol{\phi}_N^* \mathbf{x} \end{bmatrix} = F^* \mathbf{x}.$$

We use the subscript Φ only when we wish to make the frame explicit. The operator $\mathbf{x} \to F^*\mathbf{x}$, $\mathbb{C}^d \to \mathbb{C}^N$ is called the *analysis operator.* Some pictures of frame Fourier transforms of 1D signals are given in Section 6.8.

Next, given $\mathbf{Y} = \left[a_1\ a_2\ \cdots\ a_N\right]^t$, a proposed vector of frame coefficients, we may synthesize a signal from Φ:

$$\begin{aligned} &a_1\boldsymbol{\phi}_1 + a_2\boldsymbol{\phi}_1 + \cdots + a_N\boldsymbol{\phi}_N \\ &= \left[\boldsymbol{\phi}_1\ \boldsymbol{\phi}_2\ \cdots\ \boldsymbol{\phi}_N\right] \left[a_1\ a_2\ \cdots\ a_N\right]^t \\ &= F\mathbf{Y}. \end{aligned}$$

The operator $\mathbf{Y} \to F\mathbf{Y}$, $\mathbb{C}^N \to \mathbb{C}^d$ is called the *synthesis operator* with respect to Φ.

If Φ were an orthonormal basis then FF^* would be the identity and we get the reconstruction formula $\mathbf{x} = FF^*\mathbf{x} = F\mathbf{X}$ allowing us to reconstruct **x** from its Fourier transform $\mathbf{X} = \mathbf{X}_\Phi$. Seeking a reconstruction formula, let us consider the composition of analysis followed by synthesis.

$$\mathbf{x} \to F^*\mathbf{x} = \mathbf{X} \to F\mathbf{X} = FF^*\mathbf{x}.$$

The combined operator

$$S = FF^* \tag{6.7}$$

is called the *frame operator*. The frame operator S and its companion the *Gram matrix*,

$$G = F^*F = [\boldsymbol{\phi}_i^* \boldsymbol{\phi}_j] = [(\boldsymbol{\phi}_i, \boldsymbol{\phi}_j)], \tag{6.8}$$

are extremely important in the study of frames and their properties.

The frame operator S is a $d \times d$ matrix, operating on signals $\mathbf{x} \in \mathbb{C}^d$. Observe that S is self adjoint

$$S^* = (FF^*)^* = F^{**}F^* = FF^* = S. \tag{6.9}$$

If we assume that S is invertible, then we have the following reconstruction formulas, in matrix form,

$$\mathbf{x} = S^{-1}(FF^*)\mathbf{x} = (S^{-1}F)F^*\mathbf{x},$$
$$\mathbf{x} = (FF^*)S^{-1}\mathbf{x} = F(S^{-1}F)^*\mathbf{x}.$$

However, we can get much nicer statements by considering the dual frame, which we introduce in the following section.

6.4.2 Dual Frame and Perfect Reconstruction

We continue to assume that S is invertible. Considering the columns of $S^{-1}F$ we construct the (*canonical*) *dual frame* $\Phi_d = \Psi = \psi_1, \ldots, \psi_N$,

$$\psi_1 = S^{-1}\phi_1, \ldots, \psi_N = S^{-1}\phi_N,$$

so that

$$S^{-1}F = S^{-1}\left[\phi_1\ \phi_2\ \cdots\ \phi_N\right] = \left[\psi_1\ \psi_2\ \cdots\ \psi_N\right].$$

We shall show that Ψ is indeed a frame and determine its bounds in Section 6.6. To distinguish and relate the two frames, we denote the frame matrices by

$$F = \left[\phi_1\ \phi_2\ \cdots\ \phi_N\right] \tag{6.10}$$

and

$$F_d = \left[\psi_1\ \psi_2\ \cdots\ \psi_N\right]. \tag{6.11}$$

We call F_d^*, F_d and $S_d = F_dF_d^*$ the *dual analysis, synthesis and frame operators*, respectively. By construction, the operators satisfy

$$F_d = S^{-1}F,$$
$$S_d = F_dF_d^* = S^{-1}FF^*(S^{-1})^* = S^{-1}SS^{-1} = S^{-1}$$

and

$$F_dF^* = S^{-1}FF^* = S^{-1}S = I_d, \tag{6.12}$$
$$FF_d^* = (F_dF^*)^* = I_d^* = I_d. \tag{6.13}$$

The last two equations show that Φ-analysis followed by the dual Ψ-synthesis gives us a *perfect reconstruction formula*, and vice versa.

Theorem 6.1 *Let $\Phi = \phi_1, \ldots, \phi_N$ be a frame with invertible frame operator S and dual frame $\Psi = \psi_1 = S^{-1}\phi_1, \ldots, \psi_N = S^{-1}\phi_N$. Then Ψ-analysis followed by Φ-synthesis reconstructs the original signal, that is,*

$$\mathbf{x} = \sum_{n=1}^{N} (\mathbf{x}, \psi_n)\phi_n. \tag{6.14}$$

In addition, there is a dual formula

$$\mathbf{x} = \sum_{n=1}^{N} (\mathbf{x}, \boldsymbol{\phi}_n)\boldsymbol{\psi}_n. \tag{6.15}$$

Proof: Rewriting $\mathbf{x} = I_d\mathbf{x} = FF_d^*\mathbf{x}$ we get

$$\mathbf{x} = F(F_d^*\mathbf{x}) = \begin{bmatrix}\boldsymbol{\phi}_1 & \boldsymbol{\phi}_2 & \cdots & \boldsymbol{\phi}_N\end{bmatrix}\begin{bmatrix}(\mathbf{x}, \boldsymbol{\psi}_1)\\ (\mathbf{x}, \boldsymbol{\psi}_2)\\ \vdots \\ (\mathbf{x}, \boldsymbol{\psi}_N)\end{bmatrix}$$

$$= \sum_{n=1}^{N} (\mathbf{x}, \boldsymbol{\psi}_n)\boldsymbol{\phi}_n.$$

The proof of the second formula is similar.

A complication of perfect reconstruction is construction of the dual frame. But, if S is a scalar matrix $S = AI_d$ then we get a simple formula for the dual frame

$$\boldsymbol{\psi}_n = S^{-1}\boldsymbol{\phi}_n = \frac{1}{A}\boldsymbol{\phi}_n.$$

In the next section, we shall prove that for a tight frame with equality (6.3) that indeed $S = AI_d$. This also proves the following proposition.

Proposition 6.2 *Suppose that Φ is a tight frame with frame constant A. Then*

$$\mathbf{x} = \frac{1}{A}\sum_{n=1}^{N} (\mathbf{x}, \boldsymbol{\phi}_n)\boldsymbol{\phi}_n. \tag{6.16}$$

6.4.3 Partial Reconstruction

If we only have a subset of coefficients $(\mathbf{x}, \boldsymbol{\psi}_i)$, then we can reconstruct a part of the signal. In particular, suppose that we know $(\mathbf{x}, \boldsymbol{\psi}_i)$, $i = 1, \ldots, n$. Then, we get a partial Fourier sum

$$\mathbf{x}_n = \sum_{i=1}^{n} (\mathbf{x}, \boldsymbol{\psi}_i)\boldsymbol{\phi}_i$$

with "limit"

$$\lim_n \mathbf{x}_n = \mathbf{x}.$$

If we were considering infinite frames the limit would be legitimate. We may judiciously choose which partial sums to consider as we did in the progressive JPEG transmission.

The operation $\mathbf{x} \to \mathbf{x}_n$ may be written in matrix form

$$\mathbf{x}_n = \sum_{i=1}^{n} (\mathbf{x}, \boldsymbol{\psi}_i)\boldsymbol{\phi}_i = \sum_{i=1}^{n} (\boldsymbol{\psi}_i^* \mathbf{x})\boldsymbol{\phi}_i = \sum_{i=1}^{n} \boldsymbol{\phi}_i \boldsymbol{\psi}_i^* \mathbf{x} = \left(\sum_{i=1}^{n} \boldsymbol{\phi}_i \boldsymbol{\psi}_i^* \right) \mathbf{x}.$$

The *partial sum operator* is a linear combination of the rank one operators $\boldsymbol{\phi}_i \boldsymbol{\psi}_i^*$. Each of the matrices has range $\mathbb{C}\boldsymbol{\phi}_i$ so its rank is 1. The partial sum operators also have a "limit":

$$\lim_n \sum_{i=1}^{n} \boldsymbol{\phi}_i \boldsymbol{\psi}_i^* = I_d.$$

More precisely, we have a finite decomposition of the identity into rank one operators.

$$\sum_{i=1}^{N} \boldsymbol{\phi}_i \boldsymbol{\psi}_i^* = I_d. \tag{6.17}$$

6.4.4 Other Dual Frames

There are many dual frames that allow perfect reconstruction. Indeed, suppose we have another candidate for a dual-frame matrix

$$F' = \left[\boldsymbol{\psi}_1' \ \boldsymbol{\psi}_2' \ \cdots \ \boldsymbol{\psi}_N' \right], \tag{6.18}$$

where we assume $F(F')^* = I_d$, and consequently $F'F^* = I_d$. These are the analogs of equations (6.12) and (6.13). By the proof of Theorem 6.1 we get perfect reconstruction, namely:

$$\mathbf{x} = \sum_{n=1}^{N} (\mathbf{x}, \boldsymbol{\psi}_n')\boldsymbol{\phi}_n \tag{6.19}$$

and

$$\mathbf{x} = \sum_{n=1}^{N} (\mathbf{x}, \boldsymbol{\phi}_n)\boldsymbol{\psi}_n'. \tag{6.20}$$

Now $F(F')^* = I_d = FF_d^*$ so $F(F' - F_d)^* = 0$. Letting $K = F' - F_d$ we get

$$F' = F_d + K = S^{-1}F + K, \tag{6.21}$$

$$KF^* = 0. \tag{6.22}$$

Because of equation (6.19) we have some choices in determining frame coefficients. One possible use of other dual frames is to redundantly create several frame Fourier transforms $\mathbf{X}_{\Psi'}$. If one of these suffers packet loss an alternative uncorrupted transform can be used (see Section 6.8).

What is special about the canonical dual frame? The following proposition gives the answer.

Proposition 6.3 *Suppose that Φ is a frame with canonical dual frame Ψ. Given any other frame Ψ' dual to Φ the following holds:*

$$\sum_{n=1}^{N} |(\mathbf{x}, \psi_n)|^2 \leq \sum_{n=1}^{N} |(\mathbf{x}, \psi_n')|^2$$

for every $\mathbf{x}$.

The proof is left to the reader in Exercise 7.

The matrices K can be easily produced using the singular value decomposition of F. Indeed, we write $F = U \begin{bmatrix}\Sigma & Z\end{bmatrix} V^*$, where U and V are unitary $d \times d$ and $N \times N$ matrices, Σ is an invertible $d \times d$ diagonal matrix of the singular values of F, and Z is a $d \times (N - d)$ matrix of zeros. So, $KF^* = 0$ may be written

$$KV \begin{bmatrix}\Sigma \\ Z\end{bmatrix} U^* = 0.$$

Write $KV = \begin{bmatrix}L_1 & L_2\end{bmatrix}$ where L_1 and L_2 are $d \times d$ and $d \times (N - d)$ matrices, respectively. Then

$$\begin{aligned} 0 &= \begin{bmatrix}L_1 & L_2\end{bmatrix} \begin{bmatrix}\Sigma \\ Z\end{bmatrix} U^* = (L_1\Sigma + L_2 Z)U^*, \\ 0 &= L_1 \Sigma U^*, \\ L_1 &= 0. \end{aligned}$$

Thus, a typical K has the form

$$K = \begin{bmatrix}0 & L_2\end{bmatrix} V^* \tag{6.23}$$

with arbitrary matrix L_2.

6.4.5 Numerical Concerns

In a typical application of frames, either analysis or synthesis, there may be an error in the inputs or intermediate results. We are concerned about the bounding the resulting error. We shall now take a small side tour to review the condition number of a matrix, and then state a result on numerical stability of analysis and synthesis in Proposition 6.4.

6.4.5.1 Condition Number of a Matrix

Suppose we have a linear transformation $A : \mathbb{C}^m \to \mathbb{C}^n$, $\mathbf{y} = A\mathbf{x}$. If $\mathbf{x}$ is our desired input, but we make an error $\Delta\mathbf{x}$, then $\Delta\mathbf{y}$, the error of the output, is $\Delta\mathbf{y} = A(\mathbf{x} + \Delta\mathbf{x}) - A\mathbf{x} = A\Delta\mathbf{x}$. Assuming that neither $\mathbf{x}$ nor $\mathbf{y}$ equals zero, the ratio for relative errors under the transformation A is

$$\frac{\|\Delta\mathbf{y}\|/\|\mathbf{y}\|}{\|\Delta\mathbf{x}\|/\|\mathbf{x}\|} = \frac{\|\Delta\mathbf{y}\|}{\|\Delta\mathbf{x}\|}\frac{\|\mathbf{x}\|}{\|\mathbf{y}\|} = \frac{\|A\Delta\mathbf{x}\|}{\|\Delta\mathbf{x}\|}\frac{\|\mathbf{x}\|}{\|A\mathbf{x}\|}$$

$$= \sqrt{\frac{(A\Delta\mathbf{x}, A\Delta\mathbf{x})}{(\Delta\mathbf{x}, \Delta\mathbf{x})}}\sqrt{\frac{(\mathbf{x}, \mathbf{x})}{(A\mathbf{x}, A\mathbf{x})}}.$$

To get an overall upper bound, we maximize the quotient over all possible non-zero $\mathbf{x}$ and $\Delta\mathbf{x}$. There is, however, a small problem when A has a nontrivial null space. By adding a large vector $\mathbf{v}$ ($\Delta\mathbf{v}$, respectively) onto $\mathbf{x}$ ($\Delta\mathbf{x}$, respectively) with $A\mathbf{v} = \mathbf{0}$ ($A\Delta\mathbf{v} = \mathbf{0}$, respectively) the right factor becomes very large (left factor becomes very small, respectively) without actually changing $\mathbf{y}$ or $\Delta\mathbf{y}$. So we will maximize the quotient over all nonzero $\mathbf{x}$ and $\Delta\mathbf{x}$ in the orthogonal complement of the nullspace of A. We use again the singular value decomposition $A = U\Sigma V^*$, where U and V are unitary of the right size,

$$\Sigma = \begin{bmatrix} D & 0 \\ 0 & 0 \end{bmatrix},$$

where D is a $rk(A) \times rk(A)$ invertible diagonal matrix with positive real entries, and the zero matrices are of the right size. The entries of D are the singular values of A which are the square roots of the nonzero eigenvalues of A^*A or AA^*. Consider any vector $\mathbf{z}$ and set $\mathbf{w} = V^*\mathbf{z}$. We have:

$$\sqrt{\frac{(A\mathbf{z}, A\mathbf{z})}{(\mathbf{z}, \mathbf{z})}} = \sqrt{\frac{(U\Sigma V^*\mathbf{z}, U\Sigma V^*\mathbf{z})}{(\mathbf{z}, \mathbf{z})}}$$

$$= \sqrt{\frac{(\Sigma V^*\mathbf{z}, \Sigma V^*\mathbf{z})}{(V^*\mathbf{z}, V^*\mathbf{z})}}$$

$$= \sqrt{\frac{(\Sigma\mathbf{w}, \Sigma\mathbf{w})}{(\mathbf{w}, \mathbf{w})}}.$$

Now $\mathbf{z} = V\mathbf{w}$ is in the orthogonal complement of the kernel of A if and only if $\mathbf{w} = \begin{bmatrix}\mathbf{u}^t & 0\end{bmatrix}^t$ where $\mathbf{u} \in \mathbb{C}^{rk(A)}$. It follows that

$$\sqrt{\frac{(\Sigma\mathbf{w}, \Sigma\mathbf{w})}{(\mathbf{w}, \mathbf{w})}} = \sqrt{\frac{(D\mathbf{u}, D\mathbf{u})}{(\mathbf{u}, \mathbf{u})}}.$$

The maximum and minimum values of the latter quotient are $\sigma_{\max}$ and $\sigma_{\min}$ the largest and smallest singular values of A. Thus,

$$\max \frac{\|\Delta \mathbf{y}\|/\|\mathbf{y}\|}{\|\Delta \mathbf{x}\|/\|\mathbf{x}\|} \le \frac{\sigma_{\max}}{\sigma_{\min}},$$

where the maximum is restricted to nonzero vectors in the orthogonal complement of the kernel of A. This quantity is called the *condition number* of A. So, we get the following proposition on the condition numbers of analysis and synthesis operators of a frame and its dual.

Proposition 6.4 *Let Φ be a frame with frame bounds ratio $\rho = B/A$. The condition numbers of F, F^*, F_d, F_d^* are all the same and equal $\sqrt{\rho}$.*

Proof: The positive singular values of F and F^* are the same and equal the square roots of the eigenvalues of $S = FF^*$. By Proposition 6.5, $B/A = \sigma^2_{\max}/\sigma^2_{\min}$ and so the condition numbers of F and F^* both equal $\sqrt{B/A} = \sqrt{\rho}$. Since $F_d F_d^* = S^{-1}$ then the condition numbers of F_d and F_d^* also equal $\sqrt{\rho}$.

For more on the projection $\mathbb{C}^N \to \ker F$ see Exercise 8.

6.5 Initial Examples of Frames

6.5.1 Circular Frames in $\mathbb{R}^2$

Our first example generalizes the Mercedes-Benz frame. Let $N \ge 3$ and set

$$\boldsymbol{\phi}_n = \begin{bmatrix} \cos\left(\frac{2\pi n}{N}\right) \\ \sin\left(\frac{2\pi n}{N}\right) \end{bmatrix}, \quad 1 \le n \le N,$$

yielding a unit norm frame. The $\boldsymbol{\phi}_n$ are equally spaced points on the circle. The frame operator $FF^* = S = [s_{i,j}]$ satisfies:

$$s_{1,1} = \sum_{n=1}^{N} \cos^2\left(\frac{2\pi n}{N}\right) = \frac{N}{2}, \tag{6.24}$$

$$s_{2,2} = \sum_{n=1}^{N} \sin^2\left(\frac{2\pi n}{N}\right) = \frac{N}{2}, \tag{6.25}$$

$$s_{1,2} = s_{2,1} = \sum_{n=1}^{N} \sin\left(\frac{2\pi n}{N}\right) \cos\left(\frac{2\pi n}{N}\right) = 0. \tag{6.26}$$

The proofs of these formulas are left to the reader as Exercise 11. It follows that $\boldsymbol{\phi}_1, \dots, \boldsymbol{\phi}_N$ is a tight unit norm frame. The Gram matrix $F^tF = G = [g_{i,j}]$ is given by

$$\begin{aligned} g_{i,j} = (\boldsymbol{\phi}_i, \boldsymbol{\phi}_j) &= \cos\left(\frac{2\pi i}{N}\right)\cos\left(\frac{2\pi j}{N}\right) + \sin\left(\frac{2\pi i}{N}\right)\sin\left(\frac{2\pi j}{N}\right) \\ &= \cos\left(\frac{2\pi(i-j)}{N}\right). \end{aligned}$$

A way to visualize the Gram matrix is by using a scaled image matrix as in Figure 6.3, where $N = 10$. In the figure white represents 1 and black represents -1, and gray is used for intermediate values. For a real unit norm frame the entries of the Gram matrix satisfy $-1 \le g_{i,j} \le 1$.

6.5.2 Extended DFT Frames and Harmonic Frames

We generalize the previous example by tweaking the *DFT*. The *DFT* on $\mathbb{C}^d$ is constructed from discretized exponential waveforms $\exp(2\pi int)$ on $0 \le t \le 1$ with d equally spaced frequencies $n = 0, \dots, d-1$. The sample points are $t_j = \frac{j}{d}$, $j = 0, \dots, d-1$. To generalize we select times $0 = t_0 \le t_1 \le$

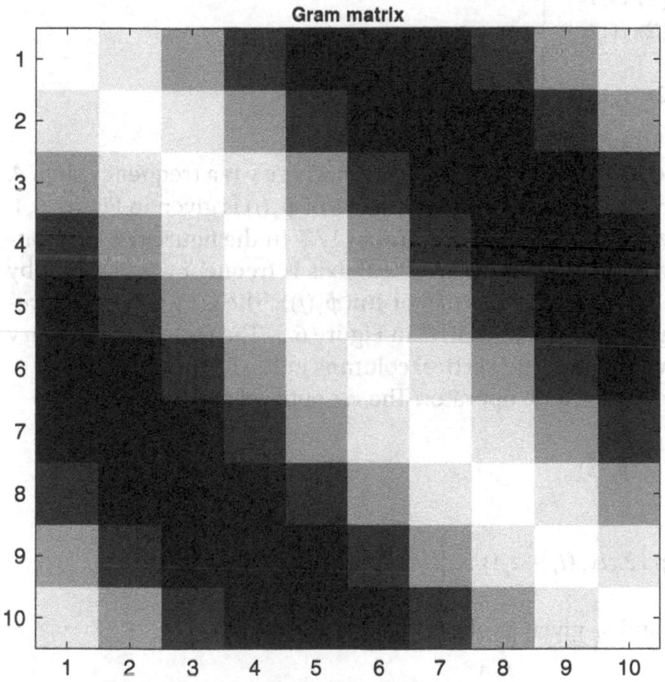

Figure 6.3 Gram matrix of circular frame.

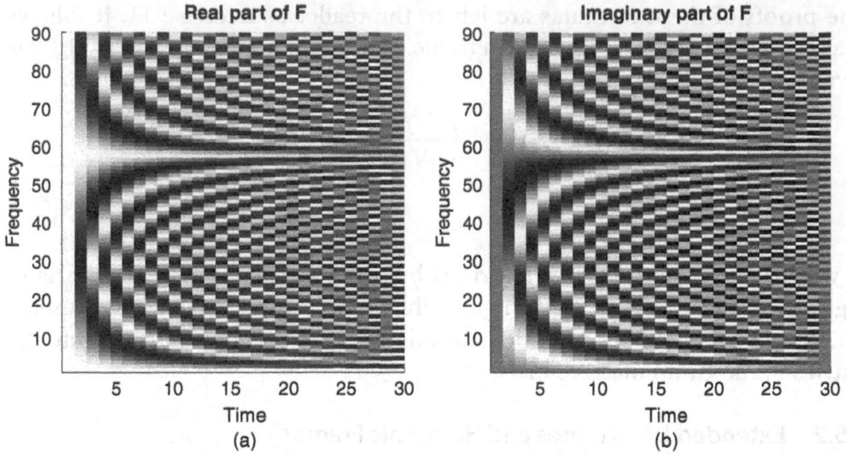

Figure 6.4 Harmonic frames.

$\cdots \leq t_{d-1} \leq 1$ and frequencies $0 = v_0 < v_1 < \cdots < v_{N-1}$ to construct a frame to analyze functions sampled at the times t_j. Define

$$\boldsymbol{\phi}_n = \begin{bmatrix} \exp\left(2\pi i v_n t_0\right) \\ \exp\left(2\pi i v_n t_1\right) \\ \vdots \\ \exp\left(2\pi i v_n t_{d-1}\right) \end{bmatrix}.$$

We shall call such a frame an *extended DFT frame.*

To make things concrete set $t_j = \frac{j}{d}$ and $v_n = vn$ where v is a frequency step. A matrix image plot of the real and imaginary parts of $\boldsymbol{\phi}_k(j)$ is given in Figure 6.4. The parameters are $d = 30, N = 90, v = d/N = 1/3$. In the figure the horizontal axis is time t_j, labeled by j, and the vertical axis is frequency v_k, labeled by k. The k, j block in the image is $\text{Re}(\boldsymbol{\phi}_k(j))$ or $\text{Im}(\boldsymbol{\phi}_k(j))$. The gray level interpretation is the same as for the Gram matrix in Figure 6.3. The real and imaginary parts of the frame vectors are the vertical columns in the images.

Now, let us analyze the frame operator. The r, s entry of $S = FF^*$ is

$$\begin{aligned} S_{r,s} &= \sum_{n=0}^{N-1} \boldsymbol{\phi}_n(r)\overline{\boldsymbol{\phi}_n(s)} \\ &= \sum_{n=0}^{N-1} \exp\left(2\pi i v_n (t_r - t_s)\right). \end{aligned}$$

If we assume the t_j and v_k given as above then

$$\sum_{n=0}^{N-1} \exp\left(2\pi i v_n (t_r - t_s)\right) = \sum_{n=0}^{N-1} \exp\left(2\pi i n \frac{v}{d}(r - s)\right)$$

$$= \sum_{n=0}^{N-1} \exp\left(2\pi i \frac{\nu}{d}(r-s)\right)^n$$

$$= \frac{1 - \exp\left(2\pi i \frac{\nu}{d}(r-s)\right)^N}{1 - \exp\left(2\pi i \frac{\nu}{d}(r-s)\right)}$$

and so,

$$S_{r,s} = \frac{1 - \exp\left(2\pi i \frac{\nu N}{d}(r-s)\right)}{1 - \exp\left(2\pi i \frac{\nu}{d}(r-s)\right)}. \tag{6.27}$$

If $\frac{\nu}{d}(r-s)$ is an integer, then the sum is N, specifically for diagonal elements $S_{r,r} = N$. Thus, S is a Hermitian, Toeplitz matrix. Write $\epsilon = \frac{\nu N}{d}$ so $\nu = \frac{\epsilon d}{N}$. The quantity ϵ gives us important information about the frame. The numerators and denominators of $S_{r,s}$ are

$$num = 1 - \exp\left(2\pi i \frac{\nu N}{d}(r-s)\right) = 1 - \exp(2\pi i \epsilon (r-s))$$

$$den = 1 - \exp\left(2\pi i \frac{\nu}{d}(r-s)\right) = 1 - \exp\left(2\pi i \epsilon \frac{(r-s)}{N}\right)$$

The formula is valid as long as $\epsilon \frac{(r-s)}{N}$ is not an integer. If $\epsilon = 1$, $\frac{(r-s)}{N}$ cannot be an integer since $|r-s| < N$. It follows that if $\epsilon = 1$ then $S = NI_d$ and the frame is tight. These frames are called *harmonic frames*. See the Matlab experiments in Section 6.9.

Remark 6.4 As a variation we can we can scale the components of a frame vectors by a mask $(a_0, \dots, a_{d-1})$ so that

$$\phi_n = \begin{bmatrix} \exp\left(2\pi i \nu_n t_0\right) a_0 \\ \exp\left(2\pi i \nu_n t_1\right) a_1 \\ \vdots \\ \exp(2\pi i \nu_n t_{d-1}) a_{d-1} \end{bmatrix}.$$

The new frame matrix will be DF where $D = diag(a_0, \dots, a_{d-1})$. This will be used in the Gabor transform in Section 6.7.

6.5.3 Canonical Tight Frame

With the simplified reconstruction formula (6.16) in mind, given any frame it is possible to construct a tight frame from it. Before proceeding, we use the fact that S is unitarily diagonalizable with positive real eigenvalues (see Proposition 6.5) to create a square root of S. So, we have: $S = U\,DU^*$, where

$U^* = U^{-1}$ and $D = diag(\lambda_1, \dots, \lambda_d)$, $0 \le \lambda_1 \le \cdots \le \lambda_d$. The square root of S, $S^{1/2}$, is defined by

$$S^{1/2} = UD^{1/2}U^*,$$

where $D^{1/2} = diag(\sqrt{\lambda_1}, \dots, \sqrt{\lambda_d})$. Obviously, $(S^{1/2})^2 = S$ and S and $S^{1/2}$ commute. If S is invertible then $S^{-1/2} = (S^{1/2})^{-1}$. We define a new frame Φ_s by means of the frame matrix

$$F_s = S^{-1/2}F.$$

The frame operator for Φ_s is

$$\begin{aligned} F_sF_s^* &= S^{-1/2}F(S^{-1/2}F)^* = S^{-1/2}FF^*(S^{-1/2})^* \\ &= S^{-1/2}SS^{-1/2} = I_d. \end{aligned}$$

It follows that Φ_s is a tight frame with frame constant $A = 1$, but it may not be nicely related to the original frame. Also, it is unlikely that it is an equal norm frame. The frame Φ_s is a Parseval frame since, in equation (6.3), $A = 1$.

6.5.4 Frames for Images

The next example creates frames for images and generalizes the two-dimensional *DFT*. All proofs are deferred to Exercises 14 and 15. The space of $d_1 \times d_2$ matrices is the vector space $\mathbb{C}^{d_1d_2}$. Let $\Phi = \{\phi_1, \dots, \phi_{N_1}\} \subset \mathbb{C}^{d_1}$ and $\Phi' = \{\phi'_1, \dots, \phi'_{N_2}\} \subset \mathbb{C}^{d_2}$ be two frames with frame matrices F_1, F_2, and frame bounds A_1, B_1, and A_2, B_2, respectively. Then an analysis operator $\mathbb{C}^{d_1d_2} \to \mathbb{C}^{N_1N_2}$ is defined by $X \to F_1^*X(F_2^*)^t$. Likewise, a synthesis operator $\mathbb{C}^{N_1N_2} \to \mathbb{C}^{d_1d_2}$, is defined by $Y \to F_1YF_2^t$. These two operators correspond to the *product frame*

$$\Phi \times \Phi' = \left\{\phi_i(\phi'_j)^t : 1 \le i \le d_1,\ 1 \le j \le d_2\right\},$$

which are matrices of the same size as X. The frame operator is $X \to S_1XS_2^t$. Now let Ψ and Ψ' be the canonical duals of Φ and Φ'. Then, the canonical dual frame of $\Phi \times \Phi'$ is the product frame

$$\Psi \times \Psi' = \left\{\psi_i(\psi'_j)^t : 1 \le i \le d_1,\ 1 \le j \le d_2\right\}.$$

The eigenvalues of the frame operator are $\alpha_i\beta_j$ where α_i is an eigenvalue of S_1 and β_j is an eigenvalue of S_2. Thus, the frame bounds of the product frame are A_1A_2, B_1B_2 and the frame bounds ratio of the product frame is $\rho(\Phi \times \Phi') = \rho(\Phi)\rho(\Phi')$. The product of tight frames is tight.

6.6 More on the Frame Operator

In this section, we shall show that the frame operator is always invertible and that for a tight frame the dual basis is easy to calculate. The main use of the frame operator is to reformulate and prove a number of consequences of the frame inequality. We start with the following equation:

$$\sum_{n=1}^{N} |(\mathbf{x}, \boldsymbol{\phi}_n)|^2 = \begin{bmatrix} \overline{(\mathbf{x}, \boldsymbol{\phi}_1)} & \overline{(\mathbf{x}, \boldsymbol{\phi}_2)} & \cdots & \overline{(\mathbf{x}, \boldsymbol{\phi}_N)} \end{bmatrix} \begin{bmatrix} (\mathbf{x}, \boldsymbol{\phi}_1) \\ (\mathbf{x}, \boldsymbol{\phi}_2) \\ \vdots \\ (\mathbf{x}, \boldsymbol{\phi}_N) \end{bmatrix}$$

$$= (F^*\mathbf{x}, F^*\mathbf{x}) = (FF^*\mathbf{x}, \mathbf{x})$$

$$= (S\mathbf{x}, \mathbf{x}). \tag{6.28}$$

Proposition 6.5 *Suppose Φ is a frame with frame bounds A and B. Then*

1. *The frame equation may be rewritten*

$$A\|\mathbf{x}\|^2 \le (S\mathbf{x}, \mathbf{x}) \le B\|\mathbf{x}\|^2. \tag{6.29}$$

2. *The frame operator is invertible.*
3. *The frame operator is Hermitian (self-adjoint) and hence unitarily diagonalizable. Namely, $S = UDU^*$, where $U^* = U^{-1}$ and $D = diag(\lambda_1, \dots, \lambda_d)$, $0 < \lambda_1 \le \cdots \le \lambda_d$.*
4. *The optimum frame bounds are the smallest and largest eigenvalues of S.*

Proof: For statement 1, we have from the definitions:

$$\sum_{n=1}^{N} |(\mathbf{x}, \boldsymbol{\phi}_n)|^2 = (F^*\mathbf{x}, F^*\mathbf{x}) = (FF^*\mathbf{x}, \mathbf{x}) = (S\mathbf{x}, \mathbf{x}).$$

For statement 2, if $S\mathbf{x}_0 = 0$ for some nonzero $\mathbf{x}_0$ then $0 = (\mathbf{0}, \mathbf{x}_0) = (S\mathbf{x}_0, \mathbf{x}_0)$. But, $0 < A\|\mathbf{x}_0\|^2 \le (S\mathbf{x}_0, \mathbf{x}_0) = 0$, a contradiction. For statement 3, we note that $S^* = (FF^*)^* = F^{**}F^* = FF^* = S$. The unitary diagonalizabilty of S then follows from a standard linear algebra theorem. For statement 4 we argue as follows. Since S is unitarily diagonalizable with eigenvalues $0 < \lambda_1 \le \cdots \le \lambda_d$ then every vector can be written in the form

$$\mathbf{x} = \mathbf{x}_1 + \cdots + \mathbf{x}_d$$

where the $\mathbf{x}_i$ are pairwise orthogonal eigenvectors of the λ_i. We know that $\lambda_1 > 0$ since S is invertible. So, by orthogonality

$$(S\mathbf{x}, \mathbf{x}) = \sum_{i,j} (S\mathbf{x}_i, \mathbf{x}_j) = \sum_{i,j} \lambda_i(\mathbf{x}_i, \mathbf{x}_j) = \sum_{i} \lambda_i(\mathbf{x}_i, \mathbf{x}_i).$$

But,

$$\lambda_1\|\mathbf{x}\|^2 = \lambda_1 \sum_{i} (\mathbf{x}_i, \mathbf{x}_i) \le \sum_{i} \lambda_i(\mathbf{x}_i, \mathbf{x}_i) \le \lambda_d \sum_{i} (\mathbf{x}_i, \mathbf{x}_i) = \lambda_d\|\mathbf{x}\|^2.$$

Thus, we may pick $A = \lambda_1$ and $B = \lambda_d$. By selecting appropriate eigenvectors we see that the bounds are sharp.

Proposition 6.6 *Let $\Phi = \{\phi_1, \dots, \phi_N\}$ be a finite frame with optimal bounds A and B, and frame operator S. Then,*

$$\Psi = \{\psi_1 = S^{-1}\phi_1, \dots, \psi_N = S^{-1}\phi_N\}$$

is a frame with optimal frame bounds $1/B$ and $1/A$. If Φ is tight, then, so is Ψ with frame constant $1/A$.

Proof: Let $S = FF^*$. The frame operator for Ψ is S^{-1} with smallest eigenvalue $1/\lambda_d$ and largest eigenvalue $1/\lambda_1$. The conclusion then follows from Proposition 6.5.

Proposition 6.7 *For tight, unit norm, and equiangular frames we have the following:*

1. *If Φ is a tight frame, then $(S\mathbf{x}, \mathbf{x}) = A(\mathbf{x}, \mathbf{x})$ and $S = AI_d$.*
2. *If Φ is a unit norm frame, then G has 1's on the diagonal.*
3. *If Φ is an equiangular unit norm frame then all off-diagonal elements of G are equal.*
4. *If Φ is a unit norm tight frame, then $A = \frac{N}{d}$.*

Proof: For statement 1, we see from Proposition 6.5 that the diagonalizable matrix S has all of its eigenvalues equal to A. It follows that $S = AI_d$. For statement 2, We see that the diagonal elements of G are $(\phi_i, \phi_i) = 1$. Statement 3 is easy. For statement 4, we observe that the Gram matrix $G = F^*F$ has $1's$ on the diagonal. So,

$$N = \text{trace}(F^*F) = \text{trace}(FF^*) = \text{trace}(AI_d) = dA$$

and $A = \frac{N}{d}$.

Finally, we make a simple remark on the frame operator of the concatenation of frames.

Lemma 6.1 *Let $\Phi_1, \dots, \Phi_r$ be frames and let $\Phi = \Phi_1 \frown \cdots \frown \Phi_r$ be the concatenation of the frames. Then, the total frame operator S of Φ is the sum of the frame operators S_i of Φ_i.*

Proof: Assuming that the elements of Φ are ordered as specified, the frame matrix of Φ has the form $\begin{bmatrix} F_1 & \cdots & F_r \end{bmatrix}$, where F_i is the frame matrix of Φ_i. Then,

$$\begin{aligned} S &= \begin{bmatrix} F_1 & \cdots & F_r \end{bmatrix} \begin{bmatrix} F_1^* \\ \vdots \\ F_r^* \end{bmatrix} \\ &= F_1F_1^* + \cdots + F_rF_r^* \\ &= S_1 + \cdots + S_r. \end{aligned}$$

Note that even if the Φ_i are not frames and the S_i are not invertible the result still holds.

6.7 Group-Based Frames

6.7.1 Unitary Matrix Groups and Frames

In this section, we are going construct examples of frames using unitary matrix groups. We would be greatly aided by using group representation theory but that takes us too far afield. We will use ad hoc methods directly related to frames. The interested reader can look in [27] for background on representations of groups.

The circular frame example from Section 6.5 may be constructed as follows. The matrix

$$R = \begin{bmatrix} \cos\left(\frac{2\pi}{N}\right) & -\sin\left(\frac{2\pi}{N}\right) \\ \sin\left(\frac{2\pi}{N}\right) & \cos\left(\frac{2\pi}{N}\right) \end{bmatrix}$$

is a rotation of finite order N. Define $\boldsymbol{\phi}_n = R^n\boldsymbol{\phi}_0$, where $\boldsymbol{\phi}_0 = \begin{bmatrix}1 & 0\end{bmatrix}^t$. The key features of the set of powers $G = \{R^n : 0 \le n \le N-1\}$ are that:

1. G is multiplicatively closed since $R^{n_1}R^{n_2} = R^{(n_1+n_2) mod\ N}$;
2. G contains the identity $I_2 = R^0$;
3. G contains inverses: $(R^n)^{-1} = R^{(N-n) mod\ N}$.

With these properties in mind, we define a *finite matrix group.*

Definition 6.2 A finite matrix group $G \subset M_d(\mathbb{C})$ or $M_d(\mathbb{R})$ is a finite set of $d \times d$ invertible matrices such that:

1. if $R_1, R_2 \in G$ then $R_1R_2 \in G$;
2. $I_d \in G$, and
3. if $R \in G$ then $R^{-1} \in G$.

From now on we will assume that $G \subset M_d(\mathbb{C})$ since this includes that case where $G \subset M_d(\mathbb{R})$. Next we define the *natural action, orbits,* and *stabilizers* of G, which we will use to construct group-based frames.

Definition 6.3 Let G be a finite matrix group in $M_d(\mathbb{C})$.

1. The map $(R, \mathbf{x}) \to R\mathbf{x}$ is called the (natural) action of G on $\mathbb{C}^d$.
2. We say the action is unitary if $(R\mathbf{x}, R\mathbf{x}) = (\mathbf{x}, \mathbf{x})$ (i.e., $R^{-1} = R^*$) for all $R \in G$ and $\mathbf{x} \in \mathbb{C}^d$.
3. Let $\mathbf{x} \in \mathbb{C}^d$. Then the G-orbit of G in $\mathbf{x}$ in $\mathbb{C}^d$ is the set

$$\mathcal{O}_{\mathbf{x}} = G\mathbf{x} = \{R\mathbf{x} : R \in G\}$$

We call $\mathbf{x}$ the *seed vector* of $\mathcal{O}_{\mathbf{x}}$.

4. The set $G_{\mathbf{x}} = \mathrm{Stab}_G(\mathbf{x}) = \{R \in G : R\mathbf{x} = \mathbf{x}\}$ is called the *stabilizer* of $\mathbf{x}$, it is a subgroup of G. Regular orbits $\mathcal{O}_{\mathbf{x}}$ are ones for which $|G_{\mathbf{x}}| = 1$ and singular orbits have $|G_{\mathbf{x}}| > 1$.

The following proposition follows from the *orbit stabilizer theorem,* which is a standard theorem in group theory.

Proposition 6.8 *Let the matrix group G act on $\mathbb{C}^d$ and $\mathbf{x} \in \mathbb{C}^d$. Then,*

$$|G| = |G_{\mathbf{x}}||\mathcal{O}_{\mathbf{x}}|,$$

$$|\mathcal{O}_{\mathbf{x}}| = \frac{|G|}{|G_{\mathbf{x}}|}.$$

Furthermore, we have these orbit sum formulas:

$$\sum_{\mathbf{y}\in\mathcal{O}_{\mathbf{x}}} \mathbf{y} = \frac{1}{|G_{\mathbf{x}}|}\sum_{n=1}^{N} R_n\mathbf{x}$$

and

$$\sum_{\mathbf{y}\in\mathcal{O}_{\mathbf{x}}} \mathbf{y}\mathbf{y}^* = \frac{1}{|G_{\mathbf{x}}|}\sum_{n=1}^{N} R_n\mathbf{x}(R_n\mathbf{x})^*. \tag{6.30}$$

Proof: We just give an idea of the proof. Write $G = \{R_1, \dots, R_N\}$ for some ordering on the elements, and for a seed vector $\mathbf{x} \in \mathbb{C}^d$ consider the sequence $R_1\mathbf{x}, \dots, R_N\mathbf{x}$. Each distinct element in the list occurs $|G_{\mathbf{x}}|$ times, and the list of distinct elements in $R_1\mathbf{x}, \dots, R_N\mathbf{x}$ is the orbit of $\mathbf{x}$. The sum formulas follow immediately.

To construct *group frames* we need the notion of *cyclic vectors,* and for tight frames we need *irreducible actions.* We define these concepts now.

Definition 6.4 Let G be a finite matrix group in $M_d(\mathbb{C})$.

1. A cyclic vector $\mathbf{x}$ for a matrix group G is vector such that the orbit $\mathcal{O}_{\mathbf{x}}$ spans $\mathbb{C}^d$.
2. We say that the group G acts irreducibly on $\mathbb{C}^d$ if every non-zero vector in $\mathbb{C}^d$ is cyclic.
3. Given $\mathbf{x}_1, \dots, \mathbf{x}_r$, if the concatenation of orbits $\mathcal{O}_{\mathbf{x}_1} \frown \cdots \frown \mathcal{O}_{\mathbf{x}_r}$ determines a frame, then we call the resulting frame a group frame.

Example 6.5 *The tetrahedral, octahedral, and icosahedral groups (defined below) act irreducibly on $\mathbb{C}^3$. The rotational group of our first example acts irreducibly on $\mathbb{R}^2$ but not on $\mathbb{C}^2$.*

Remark 6.5 Many useful frames, such as the ones we considered previously and the platonic frames, which we will discuss later, are highly symmetric and have a large group of unitary symmetries. Thus, group frames are a natural way of producing highly symmetric frames.

Proposition 6.9 *Suppose that the matrix group G acts unitarily and irreducibly on $\mathbb{C}^d$ and that $\phi_0 \in \mathbb{C}^d$ is any nonzero vector. Then $\Phi = \mathcal{O}_{\phi_0}$ is a tight frame.*

Proof: *First assume that no nontrivial element of G fixes ϕ_0. Then we know that $\Phi = \mathcal{O}_{\phi_0} = R_1\phi_0, \dots, R_N\phi_0$, without repeats, where $G = \{R_1, \dots, R_N\}$ in some order. Next, we show that the frame operator S satisfies*

$$SR = RS$$

or, equivalently

$$S = RSR^{-1} = RSR^*,$$

for every element of the group $R \in G$. To prove this note that

$$S = \sum_{n=1}^{N} \phi_n \phi_n^* = \sum_{n=1}^{N} R_n\phi_0(R_n\phi_0)^* = \sum_{n=1}^{N} R_n\phi_0\phi_0^* R_n^*.$$

So,

$$\begin{aligned} RSR^* &= \sum_{n=1}^{N} RR_n\phi_0\phi_0^* R_n^* R^* \\ &= \sum_{n=1}^{N} (RR_n)\phi_0((RR_n)\phi_0)^*. \end{aligned}$$

But, since the map $R' \to RR'$ is a bijective mapping of G onto itself, then, $\sum_{n=1}^{N} (RR_n)\phi_0((RR_n)\phi_0)^*$ is simply the sum $\sum_{n=1}^{N} R_n\phi_0(R_n\phi_0)^*$ with the terms rearranged. It follows that $S = RSR^*$ and $SR = RS$. Next, S has at least one eigenvalue–eigenvector pair $(\lambda, \mathbf{x})$ in $\mathbb{C}^d$. Then

$$S(R\mathbf{x}) = R(S\mathbf{x}) = \lambda(R\mathbf{x}).$$

It follows that the linear span of the orbit $\mathcal{O}_{\mathbf{x}}$ consists of eigenvectors with the same eigenvalue λ. But, because G acts irreducibly, the linear span of $\mathcal{O}_{\mathbf{x}}$ is $\mathbb{C}^d$ and $S = \lambda I_d$. Thus, Φ is a tight frame.

In the case where the stabilizer G_{ϕ_0} is nontrivial, we use equation (6.30) to show that for any nonzero $\mathbf{x}$

$$S = \sum_{\mathbf{y}\in\mathcal{O}_{\mathbf{x}}} \mathbf{y}\mathbf{y}^* = \frac{1}{|G_{\mathbf{x}}|} \sum_{i=1}^{N} R_i\mathbf{x}(R_i\mathbf{x})^*. \tag{6.31}$$

It then follows that $SR = RS$, and the rest of the proof is the same.

Corollary 6.1 *Suppose that the finite matrix group G acts unitarily and irreducibly on $\mathbb{C}^d$ and that the frame Φ is a disjoint concatenation of orbits $\mathcal{O}_{\mathbf{x}_1} \frown \cdots \frown \mathcal{O}_{\mathbf{x}_r}$. Then Φ is a tight frame.*

Proof: First note that orbits are either equal of disjoint, so it makes sense to assume that the orbits are disjoint. Each orbit defines a tight frame and by Proposition 6.1 the concatenated frame is also tight.

Finally, before moving onto the examples, we use the above proof to calculate the dual of a group frame. Before stating and proving our formula for the dual frame, we state and prove a lemma.

Lemma 6.2 *Suppose G acts unitarily on $\mathbb{C}^d$ and $\Phi = \mathcal{O}_{\mathbf{x}_1} \frown \cdots \frown \mathcal{O}_{\mathbf{x}_r}$ is a group frame on disjoint orbits. Then, the frame operator commutes with every element of G.*

Proof: In the proof of Proposition 6.9 we showed that for each orbit $\mathcal{O}_{\mathbf{x}_i}$ each $R \in G$ commutes with

$$S_i = \sum_{\mathbf{y} \in \mathcal{O}_{\mathbf{x}_i}} \mathbf{y}\mathbf{y}^* = \frac{1}{|G_{\mathbf{x}_i}|} \sum_{R \in G} R\mathbf{x}_i (R\mathbf{x}_i)^*.$$

Thus, each $R \in G$ commutes with $S = S_1 + \cdots + S_r$ (use Lemma 6.1).

Proposition 6.10 *Suppose G acts unitarily on $\mathbb{C}^d$ and $\Phi = \mathcal{O}_{\mathbf{x}_1} \frown \cdots \frown \mathcal{O}_{\mathbf{x}_r}$ is a group frame on disjoint orbits. Let $\mathbf{y}_i$ be the element $S^{-1}\mathbf{x}_i$. Then the dual frame is given by $\Psi = \mathcal{O}_{\mathbf{y}_1} \frown \cdots \frown \mathcal{O}_{\mathbf{y}_r}$ with the specific pairing*

$$R\mathbf{x}_i \leftrightarrow R\mathbf{y}_i, \quad R \in G.$$

Proof: We know that $R\mathbf{x}_i$ is paired with $S^{-1}R\mathbf{x}_i = R\,S^{-1}\mathbf{x}_i = R\mathbf{y}_i$.

6.7.2 Initial Examples of Group Frames

6.7.2.1 Platonic Frames

The icosahedral group is the group of symmetries of the icosahedron. A picture of a barycentrically subdivided icosahedron, projected to the unit sphere, is pictured in Figure 6.5. There are 120 small $\left(\frac{\pi}{2}, \frac{\pi}{3}, \frac{\pi}{5}\right)$ triangles. The triangles of the icosahedron can be seen by grouping the small triangles in groups of six. The pentagons of the dodecahedron can be seen by grouping small triangles together in groups of 10. There are 60 rotations in the icosahedral group. For most points $\mathbf{x}$ on the sphere the orbit $\mathcal{O}_{\mathbf{x}}$ contains 60 points. There are three types of points with smaller orbits:

- centers of equilateral triangles, which are vertices of pentagons, and have orbits of size $60/3 = 20$;

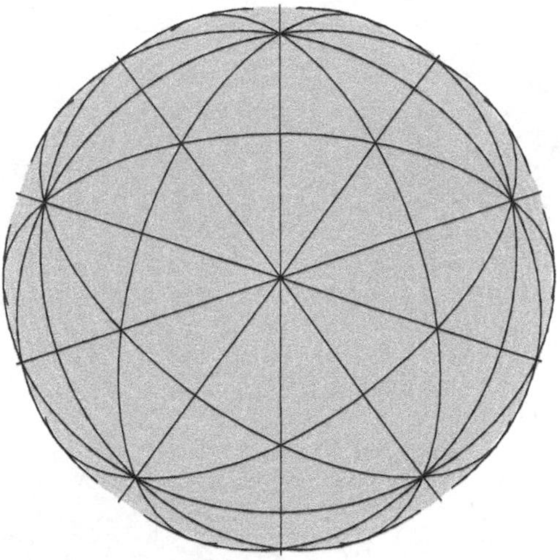

Figure 6.5 Icosahedral tiling of the sphere.

- midpoints of edges of equilateral triangles, which are midpoints of edges of pentagons, and have orbits of size $60/2 = 30$;
- vertices of equilateral triangles, which are centers of pentagons, and have orbits of size $60/5 = 12$.

The other platonic groups in $\mathbb{R}^3$ are the tetrahedral group, $|G| = 12$, and the octahedral group $|G| = 48$.

Now we construct some icosahedral frames from some randomly selected points on the sphere and generating matrices R_1 *and* R_2 (given below) for the icosahedral group. Start with a matrix J_0, whose columns are distinct points on the unit sphere, perhaps randomly selected. Then iteratively define

$$K_{n+1} = \begin{bmatrix} J_n & R_1 J_n & R_2 J_n \end{bmatrix},$$
$$J_{n+1} = K_{n+1} \text{ with duplicate columns removed.}$$

We terminate the algorithm when $|J_{n+1}| = |J_n|$. The algorithm terminates in a finite number of steps equal to the longest word required to express an element in terms of R_1 and R_2. The algorithm can be made more efficient by setting $\partial J_n = J_n \setminus J_{n-1}$, the matrix obtained by removing any columns in J_{n-1} from J_n, and following these iterative steps:

$$\partial J_n = J_n \setminus J_{n-1}$$
$$K_{n+1} = \begin{bmatrix} \partial J_n & R_1(\partial J_n) & R_2(\partial J_n) \end{bmatrix},$$
$$J_{n+1} = J_{n-1} \cup (K_{n+1} \setminus J_n) \text{ with duplicate columns removed.}$$

In the refined algorithm, we avoid creating any new points that were created in previous steps. Two generating matrices for the icosahedron (to four decimal

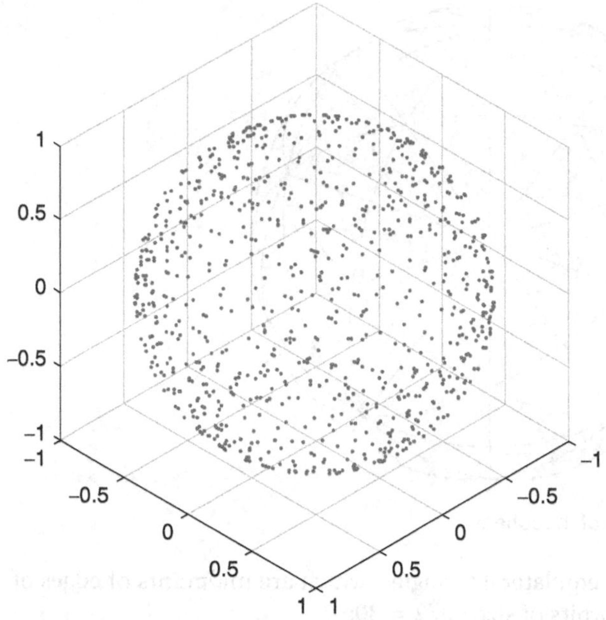

Figure 6.6 Icosahedral frame with 900 points.

places) are:

$$R_1 = \begin{bmatrix} 0.3090 & -0.9511 & 0 \\ 0.9511 & 0.3090 & 0 \\ 0 & 0 & 1.0000 \end{bmatrix},$$

$$R_2 = \begin{bmatrix} 0.8618 & -0.4253 & 0.2764 \\ 0.4253 & 0.3090 & -0.8507 \\ 0.2764 & 0.8507 & 0.4472 \end{bmatrix}.$$

The algorithm is implemented in the script platonic.m on the book website. A tight frame with 900 points is shown in Figure 6.6.

6.7.2.2 Symmetric Group Frames

Next we consider frames constructed from the *symmetric group* of permutations. The symmetric group does not act irreducibly on $\mathbb{C}^d$, so we will see how to construct frames when we have reducible actions. The group of permutations Σ_d of order $d!$ acts on signals of length d by permuting coordinates. The vector space $\mathbb{C}^d$ has two orthogonal subspaces $V_1 = \mathbb{C}(1, \dots, 1)$ and $V_2 = \{\mathbf{x} \in \mathbb{C}^d : \mathbf{x} \cdot (1, \dots, 1) = 0\}$. For each $R \in \Sigma_d$, $RV_1 = V_1$ and $RV_2 = V_2$, and V_1

and V_2 are orthogonal (by construction). It is well known that Σ_d acts irreducibly on these subspaces. If we pick $\mathbf{x}_i \in V_i$ then $\mathcal{O}_{\mathbf{x}_i}$ spans V_i so the action on all of $\mathbb{C}^d$ cannot be irreducible. Also neither of $\mathcal{O}_{\mathbf{x}_i}$ generates a frame. We need some combination of vectors from V_1 and V_2 to create frames. Some caution is required in selecting seed vectors since the number of vectors in an orbit can be rather large. Let us try to construct a frame with $d = 10$ and $\boldsymbol{\phi}_0 = (1, -2, 1, 0, 0, 0, 0, 0, 0, 0) \in V_2$. The stabilizer of $\boldsymbol{\phi}_0$ is the subgroup of arbitrary permutations of the last seven coordinates and switching the first and third coordinates. The stabilizer has order $2!7!$ and so the orbit of $\boldsymbol{\phi}_0$ has $10!/(2!7!) = 360$ elements. When $\boldsymbol{\phi}_0 = (1, 1, 1, 1, 1, 1, 1, 1, 1, 1) \in V_1$ there is a single vector. In both cases the frame operators are singular and have two eigenvalues each. The eigenvalues λ_j and multiplicities μ_j on V_j of the two different frame operators are given in the table below.

Seed vector	**Frame op**	λ_1	μ_1	λ_2	μ_2
$(1,1,1,1,1,1,1,1,1,1)$	S_1	10	1	0	9
$(1,-2,1,0,0,0,0,0,0,0)$	S_2	0	1	240	9

The complimentary dimensions are no mystery. To understand this, let $\mathbf{x}_i \in V_i$ and consider $S_i = \sum_{R \in G} R\mathbf{x}_i(R\mathbf{x}_i)^* = \sum_{R \in G} R\mathbf{x}_i\mathbf{x}_i^* R^*$. We show that each S_i is a scalar matrix on each V_j. Consider S_2, the more interesting case. Let $\mathbf{y}_1 \in V_1$, then $S_2\mathbf{y}_1 = \sum_{R \in G} R\mathbf{x}_2\mathbf{x}_2^* R^*\mathbf{y}_1$. But, $R^*\mathbf{y}_1 \in V_1$ so, $\mathbf{x}_2^* R^*\mathbf{y}_1 = (R^*\mathbf{y}_1, \mathbf{x}_2) = 0$. Thus, $S_2 V_1 = 0$. If $\mathbf{y}_2 \in V_2$ then $S_2\mathbf{y}_2 = \sum_{R \in G} R\mathbf{x}_2\mathbf{x}_2^* R^*\mathbf{y}_2$ is the linear combination $\sum_{R \in G} (R^*\mathbf{y}_2, \mathbf{x}_2)R\mathbf{x}_2$ of vectors from V_2. So, $S_2 V_2 \subseteq V_2$. By the proof of Proposition 6.9, $RS_2 = S_2 R$ for all R. Since R acts irreducibly on V_2, S_2 acts by scalar multiplication on V_2. Similar arguments apply to S_1. This explains the multiplicities in the table.

Now create a combined frame with matrix $F = [F_1\ F_2]$, where each F_i is the frame matrix created from $\mathbf{x}_i$. Then

$$S = [F_1\ F_2]\begin{bmatrix} F_1^* \\ F_2^* \end{bmatrix} = F_1F_1^* + F_2F_2^* = S_1 + S_2,$$

and S acts by the scalar 10 on V_1 and the scalar 240 on V_2. If we start with the vector pair $a\mathbf{x}_1$ and $b\mathbf{x}_2$ then $F = [aF_1\ bF_2]$ and S acts by $10|a|^2$ on V_1 and $240|b|^2$ on V_2. For the given vectors, $\sqrt{24}\mathbf{x}_1$ and $\mathbf{x}_2$ yield a tight frame with 361 vectors and frame constant of 240.

The orbits can be calculated as in the icosahedral example by observing that Σ_d is generated by the transposition $(1, 2)$ and the cycle $(1, 2, \dots, d)$ using permutation cycle format. We shall see the matrix forms of these matrices in the next example on Gabor frames. The symmetric group example works for any group of permutations of a finite set. Examples are as numerous as finite groups. The character theory of G would be helpful in these examples. The case of multiple irreducible subspaces carries over in a fairly straight forward fashion.

6.7.2.3 Harmonic Frames

If we take $\epsilon = 1$ in the extended *DFT* frame example in Section 6.5 then we get a harmonic frame. We proved that a harmonic frame is tight. We can also look at a harmonic frame as a group frame. Let $\omega = \exp\left(\frac{2\pi i}{N}\right)$, $W = diag(1, \omega, \omega^2, \dots, \omega^{d-1})$, $\boldsymbol{\phi}_0 = \begin{bmatrix}1 & \cdots & 1\end{bmatrix}^t$ and $\boldsymbol{\phi}_i = W^i\boldsymbol{\phi}_0$. Then $\boldsymbol{\phi}_0, \boldsymbol{\phi}_1, \dots, \boldsymbol{\phi}_{N-1}$ is the extended *DFT* frame with parameters, d, N, and $\epsilon = 1$. By construction, it is obviously a group frame. We may compute the transform of a vector easily using the *DFT*. Let $\mathbf{x} \in \mathbb{C}^d$, and create a vector $\mathbf{x}' \in \mathbb{C}^N$ by padding $\mathbf{x}$ with zeros at the bottom. Then $\mathbf{X} = DFT(\mathbf{x}')$ is the harmonic frame transform of $\mathbf{x}$.

6.7.3 Gabor Frames

The Gabor class of frames is related to the short-time (windowed) Fourier transform and spectrograms of Chapter 5. Considering $\mathbb{C}^d$ as a space of signals we can construct two different shift actions on $\mathbb{C}^d$. The first is the cyclic shift action in the time domain:

$$T : (x_0, x_1, \dots, x_{d-1}) \rightarrow (x_{d-1}, x_0, \dots, x_{d-2}).$$

This *time-shift operator* can be implemented by a cyclic convolution $\mathbf{x} \rightarrow \delta_1 * \mathbf{x}$ with the vector $\delta_1 = (0, 1, \dots, 0)$. As a matrix multiplication in the $d = 4$ case, T is given by:

$$T\mathbf{x} = \begin{bmatrix} 0 & 0 & 0 & 1 \\ 1 & 0 & 0 & 0 \\ 0 & 1 & 0 & 0 \\ 0 & 0 & 1 & 0 \end{bmatrix} \begin{bmatrix} x_0 \\ x_1 \\ x_2 \\ x_3 \end{bmatrix} = \begin{bmatrix} x_3 \\ x_0 \\ x_1 \\ x_2 \end{bmatrix}.$$

By the convolution theorem, the action of T and it powers can be converted to pointwise multiplication

$$\mathcal{F}(\delta_1 * \mathbf{x}) = \mathcal{F}(\delta_1) \circ \mathcal{F}(\mathbf{x}),$$

where $\mathcal{F}$ is the Fourier transform operator. Letting $\mathbf{y} = \mathcal{F}(\mathbf{x})$, the adjoint of the pointwise multiplication operator defined by $\mathcal{F}(\delta_1)$ is given by

$$W : (y_0, y_1, \dots, y_{d-1}) \rightarrow (y_0, \omega y_1, \dots, \omega^{(d-1)} y_{d-1}),$$

where $\omega = \exp\left(\frac{2\pi i}{d}\right)$. It is more convenient to work with the adjoint defined by $\mathcal{F}(\delta_1)$ than direct multiplication. As a matrix multiplication for $d = 4$ we have:

$$W\mathbf{y} = \begin{bmatrix} 1 & 0 & 0 & 0 \\ 0 & i & 0 & 0 \\ 0 & 0 & -1 & 0 \\ 0 & 0 & 0 & -i \end{bmatrix} \begin{bmatrix} y_0 \\ y_1 \\ y_2 \\ y_3 \end{bmatrix} = \begin{bmatrix} y_0 \\ iy_1 \\ -y_2 \\ -iy_3 \end{bmatrix}.$$

In terms of T and W, the convolution theorem may be written as $\mathcal{F}T = W^*\mathcal{F}$. We call W a *phase (frequency) shift operator*. The Fourier transform and its inverse turn a time shift into a phase shift and vice versa. Both shift operators T and W have order d. The time and phase shift operators do not commute but it is close:

$$WT = \omega TW,$$
$$T^{-1}WT = \omega W.$$

It follows for any i and j that

$$W^jT^i = \omega^{ij}T^iW^j,$$
$$T^{-i}W^jT^i = \omega^{ij}W^j,$$
$$T^{-i}W^jT^iW^{-j} = \omega^{ij}I_d.$$

The following proposition is easily shown.

Proposition 6.11 *The matrix group $G = \langle T, W\rangle$, generated by T and W, has order d^3 and every element has a unique representation as*

$$\omega^k W^j T^i, \quad 0 \le i, j, k \le d-1. \tag{6.32}$$

Proof: See Exercise 21.

Remark 6.6 The group G is sometimes called a finite Heisenberg group.

Before we construct Gabor frames, we prove an irreducibility result.

Proposition 6.12 *The group $G = \langle T, W\rangle$ acts irreducibly on $\mathbb{C}^d$.*

Proof: Given any vector $\mathbf{x} = \begin{bmatrix} x_0 & x_1 & \cdots & x_{d-1}\end{bmatrix}^t \in \mathbb{C}^d$ it is easy to calculate that

$$\frac{1}{d}\sum_{j=1}^{d-1} W^j\mathbf{x} = \begin{bmatrix} x_0 & 0 & \cdots & 0\end{bmatrix}^t.$$

If $\mathbf{x}$ is not the zero vector then for some power T^i, $T^i\mathbf{x}$ has a nonzero first component. It follows that

$$\mathbf{y} = \frac{1}{d}\sum_{j=1}^{d-1} W^jT^i\mathbf{x} = \begin{bmatrix} a & 0 & \cdots & 0\end{bmatrix}^t$$

for some nonzero scalar a. Then $\mathbf{y}, T\mathbf{y}, T^2\mathbf{y}, \ldots, T^{d-1}\mathbf{y}$ is a basis for $\mathbb{C}^d$. It follows that every nonzero vector is cyclic.

The Gabor frames we define will not be group frames exactly. The standard group frame determined by $\mathcal{O}_{\phi_0}$ will have redundancies because of the presence of scalar matrices in G. If aI_d is a scalar matrix in G and R any other group element then $\phi = R\phi_0$ and $\phi' = aI_dR\phi_0 = a\phi$ are frame vectors which are scalar multiples of each other. Clearly, this is not desirable, and we eliminate the redundancy as follows. The set $Z = \{aI_d : a \in \mathbb{C}, aI_d \in G\}$ is a subgroup of G. A *transversal* of Z in G is a set $T \subset G$ such that every element of $R \in G$ has a unique factorization $R = UV$ for some $U \in Z$ and $V \in T$. A transversal always exists and has cardinality $|G|/|Z|$. Given a seed vector, ϕ_0, every vector in $\{R\phi_0 : R \in G\}$ is a scalar multiple of some vector in $\{R\phi_0 : R \in T\}$.

Definition 6.5 Let G be a finite unitary matrix group with subgroup of scalars $Z = \{aI_d : a \in \mathbb{C}, aI_d \in G\}$ and suppose that T is a transversal for Z in G. Then, $\{R\phi_0 : R \in T\}$, or a concatenation of such sets, is called a *reduced group frame* for G.

The following proposition show that our group methods can be used on the reduced frame.

Proposition 6.13 *Let notation be as in Definition 6.5. Then for any* $\mathbf{x} \in \mathbb{C}^d$

$$\sum_{R\in G} R\mathbf{x}(R\mathbf{x})^* = |Z| \sum_{V\in T} V\mathbf{x}(V\mathbf{x})^*,$$

or in terms of frame operators

$$S_G = |Z| S_T,$$

where S_G and S_T denote the group frame operator and the reduced frame operator, respectively. Consequently, S_T commutes with all $R \in G$. Additionally, if the group frame is tight then the reduced group frame is also tight.

Proof: We may write

$$\begin{aligned}
\sum_{R\in G} R\mathbf{x}(R\mathbf{x})^* &= \sum_{U\in Z}\sum_{V\in T} UV\mathbf{x}(UV\mathbf{x})^* \\
&= \sum_{U\in Z}\sum_{V\in T} U(V\mathbf{x}\mathbf{x}^*V^*)U^* \\
&= \sum_{U\in Z}\sum_{V\in T} UU^*(V\mathbf{x}\mathbf{x}^*V^*) \\
&= \sum_{U\in Z}\sum_{V\in T} V\mathbf{x}\mathbf{x}^*V^* \\
&= |Z|\sum_{V\in T} V\mathbf{x}\mathbf{x}^*V^*.
\end{aligned}$$

The first line is justified by the transversal properties and the fourth line is legitimate since U^* is a scalar and commutes with $V\mathbf{x}\mathbf{x}^*V^*$. Since S_G commutes with all $R \in G$ then so does S_T. If the group frame is tight then S_G is a scalar matrix. Then S_T is scalar and the reduced group frame is also tight.

Now we construct our *Gabor frames* as follows. Select an initial seed vector $\boldsymbol{\phi}_0$ of the form $(a_0, \dots, a_{\ell-1}, 0, \dots, 0)$ with no $a_i = 0$. The number of nonzero entries ℓ is the *mask length* or the *window width*. All translates $W^jT^i\boldsymbol{\phi}_0$ have the same window width. Now let d_t, d_w divide d and consider the frame

$$\Phi = \left\{ W^{jd_w}T^{id_t}\boldsymbol{\phi}_0, \quad 1 \le i \le \frac{d}{d_t}, 1 \le j \le \frac{d}{d_w} \right\}. \tag{6.33}$$

Note that Φ is a reduced group frame for the subgroup $H = \langle T^{d_t}, W^{d_w} \rangle$. The frame has $\frac{d^2}{d_t d_w}$ vectors. The number d_t lengthens our basic shift in the time domain, if d_t is larger than the window width then we will not get a frame. So $d_t \le \ell$ and $\ell = d_t$ is a good choice. The powers of W allow us to capture frequencies. By picking $d_w > 1$ we reduce the number of detectable frequencies to $\frac{d}{d_w}$. But, the number of frequencies should be larger than the number of frequencies, ℓ, supported on a window width ℓ. Therefore, $\ell \le \frac{d}{d_w}$. Summarizing:

$$\boldsymbol{\phi}_0 = (a_0, \dots, a_{\ell-1}, 0, \dots, 0), \quad \forall a_i \ne 0, \tag{6.34}$$

$$d_t \le \ell, \quad d_w \le \frac{d}{\ell}. \tag{6.35}$$

so that

$$d_t d_w \le d, \text{and } d \le |\Phi| = \frac{d^2}{d_t d_w} \le d^2.$$

We now prove that the maximal size Gabor frames are tight.

Proposition 6.14 *Let $d_t = d_w = 1$. Then the Gabor frame given in equation (6.33) is tight.*

Proof: This follows immediately from Propositions 6.12 and 6.13.

We next examine what happens in the case where the frame is based on the subgroup $H = \langle T^{d_t}, W^{d_w} \rangle$ under the conditions in the inequalities (6.34) and (6.35).

Proposition 6.15 *Suppose $\boldsymbol{\phi}_0$, d_t, d_w satisfy the conditions in the inequalities (6.34) and (6.35). Then the set Φ given in equation (6.33) is a frame and the dual frame is given by*

$$\Psi = \left\{ W^{jd_w}T^{id_t}\boldsymbol{\psi}_0, \quad 1 \le i \le \frac{d}{d_t}, 1 \le j \le \frac{d}{d_w} \right\},$$

where $\psi_0 = S^{-1}\phi_0$, *and S is the frame operator for* Φ. *Furthermore, the frame operator is a diagonal matrix and* ψ_0 *is a mask of length* ℓ.

Proof: The proof that Φ is a frame uses the ideas of Propositions 6.12 and 6.14. We just need to prove that:

- $S^{-1}W^{jd_w}T^{id_t}\phi_0 = W^{jd_w}T^{id_t}S^{-1}\phi_0 = W^{jd_w}T^{id_t}\psi_0$ for all i and j;
- the vector ϕ_0 is H-cyclic;
- S is diagonal.

The first bullet follows from Proposition 6.13. To prove that ϕ_0 is cyclic we argue as follows. Let $r \leq \ell$, then the matrix $\omega^{-rd_w}W^{d_w}$ has a 1 in the $(r+1, r+1)$ position and all other diagonal entries in the upper left diagonal $\ell \times \ell$ block are d/d_w roots of unity, not equal to 1. Therefore,

$$\frac{1}{d/d_w}\sum_{j=1}^{d/d_w}\omega^{-rjd_w}W^{jd_w}\phi_0 = \frac{1}{d/d_w}\sum_{j=1}^{d/d_w}\left(\omega^{-rd_w}W^{d_w}\right)^j\phi_0$$
$$= (0, \ldots, 0, a_r, 0, \ldots, 0).$$

For, the k'th entry equals

$$\sum_{j=1}^{d/d_w}\left(\omega^{(k-r)d_w}\right)^j a_k,$$

and for $k \leq \ell$ the sum of roots is zero unless $k = r$. For $k > \ell$, $a_k = 0$. Thus, $(0, \ldots, 0, 1, 0, \ldots, 0)$, with 1 in the rth position, is in the linear span of the frame. Now by using a suitable power T^{id_t} of the translation T^{d_t}, of length less than ℓ, we can produce a vector

$$\frac{1}{d/d_w}\sum_{j=1}^{d/d_w}\omega^{-rjd_w}T^{id_t}W^{jd_w}\phi_0$$

in the linear span of Φ with exactly one nonzero entry in any position we choose.

Next we prove that S is diagonal. The entries of S are dot products of rows X_r and X_s of F, $0 \leq r, s \leq d-1$. We need to show that $(X_r, X_s) = 0$ when $r > s$. Set $\phi_0 = (a_0, \ldots, a_{d-1})$. The rth component of $T^{id_t}\phi_0$ is a_{r-id_t}, and the rth component of $W^{jd_w}T^{id_t}\phi_0$ is $\omega^{jrdw}a_{r-id_t}$. The indices of a are taken mod d. Then,

$$(X_r, X_s) = \sum_{j=1}^{d/d_w}\sum_{i=1}^{d/d_t}\omega^{jrd_w}a_{r-id_t}\overline{\omega^{jsd_w}a_{s-id_t}}$$
$$= \sum_{j=1}^{d/d_w}\sum_{i=1}^{d/d_t}\omega^{j(r-s)d_w}a_{r-id_t}\overline{a_{s-id_t}}$$

$$= \sum_{i=1}^{d/d_t} \sum_{j=1}^{d/d_w} \omega^{j(r-s)d_w} a_{r-id_t} \overline{a_{s-id_t}}$$

$$= \sum_{i=1}^{d/d_t} \frac{1-\omega^{(r-s)d}}{1-\omega^{(r-s)d_w}} a_{r-id_t} \overline{a_{s-id_t}}.$$

The quotient is valid if $r - s$ is not divisible by d/d_w, and then the quotient is zero. Otherwise we must replace the quotient by d/d_w. However, if $r - s$ is divisible by d/d_w then one of a_{r-id_t} or a_{s-id_t} be zero since their indices are too far apart (see Exercise 22). We do get a formula for the diagonal entry (X_r, X_r):

$$(X_r, X_r) = \frac{d}{d_w} \sum_{i=1}^{d/d_t} |a_{r-id_t}|^2.$$

Remark 6.7 There are many cases where we can relax the conditions on the size of the window and still get a frame, but the frame operator may no longer be diagonal and ψ_0 may no longer be a mask of length ℓ.

Example 6.6 *Select parameters* $d = 100,\ d_t = 10,\ d_w = 5,\ N = 200,$ *and* $\ell = 20$. *The mask and the dual mask are given in Figure 6.7.*

6.7.3.1 Flipped Gabor Frames

In the previous discussion, we time shifted first and then phase shifted (apply *DFT*). If we reverse order we get a frame that we call a *flipped Gabor frame.*

$$\Phi' = \left\{ T^{jd_t} W^{id_w} \phi_0, \quad 1 < i \leq \frac{d}{d_w},\ 1 \leq j \leq \frac{d}{d_t} \right\}. \tag{6.36}$$

The same analysis above works for the flipped frame since

$$T = \left\{ T^{jd_t} W^{id_w}, \quad 1 \leq i \leq \frac{d}{d_w},\ 1 \leq j \leq \frac{d}{d_t} \right\}$$

is an alternate transversal for the scalar subgroup. The new frame is obtained from the old by permuting vectors and multiplying by roots of unity. To see the difference, we give matrix image maps of the frame matrices in the two different cases in Figures 6.8 and 6.9.

6.8 Frame Applications

We describe some applications of oversampling via frames. A lengthy list of applications is beyond the scope of this book (see [7, 21].

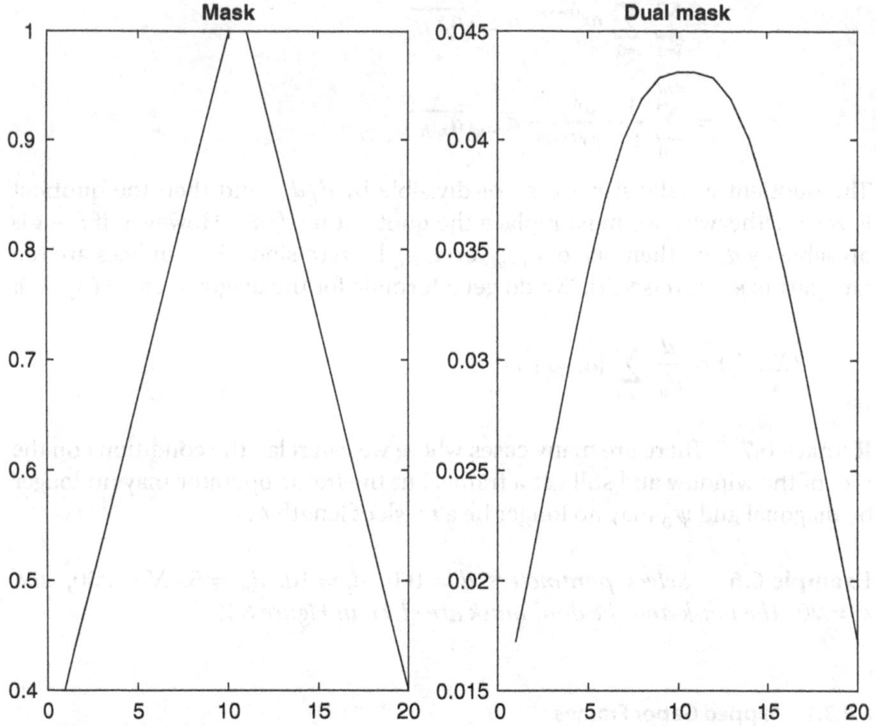

Figure 6.7 Mask and dual mask of Gabor frame.

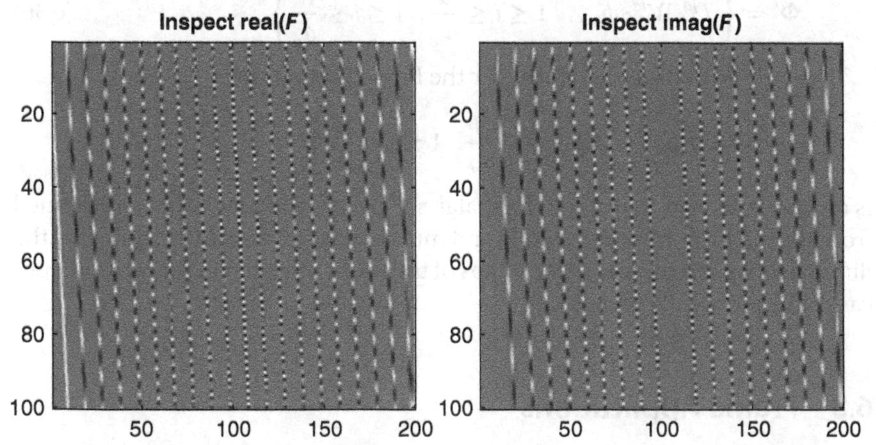

Figure 6.8 Gabor frame—time shift first then DFT.

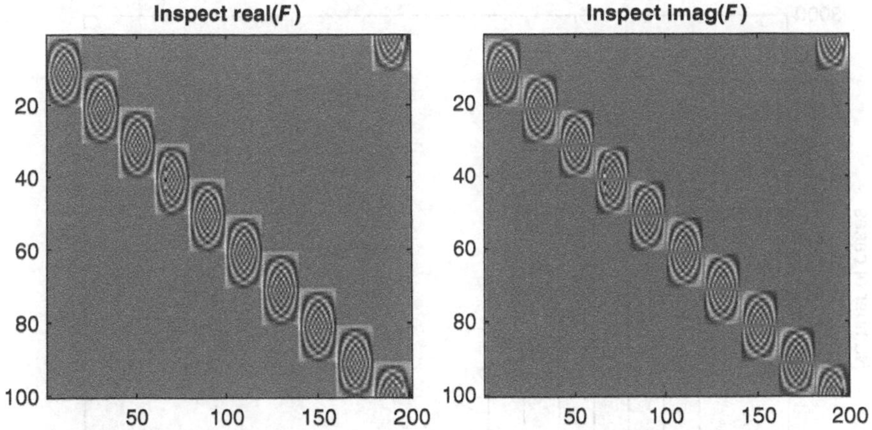

Figure 6.9 Gabor frame—DFT first then time shift.

6.8.1 Packet Loss

Suppose that Φ_1 is a synthesis frame with frame matrix F_1 and Ψ_1 is a dual analysis frame with frame matrix $F_{1,d}$. Suppose that $\Psi_2 \subset \Psi_1$ is obtained by dropping a few frame vectors and $F_{2,d}$ is the matrix obtained by dropping the columns corresponding to the dropped vectors. The new set and matrix may or may not correspond to a frame. Let $\mathbf{x}$ be any signal and $\mathbf{X}_1 = F_{1,d}^*\mathbf{x}$, $\mathbf{X}_2 = F_{2,d}^*\mathbf{x}$ be the analysis vectors of $\mathbf{x}$ with respect to Ψ_1 and Ψ_2. The vector $\mathbf{X}_2$ is simply $\mathbf{X}_1$ with some dot products dropped, that is, $\mathbf{X}_1$ with some packet loss. All is not lost, however, if Ψ_2 is a frame we can compute the dual frame Φ_2 and reconstruct the original signal $\mathbf{x}$. Specifically, in matrix form we recover $\mathbf{x}$ by

$$\mathbf{x} = S_{2,d}^{-1} F_{2,d} \mathbf{X}_2,$$

where

$$S_{2,d} = F_{2,d} F_{2,d}^*.$$

However, the nice properties of Φ_1 may be lost, for example, tightness. In particular, the frame bounds ratio of the new frame may be large, potentially causing numerical instability. The frame bounds ratio for Φ_2 may be calculated by computing the same ratio for Ψ_2. With this in mind, some frames may be better than others in repairing packet loss. Because of space limitations, we will just report the result of an experiment with packet loss. See the Matlab Project section 6.9 for more detail. For our experiment we picked a harmonic frame with $d = 100$ and $N = 200$ and used the Matlab script lostpacketdist.m from the website. In each the 10,000 runs of the experiment a random set of 20% of the frame vectors were dropped and $log_{10}(\rho)$, the base 10 logarithm of the frame bounds ratio, was recorded. A histogram of the results is given

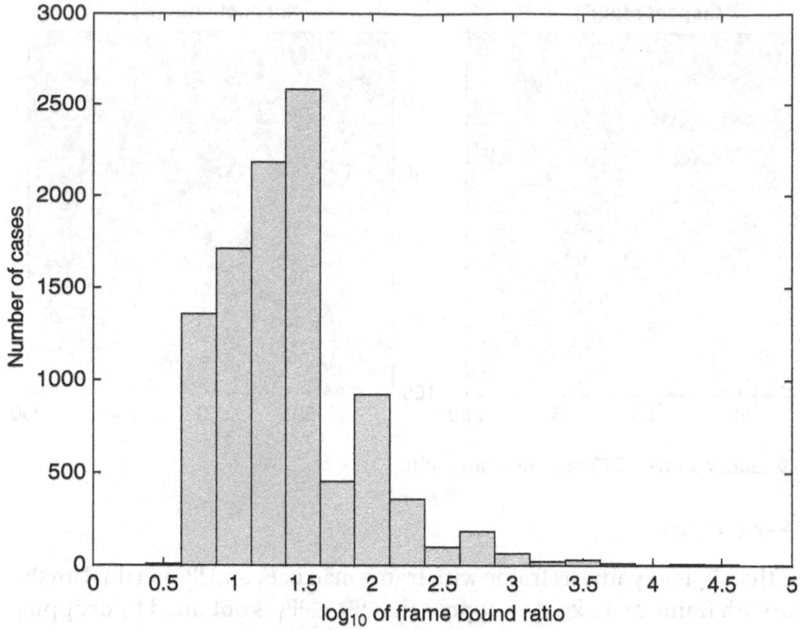

Figure 6.10 Distribution of $log_{10}(\rho)$ for lost packet frames.

in Figure 6.10. The logarithm of the condition number is $log_{10}(\rho)/2$. This number represents the number of significant digits lost on a relative basis. The histogram shows that extreme loss is rare.

6.8.2 Redundancy and other duals

In Section 6.4, we suggested that using several different dual frames would achieve some redundancy. Here, we only construct a second dual frame and compare the frame transforms. For our frame we choose a harmonic frame with $d = 100$ and $N = 200$. Our signal, defined for $0 \leq t \leq 1$ is

$$x(t) = \sin(3.5 \times 2\pi t) - 0.5\cos(6.1 \times 2\pi t) + 0.6\cos(12.5 \times 2\pi t).$$

We sample $x(t)$ at the 100 points $0, 0.01, \ldots, 0.99$ to get a sample vector $\mathbf{x}$. In Figure 6.11 we show the signal and its frame Fourier transform $\mathbf{X} = F_d\mathbf{x}$ with respect to the canonical dual frame F_d in the top two panels. Using the formulas

$$\begin{aligned} F'_d &= F_d + K, \\ F &= U\left[\Sigma\ 0\right]V^*, \\ K &= \left[0\ L\right]V^*, \end{aligned}$$

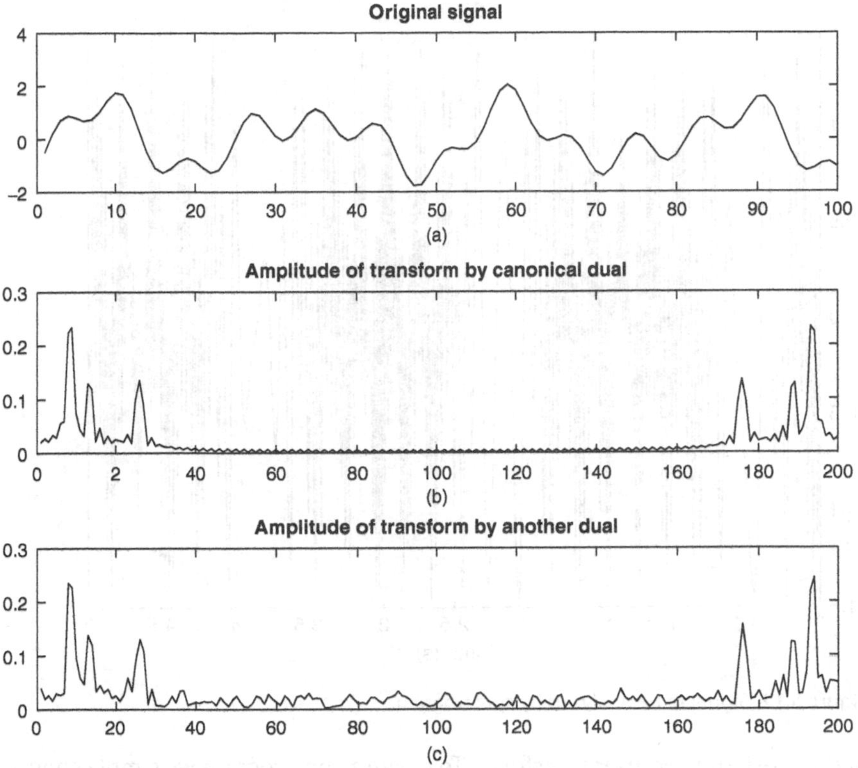

Figure 6.11 Signal and two frame transforms.

established in Section 6.4.4, we construct another dual frame matrix F'_d. We construct K by selecting a random matrix L whose entries are bounded by a small multiple of the maximum value of the entries of F_d. Thus, F'_d is a small perturbation of F_d. The alternate Fourier transform is $\mathbf{X}' = F'_d\mathbf{x}$ pictured in Figure 6.11c. There are three spikes for each of the frequencies in the signal. The canonical dual transform appears to be smoother.

6.8.3 Spectrogram

Gabor frames are an alternate way of constructing a *spectrogram* for a signal. We briefly restate the construction in terms of frames. As discussed in Chapter 5, the spectrogram is a matrix whose columns are the local frequency spectrum of a windowed portion of the signal. As the window moves left to right the corresponding column in the matrix moves left to right, recording the changing frequencies in the moving window. The flipped Gabor transform is perfect for this. Indeed, the flipped Gabor transform is simply a vector version

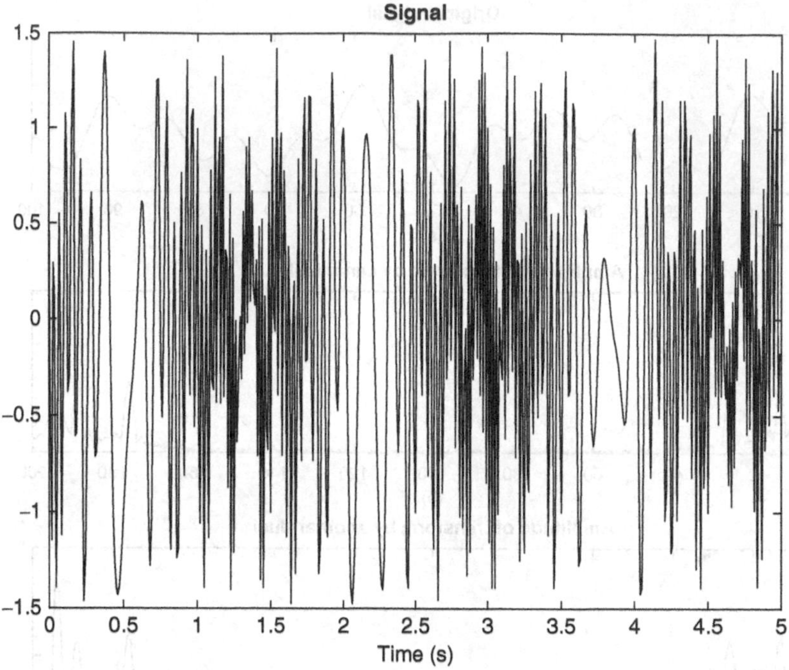

Figure 6.12 Signal to be analyzed by spectrogram.

of the short-time Fourier transform. To create a spectrogram we simply choose d, d_t, d_w and a suitable mask of length ℓ. Given a signal sample $\mathbf{x}$ we create $\mathbf{X} = G\mathbf{x}$ where G is the flipped Gabor frame. The first d/d_w entries are the harmonic transform of the first window. For the next d/d_w entries we shift the window d_t spaces to the right and compute the harmonic transform, and so on. If $d_t < \ell$ the windows overlap. If $d/d_w > \ell$ we oversample the frequencies. We break up $\mathbf{X}$ into portions of length d/d_w place them in order as columns of the spectrogram matrix. The Matlab script on the book website gaborspect.m computes and displays the spectrogram of a given signal. In Figures 6.12 and 6.13, we show the signal and its spectrogram. We picked the following parameters: $N = 500$, $d_t = 10$, $dw = 10$, with a tent-like mask similar to the one given in given in Figure 6.7.

6.9 Matlab Project

The exercises in the Matlab project are of two types: simple problems that can be executed by typing commands in at the console—to get used to the ideas—and then problems that employ more detailed scripts from the website.

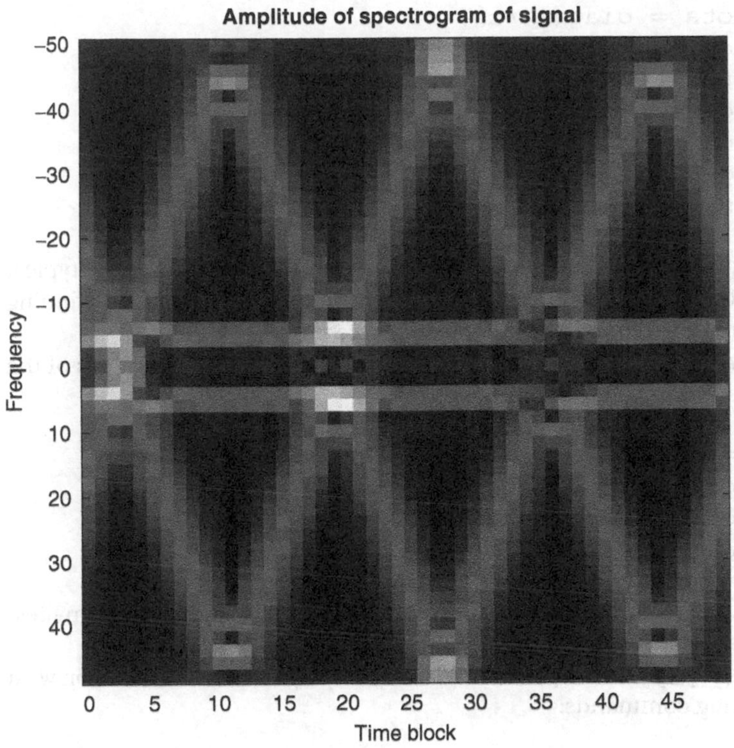

Figure 6.13 Spectrogram of signal.

6.9.1 Frames and Frame Operator

1. *Frame equation and bounds*: The frame bounds are the minimum and maximum of the quantity

$$\frac{1}{\|\mathbf{x}\|^2}\sum_{n=1}^{N}|(\mathbf{x},\boldsymbol{\phi}_n)|^2=\frac{\|F\mathbf{x}\|^2}{\|\mathbf{x}\|^2}=\frac{(F\mathbf{x})^*F\mathbf{x}}{\mathbf{x}^*\mathbf{x}}, \tag{6.37}$$

which is a dot product ratio. We will estimate the ratios for a random frame F and a large selection of random vectors $\mathbf{x}_1,\dots,\mathbf{x}_m$. Use the following commands to create a random frame and random vectors:

```
F = randn(15,30);
x = randn(15,10000);
Ftx = F'*x;
```

We may calculate the quantities in equation (6.37) and display the results in two different ways with the following Matlab commands:

```
xdots = diag(x'*x);
Fdots = diag(Fx'*Fx);
rats = Fdots./xdots;
figure(1)
plot(rats,'r.')
figure(2)
hist(rats,100)
```

The histogram commands allow us to see visually the extreme and the typical values of the dot product ratios. Calculate the approximate ratios by finding the minimum and maximum of the vector `rats`.

2. *Frame bound from eigenvalues*: Compute and display the eigenvalues of the frame operator using the commands

```
S =F*F';
eigS = sort(eig(S))
figure(3)
plot(eigS,'r*')
```

Then find the frame bounds and frame bound ratio by finding the smallest and largest eigenvalues.

3. *Visual display of frame operator*: Visually display the frame operator with the following commands:

```
surf(S)
colormap(gray)
view(90,90)
axis('square')
axis('tight')
```

Does the frame operator appear to be diagonal?

4. *Harmonic frames*: Create a harmonic frame and its frame operator with the commands below and then display the real and imaginary parts of both the frame and the frame operator using the above commands.

```
W =fft(eye(20))';
F = W(1:d,:);
S = F*F';
rF =real(F);
iF = imag(F);
rS = real(S);
```

Note that Matlab may complain if you try to display the matrices F and S directly since they may be complex valued.

5. *Scaled harmonic frames*: Using the commands below, create a scaled harmonic frame. Display the frame operator, compute the eigenvalues and find the frame bound ratio.

```
mask = [1:9,10,10,9:(-1):1]/10; % length 20
D= diag(mask);
F = D*W(1:d,:);
S = F*F';
```

   Check that S is diagonal with this command.

```
checkdiag = sum(sum(abs(S)))-sum(diag(abs(S)));
```

6.9.2 Analysis and Synthesis

1. *Analyze a signal*: Let

$$x(t) = \sin(2.5 \times 2\pi t) - 0.4\cos(10.5 \times 2\pi t) - 0.2\sin(25.5 \times 2\pi t), \quad 0 \le t \le 1.$$

   Compute the frame Fourier transform with three different harmonic frames with parameters $d_1 = d_2 = d_3 = 128$ and $N_1 = 128, N_2 = 2 \times 128$ and $N_2 = 3 \times 128$. The harmonic frames are constructed as in Problem 4. The discretized signal $\mathbf{x}$ and its frame transform $\mathbf{X} = F^*\mathbf{x}$ may be constructed by:

```
T = (0:(127))'/128;
x = sin(2.5*2*pi*T)-0.4*cos(10.5*2*pi*T)...
-0.2*sin(25.5*2*pi*T);
X = F'*x;
```

   Plot both the signal and its frame transform, noting the character of the frame transform. What happens when you make N bigger?
2. *Reconstructing signals*: Since F is tight frame with frame constant N we may reconstruct the signal via $\mathbf{x}_r = \frac{1}{N}F\mathbf{X}$. Verify that the reconstructed $\mathbf{x}_r$ and the original signal $\mathbf{x}$ are the same.
3. *Thresholding*: Repeat the above two experiments when thresholding is applied to the transform $\mathbf{X}$. Use this Matlab code to threshold the transform.

```
eps = 0.01*max(abs(X))
cX = X;
cX(find(abs(X)<eps)) = 0;
```

   Compute the reconstructed signal and the distortion.

```
cx = (1/N)*F*cX;
percentdist = 100*(x-cx)'*(x-cx)/(x'*x)
```

6.9.3 Condition Number

1. *Analysis errors*: We experiment with the condition number of the analysis operator $\mathbf{x} \to F^*\mathbf{x}$ as in the exercise above on frame bounds. The relative error ratio is

$$\sqrt{\frac{(F^*\Delta\mathbf{x})^*F^*\Delta\mathbf{x}}{(F^*\mathbf{x})^*F^*\mathbf{x}}\frac{\mathbf{x}^*\mathbf{x}}{\Delta\mathbf{x}^*\Delta\mathbf{x}}}.$$

   Modify the code in the first exercise to compute and display the relative error ratio for a large number of random $\mathbf{x}$ and $\Delta\mathbf{x}$. Replace the command

```
rats = Fdots./xdots
```

   with

```
rats = sqrt((delFdots./Fdots).*(xdots./
delxdots)).
```

   Experiment with d and N to see that for large d and N we rarely get near the upper bound on the condition number of F.
2. *Synthesis errors*: Do the same experiment for the synthesis operator $\mathbf{y} \to F\mathbf{y}$. In this case, you will need to project onto the orthogonal complement of the kernel of F with a command like `y = Q*randn(30,10000);` where Q is the desired projection operator. See Exercise 8 to see how to construct Q from the singular value decomposition of F.

6.9.4 Packet Loss

1. Experiment with the script lostpacketdemo.m, paying particular attention to the eigenvalue displays. Vary d, N, and the number of lost vectors. Try the same experiments with a random frame.
2. Experiment with the script lostpacketdist.m, paying particular attention to the histogram of condition numbers. Vary d, N, and the number of dropped frame vectors. Try the same experiments with a random frame.

6.9.5 Gabor Frames

1. Experiment with the script gabor.m, paying particular attention to the masks, frame and frame operator matrix image maps, eigenvalue structure and whether the frame operator is diagonal. Vary d, d_t, d_w and the mask.

2. Do the same with gaborflip.m.The conclusions should be the same except that the matrix image maps will be different.
3. Experiment with the script gaborspect.m.Vary the signal and the mask to see the effect on the dual mask and the spectrogram. Determine the maximum frequency that the spectrogram will detect without aliasing.

Exercises

1. Show that $\Phi = \boldsymbol{\phi}_1, \dots, \boldsymbol{\phi}_N$ is a frame if and only if contains a basis of $\mathbb{C}^d$. Alternatively, the frame matrix F has d linearly independent columns.

2. Show that if we add a finite number of vectors to an existing frame then we still get a frame.

3. Let $\Phi = \boldsymbol{\phi}_1, \dots, \boldsymbol{\phi}_N$ be any frame with optimal bounds A, B. Show that $c\Phi = c\boldsymbol{\phi}_1, \dots, c\boldsymbol{\phi}_N$ has frame bounds $|c|^2A$ and $|c|^2B$.

4. If $\Phi = \boldsymbol{\phi}_1, \dots, \boldsymbol{\phi}_N$ is an equal norm tight, frame show that

$$S = \frac{N}{d}\|\boldsymbol{\phi}_1\|^2 I_d.$$

Suggestion: Use the previous exercise and Proposition 6.7.

5. Follow all the steps of analysis and synthesis on the arbitrary vector $\mathbf{x} = \left[x_1\ x_2\right]^t$ using the Mercedes-Benz frame. Determine F, S, F_d, $\mathbf{X} = F_d^*\mathbf{x}$, and verify that $F\mathbf{X} = \mathbf{x}$.

6. Show for an $n \times m$ matrix A, an $m \times n$ matrix B, and vectors $\mathbf{x} \in \mathbb{C}^m$, $\mathbf{y} \in \mathbb{C}^n$, that $A\mathbf{x}$, $A^*\mathbf{y}$, $B^*\mathbf{x}$, and $B\mathbf{y}$ are all defined (easy) and that

$$(A\mathbf{x}, \mathbf{y}) = (\mathbf{x}, A^*\mathbf{y}),\ (\mathbf{x}, B\mathbf{y}) = (B^*\mathbf{x}, \mathbf{y})$$
$$(A\mathbf{x}, B\mathbf{y}) = (B^*A\mathbf{x}, \mathbf{y}) = (\mathbf{x}, A^*B\mathbf{y})$$
$$(B^*\mathbf{x}, A^*\mathbf{y}) = (AB^*\mathbf{x}, \mathbf{y}) = (\mathbf{x}, BA^*\mathbf{y}).$$

Suggestion: The scalar product $(A\mathbf{x}, \mathbf{y}) = \mathbf{y}^*A\mathbf{x}$ when $\mathbf{x}$ and $\mathbf{y}$ are considered as column vectors.

7. Prove Proposition 6.3. Suggestion: Set $F' = S^{-1}F + K$ and then expand $(F'^*x, F'^*x) = \sum_{n=1}^{N} |(\mathbf{x}, \boldsymbol{\psi}'_n)|^2$ (see equation (6.28)).

8. Find the projection operators onto ker(F) and the orthogonal complement of ker(F) as follows. Using the singular value decomposition of

$F = U\,[\Sigma\ 0]\, V^*$, and the projection operators

$$P_0 = \begin{bmatrix} 0 & 0 \\ 0 & I_{N-d} \end{bmatrix},\ Q_0 = \begin{bmatrix} I_d & 0 \\ 0 & 0 \end{bmatrix},$$

define $P = VP_0V^*$, $Q = VQ_0V^*$. Verify the following properties:

$$P^* = P,\ Q^* = Q$$
$$P^2 = P,\ Q^2 = Q$$
$$P + Q = I_N,\ PQ = QP = 0$$
$$FP = 0, FQ = F.$$

9. *Projection operators continued*: Let $\mathbf{x} \in \mathbb{C}^N$. Show that $\mathbf{x} \in \ker F$, if and only if $P\mathbf{x} = \mathbf{x}$ and $\mathbf{x} \perp \ker F$ if and only if $Q\mathbf{x} = \mathbf{x}$.

10. Show that Q is the projection of $\mathbb{C}^N$ onto the range of F^*.

11. Prove equations (6.24)– (6.26). Suggestion: Use these trigonometric identities

$$\cos^2(x) = \frac{1+\cos(2x)}{2},\quad \sin^2(x) = \frac{1-\cos(2x)}{2},\quad \cos(x)\sin(x) = \frac{\sin(2x)}{2}$$

and realize that for any integer M

$$\sum_{n=0}^{M-1} (\cos(nx) + i\sin(nx)) = \sum_{n=0}^{M-1} e^{inx}.$$

12. For a harmonic frame of type d, N show that $S = NI_d$. Suggestion: Review the discussion in Section 6.5.

13. For a harmonic frame with mask (Remark 6.4), show that the frame operator equals NDD^*. What are the eigenvalues of S and what is the frame bounds ratio?

14. Prove the assertions about product frames at the end of Section 6.5 except for the eigenvalues of the frame operator and the frame bounds.

15. Prove the assertions about eigenvalues of the frame operator and frame bounds for product frames. Suggestion: If $S_1\mathbf{x}_i = \alpha_i\mathbf{x}_i$ and $S_2\mathbf{y}_j = \beta_j\mathbf{y}_j$, then consider $S_1(\mathbf{x}_i\mathbf{y}_j^t)S_2^t$.

The next two exercises demonstrate that a tight frame operator is scalar multiple of the identity without using diagonalizabilty of the frame operator.

16. Suppose that we have a real, tight frame with frame operator S. Using a version of the polarization identity:

$$2(S\mathbf{x},\mathbf{y}) = (S(\mathbf{x}+\mathbf{y}),\mathbf{x}+\mathbf{y}) - (S\mathbf{x},\mathbf{x}) - (S\mathbf{y},\mathbf{y}),$$

show that

$$(S\mathbf{x},\mathbf{y}) = A(\mathbf{x},\mathbf{y}).$$

Now compute $(S\mathbf{x} - A\mathbf{x},\mathbf{y})$ to show that $S = AI_d$.

17. Rework the above exercise for complex frames. First, show that

$$(S\mathbf{x},\mathbf{y}) + \overline{(S\mathbf{x},\mathbf{y})} = (S(\mathbf{x}+\mathbf{y}),\mathbf{x}+\mathbf{y}) - (S\mathbf{x},\mathbf{x}) - (S\mathbf{y},\mathbf{y})$$

to prove that

$$(S\mathbf{x},\mathbf{y}) + \overline{(S\mathbf{x},\mathbf{y})} = A((\mathbf{x},\mathbf{y}) + \overline{(\mathbf{x},\mathbf{y})})$$

Replace $\mathbf{y}$ by $i\mathbf{y}$ and use the resulting equations to show that $(S\mathbf{x},\mathbf{y}) = A(\mathbf{x},\mathbf{y})$.

18. Show that the frame operator of Φ is given by

$$S = \sum_{n=1}^{N} \phi_n \phi_n^*$$

and hence is independent of the ordering of the frame vectors.

19. Prove that $(R\mathbf{x}, R\mathbf{x}) = (\mathbf{x},\mathbf{x})$ if and only if $R^* = R^{-1}$. See the suggestions for Exercises 16 and 17.

20. Supply the missing details of Proposition 6.8.

21. Prove Proposition 6.11.

22. Supply the missing details of Proposition 6.15.

7

Filter Banks

7.1 Overview

In previous chapters, we encountered various signal models, continuous and discrete, for signals of both finite and infinite duration. We then developed notions of Fourier analysis appropriate to several of these models. Wavelet theory is similar in that it can be approached from both the continuous point of view and the discrete point of view, for signals of either finite or infinite length. As with Fourier analysis, wavelet analysis in the continuous setting involves more technical details and does not allow one to get to the computational applications very quickly. In this chapter, we will thus focus on discrete signals only. We will examine continuous models in Chapter 9.

However, it is easier to first carry out the analysis for signals that are bi-infinite in extent, for example, signals in $L^2(\mathbb{Z})$ or $L^\infty(\mathbb{Z})$ rather than $\mathbb{R}^N$. In specific computational examples where we have a finite length signal $\mathbf{x}$, we will embed $\mathbf{x}$ in $L^2(\mathbb{Z})$ by zero extension or some other technique, and then apply techniques for bi-infinite signals.

Wavelet analysis is in part motivated by the need to analyze signals whose frequency content varies over the duration of the signal. This is one application where Fourier methods do not perform well; for the basic, Fourier waveforms are global in nature. As a result, standard Fourier techniques do not really recognize or exploit the fact that the frequency content of a signal or image can vary considerably from one point to another. Wavelet techniques are an alternative to the windowing approach of Chapter 5, and they provide an elegant, flexible, computationally efficient solution to the problem of "localizing" signal frequency analysis. The ideas have found use in many areas of mathematics, science, and engineering, and remain a very active area of research.

One simple approach to discrete wavelet analysis is through the mechanism of a *filter bank*. A properly designed filter bank naturally gives rise to a discrete wavelet transform (DWT) similar to the discrete Fourier transform (DFT). The "discrete wavelets" appear as basic waveforms associated to the DWT, just as the exponential basic waveforms arise from the DFT matrix. In order to get

Discrete Fourier Analysis and Wavelets: Applications to Signal and Image Processing, Second Edition.
S. Allen Broughton and Kurt Bryan.

Companion Website: www.wiley.com/go/Broughton/Discrete_Fourier_Analysis_and_Wavelets

to applications and Matlab experimentation sooner, we defer the topic of filter design for filter banks until Section 7.7. Indeed, if the reader is content to use "off-the-shelf" filters, that section can be omitted, though the mathematics of filter design ties in nicely with the wavelet analysis of Chapter 9. Moreover, the material of Section 7.7 is also necessary to understand the modern "lifting" implementation of filter banks discussed in Chapter 8. For additional reading on filter banks see [33].

To build up some intuition, we will begin by examining the very simplest kind of filter bank, the Haar filter bank.

7.2 The Haar Filter Bank

One of the key features of the Fourier transform is that it allows us to decompose a signal into constituent frequencies and then deal with the signal one frequency at a time. The filter bank method also adopts this approach by splitting the signal into various frequency bands. In the simplest case, that of a single-stage two-channel filter bank, we merely separate the signal into high and low frequencies by filtering with the methods of Chapter 4.

7.2.1 The One-Stage Two-Channel Filter Bank

In what follows, we will consider bi-infinite signals in the space $L^2(\mathbb{Z})$, although many of the ideas extend to signals in $L^\infty(\mathbb{Z})$.

Consider a pair of vectors $\boldsymbol{\ell}$ and $\mathbf{h}$ in $L^2(\mathbb{Z})$, to be used as low- and high-pass filters for a signal $\mathbf{x} \in L^2(\mathbb{Z})$. The vectors $\boldsymbol{\ell}$ and $\mathbf{h}$ act by convolution with $\mathbf{x}$, via equation (4.22). We will assume that $\boldsymbol{\ell}$ and $\mathbf{h}$ are finite impulse response (FIR) filters as defined in Section 4.5.6. In this case, the infinite sum defining the convolution of either filter with $\mathbf{x} \in L^2(\mathbb{Z})$ converges, and in fact both $\mathbf{x} * \boldsymbol{\ell}$ and $\mathbf{x} * \mathbf{h}$ will also be elements of $L^2(\mathbb{Z})$ (recall Exercise 30 of Chapter 4).

For simplicity, let us first consider the familiar two-point averaging and differencing filters with components

$$\begin{aligned} \ell_0 &= \frac{1}{2}, \quad \ell_1 = \frac{1}{2}, \quad \ell_r = 0, \text{ otherwise,} \\ h_0 &= \frac{1}{2}, \quad h_1 = -\frac{1}{2}, \quad h_r = 0, \text{ otherwise.} \end{aligned} \tag{7.1}$$

It was shown in Example 4.5 that $\boldsymbol{\ell}$ acts as a low-pass filter, whereas Exercise 35 of Chapter 4 shows that $\mathbf{h}$ is a high-pass filter. In the context of filter banks, these are called the *Haar filters,* although the definition may vary a bit.

Example 7.1 *In Figure 7.1, the signal in the upper pane is a portion of the vector* $\mathbf{x} \in L^2(\mathbb{Z})$ *with components*

$$x_k = \frac{1}{2}\sin\left(\frac{2\pi \cdot 3k}{128}\right) + \frac{1}{2}\sin\left(\frac{2\pi \cdot 49k}{128}\right),$$

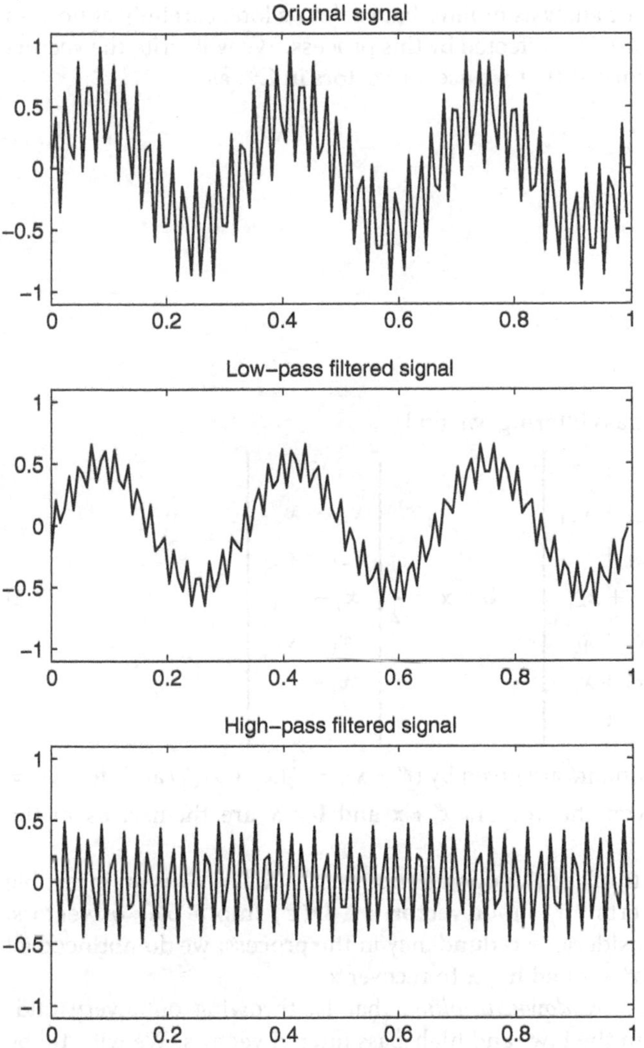

Figure 7.1 Original signal with low- and high-pass filtered versions.

for $0 \leq k \leq 128$*; that is, we sample* $\frac{1}{2}\sin(2\pi \cdot 3t) + \frac{1}{2}\sin(2\pi \cdot 49t)$ *at* 128 *points* $t = k/128$ *for* $0 \leq k \leq 127$*. Outside this range, we extend* $\mathbf{x}$ *by zero, although this of no particular importance at the moment.*

The lower panes show the nonzero portion of the convolutions $\mathbf{x} * \boldsymbol{\ell}$ *and* $\mathbf{x} * \mathbf{h}$*. The low-pass filtered version* $\mathbf{x} * \boldsymbol{\ell}$ *is a smoothed approximation to the original signal, and shows the overall "trend" of the signal. The high-pass version* $\mathbf{x} * \mathbf{h}$ *quantifies "detail" information, the deviation of the signal from the trend. It is worth noting that because* $\boldsymbol{\ell} + \mathbf{h} = \mathbf{e}$*, where* $e_0 = 1$ *and* $e_k = 0$ *for all other k, we have* $\mathbf{x} = \boldsymbol{\ell} * \mathbf{x} + \mathbf{h} * \mathbf{x}$*. We can thus recover* $\mathbf{x}$ *by adding the low- and high-pass filtered vectors. No information has been lost in the course of filtering.*

To continue with our analysis of filter banks, let us look carefully at how an arbitrary vector $\mathbf{x} \in L^2(\mathbb{Z})$ is affected by this process. We will write the vectors in the same column format that we use for vectors in $\mathbb{C}^N$, as

$$\mathbf{x} = \begin{bmatrix} \vdots \\ x_{-2} \\ x_{-1} \\ x_0 \\ x_1 \\ x_2 \\ \vdots \end{bmatrix}.$$

After low- and high-pass filtering, we find

$$\boldsymbol{\ell} * \mathbf{x} = \frac{1}{2}\begin{bmatrix} \vdots \\ x_{-2}+x_{-3} \\ x_{-1}+x_{-2} \\ x_0+x_{-1} \\ x_1+x_0 \\ x_2+x_1 \\ \vdots \end{bmatrix}, \quad \mathbf{h} * \mathbf{x} = \frac{1}{2}\begin{bmatrix} \vdots \\ x_{-2}-x_{-3} \\ x_{-1}-x_{-2} \\ x_0-x_{-1} \\ x_1-x_0 \\ x_2-x_1 \\ \vdots \end{bmatrix}. \tag{7.2}$$

In general, the components are given by $(\boldsymbol{\ell} * \mathbf{x})_k = \frac{1}{2}(x_k + x_{k-1})$ and $(\mathbf{h} * \mathbf{x})_k = \frac{1}{2}(x_k - x_{k-1})$. Moreover, the vectors $\boldsymbol{\ell} * \mathbf{x}$ and $\mathbf{h} * \mathbf{x}$ are themselves in the space $L^2(\mathbb{Z})$.

We could actually think of the operation $\mathbf{x} \to (\boldsymbol{\ell} * \mathbf{x}, \mathbf{h} * \mathbf{x})$ as an invertible transform that converts any input vector $\mathbf{x} \in L^2(\mathbb{Z})$ into a pair of vectors. However, there is considerable redundancy in the process; we do not need all components of both $\boldsymbol{\ell} * \mathbf{x}$ and $\mathbf{h} * \mathbf{x}$ to recover $\mathbf{x}$.

We can economize by *downsampling*, that is, throwing out every odd-indexed component in the low- and high-pass filtered vectors. We will still be able to reconstruct $\mathbf{x}$. In the present case, we obtain from equation (7.2) the vectors $\mathbf{X}_\ell$ and $\mathbf{X}_h$ defined by

$$\mathbf{X}_\ell := D(\boldsymbol{\ell} * \mathbf{x}) = \frac{1}{2}\begin{bmatrix} \vdots \\ x_{-2}+x_{-3} \\ x_0+x_{-1} \\ x_2+x_1 \\ x_4+x_3 \\ \vdots \end{bmatrix}, \tag{7.3}$$

$$\mathbf{X_h} := D(\mathbf{h} * \mathbf{x}) = \frac{1}{2} \begin{bmatrix} \vdots \\ x_{-2} - x_{-3} \\ x_0 - x_{-1} \\ x_2 - x_1 \\ x_4 - x_3 \\ \vdots \end{bmatrix}. \tag{7.4}$$

where D is the *downsampling* operator:

Definition 7.1 The downsampling operator $D : L^2(\mathbb{Z}) \to L^2(\mathbb{Z})$ is defined by

$$D : (\ldots, x_{-2}, x_{-1}, x_0, x_1, x_2, \ldots) \to (\ldots, x_{-2}, x_0, x_2, \ldots),$$

that is, $(D(\mathbf{x}))_k = x_{2k}$.

Let us define the transform $W(\mathbf{x}) = \mathbf{X}$, where $\mathbf{X} = (\mathbf{X}_\ell, \mathbf{X}_h)$. The transform W is a linear mapping from $L^2(\mathbb{Z})$ to $L^2(\mathbb{Z}) \times L^2(\mathbb{Z})$. The components of $\mathbf{X}_\ell$ are called the *approximation coefficients*, and the components of $\mathbf{X}_h$ are called the *detail coefficients*. The vector $\mathbf{X}_\ell$ is a low-pass filtered (hence smoothed) version of $\mathbf{x}$, but with every other sample omitted. The vector $\mathbf{X}_h$ is similar, but high-pass filtered. We may also refer to $\mathbf{X}_\ell$ as the signal *trend* and $\mathbf{X}_h$ as the signal *detail*.

The transform W is represented by the schematic in Figure 7.2 and is referred to as an *analysis filter bank*. More precisely, W is a *one-stage two-channel analysis filter bank* since it splits the input signal into two channels and filters each. The "one-stage" stems from the fact that the splitting and filtering is performed only once; we will soon encounter multi-stage filter banks in which the split/filter operation is cascaded.

In Figure 7.2, we have subscripted the low- and high-pass filters with the letter "a" to indicate that these are the *analysis* filters, used to break the input signal into frequency subbands. The blocks $[\boldsymbol{\ell}_a]$ and $[\mathbf{h}_a]$ in the diagram represent $L^2(\mathbb{Z})$ convolution with the low- and high-pass filters, respectively, while the $[\downarrow 2]$ blocks represent the operation of downsampling. We will refer to the transform W as the *Haar filter bank transform* for now.

As it turns out, there is a simple and elegant method for inverting this transform.

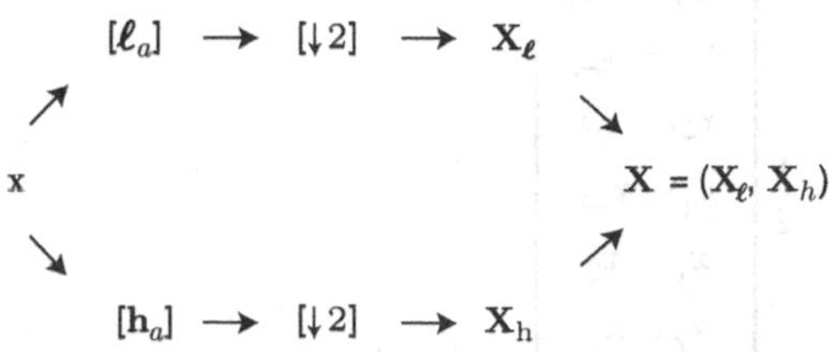

Figure 7.2 Schematic of one-stage analysis filter bank.

7.2.2 Inverting the One-stage Transform

The inverse transform can be computed in a structured way that is quite similar to the forward transform. We begin by *upsampling* each piece $\mathbf{X}_\ell$ and $\mathbf{X}_h$ of the transform $\mathbf{X}$.

Definition 7.2 The upsampling operator $U : L^2(\mathbb{Z}) \to L^2(\mathbb{Z})$ is defined by

$$U : (\dots, x_{-2}, x_{-1}, x_0, x_1, x_2, \dots) \to (\dots, x_{-2}, 0, x_{-1}, 0, x_0, 0, x_1, 0, x_2, \dots).$$

The components of $U(\mathbf{x})$ are given by

$$(U(\mathbf{x}))_k = \begin{cases} x_{k/2}, & k \text{ even}, \\ 0, & k \text{ odd}. \end{cases}$$

If we upsample the vectors $\mathbf{X}_\ell$ and $\mathbf{X}_h$ in equations (7.3) and (7.4), we obtain

$$U(\mathbf{X}_\ell) = \frac{1}{2}\begin{bmatrix} \vdots \\ 0 \\ x_{-2} + x_{-3} \\ 0 \\ x_0 + x_{-1} \\ 0 \\ x_2 + x_1 \\ 0 \\ \vdots \end{bmatrix}, \quad U(\mathbf{X}_h) = \frac{1}{2}\begin{bmatrix} \vdots \\ 0 \\ x_{-2} - x_{-3} \\ 0 \\ x_0 - x_{-1} \\ 0 \\ x_2 - x_1 \\ 0 \\ \vdots \end{bmatrix}. \tag{7.5}$$

The next step in inverting the transform is to convolve $U(\mathbf{X}_\ell)$ and $U(\mathbf{X}_h)$ with appropriate *synthesis filters*. In this example, we convolve $U(\mathbf{X}_\ell)$ with the low-pass synthesis filter $\boldsymbol{\ell}_s$ with components $(\boldsymbol{\ell}_s)_{-1} = 1, (\boldsymbol{\ell}_s)_0 = 1, (\boldsymbol{\ell}_s)_k = 0$ for all other indexes k. We convolve $U(\mathbf{X}_h)$ with the high-pass synthesis filter $\mathbf{h}_s$ that has components $(\mathbf{h}_s)_{-1} = -1, (\mathbf{h}_s)_0 = 1$, and $(\mathbf{h}_s)_k = 0$ for all other k. Note that $\boldsymbol{\ell}_s$ and $\mathbf{h}_s$ are not causal, but let us not worry about that for the moment. This filtering process yields vectors $\mathbf{v}_\ell = \boldsymbol{\ell}_s * (U(\mathbf{X}_\ell))$ and $\mathbf{v}_h = \mathbf{h}_s * (U(\mathbf{X}_h))$ given by

$$\mathbf{v}_\ell = \frac{1}{2}\begin{bmatrix} \vdots \\ x_{-2} + x_{-3} \\ x_{-2} + x_{-3} \\ x_0 + x_{-1} \\ x_0 + x_{-1} \\ x_2 + x_1 \\ x_2 + x_1 \\ x_4 + x_3 \\ \vdots \end{bmatrix}, \quad \mathbf{v}_h = \frac{1}{2}\begin{bmatrix} \vdots \\ x_{-3} - x_{-2} \\ x_{-2} - x_{-3} \\ x_{-1} - x_0 \\ x_0 - x_{-1} \\ x_1 - x_2 \\ x_2 - x_1 \\ x_3 - x_4 \\ \vdots \end{bmatrix}.$$

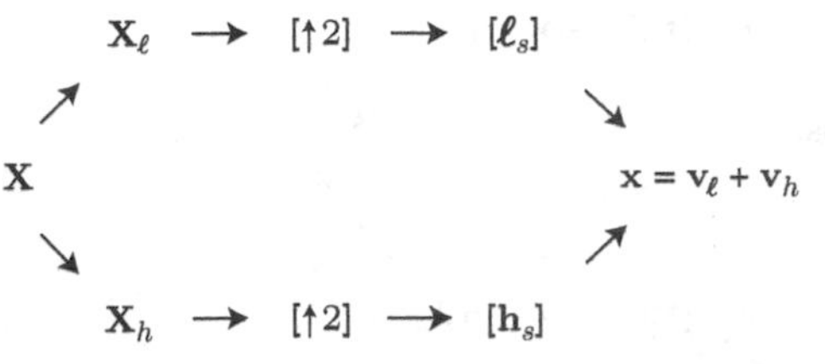

Figure 7.3 Schematic of one-stage synthesis filter bank.

In general, the kth component of $\mathbf{v}_\ell$ is $\frac{1}{2}(x_{k-1}+x_k)$ if k is even and $\frac{1}{2}(x_k+x_{k+1})$ if k is odd. The kth component of $\mathbf{v}_h$ is $\frac{1}{2}(x_k-x_{k-1})$ if k is even and $\frac{1}{2}(x_k-x_{k+1})$ if k is odd. In general, the synthesis filters $\boldsymbol{\ell}_s$ and $\mathbf{h}_s$ must be chosen to satisfy certain constraints that, not surprisingly, depend on the analysis filters.

Finally, we recover the original vector as $\mathbf{x} = \mathbf{v}_\ell + \mathbf{v}_h$. All operations above are linear. The combination of upsampling and convolution with the synthesis filters is called the *synthesis filter bank*.

The inverse transform/synthesis process can be summarized graphically as in Figure 7.3. In contrast to the analysis filter bank, the downsampling operator is replaced by the upsampling operator $[\uparrow 2]$, followed by convolution with the synthesis filters $\boldsymbol{\ell}_s$ and $\mathbf{h}_s$.

Remark 7.1 In the discussion above, the synthesis filters are not causal. If causal filters are required, we can simply shift the filter vectors $\boldsymbol{\ell}_s$ and $\mathbf{h}_s$ to obtain causal filters. Specifically, let us define the shift operator $S : L^2(\mathbb{Z}) \to L^2(\mathbb{Z})$ as

$$(S(\mathbf{y}))_k = y_{k-1}.$$

The S operator shifts $\mathbf{y}$ one index to the right, which we can interpret as delaying $\mathbf{y}$ one index in time (yes, delaying; e.g., $S(\mathbf{y})$ attains at time index 1 the value $\mathbf{y}$ attained one unit earlier, since $(S(\mathbf{y}))_1 = y_0$). If we replace the synthesis filters $\boldsymbol{\ell}_s$ and $\mathbf{h}_s$ by the causal filters $S(\boldsymbol{\ell}_s)$ and $S(\mathbf{h}_s)$ in the computations above, then the output of the synthesis filter bank is $S(\mathbf{x})$, a one-index delayed version of $\mathbf{x}$. We consider this shifted output a legitimate inversion of the analysis filter bank.

More generally, if $\mathbf{g}$ is any filter, then it is easy to show that $(S^m(\mathbf{g})) * \mathbf{x} = S^m(\mathbf{g} * \mathbf{x})$ for any m and $\mathbf{x}$ (see Exercise 7). Given any noncausal FIR filter vector $\mathbf{g}$, we can always shift $\mathbf{g}$ so that it gives a causal filter, if we are willing to accept the resulting delay in the output. However, in image processing causality is not a concern.

7.2.3 Summary of Filter Bank Operation

The overall operation of a filter bank in a typical application is illustrated in Figure 7.4. The output $\tilde{\mathbf{x}}$ denotes an altered version (denoised, compressed) of the input vector $\mathbf{x}$. In the event that no compression, denoising, or other alteration is performed between the analysis and synthesis banks, we would

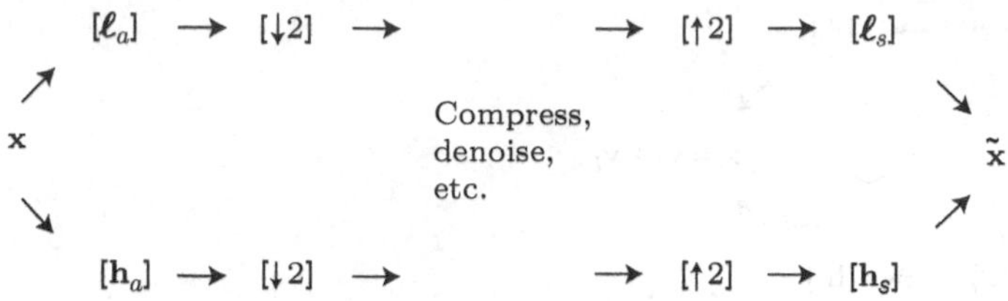

Figure 7.4 Analysis/synthesis filter bank.

like $\tilde{\mathbf{x}} = S^m(\mathbf{x})$ for some m. This means that the output of the filter bank is a perfect reconstruction of the input, with a possible m-index delay.

The analysis–synthesis filter bank pair is yet another specific instance of the ubiquitous "transform, process, reverse transform" operation in mathematics as was illustrated in Figure 2.1.

Example 7.2 *Before proceeding to more general filter banks, it will be useful to see the Haar filter bank applied to a specific signal. We also examine how this approach might be used in a crude compression application, and why its more local nature can give superior performance over a DFT-based approach.*

Let $\mathbf{x} \in L^2(\mathbb{Z})$ *denote the sampled piecewise constant signal shown in Figure 7.5 for the index range* $0 \le k \le 1023$, *with* $x_k = 0$ *outside this range. The signal consists of an oscillatory section followed by large stretches of constant values interspersed with abrupt jumps; the jumps contain a lot of (localized) high-frequency energy. This is just the kind of signal on which a global DFT will perform poorly when used for compression. It is also similar to two-dimensional images, which often have large areas of constant value with abrupt edges.*

We run the signal through a one-stage Haar filter bank with analysis filters defined by equation (7.1), to produce approximation coefficients $\mathbf{X}_\ell$ *and detail coefficients* $\mathbf{X}_h$ *via equation (7.3) or (7.4), both in* $L^2(\mathbb{Z})$. *Since* $\mathbf{x}$ *is zero outside a finite range of indexes, so are* $\mathbf{X}_\ell$ *and* $\mathbf{X}_h$ *(Exercise 20). In the present case, it*

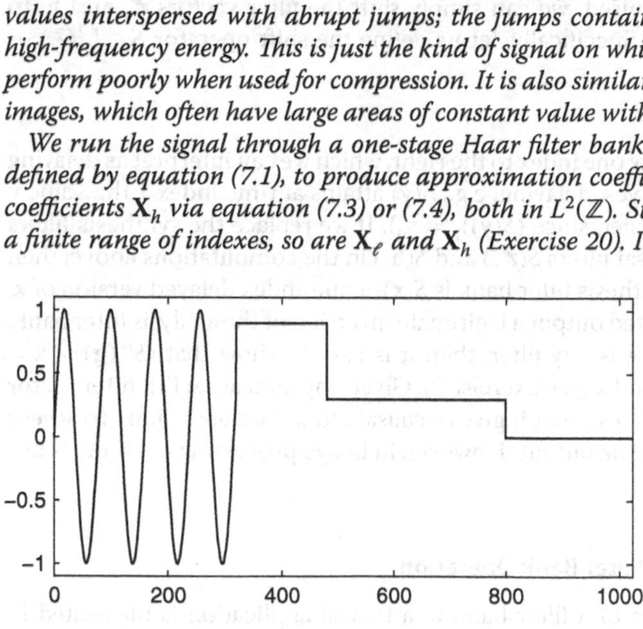

Figure 7.5 Signal to be filtered.

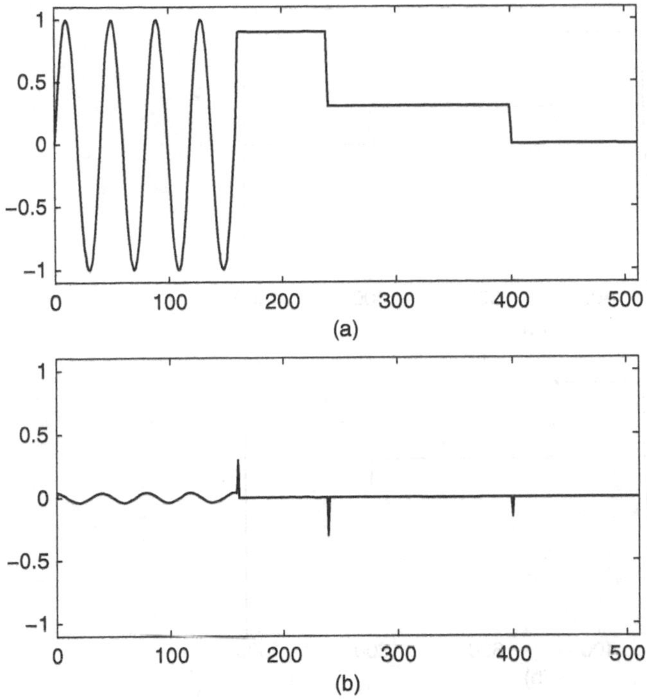

Figure 7.6 Approximation (a) and detail (b) coefficients for sample signal.

suffices to display only components indexed in the range $0 \leq k \leq 512$ *for* $\mathbf{X}_\ell$ *and* $\mathbf{X}_h$*. Plots of both are shown in Figure 7.6. The vector* $\mathbf{X}_\ell$ *is shown in Figure 7.6a. Note that the beginning of* $\mathbf{X}_\ell$ *corresponds to the beginning of the signal* $\mathbf{x}$ *and the end of* $\mathbf{X}_\ell$ *corresponds to the end of* $\mathbf{x}$*. The same holds for* $\mathbf{X}_h$ *in Figure 7.6b. The approximation coefficients strongly resemble the original signal. The detail coefficients are large where the original signal* $\mathbf{x}$ *is highly oscillatory (during the first third of the signal, and at the jumps) but otherwise relatively small.*

We can obtain an "easy" 50% compression of $\mathbf{x}$ *by zeroing out all components of* $\mathbf{X}_h$ *and then inverse transforming with the appropriate synthesis filter bank. The resulting vector* $\tilde{\mathbf{x}}$ *is a compressed version of* $\mathbf{x}$ *and is shown in Figure 7.7a.*

Contrast this to the compression obtained using a DFT on the nonzero portion of $\mathbf{x}$*. Specifically, we compute the DFT of* $\mathbf{x}$ *and zero out the top half of the frequencies (alternatively, a threshold at a level that zeros out half the frequencies; it does not make much difference). We then inverse transform. The resulting vector is shown in Figure 7.7b. The first third of the signal is very local in the frequency domain and compresses well with this approach. The jumps, however, require many frequencies to synthesize. As a result, the DFT deals with the jumps less gracefully, and the difficulty is somewhat evident. In a compressed image, this phenomenon can be pretty objectionable.*

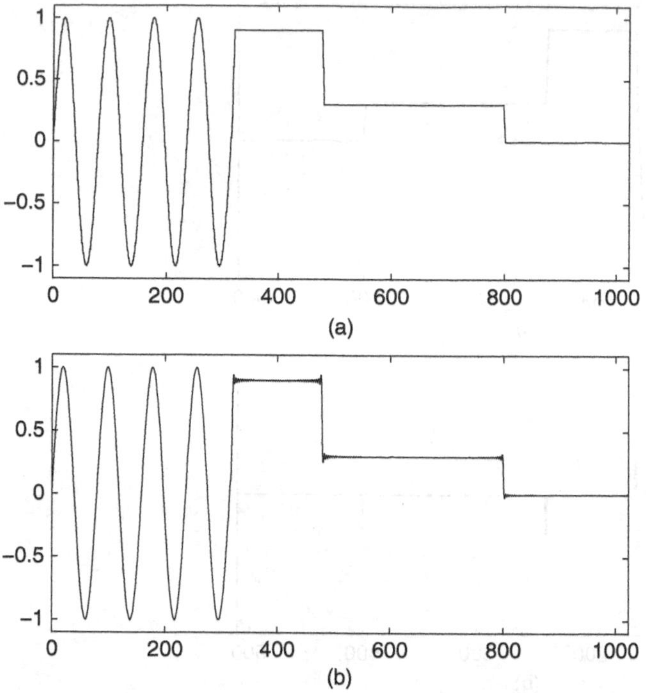

Figure 7.7 Signal **x** compressed via Haar transform (a) and DFT (b).

7.3 The General One-stage Two-channel Filter Bank

7.3.1 Formulation for Arbitrary FIR Filters

The procedure above can be generalized beyond the simple two-point averaging filters. For an input signal vector $\mathbf{x} \in L^2(\mathbb{Z})$, we begin with a pair $\boldsymbol{\ell}_a$ and $\mathbf{h}_a$ of FIR *analysis filters* (conventionally low-pass/high-pass, but not necessarily so). To compute the transform pair $\mathbf{X}_\ell$ and $\mathbf{X}_h$, we proceed as illustrated by Figure 7.2. Specifically, we proceed in two steps:

1. Compute $\mathbf{x} * \boldsymbol{\ell}_a$ and $\mathbf{x} * \mathbf{h}_a$.
2. Downsample each filtered signal to obtain

$$\mathbf{X}_\ell = D(\mathbf{x} * \boldsymbol{\ell}_a), \quad \mathbf{X}_h = D(\mathbf{x} * \mathbf{h}_a).$$

If $\boldsymbol{\ell}_a$ is a low-pass filter, then $\mathbf{X}_\ell$ is a smoothed approximation to the original signal that gives the overall trend in **x**. If **h** is a high-pass filter, then $\mathbf{X}_h$ contains details that show how **x** deviates from the overall signal trend.

Note also that since the filters are FIR, each component of $\mathbf{X}_h$ is computed from a relatively small number of components of **x**. Specifically, $(\mathbf{X}_h)_k$ is computed from x_{2k} and nearby samples (where "nearby" is dictated by the filter length). Similar remarks apply to $\mathbf{X}_\ell$. Contrast this to the DFT, where each transform coefficient depends on *all* of the sample values. In this sense, the filter bank transform is more local than Fourier-based methods.

To invert the process, we essentially reverse the steps:

1. Upsample each of $\mathbf{X}_\ell$ and $\mathbf{X}_h$, to obtain signals $U(\mathbf{X}_\ell)$ and $U(\mathbf{X}_h)$.
2. Compute $\mathbf{v}_\ell = \boldsymbol{\ell}_s * (U(\mathbf{X}_\ell))$ and $\mathbf{v}_h = \mathbf{h}_s * (U(\mathbf{X}_h))$, where $\boldsymbol{\ell}_s$ and $\mathbf{h}_s$ are certain low- and high-pass filters, called the *synthesis* filters.
3. Add the vectors $\mathbf{v}_\ell$ and $\mathbf{v}_h$ to recover $\mathbf{x}$, or perhaps $S^m(\mathbf{x})$, the signal $\mathbf{x}$ shifted m indexes as per Remark 7.1.

The synthesis filters depend on the analysis filters. There are other hidden constraints too—we cannot take just anything for the analysis filters, there may not be appropriate matching synthesis filters. And even if there are, the synthesis filters may be too complicated or have too many taps.

Remark 7.2 As previously noted, the scheme above is called a *two-channel* filter bank, since the incoming signal is split into two channels and each is subjected to a different filter, to isolate a different frequency portion of the signal. More generally, we can split a signal into M channels, filter each to extract information in a different frequency range, and then downsample (downsampling each channel here usually consists of keeping only every Mth filtered component). This is an example of an M-channel filter bank. We will not examine such filter banks in this text, but see [28] for more information. This general strategy of splitting a signal into various frequency bands and processing each separately is also known as *subband coding*.

Below we examine a bit more carefully what is required for the analysis/synthesis filter pair in the time domain if the process above is to perfectly reconstruct $\mathbf{x}$. Later, it will be more convenient to look at the filter requirements via the z-transform.

7.3.2 Perfect Reconstruction

The requirement that the filter bank output be a (possibly delayed) copy of the input signal $\mathbf{x}$ can be expressed as

$$S^m(\mathbf{x}) = \boldsymbol{\ell}_s * (U(D(\boldsymbol{\ell}_a * \mathbf{x}))) + \mathbf{h}_s * (U(D(\mathbf{h}_a * \mathbf{x}))). \tag{7.6}$$

We want to choose the various filters so that equation (7.6) holds for all $\mathbf{x} \in L^2(\mathbb{Z})$.

Definition 7.3 A "perfect reconstruction" or "biorthogonal" filter bank is one for which analysis followed by synthesis on any signal $\mathbf{x}$ yields $S^m(\mathbf{x})$ for some m.

The terminology "perfect reconstruction" is very sensible; the term "biorthogonal" will be explained later.

Unfortunately, equation (7.6) is not particularly helpful when it comes to actually designing filters, even when written out explicitly in summation notation. The equation does, however, make it clear that if one specifies the analysis filters, then the synthesis filters are determined from an infinite set of linear equations in the filter coefficients. Let us look at a simple example.

Example 7.3 *Let us consider the case $m = 0$, where the analysis filters have already been chosen according to equation (7.1), the Haar filters. Are the synthesis filters uniquely determined? To find out, we will make use of equation (7.6).*

Let $\mathbf{x}$ be an arbitrary vector in $L^2(\mathbb{Z})$. The result of filtering, downsampling, and then upsampling $\mathbf{x}$ was worked out in equation (7.5). In general, the vector $\mathbf{v} = U(D(\boldsymbol{\ell}_a * \mathbf{x}))$ has components $v_k = (x_k + x_{k-1})/2$ for k even, and $v_k = 0$ for k odd. Similarly, the vector $\mathbf{w} = U(D(\mathbf{h}_a * \mathbf{x}))$ has components $w_k = (x_k - x_{k-1})/2$ for k even, and $w_k = 0$ for k odd.

We convolve $\mathbf{v}$ with the synthesis low-pass filter $\boldsymbol{\ell}_s$ and convolve $\mathbf{w}$ with the synthesis high-pass filter $\mathbf{h}_s$; then we require that the nth component of the sum $\mathbf{v} + \mathbf{w}$ equal x_n, for perfect reconstruction. This yields

$$\frac{1}{2}\sum_{k\ even} ((x_k + x_{k-1})(\boldsymbol{\ell}_s)_{n-k} + (x_k - x_{k-1})(\mathbf{h}_s)_{n-k}) = x_n,$$

since odd components of $\mathbf{v}$ and $\mathbf{w}$ are zero. A little regrouping yields

$$\frac{1}{2}\sum_{k\ even} [((\boldsymbol{\ell}_s)_{n-k} + (\mathbf{h}_s)_{n-k})x_k + ((\boldsymbol{\ell}_s)_{n-k} - (\mathbf{h}_s)_{n-k})x_{k-1}] = x_n, \tag{7.7}$$

which must hold for all $\mathbf{x} \in L^2(\mathbb{Z})$ and all n.

Choose $\mathbf{x}$ to have components $x_0 = 1$ and $x_k = 0$ for $k \neq 0$. Equation (7.7) yields (only the $k = 0$ term contributes)

$$(\boldsymbol{\ell}_s)_n + (\mathbf{h}_s)_n = \begin{cases} 2, & n = 0, \\ 0, & n \neq 0. \end{cases} \tag{7.8}$$

Next, choose $\mathbf{x}$ to have components $x_1 = 1$ and $x_k = 0$ for $k \neq 1$. In this case, equation (7.7) yields (only $k = 2$ contributes here)

$$(\boldsymbol{\ell}_s)_{n-2} - (\mathbf{h}_s)_{n-2} = \begin{cases} 2, & n = 1, \\ 0, & n \neq 1, \end{cases} \tag{7.9}$$

for all n. From equation (7.8) with $n = 0$ and (7.9) with $n = 2$, we obtain a pair of linear equations $(\boldsymbol{\ell}_s)_0 + (\mathbf{h}_s)_0 = 2$ and $(\boldsymbol{\ell}_s)_0 - (\mathbf{h}_s)_0 = 0$ for $(\boldsymbol{\ell}_s)_0$ and $(\mathbf{h}_s)_0$, from which we can deduce that $(\boldsymbol{\ell}_s)_0 = (\mathbf{h}_s)_0 = 1$. Choosing $n = -1$ in (7.8) and $n = 1$ in (7.9) leads to $(\boldsymbol{\ell}_s)_{-1} = 1$ and $(\mathbf{h}_s)_{-1} = -1$. By considering $n = n_0$ in equation (7.8) and $n = n_0 + 2$ in equation (7.9) for $n_0 \neq -1, 0$, it can be shown that all other synthesis filter coefficients are zero. These are, of course, *the synthesis filters we used in Section 7.2.2. We now know that they are the only filters that will work with the chosen analysis filters (for zero output delay).*

Example 7.4 *Another set of biorthogonal filters is the Le Gall 5/3 filters, with coefficients*

$$\boldsymbol{\ell}_a = \left(\dots, 0, -\frac{1}{8}, \frac{1}{4}, \frac{3}{4}, \frac{1}{4}, -\frac{1}{8}, 0, \dots\right), \quad \mathbf{h}_a = \left(\dots, 0, -\frac{1}{2}, 1, -\frac{1}{2}, 0, \dots\right),$$
$$\boldsymbol{\ell}_s = \left(\dots, 0, \frac{1}{2}, 1, \frac{1}{2}, 0, \dots\right), \quad \mathbf{h}_s = \left(\dots, 0, -\frac{1}{8}, -\frac{1}{4}, \frac{3}{4}, -\frac{1}{4}, -\frac{1}{8}, 0, \dots\right),$$

used in the JPEG 2000 compression standard. The derivation of these filters is given in Section 7.7, where we look at a variety of techniques for designing filters.

7.3.3 Orthogonal Filter Banks

As we will see, there are infinitely many analysis/synthesis FIR filters that satisfy equation (7.6). It is somewhat helpful to narrow the types of filters under consideration by imposing additional constraints. For example, in the simple Haar two-point averaging/differencing filters above, the synthesis filters were merely time-reversed and rescaled versions of the analysis filters. More specifically, $(\boldsymbol{\ell}_s)_k = C(\boldsymbol{\ell}_a)_{-k}$ and $(\mathbf{h}_s)_k = C(\mathbf{h}_a)_{-k}$ for some constant C. This turns out to be a common and useful arrangement, and in fact by properly rescaling, we can also arrange $C = 1$. In this case, the analysis and synthesis filters satisfy

$$(\boldsymbol{\ell}_s)_k = (\boldsymbol{\ell}_a)_{-k}, \quad (\mathbf{h}_s)_k = (\mathbf{h}_a)_{-k}. \tag{7.10}$$

Filters in this relation are said to be *adjoints* or *time-reversals* of each other (previously discussed in Exercise 9). The resulting filter bank is said to be *orthogonal*, for reasons that will be explained in Section 7.5.3. In this case, if the analysis filters are causal, then the synthesis filters cannot be, but by Remark 7.1, if desired, we can always shift the synthesis filters (by the same amount) to obtain causality at the expense of introducing a delay in the filter bank output. Then equation (7.10) will not hold, but we will still call the filter bank "orthogonal."

Example 7.5 *A simple rescaling of the Haar filters above yields an orthogonal filter bank. Specifically, multiply both analysis filters by $\sqrt{2}$ and compensate by dividing the synthesis filters by $\sqrt{2}$. We obtain filters with nonzero coefficients $(\boldsymbol{\ell}_a)_0 = 1/\sqrt{2}, (\boldsymbol{\ell}_a)_1 = 1/\sqrt{2}$ and $(\mathbf{h}_a)_0 = 1/\sqrt{2}, (\mathbf{h}_a)_1 = -1/\sqrt{2}$. The synthesis filters become $(\boldsymbol{\ell}_s)_0 = 1/\sqrt{2}, (\boldsymbol{\ell}_s)_{-1} = 1/\sqrt{2}$ and $(\mathbf{h}_s)_0 = 1/\sqrt{2}, (\mathbf{h}_s)_{-1} = -1/\sqrt{2}$, merely the time reversals of the analysis filters. All other filter coefficients are zero.*

7.4 Multistage Filter Banks

It can be extremely useful to iterate or nest filter banks. For example, if an input vector $\mathbf{x}$ is passed through an analysis filter bank to produce filtered and downsampled vectors $\mathbf{X}_\ell = D(\boldsymbol{\ell}_a * \mathbf{x})$ and $\mathbf{X}_h = D(\mathbf{h}_a * \mathbf{x})$, it can be useful to pass $\mathbf{X}_\ell$ through another filter bank to further separate low- and high-pass components. Or, we might choose to pass both $\mathbf{X}_\ell$ and $\mathbf{X}_h$ through another filter bank, or even just $\mathbf{X}_h$. A couple of possibilities are illustrated in Figures 7.8 and 7.9.

The subscripts indicate the order in which the filtering operations are applied; for example, $\mathbf{X}_{\ell h}$ means the low-pass filtering is done first and then the high-pass filtering. In other words, $\mathbf{X}_{\ell h} = (\mathbf{X}_\ell)_h$. The filter bank in Figure 7.8 and others similarly obtained by successively refining the low-pass

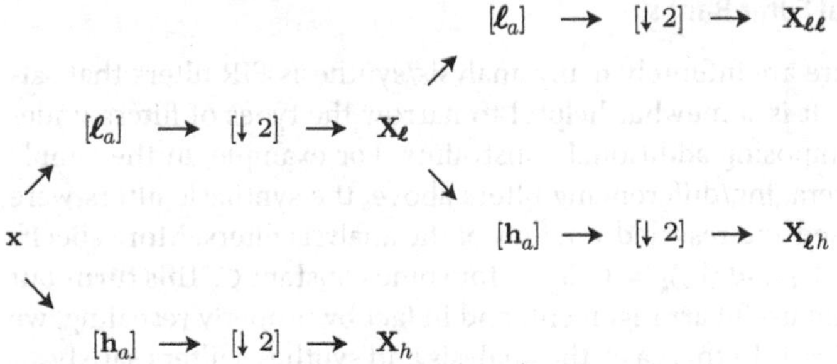

Figure 7.8 Filter bank possibility (leads to wavelets).

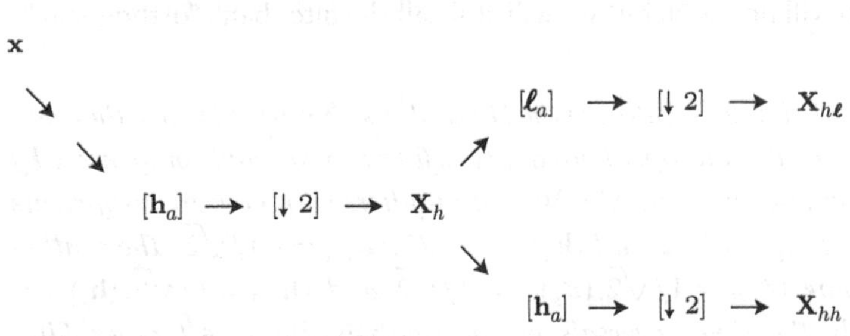

Figure 7.9 Full filter bank tree.

output are the filter banks usually associated to wavelets. This is the case that we will focus on for the rest of the chapter. The filter bank of Figure 7.9 is the two-stage full filter bank tree, which leads to the theory of wavelet packets. We do not pursue this here, though it is implemented in Matlab. It is even possible to use different filter pairs at each stage in a filter bank.

Let us consider more carefully the process of filtering a signal using the structure of the first two-stage filter bank above, in which we repeatedly put the low-pass portion through a filter bank. The two-stage filter bank operation might also be illustrated as below:

$$\mathbf{x} \to \begin{bmatrix} \mathbf{X}_{\ell} \\ \mathbf{X}_{h} \end{bmatrix} \to \begin{bmatrix} \mathbf{X}_{\ell\ell} \\ \mathbf{X}_{\ell h} \\ \mathbf{X}_{h} \end{bmatrix}.$$

Each of $\mathbf{X}_{\ell}, \mathbf{X}_{h}, \mathbf{X}_{\ell\ell}, \mathbf{X}_{\ell h}$, and $\mathbf{X}_{h}$ is an element of the same space, $L^2(\mathbb{Z})$, as $\mathbf{x}$.

Of course, we can iterate again, by passing $\mathbf{X}_{\ell\ell}$ through the analysis bank to produce vectors $\mathbf{X}_{\ell\ell\ell}$ and $\mathbf{X}_{\ell\ell h}$. The process can be continued in an obvious manner, repeatedly passing the $\mathbf{X}_{\ell\ldots\ell}$ subvector through the filter bank to produce $\mathbf{X}_{\ell\ell\ldots\ell}$ and $\mathbf{X}_{\ell\ldots\ell h}$. For example, if we use four stages, the net result can be written in the form

$$\begin{bmatrix} \mathbf{X}_{\ell\ell\ell\ell} \\ \mathbf{X}_{\ell\ell\ell h} \\ \mathbf{X}_{\ell\ell h} \\ \mathbf{X}_{\ell h} \\ \mathbf{X}_{h} \end{bmatrix}.$$

It is clear how this multistage analysis filter bank can be inverted. We upsample and apply the appropriate synthesis filters to $\mathbf{X}_{\ell\ell\ell\ell}$ and $\mathbf{X}_{\ell\ell\ell h}$, to produce $\mathbf{X}_{\ell\ell\ell}$. We repeat with $\mathbf{X}_{\ell\ell\ell}$ and $\mathbf{X}_{\ell\ell h}$ to produce $\mathbf{X}_{\ell\ell}$, and then use $\mathbf{X}_{\ell\ell}$ and $\mathbf{X}_{\ell h}$ to synthesize $\mathbf{X}_{\ell}$. Finally, we synthesize the output of the filter bank with $\mathbf{X}_{\ell}$ and $\mathbf{X}_{h}$.

Example 7.6 *Let $\mathbf{x} \in L^2(\mathbb{Z})$ be the signal in the upper left of Figure 7.5, considered in Example 7.2. We pass $\mathbf{x}$ through a three-stage orthogonal Haar filter bank with analysis filters $\boldsymbol{\ell} = (\ldots, 0, 1/\sqrt{2}, 1/\sqrt{2}, 0, \ldots)$ and $\mathbf{h} = (\ldots, 0, 1/\sqrt{2}, -1/\sqrt{2}, 0, \ldots)$ as computed in Example 7.5. The quantity $\mathbf{X}_h$ was shown in Figure 7.6. In Figure 7.10, we show $\mathbf{X}_{\ell h}$ in panel (a) and $\mathbf{X}_{\ell\ell}$ in panel (b). In each case, only the nonzero components of the relevant signal are shown, so $\mathbf{X}_{\ell h}$ and $\mathbf{X}_{\ell\ell}$ span 0–256.*

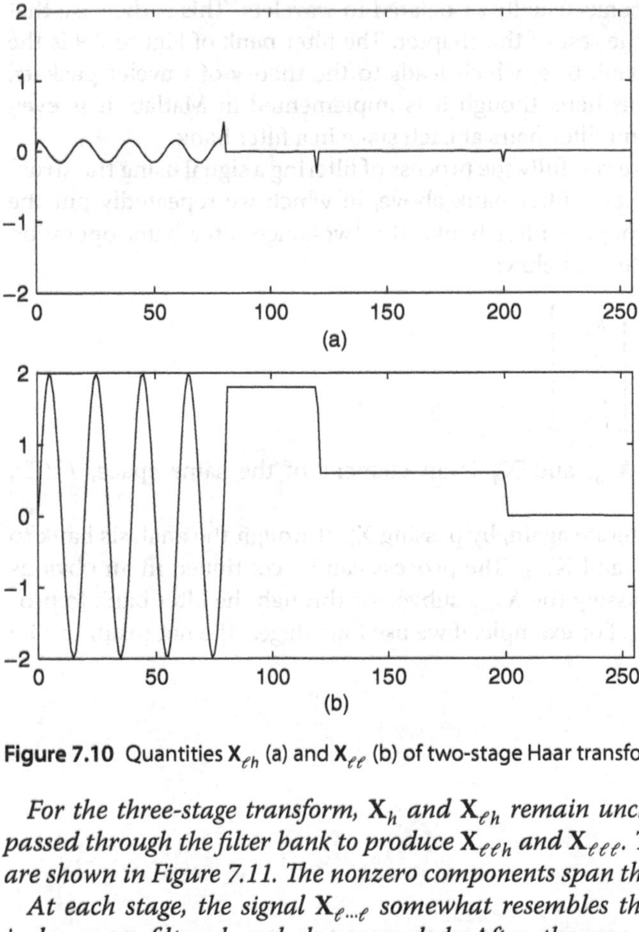

Figure 7.10 Quantities $\mathbf{X}_{\ell h}$ (a) and $\mathbf{X}_{\ell\ell}$ (b) of two-stage Haar transform.

For the three-stage transform, $\mathbf{X}_h$ and $\mathbf{X}_{\ell h}$ remain unchanged, while $\mathbf{X}_{\ell\ell}$ is passed through the filter bank to produce $\mathbf{X}_{\ell\ell h}$ and $\mathbf{X}_{\ell\ell\ell}$. The latter two vectors are shown in Figure 7.11. The nonzero components span the index range 0–128.

At each stage, the signal $\mathbf{X}_{\ell\ldots\ell}$ somewhat resembles the original signal but is low-pass filtered and downsampled. After the one-stage transform, the high-frequency portion $\mathbf{X}_h$ is nonzero only in those parts of the signal where high-frequency energy is present, such as the sinusoidal portion at the start and the two discontinuities. To a large extent, the same observation applies to $\mathbf{X}_{\ell h}$ and even $\mathbf{X}_{\ell\ell h}$.

Let us examine what happens if we compress the signal by "zeroing out" selective high-frequency portions of the three-stage filter bank output and then inverting the transform. For example, if we zero out $\mathbf{X}_h$ and then resynthesize the signal with the remaining components (by resynthesizing $\mathbf{X}_{\ell\ell h}$ and $\mathbf{X}_{\ell\ell\ell}$ into $\mathbf{X}_{\ell h}$, then $\mathbf{X}_{\ell h}$ and $\mathbf{X}_{\ell\ell}$ to $\mathbf{X}_\ell$, and finally use $\mathbf{X}_\ell$ with $\mathbf{X}_h = \mathbf{0}$ to synthesize the output), we obtain the signal that was shown in Figure 7.7a. The right side of Figure 7.12a shows the reconstruction from $\mathbf{X}_{\ell\ell}$ alone (implicitly setting $\mathbf{X}_h$ and $\mathbf{X}_{\ell h}$ to zero). The vector $\mathbf{X}_{\ell\ell}$ has at most 257 *nonzero coefficients. Finally, Figure 7.12b shows the reconstruction from $\mathbf{X}_{\ell\ell\ell}$ alone (at most* 129 *nonzero coefficients).*

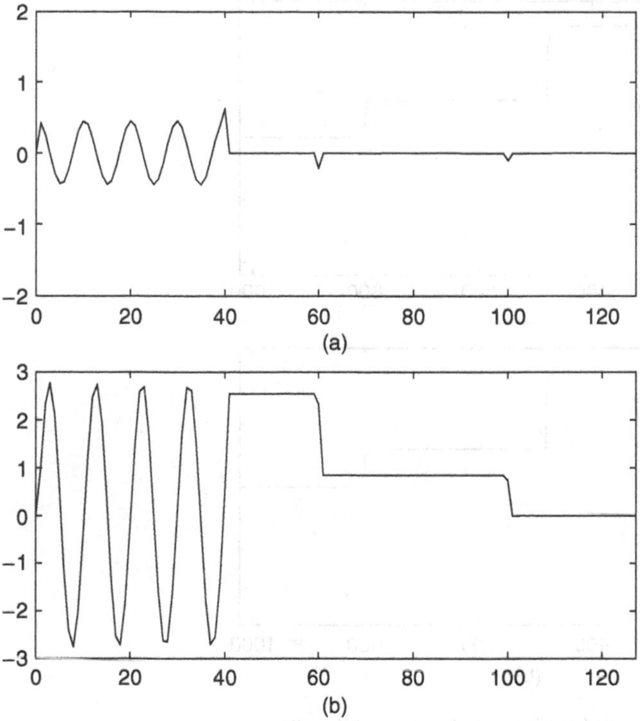

Figure 7.11 Portions $\mathbf{X}_{\ell\ell h}$ (a) and $\mathbf{X}_{\ell\ell\ell}$ (b) of three-stage Haar transform.

7.5 Filter Banks for Finite Length Signals

7.5.1 Extension Strategy

The theory developed thus far is for signals of infinite length, for example, signals in $L^2(\mathbb{Z})$. Signals of finite length present a certain difficulty that we must overcome: the convolution of a filter $\mathbf{g} \in L^2(\mathbb{Z})$ and a finite length signal $\mathbf{x} \in \mathbb{C}^N$ is undefined. The obvious fix is to extend $\mathbf{x}$ to be an element of $L^2(\mathbb{Z})$, and then convolve. This is exactly what we will do.

Specifically, we use the following conceptual approach: we first extend $\mathbf{x}$ to $L^2(\mathbb{Z})$ in some manner, filter the extended signal, and then truncate the filtered signal to some finite length. We call this a "conceptual approach," since we will not actually perform numerical computations on signals in $L^2(\mathbb{Z})$.

There are many ways to extend a signal $\mathbf{x} \in \mathbb{C}^N$ to a signal $\widetilde{\mathbf{x}} \in L^2(\mathbb{Z})$. Some common choices are listed below.

1. We can extend $\mathbf{x}$ by zero-padding; that is, set $\widetilde{x}_k = x_k$ for $0 \leq k \leq N-1$ and $\widetilde{x}_k = 0$ for k outside this range.

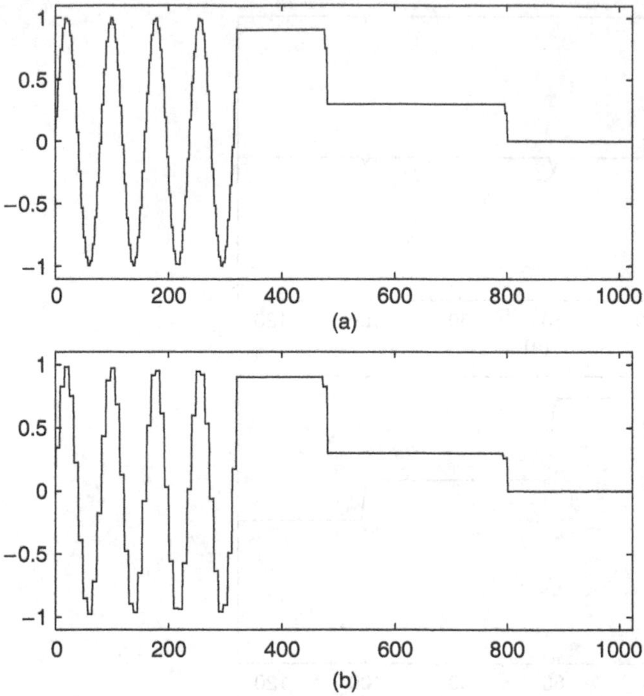

Figure 7.12 Reconstruction of **x** using only $\mathbf{X}_{\ell\ell}$ (a) and $\mathbf{X}_{\ell\ell\ell}$ (b).

2. We can extend **x** periodically, by setting $\widetilde{x}_k = x_{k \bmod N}$. But note that if we do this for all k, then $\widetilde{\mathbf{x}}$ will not be in $L^2(\mathbb{Z})$. In fact, since we are working with FIR filters, we will only need to extend **x** a few cycles in either direction, and then we can extend by zero. In this case, the extension is in $L^2(\mathbb{Z})$. As it turns out, this results in a simple modification of the $L^2(\mathbb{Z})$ theory in which all convolutions in $L^2(\mathbb{Z})$ are replaced by circular convolutions in $\mathbb{C}^N$. We will look at this more closely below.
3. Periodic extension as in scheme 2 above has caused difficulty in using the DFT for compression, since this extension introduces artificial discontinuities at the signal edges. The same thing can happen here. For most analyses, it is better to extend **x** by even reflection as in Chapter 3, first to $\mathbf{C}^{2N}$ as

$$\widetilde{x}_k = \begin{cases} x_k, & 0 \le k \le N-1, \\ x_{2N-k-1}, & N \le k \le 2N-1, \end{cases} \tag{7.11}$$

and then via $\widetilde{x}_k = \widetilde{x}_{k \bmod 2N}$. This type of extension duplicates the boundary points x_0 and x_{N-1}, so the extended version $\widetilde{\mathbf{x}}$ looks like

$$\ldots, x_3, x_2, x_1, x_0, x_0, x_1, \ldots, x_{N-1}, x_{N-1}, x_{N-2}, \ldots$$

and is periodic with period $2N$. In conjunction with the DFT, this is the extension that led to the discrete cosine transform. This type of extension is called a *half-point symmetric* extension.

Alternatively, we may extend $\mathbf{x}$ to $\mathbb{C}^{2N-2}$ as

$$\tilde{x}_k = \begin{cases} x_k, & 0 \le k \le N-1, \\ x_{2N-k-2}, & N \le k \le 2N-3, \end{cases} \tag{7.12}$$

and then via $\tilde{x}_k = \tilde{x}_{k \bmod 2N-2}$. This type of extension does not duplicate the boundary points x_0 and x_{N-1}, so the extended version $\tilde{\mathbf{x}}$ looks like

$$\ldots, x_3, x_2, x_1, x_0, x_1, \ldots, x_{N-2}, x_{N-1}, x_{N-2}, \ldots$$

and is periodic with period $2N - 2$. This is called a *whole-point symmetric* extension.

As with simple periodic extension, whatever method we use only requires that we extend a few cycles in each direction. We can then extend by zero to obtain an extension in $L^2(\mathbb{Z})$.

There are other methods for extending signals (many built into Matlab's wavelet commands), but we will not concern ourselves with them here.

After extending the signal $\mathbf{x}$ to $\tilde{\mathbf{x}}$ and filtering, we obtain a vector $\mathbf{y} = \tilde{\mathbf{x}} * \mathbf{g}$ in $L^2(\mathbb{Z})$. We then truncate $\mathbf{y}$ to a finite length vector. It is not clear how this last step should be done, or how it will affect the theory we have developed thus far. Thus to better understand this approach, let us take a closer look at extension approach (list item 2) above to see how the computations proceed, and obtain a "discrete filter bank transform" for signals in $\mathbb{R}^N$.

7.5.2 Analysis of Periodic Extension

Recall the overall operation of a filter bank for bi-infinite signals: A signal $\mathbf{x} \in L^2(\mathbb{Z})$ is filtered with analysis low- and high-pass filters, and then downsampled to produce vectors $\mathbf{X}_\ell = D(\mathbf{x} * \boldsymbol{\ell}_a)$ and $\mathbf{X}_h = D(\mathbf{x} * \mathbf{h}_a)$. To invert the process, we upsample, convolve with appropriate synthesis filters, and add. With perfect reconstruction, we have $U(\mathbf{X}_\ell) * \boldsymbol{\ell}_s + U(\mathbf{X}_h) * \mathbf{h}_s = S^m(\mathbf{x})$ for some m.

The punch line of the analysis in this section is that this same procedure works for signals of finite length provided that the filters involved are part of a perfect reconstruction filter bank for signals in $L^2(\mathbb{Z})$. Specifically, we have the following theorem.

Theorem 7.1 *Let $\mathbf{x} \in \mathbb{C}^N$. Let $\boldsymbol{\ell}_a$, $\mathbf{h}_a$, $\boldsymbol{\ell}_s$, and $\mathbf{h}_s$ be the analysis and synthesis FIR filters for a perfect reconstruction filter bank, each with N or fewer taps. If we interpret these filters as elements of $\mathbb{C}^N$ by taking all indexes modulo N and all convolutions as circular convolutions, then the linear transform $\mathbf{x} \to \mathbf{X} :=$*

$(\mathbf{X}_\ell, \mathbf{X}_h)$ from $\mathbb{C}^N$ to $\mathbb{C}^N$ is invertible. The inverse is given by the mapping $\mathbf{X} \to U(\mathbf{X}_\ell) * \boldsymbol{\ell}_s + U(\mathbf{X}_h) * \mathbf{h}_s$, with all convolutions as circular convolutions in $\mathbb{C}^N$.

The next couple subsections explain why this works.

7.5.2.1 Adapting the Analysis Transform to Finite Length

Given a signal $\mathbf{x} \in \mathbb{C}^N$, we begin by extending to a signal $\widetilde{\mathbf{x}} \in L^2(\mathbb{Z})$ according to scheme 2 above. Specifically, suppose that we set

$$\widetilde{x}_k = x_{k \bmod N}$$

in the range $-qN \le k \le qN$ for some "large" integer q (thus extending $\mathbf{x}$ periodically with period N, to at least $2q$ cycles); we set $\widetilde{x}_k = 0$ for k outside this range. We can take q as large as we like, so for all practical purposes $\widetilde{x}_k$ is periodic in k for "all" k.

In what follows, we assume that N is even. We assume that all filters involved have N or fewer taps, but no other special properties. We will assume that the filters (elements of $L^2(\mathbb{Z})$) have nonzero coefficients indexed in the range $-N/2$ to $N/2$, though this is primarily for convenience. Because we will ultimately interpret the filters as elements of $\mathbb{C}^N$ by considering indices modulo N, the range does not matter. In the periodic case, the notion of "causal filter" is not really meaningful.

We begin by examining the action of the filter bank on the signal $\widetilde{\mathbf{x}} \in L^2(\mathbb{Z})$. Let $\mathbf{y} = \widetilde{\mathbf{x}} * \boldsymbol{\ell}_a \in L^2(\mathbb{Z})$ so that

$$y_m = \sum_{k=-N/2}^{N/2} \widetilde{x}_{m-k}(\boldsymbol{\ell}_a)_k.$$

The component y_m depends on $\widetilde{x}_j$ only for the range $m - N/2 \le j \le m + N/2$. It is easy to see that y_m is periodic in the index m with period N, that is,

$$y_m = y_{m+N}$$

for any integer m sufficiently close to 0, such as $-(q-1)N \le m \le (q-1)N$. The reason is that

$$y_{m+N} = \sum_{k=-N/2}^{N/2} \widetilde{x}_{m+N-k}(\boldsymbol{\ell}_a)_k = \sum_{k=-N/2}^{N/2} \widetilde{x}_{m-k}(\boldsymbol{\ell}_a)_k = y_m,$$

since $\widetilde{x}_{m+N-k} = \widetilde{x}_{m-k}$ (because $\widetilde{x}_j$ is periodic with period N). Thus knowledge of y_m for $0 \le m \le N-1$ determines y_m in the larger range $-(q-1)N \le m \le (q-1)N$.

Indeed, by Definition 4.1, it is clear that the vector $(y_0, y_1, \ldots, y_{N-1})$ is just the circular convolution of $\mathbf{x}$ and $\boldsymbol{\ell}_a$ as vectors in $\mathbb{C}^N$, with all indexes interpreted modulo N.

The next step in the filter bank is downsampling. Let

$$\widetilde{\mathbf{X}}_\ell = D(\widetilde{\mathbf{x}} * \boldsymbol{\ell}_a),$$

where D is the downsampling operator on $L^2(\mathbb{Z})$. It is easy to see that $(\widetilde{\mathbf{X}}_\ell)_k$ is periodic in the index k with period $N/2$, for at least $q-1$ cycles in either direction from $k=0$. Thus knowledge of $(\widetilde{\mathbf{X}}_\ell)_k$ for the range $0 \le k \le N/2-1$ determines $(\widetilde{\mathbf{X}}_\ell)_k$ on the larger range $-(q-1)N/2 \le k \le (q-1)N/2$.

Similar remarks apply to the high-pass analysis filter $\mathbf{h}_a$, and we can produce the analogous quantity

$$\widetilde{\mathbf{X}}_h = D(\widetilde{\mathbf{x}} * \mathbf{h}_a).$$

Define truncated versions $\mathbf{X}_\ell$ and $\mathbf{X}_h$ of the downsampled filtered signals as

$$(\mathbf{X}_\ell)_k = (\widetilde{\mathbf{X}}_\ell)_k, \quad (\mathbf{X}_h)_k = (\widetilde{\mathbf{X}}_h)_k,$$

for $0 \le k \le N/2-1$, so both $\mathbf{X}_\ell$ and $\mathbf{X}_h$ are elements of $\mathbb{C}^{N/2}$. Concatenate these vectors by defining $\mathbf{X} = (\mathbf{X}_\ell, \mathbf{X}_h) \in \mathbb{C}^N$. We will refer to the mapping

$$\mathbf{x} \to \mathbf{X} \tag{7.13}$$

obtained via this procedure as a discrete wavelet transform (DWT). This transform is *linear* (because the extension procedure, convolution, and truncation are all linear). As such, the mapping in (7.13) is represented by an $N \times N$ matrix.

Example 7.7 *Let us consider signals $\mathbf{x} = (x_0, x_1, x_2, x_3) \in \mathbb{C}^4$, with the orthogonal Haar filter bank ($\boldsymbol{\ell}_a = (1/\sqrt{2}, 1/\sqrt{2}, 0, 0), \mathbf{h}_a = (1/\sqrt{2}, -1/\sqrt{2}, 0, 0)$) constructed in Example 7.5. The extension $\widetilde{\mathbf{x}}$ looks like*

$$\widetilde{\mathbf{x}} = (\ldots, x_0, x_1, x_2, x_3, x_0, x_1, x_2, x_3, x_0, x_1, x_2, x_3, \ldots)$$

with period 4, at least for some distance. The convolution $\mathbf{x} * \boldsymbol{\ell}_a$ is periodic with period 4 and looks like

$$\left(\ldots, \frac{x_0}{\sqrt{2}} + \frac{x_3}{\sqrt{2}}, \frac{x_1}{\sqrt{2}} + \frac{x_0}{\sqrt{2}}, \frac{x_2}{\sqrt{2}} + \frac{x_1}{\sqrt{2}}, \frac{x_3}{\sqrt{2}} + \frac{x_2}{\sqrt{2}}, \ldots\right).$$

Downsampling this and then truncating (take only the first $N/2 = 2$ components) produces

$$\mathbf{X}_\ell = \left(\frac{x_0 + x_3}{\sqrt{2}}, \frac{x_2 + x_1}{\sqrt{2}}\right).$$

Similar computations show that

$$\mathbf{X}_h = \left(\frac{x_0 - x_3}{\sqrt{2}}, \frac{x_2 - x_1}{\sqrt{2}}\right).$$

All in all, the DWT looks like

$$(x_0, x_1, x_2, x_3) \to \left(\frac{x_0 + x_3}{\sqrt{2}}, \frac{x_2 + x_1}{\sqrt{2}}, \frac{x_0 - x_3}{\sqrt{2}}, \frac{x_2 - x_1}{\sqrt{2}} \right).$$

This linear transform is embodied by the matrix

$$\mathbf{W}_4^a = \begin{bmatrix} \frac{1}{\sqrt{2}} & 0 & 0 & \frac{1}{\sqrt{2}} \\ 0 & \frac{1}{\sqrt{2}} & \frac{1}{\sqrt{2}} & 0 \\ \frac{1}{\sqrt{2}} & 0 & 0 & -\frac{1}{\sqrt{2}} \\ 0 & -\frac{1}{\sqrt{2}} & \frac{1}{\sqrt{2}} & 0 \end{bmatrix}. \tag{7.14}$$

The **W** *is for "wavelet," the 4 for the size of the transform, and a for "analysis."*

7.5.2.2 Adapting the Synthesis Transform to Finite Length

To invert the DWT, we proceed as follows. First, recall that we already know how to invert the transform for signals in $L^2(\mathbb{Z})$. Specifically, given the vectors $\widetilde{\mathbf{X}}_\ell$ and $\widetilde{\mathbf{X}}_h$ in $L^2(\mathbb{Z})$, we upsample each and convolve with the synthesis filters, to obtain vectors $\widetilde{\mathbf{v}}_\ell = \boldsymbol{\ell}_s * U(\widetilde{\mathbf{X}}_\ell)$ and $\widetilde{\mathbf{v}}_h = \mathbf{h}_s * U(\widetilde{\mathbf{X}}_h)$ in $L^2(\mathbb{Z})$ with components

$$(\widetilde{\mathbf{v}}_\ell)_m = \sum_{k=-N/2}^{N/2} U(\widetilde{\mathbf{X}}_\ell)_{m-k}(\boldsymbol{\ell}_s)_k, \quad (\widetilde{\mathbf{v}}_h)_m = \sum_{k=-N/2}^{N/2} U(\widetilde{\mathbf{X}}_h)_{m-k}(\mathbf{h}_s)_k. \tag{7.15}$$

We then recover $\widetilde{\mathbf{x}} = \widetilde{\mathbf{v}}_\ell + \widetilde{\mathbf{v}}_h$. Since $x_m = \widetilde{x}_m$ for $0 \le m \le N-1$, we also recover the vector $\mathbf{x}$. However, in the finite length case, we have only the truncated versions $\mathbf{X}_\ell$ and $\mathbf{X}_h$ in $\mathbb{C}^{N/2}$ to work with, not the full $L^2(\mathbb{Z})$ versions $\widetilde{\mathbf{X}}_\ell$ and $\widetilde{\mathbf{X}}_h$.

But within the range $0 \le m \le N-1$ equation (7.15) only requires knowledge of $U(\widetilde{\mathbf{X}}_\ell)_j$ and $U(\widetilde{\mathbf{X}}_h)_j$ on at most the range $-N/2 \le j \le 3N/2$. Since both $\widetilde{\mathbf{X}}_\ell$ and $\widetilde{\mathbf{X}}_h$ are periodic with period $N/2$ on the index range $-(q-1)N/2 \le k \le (q-1)N/2$ for some "large" q, knowledge of the truncated versions $\mathbf{X}_\ell$ and $\mathbf{X}_h$ allows us to reconstruct $\widetilde{\mathbf{X}}_\ell$ and $\widetilde{\mathbf{X}}_h$ on this range. As a result we can reconstruct $U(\widetilde{\mathbf{X}}_\ell)$ and $U(\widetilde{\mathbf{X}}_\ell)$ (both period N) on the index range $-(q-1)N \le k \le (q-1)N$ from knowledge of $U(\mathbf{X}_\ell)$ and $U(\mathbf{X}_h)$, by extending the latter two vectors periodically with period N. Indeed both $U(\widetilde{\mathbf{X}}_\ell)_j$ and $U(\widetilde{\mathbf{X}}_h)_j$ are just the truncated versions $U(\mathbf{X}_\ell)_j$ and $U(\mathbf{X}_h)_j$ with indexes interpreted modulo N.

With these observations equation (7.15) can be written

$$(\widetilde{\mathbf{v}}_\ell)_m = \sum_{k=-N/2}^{N/2} U(\mathbf{X}_\ell)_{m-k}(\boldsymbol{\ell}_s)_k, \quad (\widetilde{\mathbf{v}}_h)_m = \sum_{k=-N/2}^{N/2} U(\mathbf{X}_h)_{m-k}(\mathbf{h}_s)_k \tag{7.16}$$

on (at least) the index range $0 \le m \le N-1$. In particular, define truncated vectors $\mathbf{v}_\ell$ and $\mathbf{v}_h$ in $\mathbb{C}^N$ as

$$\mathbf{v}_\ell = ((\widetilde{\mathbf{v}}_\ell)_0, (\widetilde{\mathbf{v}}_\ell)_1, \dots, (\widetilde{\mathbf{v}}_\ell)_{N-1}),$$
$$\mathbf{v}_h = ((\widetilde{\mathbf{v}}_h)_0, (\widetilde{\mathbf{v}}_h)_1, \dots, (\widetilde{\mathbf{v}}_h)_{N-1}).$$

The equations (7.16) are just

$$\mathbf{v}_\ell = U(\mathbf{X}_\ell) * \boldsymbol{\ell}_s, \quad \mathbf{v}_h = U(\mathbf{X}_h) * \mathbf{h}_s,$$

where "$*$" denotes circular convolution in $\mathbb{C}^N$. But $\mathbf{v}_\ell + \mathbf{v}_h = \mathbf{x}$, and thus the linear operation $\mathbf{X} \to U(\mathbf{X}_\ell) * \boldsymbol{\ell}_s + U(\mathbf{X}_h) * \mathbf{h}_s$ inverts the original transform.

The discussion above is a proof of Theorem 7.1.

Example 7.8 *Let us work out the inverse transform corresponding to Example 7.7. Let* $\mathbf{X} = (X_0, X_1, X_2, X_3) \in \mathbb{C}^4$ *so that* $\mathbf{X}_\ell = (X_0, X_1)$ *and* $\mathbf{X}_h = (X_2, X_3)$. *Recall that the orthogonal Haar synthesis filters in* $L^2(\mathbb{Z})$ *had coefficients* $(\boldsymbol{\ell}_s)_0 = 1/\sqrt{2}, (\boldsymbol{\ell}_s)_{-1} = 1/\sqrt{2}, (\mathbf{h}_s)_0 = 1/\sqrt{2}$, *and* $(\mathbf{h}_s)_{-1} = -1/\sqrt{2}$. *However, if we interpret these as vectors in* $\mathbb{C}^4$ *by considering indexes modulo 4 we have*

$$\boldsymbol{\ell}_s = \left(\frac{1}{\sqrt{2}}, 0, 0, \frac{1}{\sqrt{2}}\right), \quad \mathbf{h}_s = \left(\frac{1}{\sqrt{2}}, 0, 0, \frac{-1}{\sqrt{2}}\right).$$

Upsampling produces the vectors $U(\mathbf{X}_\ell) = (X_0, 0, X_1, 0)$ *and* $U(\mathbf{X}_h) = (X_2, 0, X_3, 0)$. *Convolution with the synthesis filters yields*

$$U(\mathbf{X}_\ell) * \boldsymbol{\ell}_s = \left(\frac{X_0}{\sqrt{2}}, \frac{X_1}{\sqrt{2}}, \frac{X_1}{\sqrt{2}}, \frac{X_0}{\sqrt{2}}\right),$$
$$U(\mathbf{X}_h) * \mathbf{h}_s = \left(\frac{X_2}{\sqrt{2}}, \frac{-X_3}{\sqrt{2}}, \frac{X_3}{\sqrt{2}}, \frac{-X_2}{\sqrt{2}}\right).$$

Adding produces the linear transform $\mathbf{X} \to U(\mathbf{X}_\ell) * \boldsymbol{\ell}_s + U(\mathbf{X}_h) * \mathbf{h}_s$ *or*

$$\mathbf{X} \to \left(\frac{X_0 + X_2}{\sqrt{2}}, \frac{X_1 - X_3}{\sqrt{2}}, \frac{X_1 + X_3}{\sqrt{2}}, \frac{X_0 - X_2}{\sqrt{2}}\right).$$

This transformation is embodied in the matrix

$$\mathbf{W}_4^s = \begin{bmatrix} \frac{1}{\sqrt{2}} & 0 & \frac{1}{\sqrt{2}} & 0 \\ 0 & \frac{1}{\sqrt{2}} & 0 & -\frac{1}{\sqrt{2}} \\ 0 & \frac{1}{\sqrt{2}} & 0 & \frac{1}{\sqrt{2}} \\ \frac{1}{\sqrt{2}} & 0 & -\frac{1}{\sqrt{2}} & 0 \end{bmatrix}. \tag{7.17}$$

Of course $\mathbf{W}_4^a = (\mathbf{W}_4^s)^{-1}$*, as is easy to check directly. In fact, both* $\mathbf{W}_4^a$ *and* $\mathbf{W}_4^s$ *are orthogonal matrices, so* $(\mathbf{W}_4^a)^T = \mathbf{W}_4^s$*. This is one reason behind the terminology "orthogonal" for certain filter banks—the analysis and synthesis matrices obtained in the finite-length filter bank are orthogonal, which we show more generally in Section 7.5.3.*

7.5.2.3 Other Extensions

It is also possible to construct a DWT and corresponding inverse for vectors in $\mathbb{C}^N$ based on the symmetric extension technique defined by equation (7.11) or (7.12). This is especially easy when the filters involved possess even or odd symmetries, such as the Le Gall filters of Example 7.4. Similar transforms can be constructed for other extension techniques too, and many are built into Matlab's Wavelet Toolbox. See [28] and [5] for more information, and other approaches to adapting filter banks to finite-length signals.

7.5.3 Matrix Formulation of the Periodic Case

If all convolutions in the finite case are circular, then $\mathbf{X}_\ell = \mathbf{D}\mathbf{M}_{\ell_a}\mathbf{x}$ and $\mathbf{X}_h = \mathbf{D}\mathbf{M}_{h_a}\mathbf{x}$, where $\mathbf{M}_{\ell_a}$ and $\mathbf{M}_{h_a}$ are the circulant matrices for the analysis filters and $\mathbf{D}$ is a matrix that represents downsampling. The matrix $\mathbf{D}$ has dimensions $N/2 \times N$ with entries

$$D_{j,k} = \begin{cases} 1, & k = 2j,\ 0 \le j \le N/2 - 1, \\ 0, & \text{else}, \end{cases} \tag{7.18}$$

with rows and columns indexed from 0. It is not hard to check that $(\mathbf{D}\mathbf{x})_k = x_{2k}$ for any $\mathbf{x} \in \mathbb{C}^N$ (see Exercise 13). In this case, the N-point filter bank analysis transform has the form $\mathbf{x} \to \mathbf{W}_N^a\mathbf{x}$, where

$$\mathbf{W}_N^a = \begin{bmatrix} \mathbf{D}\mathbf{M}_{\ell_a} \\ \mathbf{D}\mathbf{M}_{h_a} \end{bmatrix} \tag{7.19}$$

is the $N \times N$ matrix obtained by "stacking" the $N/2 \times N$ matrices $\mathbf{D}\mathbf{M}_{\ell_a}$ and $\mathbf{D}\mathbf{M}_{h_a}$.

By Theorem 7.1, the inverse transform is obtained by upsampling each of $\mathbf{X}_\ell$ and $\mathbf{X}_h$, and then circularly convolving each with the synthesis filters and adding. Let us define the $N \times N/2$ upsampling matrix $\mathbf{U}$ with entries

$$U_{j,k} = \begin{cases} 1, & 2k = j,\ 0 \le k \le N/2 - 1, \\ 0, & \text{else}. \end{cases} \tag{7.20}$$

The synthesis bank operation can be then expressed as $\mathbf{X} \to \mathbf{W}_N^s\mathbf{X}$, where

$$\mathbf{W}_N^s = [\mathbf{M}_{\ell_s}\mathbf{U} | \mathbf{M}_{h_s}\mathbf{U}] \tag{7.21}$$

is the $N \times N$ matrix obtained by concatenating the $N \times N/2$ matrices $\mathbf{M}_{\ell_s}\mathbf{U}$ and $\mathbf{M}_{h_s}\mathbf{U}$. Here $\mathbf{M}_{\ell_s}$ and $\mathbf{M}_{h_s}$ are the circulant matrices for the synthesis filters. It can be shown that $\mathbf{U} = \mathbf{D}^T$ (again, see Exercise 13).

For a perfect reconstruction filter bank with zero delay, we have $\mathbf{W}_N^s \mathbf{W}_N^a = \mathbf{I}_N$, where $\mathbf{I}_N$ is the $N \times N$ identity matrix. In the case, that the filter bank is orthogonal (so that the filters satisfy equation (7.10)) a little matrix algebra yields

$$\begin{aligned}(\mathbf{W}_N^a)^T &= \begin{bmatrix}\mathbf{D}\mathbf{M}_{\ell_a}\\ \mathbf{D}\mathbf{M}_{h_a}\end{bmatrix}^T\\ &= \left[(\mathbf{D}\mathbf{M}_{\ell_a})^T \mid (\mathbf{D}\mathbf{M}_{h_a})^T\right]\\ &= \left[\mathbf{M}_{\ell_a}^T\mathbf{D}^T \mid \mathbf{M}_{h_a}^T\mathbf{D}^T\right]\\ &= \left[\mathbf{M}_{\ell_a}^T\mathbf{U} \mid \mathbf{M}_{h_a}^T\mathbf{U}\right]\\ &= \left[\mathbf{M}_{\ell_s}\mathbf{U} \mid \mathbf{M}_{h_s}\mathbf{U}\right]\\ &= \mathbf{W}_N^s,\end{aligned} \tag{7.22}$$

where we have used $\mathbf{M}_{\ell_a}^T = \mathbf{M}_{\ell_s}$ and $\mathbf{M}_{h_a}^T = \mathbf{M}_{h_s}$, both of which are consequences of equation (7.10) and Exercise 9.

From equation (7.22) and $\mathbf{W}_N^s \mathbf{W}_N^a = \mathbf{I}_N$, we conclude that for a perfect reconstruction orthogonal filter bank we have $(\mathbf{W}_N^a)^T \mathbf{W}_N^a = \mathbf{W}_N^s (\mathbf{W}_N^s)^T = \mathbf{I}_N$. Thus both the analysis and synthesis transform matrices are in fact orthogonal. This is one motivation for the terminology "orthogonal filter bank," at least in the finite-dimensional case.

7.5.4 Multistage Transforms

The ideas that follow do not depend on the particular method by which we obtain a wavelet transform for finite length signals, such as periodic extension or symmetric extension. All we need is a well defined method for performing a DWT on a signal of a given length. We thus assume that for signals of any (even) length k we can construct matrices $\mathbf{W}_k^a$ and $\mathbf{W}_k^s$ that govern a forward and inverse DWT, for example, as equations (7.19) and (7.21) do in the periodic extension case.

7.5.4.1 Iterating the One-stage Transform

Just as for signals in $L^2(\mathbb{Z})$ we can iterate DWTs. The typical procedure is to transform a signal $\mathbf{x} \in \mathbb{C}^N$ to produce $\mathbf{X} = (\mathbf{X}_\ell, \mathbf{X}_h) \in \mathbb{C}^N$ consisting of approximation and detail coefficients $\mathbf{X}_\ell$ and $\mathbf{X}_h$, each elements of $\mathbf{C}^{N/2}$, and then to pass the approximation coefficient vector $\mathbf{X}_\ell$ through another filter bank to produce $N/4$-dimensional vectors $\mathbf{X}_{\ell\ell}$ and $\mathbf{X}_{\ell h}$. The entire two-stage transform can be amalgamated into the single vector $(\mathbf{X}_{\ell\ell}, \mathbf{X}_{\ell h}, \mathbf{X}_h) \in \mathbf{C}^N$.

The process can be iterated further. For example, we can pass $\mathbf{X}_{\ell\ell}$ through the filter bank again to produce $N/8$-dimensional vectors $\mathbf{X}_{\ell\ell\ell}$ and $\mathbf{X}_{\ell\ell h}$, and so obtain a three-stage DWT

$$\mathbf{X} = (\mathbf{X}_{\ell\ell\ell}, \mathbf{X}_{\ell\ell h}, \mathbf{X}_{\ell h}, \mathbf{X}_h),$$

an element of $\mathbb{C}^N$. The appropriate synthesis filters can be used to reconstruct $\mathbf{X}_{\ell\ell}$ from the vectors $\mathbf{X}_{\ell\ell\ell}$ and $\mathbf{X}_{\ell\ell h}$. We can then reconstruct $\mathbf{X}_\ell$ from $\mathbf{X}_{\ell\ell}$ and $\mathbf{X}_{\ell h}$. Finally, we recover $\mathbf{x}$ from $\mathbf{X}_\ell$ and $\mathbf{X}_h$.

This process can be carried out to any stage provided that N is divisible by a suitably large power of 2.

Example 7.9 *Consider the* 1024*-point signal from Example 7.2, shown in Figure 7.5. In Figure 7.13, we show the one-, two-, four-, and six-stage orthogonal Haar transforms of that signal. The one-stage transform yields vectors* $\mathbf{X}_\ell$ *and* $\mathbf{X}_h$*, each an element of* $\mathbb{R}^{512}$*, which we display as the vector* $(\mathbf{X}_\ell, \mathbf{X}_h) \in \mathbb{R}^{1024}$ *in Figure 7.13a. The two-stage transform consists of the* 256*-dimensional vectors*

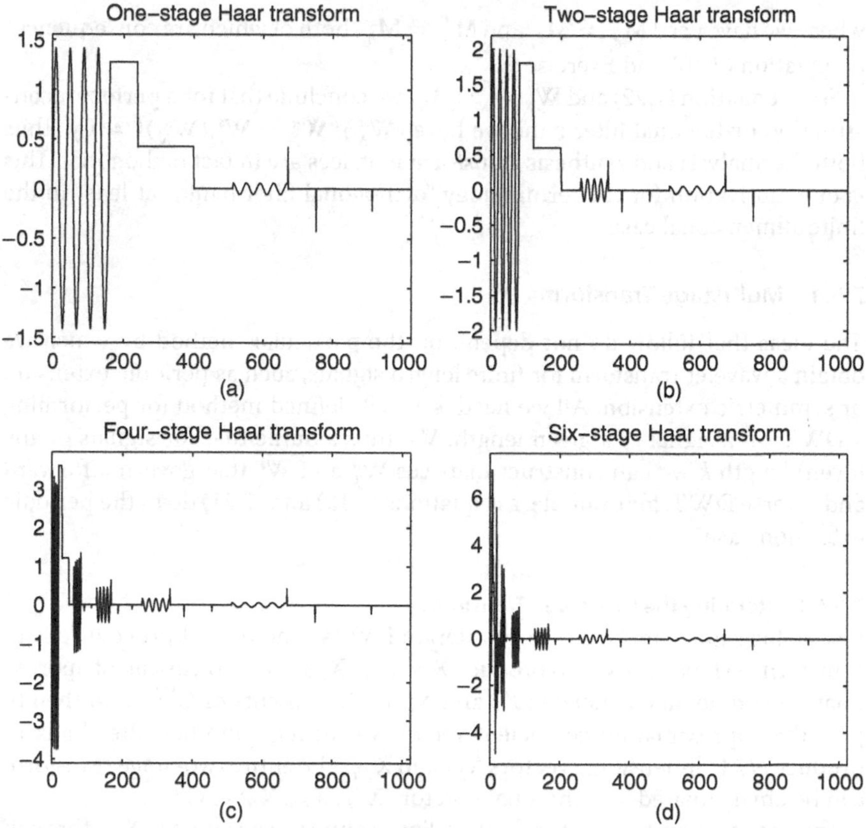

Figure 7.13 One, two, four, and six stage Haar discrete wavelet transforms.

$\mathbf{X}_{\ell\ell}, \mathbf{X}_{\ell h}$ and the same vector $\mathbf{X}_h$ from the one-stage transform. The overall two-stage transform is displayed as the 1024-dimensional vector $(\mathbf{X}_{\ell\ell}, \mathbf{X}_{\ell h}, \mathbf{X}_h)$. The same general scheme is used to plot the other transforms. For example, the six-stage transform is a plot of the 1024-component vector

$$(\mathbf{X}_{\ell\ell\ell\ell\ell\ell}, \mathbf{X}_{\ell\ell\ell\ell\ell h}, \mathbf{X}_{\ell\ell\ell\ell h}, \mathbf{X}_{\ell\ell\ell h}, \mathbf{X}_{\ell\ell h}, \mathbf{X}_{\ell h}, \mathbf{X}_h)$$

consisting of subvectors with 16, 16, 32, 64, 128, 256, and 512 components. At each stage the previously high-pass filtered portions remain unchanged. The portion $\mathbf{X}_{\ell\cdots\ell}$ is a progressively smoother and smoother (and downsampled) version of the original signal.

7.5.4.2 Matrix Formulation of Multistage Transform

The multistage transform for finite length signals has a simple matrix formulation. Let $\mathbf{W}_m^a$ denote the analysis matrix for a single-stage m-point DWT. A two-stage transform consisting of an N-point transform of $\mathbf{x}$ followed by an $N/2$ point transform of $\mathbf{X}_\ell$ can be implemented as $\mathbf{x} \to \mathcal{W}_2^a\mathbf{x}$, where

$$\mathcal{W}_2^a = \begin{bmatrix} \mathbf{W}_{N/2}^a & 0 \\ 0 & \mathbf{I}_{N/2} \end{bmatrix} \mathbf{W}_N^a. \tag{7.23}$$

The subscript "2" in $\mathcal{W}_2^a$ denotes a two-stage transform. Notice that $\mathcal{W}_2^a$ also depends on N, but we do not explicitly notate this. Of course, this construction requires that N be divisible by 4 in order to perform the $N/2$ point transform. The corresponding inverse transform $\mathbf{X} \to \mathcal{W}_2^s\mathbf{X}$ is implemented via the matrix

$$\mathcal{W}_2^s = \mathbf{W}_N^s \begin{bmatrix} \mathbf{W}_{N/2}^s & 0 \\ 0 & \mathbf{I}_{N/2} \end{bmatrix}, \tag{7.24}$$

which inverts each transform in reverse order.

Generally, we can perform an r-stage DWT, governed by the matrix product of the r matrices

$$\mathcal{W}_r^a = \begin{bmatrix} \mathbf{W}_{N/2^{r-1}}^a & 0 \\ 0 & \mathbf{I}_{N(1-1/2^{r-1})} \end{bmatrix} \cdots \begin{bmatrix} \mathbf{W}_{N/2}^a & 0 \\ 0 & \mathbf{I}_{N/2} \end{bmatrix} \mathbf{W}_N^a,$$

where each factor is an $N \times N$ matrix of the form

$$\begin{bmatrix} \mathbf{W}_{N/2^{k-1}}^a & 0 \\ 0 & \mathbf{I}_{N(1-1/2^{k-1})} \end{bmatrix}$$

for $1 \le k \le r$. All this requires is that N be divisible by 2^r. The one-stage transform matrix $\mathcal{W}_1^a$ is just the matrix $\mathbf{W}_N^a$.

The inverse transform is governed by the matrix

$$\mathcal{W}_r^s = \mathbf{W}_N^s \begin{bmatrix} \mathbf{W}_{N/2}^s & 0 \\ 0 & \mathbf{I}_{N/2} \end{bmatrix} \cdots \begin{bmatrix} \mathbf{W}_{N/2^{r-1}}^s & 0 \\ 0 & \mathbf{I}_{N(1-1/2^{r-1})} \end{bmatrix},$$

where each factor is an $N \times N$ matrix of the form

$$\begin{bmatrix} \mathbf{W}^s_{N/2^{k-1}} & 0 \\ 0 & \mathbf{I}_{N(1-1/2^{k-1})} \end{bmatrix}.$$

7.5.4.3 Reconstruction from Approximation Coefficients

A crude form of compression would be to store only the approximation coefficient $\mathbf{X}_\ell$ vector, which is half as long as $\mathbf{x}$. We could then approximately reconstruct $\mathbf{x}$ from $\mathbf{X}_\ell$ alone. Specifically, we can write the one-stage DWT of a signal $\mathbf{x}$ as

$$\begin{bmatrix} \mathbf{X}_\ell \\ \mathbf{X}_h \end{bmatrix} = \mathcal{W}_1^a(\mathbf{x}).$$

Since the DWT and its inverse are linear, we have

$$\begin{aligned} \mathbf{x} &= \mathcal{W}_1^s(\mathcal{W}_1^a(\mathbf{x})) = \mathcal{W}_1^s\left(\begin{bmatrix} \mathbf{X}_\ell \\ \mathbf{X}_h \end{bmatrix}\right) \\ &= \mathcal{W}_1^s\left(\begin{bmatrix} \mathbf{X}_\ell \\ 0 \end{bmatrix}\right) + \mathcal{W}_1^s\left(\begin{bmatrix} 0 \\ \mathbf{X}_h \end{bmatrix}\right) \\ &:= \alpha_1(\mathbf{x}) + \delta_1(\mathbf{x}). \end{aligned} \tag{7.25}$$

The vector $\alpha_1(\mathbf{x})$ is called the *level-1* or *stage-1 approximation* to $\mathbf{x}$ and $\delta_1(\mathbf{x})$ is called the *stage 1 detail.* Note that $\alpha_1(\mathbf{x})$ and $\delta_1(\mathbf{x})$ are vectors of the same size as $\mathbf{x}$, even though $\mathbf{X}_\ell$ and $\mathbf{X}_h$ are half that size. The vector $\alpha_1(\mathbf{x})$ is an approximation to the original signal $\mathbf{x}$ and $\delta_1(\mathbf{x})$ constitutes the (probably small) corrections necessary to perfectly reconstruct the original signal. Figure 7.14a shows $\alpha_1(\mathbf{x})$ for the 1024-point signal of Examples 7.2 and 7.9, using the Le Gall 5/3 filters and circular convolution.

In the two-stage case, we have

$$\begin{aligned} \mathbf{x} &= \mathcal{W}_2^s(\mathcal{W}_2^a(\mathbf{x})) = \mathcal{W}_2^s\left(\begin{bmatrix} \mathbf{X}_{\ell\ell} \\ \mathbf{X}_{\ell h} \\ \mathbf{X}_h \end{bmatrix}\right) \\ &= \mathcal{W}_2^s\left(\begin{bmatrix} \mathbf{X}_{\ell\ell} \\ 0 \\ 0 \end{bmatrix}\right) + \mathcal{W}_2^s\left(\begin{bmatrix} 0 \\ \mathbf{X}_{\ell h} \\ 0 \end{bmatrix}\right) + \mathcal{W}_2^s\left(\begin{bmatrix} 0 \\ 0 \\ \mathbf{X}_h \end{bmatrix}\right). \end{aligned} \tag{7.26}$$

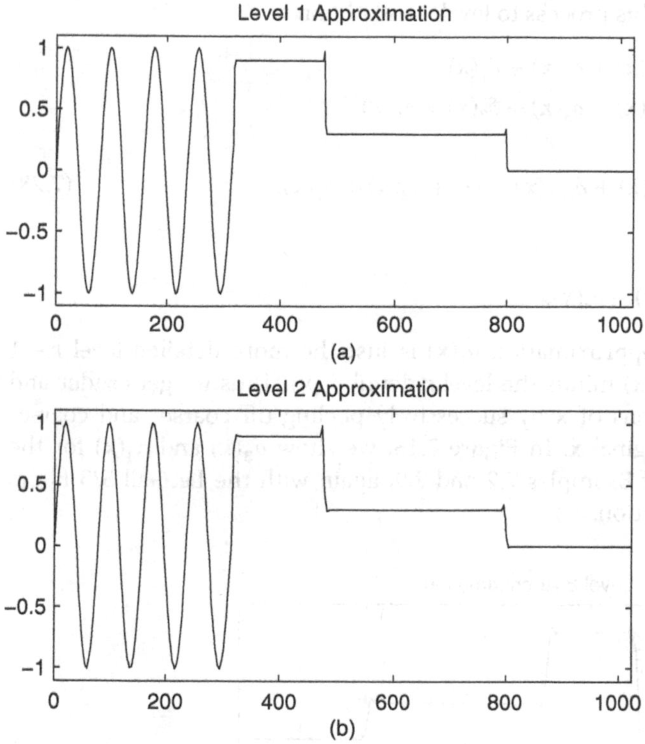

Figure 7.14 Reconstructions $\alpha_1(\mathbf{x})$ (a) and $\alpha_2(\mathbf{x})$ (b).

Now by the construction of the second stage of the *DWT* we have

$$\mathcal{W}_2^s\left(\begin{bmatrix} 0 \\ 0 \\ \mathbf{X}_h \end{bmatrix}\right) = \mathcal{W}_1^s\left(\begin{bmatrix} 0 \\ 0 \\ \mathbf{X}_h \end{bmatrix}\right)$$

(look back at equation (7.24) to see why), and thus we can express equation (7.26) as

$$\mathbf{x} = \alpha_2(\mathbf{x}) + \delta_2(\mathbf{x}) + \delta_1(\mathbf{x}), \tag{7.27}$$

where $\alpha_2(\mathbf{x})$ and $\delta_2(\mathbf{x})$ are called the *level-2 approximation and details* for $\mathbf{x}$, respectively; the vector $\delta_1(\mathbf{x})$ is unchanged from the one-stage case. The quantity $\alpha_2(\mathbf{x})$ is computable from $\mathbf{X}_{\ell\ell}$ alone. Figure 7.14b shows $\alpha_2(\mathbf{x})$ for the 1024-point signal of Examples 7.2 and 7.9, again with the Le Gall 5/3 filters and circular convolution.

We can continue this process to level r and obtain

$$\begin{aligned}
\mathbf{x} &= \alpha_3(\mathbf{x}) + \delta_3(\mathbf{x}) + \delta_2(\mathbf{x}) + \delta_1(\mathbf{x}) \\
&= \alpha_4(\mathbf{x}) + \delta_4(\mathbf{x}) + \delta_3(\mathbf{x}) + \delta_2(\mathbf{x}) + \delta_1(\mathbf{x}) \\
&= \vdots \\
&= \alpha_r(\mathbf{x}) + \delta_r(\mathbf{x}) + \delta_{r-1}(\mathbf{x}) + \cdots + \delta_2(\mathbf{x}) + \delta_1(\mathbf{x}).
\end{aligned} \tag{7.28}$$

It is easy to see that

$$\alpha_r(\mathbf{x}) = \alpha_{r-1}(\mathbf{x}) - \delta_r(\mathbf{x}),$$

that is, the level-r approximation $\alpha_r(\mathbf{x})$ is just the more detailed level $r-1$ approximation $\alpha_{r-1}(\mathbf{x})$ minus the level r detail $\delta_r(\mathbf{x})$. Thus we get cruder and cruder approximations of $\mathbf{x}$ by successively peeling off coarser and coarser details from the original $\mathbf{x}$. In Figure 7.15, we show $\alpha_3(\mathbf{x})$ and $\alpha_4(\mathbf{x})$ for the 1024-point signal of Examples 7.2 and 7.9, again with the Le Gall 5/3 filters and circular convolution.

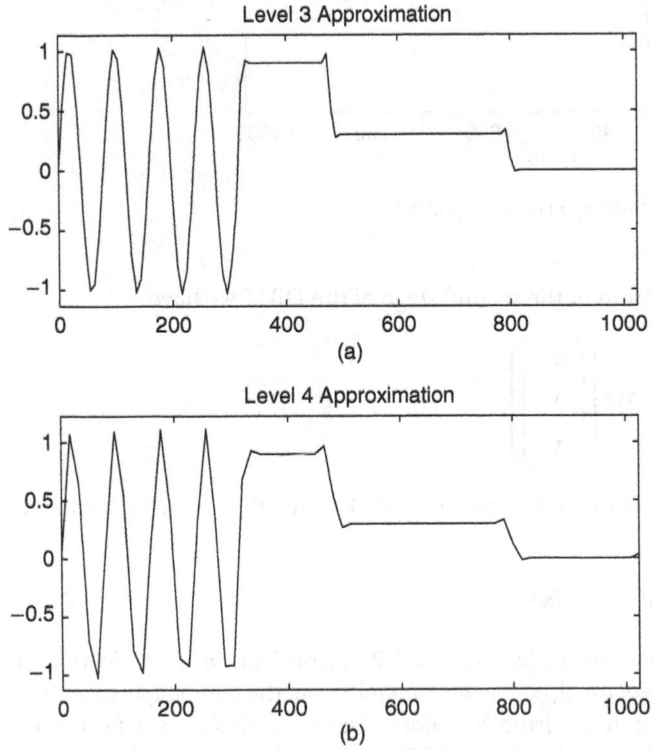

Figure 7.15 Reconstructions $\alpha_3(\mathbf{x})$ (a) and $\alpha_4(\mathbf{x})$ (b).

If $\mathbf{x}$ is a finite-dimensional vector, we can continue this for only finitely many stages. However, in the $L^2(\mathbb{Z})$ case, it can be continued indefinitely.

7.5.5 Matlab Implementation of Discrete Wavelet Transforms

Matlab's wavelet decomposition command `dwt` implements the DWT for finite length signals. The command can be tailored to use any of the extension methods described above, as well as many more. We will explore this in the Matlab project at the end of this chapter.

7.6 The 2D Discrete Wavelet Transform and JPEG 2000

7.6.1 Two-dimensional Transforms

We will restrict our wavelet analysis in two dimensions to the case of finite-extent signals or images rather than signals in $L^2(\mathbb{Z} \times \mathbb{Z})$. Let $(\boldsymbol{\ell}_a, \mathbf{h}_a)$ be the low- and high-pass analysis filters for a filter bank for $L^2(\mathbb{Z})$ signals. We assume that the analysis filters have been adapted for use in a filter bank for finite length signals, either via periodic extension as in Section 7.5.2 or some other technique, and hence we can define a corresponding analysis matrix $\mathbf{W}_k^a$ for signals of any length k. We also assume that suitable perfect reconstruction synthesis filters exist with corresponding matrix $\mathbf{W}_k^s$.

Let $\mathbf{A}$ be an $M \times N$ matrix or grayscale image, where we assume that both M and N are divisible by reasonably large powers of 2. There are many ways we can use a filter bank for one-dimensional signals to define a 2D wavelet transform. The simplest is to do precisely what we did with the discrete Fourier and cosine transforms: transform each column of $\mathbf{A}$ and then transform each row of the resulting matrix. We thus make the following definition:

Definition 7.4 Let $\mathbf{A}, M, N, \boldsymbol{\ell}_a, \mathbf{h}_a$ be as above. The one-stage, two-dimensional DWT determined by $(\boldsymbol{\ell}_a, \mathbf{h}_a)$ is the linear mapping given by

$$\mathcal{W}_1^a(\mathbf{A}) = \mathbf{W}_M^a \mathbf{A} (\mathbf{W}_N^a)^T,$$

where $\mathbf{W}_M^a$ and $\mathbf{W}_N^a$ are the $M \times M$ and $N \times N$ analysis matrices determined by $(\boldsymbol{\ell}_a, \mathbf{h}_a)$. We define the inverse one-stage transform as

$$\mathcal{W}_1^s(\hat{\mathbf{A}}) = \mathbf{W}_M^s \hat{\mathbf{A}} (\mathbf{W}_N^s)^T.$$

To graphically present a one-stage DWT for an image we will use the following scheme: if the image $\mathbf{A}$ is $M \times N$, then the transform $\mathcal{W}_1^a(\mathbf{A})$ will consist of four distinct parts, each an array of size $M/2 \times N/2$. One such part is that in which each row and column of the original image has been low-pass filtered and downsampled (downsampling the rows cuts them to length $N/2$ and

Table 7.1 Arrangement of low and high-pass components.

	low	high
low	*LL* (approximation)	*HL* (vertical details)
high	*LH* (horizontal details)	*HH* (diagonal details)

downsampling columns cuts them to length $M/2$). Another part of the transform consists of rows that are low-pass filtered with columns that are high-pass filtered, both downsampled. Similarly, we obtain high-pass row, low-pass column, and high-pass rows and columns. We present the transform as a grayscale image according to the scheme of Table 7.1. The *LL* piece (low-low) is in the upper left-hand corner and comes from low-pass filtering in both directions. Of the four pieces it is the most like the original picture and so is called the *stage 1 approximation coefficients.* The remaining three pieces are called *stage 1 detail components.* The upper right corner comes from high-pass filtering in the horizontal direction (rows) and low-pass filtering in the vertical direction (columns), and so has the label *HL*. The visible details in this subimage, such as edges, have an overall vertical orientation, since their alignment is orthogonal to the direction of the high pass filtering. The remaining components have analogous explanations.

In what follows, if $\mathbf{C} = \mathcal{W}_1^a(\mathbf{A})$ (so that $\mathbf{C}$ is an $M \times N$ matrix), then we write

$$\mathbf{C} = \begin{bmatrix} \mathbf{C}_A & \mathbf{C}_V \\ \mathbf{C}_H & \mathbf{C}_D \end{bmatrix} \tag{7.29}$$

for the various portions of the transform, where each of the submatrices is an $M/2 \times N/2$ matrix.

Example 7.10 *Let us show before and after pictures for an image using the 1-stage DWT. The original picture is in Figure 7.16a, 600 by 800 pixels, and the one-stage DWT is in panel (b), displayed according to the scheme of equation (7.29). The analysis filters are the Le Gall 5/3 filters of Example 7.4. The DWT is displayed on a logarithmic scale. In the upper right portion of the transform, the tree trunks are most visible, since this is the vertical detail section of the transform. The relatively flat and featureless portions of the image (e.g., the water) ends up appearing black in all but the approximation coefficients, since high-pass filtering in either the vertical or the horizontal direction zeros out dc components.*

7.6.2 Multistage Transforms for Two-dimensional Images

We may of course do a multistage transform. In Figure 7.17, we show two-stage and three-stage transforms of the image from Example 7.10, both displayed on a logarithmic scale.

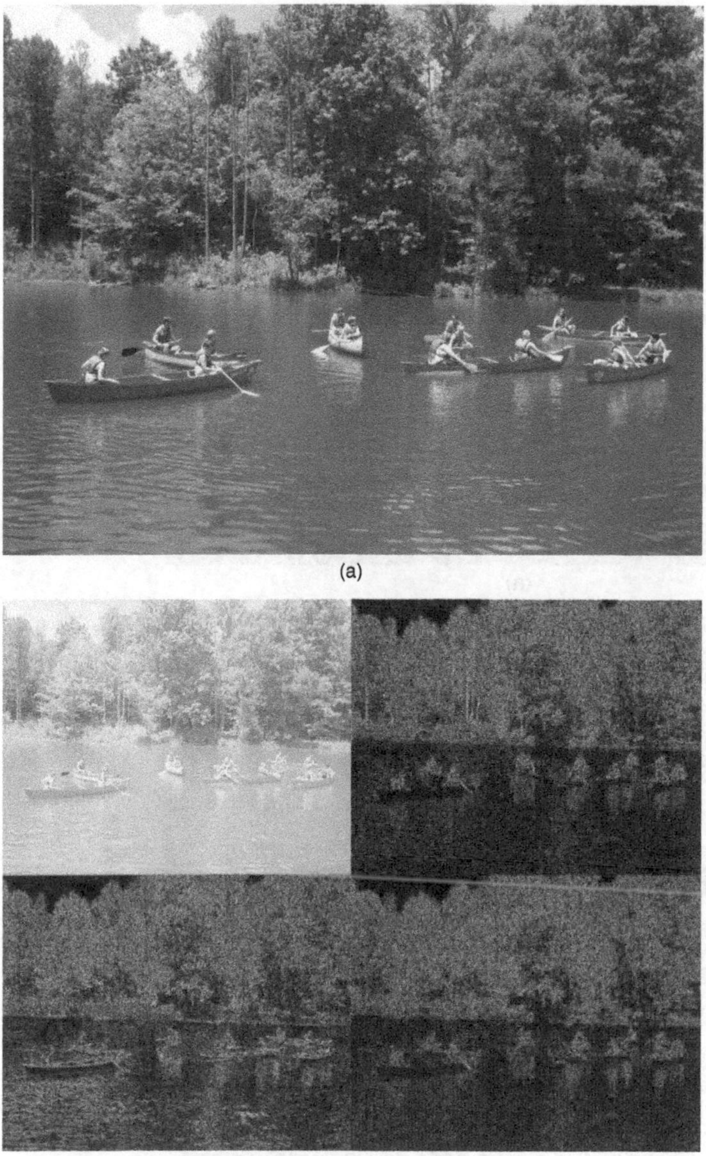

(a)

(b)

Figure 7.16 Original image (a) and one-stage DWT (b).

(a)

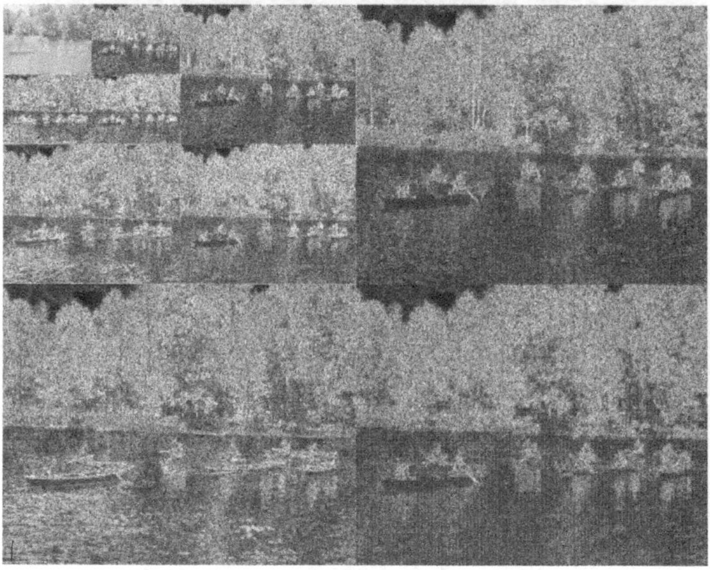

(b)

Figure 7.17 Two-stage (a) and three-stage (b) DWT of image from Example 7.10.

Remark 7.3 For image analysis, one usually goes to $r = 4$ or 5 levels. There is always one approximation submatrix and $3r$ detail submatrices.

Let us examine the multistage procedure more carefully. Let $(\boldsymbol{\ell}_a, \mathbf{h}_a)$ be the analysis filters and $(\boldsymbol{\ell}_s, \mathbf{h}_s)$ the synthesis filters for a biorthogonal filter bank with DWT analysis and synthesis matrices $\mathbf{W}_k^a$ or $\mathbf{W}_k^s$ for one-dimensional transforms of size k. Let $\mathbf{A}$ be an $M \times N$ array. Let $\mathbf{C}_1, \mathbf{C}_2, \ldots, \mathbf{C}_r$ be the matrices formed as follows: the matrix $\mathbf{C}_1$ is obtain by computing the DWT of $\mathbf{A}$ via

$$\mathbf{C}_1 = \mathbf{W}_M^a \mathbf{A} (\mathbf{W}_N^a)^T = \begin{bmatrix} \mathbf{C}_{1,A} & \mathbf{C}_{1,V} \\ \mathbf{C}_{1,H} & \mathbf{C}_{1,D} \end{bmatrix},$$

where $\mathbf{C}_{1,A}$ denotes the $M/2 \times N/2$ upper left quadrant of $\mathbf{C}_1$ (the approximation subimage; this is the same as equation (7.29)). Let $\mathbf{C}_2$ be the matrix obtained from $\mathbf{C}_1$ by replacing $\mathbf{C}_{1,A}$ by the DWT of $\mathbf{C}_{1,A}$,

$$\mathbf{C}_2 = \begin{bmatrix} \mathbf{W}_{M/2}^a \mathbf{C}_{1,A} (\mathbf{W}_{N/2}^a)^T & \mathbf{C}_{1,V} \\ \mathbf{C}_{1,H} & \mathbf{C}_{1,D} \end{bmatrix}$$

$$= \begin{bmatrix} \begin{bmatrix} (\mathbf{C}_{1,A})_A & (\mathbf{C}_{1,A})_V \\ (\mathbf{C}_{1,A})_H & (\mathbf{C}_{1,A})_D \end{bmatrix} & \mathbf{C}_{1,V} \\ \mathbf{C}_{1,H} & \mathbf{C}_{1,D} \end{bmatrix}$$

$$= \begin{bmatrix} \begin{bmatrix} \mathbf{C}_{2,A} & \mathbf{C}_{2,V} \\ \mathbf{C}_{2,H} & \mathbf{C}_{2,D} \end{bmatrix} & \mathbf{C}_{1,V} \\ \mathbf{C}_{1,H} & \mathbf{C}_{1,D} \end{bmatrix}. \tag{7.30}$$

Compare the array in (7.30) with the picture in Figure 7.17a, to build up some intuition about the various pieces of $\mathbf{C}_2$.

In general, having defined $\mathbf{C}_r$, let $\mathbf{C}_{r,A}$ be the $M/2^r \times N/2^r$ upper left approximation submatrix in the DWT of $\mathbf{C}_r$. Then $\mathbf{C}_{r+1}$ is the matrix obtained from $\mathbf{C}_r$ by replacing $\mathbf{C}_{r,A}$ by its one-stage DWT, which has four component pieces:

$$\begin{bmatrix} \mathbf{C}_{r+1,A} & \mathbf{C}_{r+1,V} \\ \mathbf{C}_{r+1,H} & \mathbf{C}_{r+1,D} \end{bmatrix}.$$

For example, at stage 3, we have

$$\mathbf{C}_3 = \begin{bmatrix} \begin{bmatrix} \begin{bmatrix} \mathbf{C}_{3,A} & \mathbf{C}_{3,V} \\ \mathbf{C}_{3,H} & \mathbf{C}_{3,D} \end{bmatrix} & \mathbf{C}_{2,V} \\ \mathbf{C}_{2,H} & \mathbf{C}_{2,D} \end{bmatrix} & \mathbf{C}_{1,V} \\ \mathbf{C}_{1,H} & \mathbf{C}_{1,D} \end{bmatrix}.$$

Compare this result with the image in Figure 7.17b.

Definition 7.5 We define the r-stage (or r-level) two-dimensional DWT of $\mathbf{A}$ as the matrix $\mathbf{C}_r$. Symbolically we denote this r-stage transform as

$$\mathbf{A} \to \mathcal{W}_r^a(\mathbf{A}).$$

The transform is linear.

Analogously the r-stage synthesis transform is denoted

$$\hat{\mathbf{A}} \to \mathcal{W}_r^s(\hat{\mathbf{A}})$$

and is obtained by reversing the above steps with the synthesis filters and corresponding matrices. The inverse transform is linear. For a biorthogonal (perfect reconstruction) filter bank, we have $\mathcal{W}_r^s(\mathcal{W}_r^a(\mathbf{A})) = \mathbf{A}$.

Matlab implements DWTs in two dimensions using the command `dwt2`. As in the one-dimensional case, there are a variety of options for extending the images past their natural boundaries in order to apply the analysis filters, including, periodic extension, whole and half-point symmetric extension, and many others. We explore them in the project sections 7.8 and 7.9.

7.6.3 Approximations and Details for Images

As for one-dimensional signals, we can perform a multistage decomposition into approximation and details for images. For an image $\mathbf{A}$, for example, we can write

$$\begin{aligned}
\mathbf{A} &= \mathcal{W}_1^s(\mathcal{W}_1^a(\mathbf{A})) \\
&= \mathcal{W}_1^s\left(\begin{bmatrix} \mathbf{C}_{1,A} & \mathbf{C}_{1,V} \\ \mathbf{C}_{1,H} & \mathbf{C}_{1,D} \end{bmatrix}\right) \\
&= \mathcal{W}_1^s\left(\begin{bmatrix} \mathbf{C}_{1,A} & 0 \\ 0 & 0 \end{bmatrix}\right) + \mathcal{W}_1^s\left(\begin{bmatrix} 0 & \mathbf{C}_{1,V} \\ \mathbf{C}_{1,H} & \mathbf{C}_{1,D} \end{bmatrix}\right) \\
&:= \alpha_1(\mathbf{A}) + \delta_1(\mathbf{A}),
\end{aligned}$$

analogous to equation (7.25). The quantity $\alpha_1(\mathbf{A})$ is the *level-1 approximation* to $\mathbf{A}$ and $\delta_1(\mathbf{A})$ is the *level-1 detail.* Note that $\alpha_1(\mathbf{A})$ is a full $M \times N$ image constructed from the $M/2 \times N/2$ approximation coefficient matrix $\mathbf{C}_{1,A}$. In Figure 7.18a, we show $\alpha_1(\mathbf{A})$ for the image from Figure 7.16.

As with signals, the decomposition process can be iterated, for example, as

$$\begin{aligned}
\mathbf{A} &= \mathcal{W}_2^s(\mathcal{W}_2^a(\mathbf{A})) \\
&= \mathcal{W}_2^s\left(\begin{bmatrix} \begin{bmatrix} \mathbf{C}_{2,A} & \mathbf{C}_{2,V} \\ \mathbf{C}_{2,H} & \mathbf{C}_{2,D} \end{bmatrix} & \mathbf{C}_{1,V} \\ \mathbf{C}_{1,H} & \mathbf{C}_{1,D} \end{bmatrix}\right)
\end{aligned}$$

(a)

(b)

Figure 7.18 Reconstructions $\alpha_1(\mathbf{A})$ (a) and $\alpha_2(\mathbf{A})$ (b).

$$
\begin{aligned}
&= \mathcal{W}_2^s\left(\begin{bmatrix} \begin{bmatrix} \mathbf{C}_{2,A} & \mathbf{0} \\ \mathbf{0} & \mathbf{0} \end{bmatrix} & \mathbf{0} \\ \mathbf{0} & \mathbf{0} \end{bmatrix}\right) + \mathcal{W}_2^s\left(\begin{bmatrix} \begin{bmatrix} \mathbf{0} & \mathbf{C}_{2,V} \\ \mathbf{C}_{2,H} & \mathbf{C}_{2,D} \end{bmatrix} & \mathbf{0} \\ \mathbf{0} & \mathbf{0} \end{bmatrix}\right) \\
&+ \mathcal{W}_1^s\left(\begin{bmatrix} \begin{bmatrix} \mathbf{0} & \mathbf{0} \\ \mathbf{0} & \mathbf{0} \end{bmatrix} & \mathbf{C}_{1,V} \\ \mathbf{C}_{1,H} & \mathbf{C}_{1,D} \end{bmatrix}\right) \\
&:= \alpha_2(\mathbf{A}) + \delta_2(\mathbf{A}) + \delta_1(\mathbf{A}),
\end{aligned}
$$

analogous to equation (7.28), where we have made use of equation (7.30) and the fact that $\mathbf{C}_2 = \mathcal{W}_2^a(\mathbf{A})$. We can thus approximate $\mathbf{A}$ "crudely," as $\alpha_2(\mathbf{A})$, or with slightly more detail, as $\alpha_2(\mathbf{A}) + \delta_2(\mathbf{A})$ (which is just $\alpha_1(\mathbf{A})$). The level-2 approximation $\alpha_2(\mathbf{A})$ for the image from Figure 7.16 is shown in Figure 7.18b; the approximation $\alpha_3(\mathbf{A})$ is shown in Figure 7.19. Not surprisingly, the level-r approximation degrades as r increases. The decomposition can be carried to any level as long as M and N are divisible by suitable powers of two.

7.6.4 JPEG 2000

The JPEG 2000 image compression standard was developed in several stages, beginning in 1997. The goal is to improve upon the traditional JPEG standard,

Figure 7.19 Reconstruction $\alpha_3(\mathbf{A})$.

not merely by increasing compression performance but by also providing greater flexibility and a wider range of options in compressing images. Thus, for example, the possibility exists for compressing different parts of an image with different parameters, or specifying that part (or all) of an image be compressed with no information loss. Many other features are built into the standard.

We do not intend to give a comprehensive or even partial survey of the many facets of the standard, but merely to indicate how the DWT comes into play. See [26] for additional information, and the official JPEG 2000 web page [17].

We should mention that image compression is important for more than just personal photos and transmission of Internet images. Image compression, and more generally image processing, plays an important role in printing, medical imaging, nondestructive testing and remote sensing, image archiving, and other areas. The JPEG 2000 standard is an attempt to serve many of these needs with a coherent, flexible standard.

The overall structure of the JPEG 2000 standard is similar to the traditional JPEG standard, but at the core of the JPEG 2000 is the DWT. Preprocessing steps similar to those of the traditional JPEG algorithm take place, such as, subtraction of dc levels or transformations to separate color components. After this the relevant data are subjected to a DWT using either the Daubechies 9/7 filters or the Le Gall 5/3 filters (the latter are used for lossless compression). The transformed image components are then quantized using one or more schemes like those described in Chapter 1, and then the results are compressed using an entropy encoding algorithm. Reconstruction of the image consists of applying the appropriate decoding algorithm and inverse quantization. The appropriate synthesis filter (inverse wavelet transform) is then applied to reconstruct the image.

The Matlab project at the end of this chapter provides the opportunity to use Matlab's Wavelet Toolbox to process some signals and grayscale images.

7.7 Filter Design

This section focuses on filter design for perfect reconstruction filter banks. It is not all that essential for using filter banks if you are content to use off-the-shelf filters. But it is useful for those readers who seek to understand the essential quantitative relationship between filter banks and wavelets.

It is natural to ask whether we can choose any analysis filters we want and then design appropriate synthesis filters for perfect reconstruction. A moment's thought should show that this is not the case—taking both analysis filters to be identically zero provides an obvious counterexample. Therefore, what constraints exist on the analysis filters? If suitable synthesis filters exist, how do we find them? Some information can be gleaned from the time-domain relations that come from equation (7.6), but it is easier to carry out the analysis via z-transforms. Our goal here is not an exhaustive treatment of filter design techniques, but a few representative and important examples.

7.7.1 Filter Banks in the z-domain

We begin by considering the effect of downsampling, upsampling and filtering in the "frequency domain" of the z-transform. Let $\mathbf{x}$ be a signal in $L^2(\mathbb{Z})$, and define

$$X(z) = \sum_{k=-\infty}^{\infty} x_k z^{-k},$$

the z-transform of $\mathbf{x}$. We suppose that $\mathbf{x}$ is passed through a two-channel filter bank with FIR analysis filters $\boldsymbol{\ell}_a$, $\mathbf{h}_a$ and synthesis filters $\boldsymbol{\ell}_s$, $\mathbf{h}_s$ as in Figure 7.4 (with no intermediate compression, denoising, etc.) We seek perfect reconstruction.

7.7.1.1 Downsampling and Upsampling in the z-domain

The first and last operations performed in a filter bank are convolutions with the filters. In between, each filtered portion of the signal is downsampled and then upsampled. The operation of downsampling and immediately upsampling a signal is simple to quantify with z-transforms.

Let $\mathbf{y}$ be the result of downsampling and then upsampling some signal $\mathbf{w}$. It is easy to see that $y_j = w_j$ for j even and $y_j = 0$ for j odd. This observation leads to the following proposition:

Proposition 7.1 *Let $W(z)$ denote the z-transform of $\mathbf{w} \in L^2(\mathbb{Z})$ and $Y(z)$ the z-transform of the vector $\mathbf{y} = U(D(\mathbf{w}))$ obtained by downsampling and then upsampling $\mathbf{w}$. The function $Y(z)$ is given by*

$$Y(z) = \frac{1}{2}W(z) + \frac{1}{2}W(-z).$$

The proof is left as Exercise 21.

7.7.1.2 Filtering in the Frequency Domain

If $\mathbf{y} = \mathbf{x} * \mathbf{g}$ for some FIR filter $\mathbf{g}$, then the equivalent frequency domain statement is the Convolution Theorem 4.6, which asserts that $Y(z) = X(z)G(z)$. By assumption $\mathbf{g}$ is FIR, so there is no difficulty in summing the terms involved in the product $X(z)G(z)$: the sums will converge for any $\mathbf{x} \in L^2(\mathbb{Z})$ (or even $L^\infty(\mathbb{Z})$).

7.7.2 Perfect Reconstruction in the z-frequency Domain

Let $\mathbf{x}$ be passed through a two-channel filter bank with analysis filters $\boldsymbol{\ell}_a$ and $\mathbf{h}_a$. Let L_a and H_a denote the z-transform of these filters. By the Convolution Theorem 4.6, the z-transform of $\mathbf{x} * \boldsymbol{\ell}_a$ is $L_a(z)X(z)$. By Proposition 7.1, when

$\mathbf{x} * \boldsymbol{\ell}_a$ is downsampled and then upsampled, the resulting signal $U(D(\mathbf{x} * \boldsymbol{\ell}_a))$ has z-transform

$$\frac{1}{2}(X(z)L_a(z) + X(-z)L_a(-z)).$$

Again, by the Convolution Theorem 4.6, if we convolve $U(D(\mathbf{x} * \boldsymbol{\ell}_a))$ with the synthesis low-pass filter, we obtain a signal with z-transform

$$\frac{1}{2}(X(z)L_a(z) + X(-z)L_a(-z))L_s(z).$$

This is the z-transform of the output from the low-pass portion of the filter bank. Performing the analogous operation with the high-pass portion of the filter bank yields $\frac{1}{2}(X(z)H_a(z) + X(-z)H_a(-z))H_s(z)$. The final step in the filter bank is to add the results, which yields an output vector $\tilde{\mathbf{x}}$ with z-transform

$$\begin{aligned}\tilde{X}(z) &= \frac{1}{2}(X(z)L_a(z) + X(-z)L_a(-z))L_s(z) \\ &\quad + \frac{1}{2}(X(z)H_a(z) + X(-z)H_a(-z))H_s(z) \\ &= \frac{1}{2}[L_a(z)L_s(z) + H_a(z)H_s(z)]X(z) \\ &\quad + \frac{1}{2}[L_a(-z)L_s(z) + H_a(-z)H_s(z)]X(-z). \end{aligned} \tag{7.31}$$

Perfect reconstruction dictates that the output signal should be a copy of the input signal, possibly delayed, so that $\tilde{X}(z) = z^{-m}X(z)$ for some $m \geq 0$; recall Exercise 37 of Chapter 4. (We could also allow $m < 0$, corresponding to an advance; it will not affect the algebra to follow.) In light of equation (7.31), if $\tilde{X}(z) = z^{-m}X(z)$ and if we require

$$L_a(z)L_s(z) + H_a(z)H_s(z) = 2z^{-m}, \tag{7.32}$$

$$L_a(-z)L_s(z) + H_a(-z)H_s(z) = 0, \tag{7.33}$$

then we will obtain perfect reconstruction. Equations (7.32) and (7.33) are thus sufficient conditions on the analysis and synthesis filter z-transforms for a perfect reconstruction filter bank with m-step delay. As it turns out, they are also necessary (see Exercise 26).

There are many approaches to designing suitable filters for use in a biorthogonal or orthogonal filter bank. We will focus on a few illustrative examples (see [28] for more information).

Remark 7.4 For any perfect reconstruction filter bank, we can make numerous minor alterations to the filters, and equations (7.32) and (7.33) will still hold. We can, for example, multiply one low-pass filter by a constant c while dividing the other by c, or do the same to the high-pass filters. We can also introduce shifts or delays of the same magnitude to both analysis or to both synthesis filters.

7.7.3 Filter Design I: Synthesis from Analysis

Suppose that the analysis filters are given and we seek suitable synthesis filters. Equations (7.32) and (7.33) are linear in the quantities $L_s(z), H_s(z)$ if we consider the analysis filters as given. We can solve as

$$L_s(z) = \frac{2z^{-m}H_a(-z)}{L_a(z)H_a(-z) - L_a(-z)H_a(z)}, \tag{7.34}$$

$$H_s(z) = \frac{-2z^{-m}L_a(-z)}{L_a(z)H_a(-z) - L_a(-z)H_a(z)}. \tag{7.35}$$

However, this is a bit suspect: *we have not defined what it means to divide z-transforms, in the formal sense.* In particular, there may not exist Laurent polynomials $L_s(z)$ and $H_s(z)$ so that equations (7.32) and (7.33) are satisfied. However, if the denominator $L_a(z)H_a(-z) - L_a(-z)H_a(z)$ is a monomial z^q, then division by z^q is formally well defined: $z^qX(z)$ just corresponds to a shift of $\mathbf{x}$, q indexes to the right. In this case, L_s and H_s will correspond to FIR filters. And indeed, this is not only necessary, it is also sufficient, a consequence of the following lemma:

Lemma 7.1 *If $p(z)$ and $q(z)$ are Laurent polynomials such that $p(z)q(z)$ is a monomial, then $p(z)$ and $q(z)$ are both monomials.*

Proof: Let

$$p(z) = \sum_{k=m_1}^{n_1} a_k z^k \text{ and } q(z) = \sum_{k=m_2}^{n_2} b_k z^k$$

where $m_1 \leq n_1$, $m_2 \leq n_2$, with $a_{m_1}, a_{n_1}, b_{m_2}, b_{n_2}$ all nonzero. Then

$$p(z)q(z) = a_{m_1}b_{m_2}z^{m_1+m_2} + \cdots + a_{n_1}b_{n_2}z^{n_1+n_2}$$

where all intermediate terms involve z^k with $m_1 + m_2 < k < n_1 + n_2$. If $p(z)q(z)$ is a monomial, say $p(z)q(z) = cz^m$, then we must have

$$m_1 + m_2 = n_1 + n_2,$$

both of which equal m. A little rearrangement of the last equation yields $n_1 - m_1 = m_2 - n_2$. But $n_1 - m_1 \geq 0$, while $m_2 - n_2 \leq 0$. We conclude that $m_1 = n_1$ and $m_2 = n_2$, so that $p(z)$ and $q(z)$ are monomials. This completes the proof of the lemma. In particular, an easy consequence is that if $p(z)q(z) = 0$ then either $p \equiv 0$ or $q \equiv 0$ (or both). That is, the set of Laurent polynomials is an "integral domain."

We can now show that

Lemma 7.2 *There exist Laurent polynomials $L_s(z), H_s(z)$ so that equations (7.32) and (7.33) are satisfied if and only if $L_a(z)H_a(-z) - L_a(-z)H_a(z)$ is a monomial.*

Proof: The "if" part we argued above. For the converse, suppose that $L_s(z)$ and $H_s(z)$ are Laurent polynomials such that (7.32) and (7.33) are satisfied. For convenience, define

$$d_1(z) = L_a(z)L_s(z) + H_a(z)H_s(z),$$
$$d_2(z) = L_a(-z)L_s(z) + H_a(-z)H_s(z)$$

so that (7.32) becomes $d_1(z) = 2z^{-m}$ while (7.33) becomes $d_2(z) = 0$. Then

$$d_1(z)d_1(-z) - d_2(z)d_2(-z) = 4(-1)^m z^{-2m}. \tag{7.36}$$

A bit of tedious algebra shows that in fact

$$d_1(z)d_1(-z) - d_2(z)d_2(-z) = (L_a(z)H_a(-z) - L_a(-z)H_a(z)) \times (L_s(z)H_s(-z) - L_s(-z)H_s(z))$$

so that from equation (7.36) we have

$$(L_a(z)H_a(-z) - L_a(-z)H_a(z)) \times (L_s(z)H_s(-z) - L_s(-z)H_s(z)) = 4(-1)^m z^{-2m}.$$

The left side above is, by assumption, a Laurent polynomial. From Lemma 7.1, we conclude that both $L_a(-z)H_a(z) - L_a(z)H_a(-z)$ and $L_s(-z)H_s(z) - L_s(z)H_s(-z)$ are monomials.

Remark 7.5 Note that since $L_a(-z)H_a(z) - L_a(z)H_a(-z)$ is an odd function of z, it is in fact a monomial of odd degree.

We can thus choose any analysis filters for which the common right side denominator $L_a(z)H_a(-z) - L_a(-z)H_a(z)$ in equations (7.34) and (7.35) is a monomial and then read off the synthesis filters. It is worth noting that the z^{-m} factor in each numerator on the right in (7.34) and (7.35) merely corresponds to an m-index translate of the filters obtained using $m = 0$. We will thus assume $m = 0$ for now.

Example 7.11 *Let the analysis filters be given by the Haar filters $\boldsymbol{\ell}_a = \left(\ldots, 0, \frac{1}{2}, \frac{1}{2}, 0, \ldots\right)$ and $\mathbf{h}_a = \left(\ldots, 0, \frac{1}{2}, -\frac{1}{2}, 0, \ldots\right)$ as in Example 7.3, with z-transforms*

$$L_a(z) = \frac{1}{2} + \frac{1}{2}z^{-1}, \quad H_a(z) = \frac{1}{2} - \frac{1}{2}z^{-1}.$$

In this case, the denominator in equations (7.34) and (7.35) turns out to be z^{-1}*, and we obtain (with* $m = 0$*) that*

$$L_s(z) = 1 + z, \quad H_s(z) = 1 - z,$$

the same filters we obtained in Example 7.3. These correspond to filters that are not causal, but by shifting the filter coefficients 1 index (equivalently, use $m = 1$ *in (7.34)–(7.35)), we obtain causal filters with transforms* $L_s(z) = 1 + z^{-1}, H_s(z) = -1 - z^{-1}$*. The resulting filter bank perfectly reconstructs the input, but with a one index delay.*

Example 7.12 *Let the analysis filters be FIR with coefficients* $(\ell_a)_0 = \frac{1}{2}, (\ell_a)_1 = 0, (\ell_a)_2 = \frac{1}{2}$ *and all other coefficients zero, and take* $\mathbf{h}_a$ *as in the last example. In this case, we have* $L_a(z) = \frac{1}{2} + \frac{1}{2}z^{-2}, H_a(z) = \frac{1}{2} - \frac{1}{2}z^{-1}$*, and equations (7.34) and (7.35) (with* $m = 0$*) yield*

$$L_s(z) = \frac{2z^2(z+1)}{z^2+1}, \quad H_s(z) = -2z.$$

In this case, the denominator is not a monomial. The transform $H_s(z)$ *corresponds to an FIR filter, but* $L_s(z)$ *does not. Why? If* $L_s(z)$ *did correspond to an FIR filter, then* $z^n L_s(z)$ *would be a polynomial in* z *for* n *sufficiently large, which is not the case (see Exercise 23).*

Remark 7.6 One way to guarantee that the required filters are FIR is to obtain all the filters as simple permutations of one or more specified FIR filters, a technique we use in the sections that follow. In this case, it is helpful to note the relation between certain filter coefficient permutations in the time domain and the resulting effect in the z-domain. In particular, let $\mathbf{g}$ be a causal N-tap FIR filter with nonzero coefficients g_k only in the range $0 \le k \le N-1$, and with z-transform $G(z) = \sum_{k=0}^{N-1} g_k z^{-k}$. If $\tilde{\mathbf{g}}$ is one of the permutations of $\mathbf{g}$ in Table 7.2, then the z-transform of $\tilde{\mathbf{g}}$ is given in the rightmost column. The proofs are left to Exercise 19.

Table 7.2 Filter permutations and corresponding z-transforms.

$\tilde{g}$ Coefficients	z Transform $\tilde{G}(z)$
$\tilde{g}_k = (-1)^k g_k$	$G(-z)$
$\tilde{g}_k = g_{-k}$	$G(z^{-1})$
$\tilde{g}_k = g_{N-1-k}$	$z^{-N+1}G(z^{-1})$
$\tilde{g}_k = (-1)^k g_{N-1-k}$	$(-z)^{-N+1}G(-z^{-1})$

Remark 7.7 A good "low-pass" filter $\mathbf{g}$ ought to at least pass dc (perhaps scaled) but zero out the Nyquist frequency. Thus we want $\mathbf{g} * \mathbf{x} = C\mathbf{x}$ for some nonzero constant C if $x_k = 1$ for all k, and $\mathbf{g} * \mathbf{x} = \mathbf{0}$ when $x_k = (-1)^k$. If $\mathbf{g}$ is an FIR filter with coefficients g_k for $-m \leq k \leq n$, this means that

$$\sum_{k=-m}^{n} g_k \neq 0 \quad \text{and} \quad \sum_{k=-m}^{n} g_k(-1)^k = 0.$$

The equations above can be stated even more compactly: if $G(z)$ is the z-transform of $\mathbf{g}$, then $G(1) \neq 0$ and $G(-1) = 0$. This condition will be our rough characterization of low-pass filters. We call $\mathbf{g}$ a "high-pass filter" if convolution with $\mathbf{g}$ zeros out dc and passes the Nyquist frequency, again perhaps rescaled or shifted. Similar reasoning to the low-pass case shows that this is equivalent to $G(1) = 0$ and $G(-1) \neq 0$.

These characterizations make it easy to see that the permutations of the first and fourth rows in Table 7.2 turn low-pass filters into high-pass filters, and vice versa. It also is clear that if the analysis filters really are low- and high-pass, then the synthesis filters obtained from equations (7.34) and (7.35) will be of the appropriate type, after possible rescaling.

7.7.4 Filter Design II: Product Filters

The fact that we have four filters to play with leaves considerable leeway in designing a filter bank. One reasonable way to add more structure and ensure that the synthesis filters are FIR is to obtain them as simple modifications of the analysis filters. In particular, if we choose the synthesis filters so that

$$L_s(z) = H_a(-z), \quad H_s(z) = -L_a(-z), \tag{7.37}$$

then equation (7.33) is automatically satisfied. From Remark 7.7, this will also ensure that $\boldsymbol{\ell}_s$ is low-pass if $\mathbf{h}_a$ is high-pass, and similarly that $\mathbf{h}_s$ is high-pass if $\boldsymbol{\ell}_a$ is low-pass.

According to Remark 7.6, equation (7.37) correspond to taking the synthesis filters as

$$(\boldsymbol{\ell}_s)_k = (-1)^k(\mathbf{h}_a)_k, \quad (\mathbf{h}_s)_k = (-1)^{k+1}(\boldsymbol{\ell}_a)_k, \tag{7.38}$$

so that, for example, the synthesis low-pass filter is just the analysis high-pass filter with alternating signs. Clearly, the synthesis filters are FIR if the analysis filters are FIR.

If we use (7.37) to substitute $H_a(z) = L_s(-z)$ and $H_s(z) = -L_a(-z)$ in equation (7.32), we obtain

$$L_a(z)L_s(z) - L_a(-z)L_s(-z) = 2z^{-m}. \tag{7.39}$$

Define $P(z) = L_a(z)L_s(z)$ so that (7.39) becomes

$$P(z) - P(-z) = 2z^{-m}. \tag{7.40}$$

The expression $P(z) - P(-z)$ is an odd function of z; if filters with this structure exist, m must be odd. In addition, $P(z)$ will itself correspond to a low-pass filter, called the *product filter*, since $P(z)$ corresponds to concatenating the low-pass filters ℓ_a and ℓ_s. This yields a prospective approach for finding suitable filters: choose an odd value for m and design a low-pass filter $\mathbf{p}$ with z-transform $P(z)$ satisfying equation (7.40); then factor

$$P(z) = L_a(z)L_s(z) \tag{7.41}$$

to obtain low-pass analysis and synthesis filters. We obtain the high-pass filters from (7.38).

It is easy to see that equation (7.40) requires that the function $P(z)$ contain no odd powers of z except for z^{-m}. However, the equation imposes no conditions at all on the coefficients of the even powers of z in $P(z)$.

Example 7.13 *Let us design a causal filter* $\mathbf{p}$ *with nonzero coefficients* p_0, p_1, p_2, *and* p_3 *whose z-transform* $P(z)$ *satisfies equation (7.40) with* $m = 1$. *Start with*

$$P(z) = p_0 + p_1 z^{-1} + p_2 z^{-2} + p_3 z^{-3}.$$

Equation (7.40) becomes

$$P(z) - P(-z) = 2p_1 z^{-1} + 2p_3 z^{-3} = 2z^{-1},$$

which forces $p_1 = 1$ *and* $p_3 = 0$. *The coefficients* p_0 *and* p_2 *can be "anything." However, according to Remark 7.7 a low-pass filter ought to satisfy* $P(-1) = 0$. *This condition forces*

$$p_0 - 1 + p_2 = 0.$$

One obvious choice is $p_0 = p_2 = \frac{1}{2}$, *which gives*

$$P(z) = \frac{1}{2} + z^{-1} + \frac{1}{2}z^{-2}.$$

We can then factor

$$P(z) = \frac{(1+z)^2}{2z^2}$$

and obtain $L_a(z) = L_s(z) = \frac{1}{\sqrt{2}}(1 + z^{-1})$. *This is only one possible factorization into* $L_a(z)$ *and* $L_s(z)$. *The analysis filter transforms are, from equation (7.37),* $H_a(z) = \frac{1}{\sqrt{2}}(-1 + z^{-1})$ *and* $H_s(z) = \frac{1}{\sqrt{2}}(1 - z^{-1})$, *which correspond to the orthogonal Haar filters from Example 7.5. The low-pass filters really are low-pass since* $L_a(-1) = L_s(-1) = 0$.

7.7.5 Filter Design III: More Product Filters

In this subsection, we will continue with the choice of equations (7.37) and (7.38) that led to the design strategy based on equation (7.41).

In Example 7.13, we could just as well have chosen $p_0 = \frac{3}{4}$ and $p_2 = \frac{1}{4}$, to obtain $P(z) = \frac{3}{4} + z^{-1} + \frac{1}{4}z^{-2}$. Then we can factor to obtain, for example, $L_a(z) = \frac{1}{2} + \frac{1}{2}z^{-1}$ and $L_s(z) = \frac{3}{2} + \frac{1}{2}z^{-1}$. In this case, however, $L_s(-1) = 1$, so the synthesis filter $\boldsymbol{\ell}_s$ is not a low-pass filter (though the filter bank will still give perfect reconstruction). As a result the analysis filter $\mathbf{h}_a$ is not high-pass. If it is desired to obtain true low-pass filters, then we need to make sure we can factor $P(z)$ into pieces $L_a(z)$ and $L_s(z)$ such that $L_a(-1) = L_s(-1) = 0$. One way to do this is to build a factor $(1 + z^{-1})^n$ with $n \geq 2$ into $P(z)$ so that we can arrange for both L_a and L_s to contain some positive power of $(1 + z^{-1})$ and obtain $L_a(-1) = L_s(-1) = 0$.

To accomplish this, we take

$$P(z) = (1 + z^{-1})^{2r}Q(z), \tag{7.42}$$

where $Q(z)$ is a polynomial to be chosen and $r \geq 1$. We can determine the appropriate degree for Q as follows: The binomial $(1 + z^{-1})^{2r}$ contains powers of z from z^{-2r} to z^0. Therefore, if Q is of nth degree, then $P(z)$ will contain powers of z from z^{-2r} to z^n, $2r + n + 1$ powers in all. Thus $P(z) - P(-z)$ will contain all odd powers of z between z^{-2r} to z^n. If n is even, this is a total of $r + n/2$ powers of z. Then the equation $P(z) - P(-z) = 2z^{-m}$ yields $r + n/2$ linear conditions on the $n + 1$ coefficients of Q. In order to solve uniquely, we require $r + n/2 = n + 1$, or $n = 2r - 2$. We thus seek a polynomial Q of degree $2r - 2$ to use in equation (7.42) so that $P(z) - P(-z) = 2z^{-m}$ is satisfied.

Example 7.14 *Let $m = 1$ in $P(z) - P(-z) = 2z^{-m}$ and take $r = 2$. We want a quadratic polynomial $Q(z) = q_0 + q_1 z + q_2 z^2$ so that if $P(z) = (1 + z^{-1})^4 Q(z)$, then $P(z) - P(-z) = 2z^{-1}$. This last equation, when expanded, leads to*

$$(2q_1 + 8q_2)z + (8q_0 + 12q_1 + 8q_2)z^{-1} + (8q_0 + 2q_1)z^{-3} = 2z^{-1}.$$

Matching powers of z leads to three linear equations, $2q_1 + 8q_2 = 0, 8q_0 + 12q_1 + 8q_2 = 2, 8q_0 + 2q_1 = 0$ in three unknowns q_0, q_1, q_2. The solution is $q_0 = -\frac{1}{16}, q_1 = \frac{1}{4}, q_2 = -\frac{1}{16}$. We find that

$$P(z) = \frac{1}{16}(-z^2 + 9 + 16z^{-1} + 9z^{-2} - z^{-4}).$$

From (7.42), we see that $P(z)$ factors completely as

$$P(z) = -\frac{1}{16}(1 + z^{-1})^4(z^2 - 4z + 1) = -\frac{1}{16}(1 + z^{-1})^4(z - r_1)(z - r_2) \tag{7.43}$$

where $r_1 = 2 - \sqrt{3}, r_2 = 1/r_1 = 2 + \sqrt{3}$.

There are now many possible ways to allocate the various linear factors (and constant out front) to $L_a(z)$ and $L_s(z)$ so that $L_a(z)L_s(z) = P(z)$ in equation (7.43). For example, we might take

$$L_a(z) = -\frac{1}{8}(1+z^{-1})^2(z^2-4z+1) = -\frac{1}{8}z^2 + \frac{1}{4}z + \frac{3}{4} + \frac{1}{4}z^{-1} - \frac{1}{8}z^{-2},$$
$$L_s(z) = \frac{1}{2}(1+z^{-1})^2 = \frac{1}{2} + z^{-1} + \frac{1}{2}z^{-2}.$$

From equation (7.37), then

$$H_a(z) = L_s(-z) = \frac{1}{2} - z^{-1} + \frac{1}{2}z^{-2},$$
$$H_s(z) = -L_a(-z) = \frac{1}{8}z^2 + \frac{1}{4}z - \frac{3}{4} + \frac{1}{4}z^{-1} + \frac{1}{8}z^{-2}.$$

The corresponding filter coefficients are given in Table 7.3. The function $L_s(z)$ (and hence $H_a(z)$) does not correspond to a causal filter, but if we redefine $L_s(z)$ by multiplying z^{-2} (which also multiplies $H_a(z)$ by z^{-2}), we obtain causal filters, and also see that now $P(z) - P(-z) = z^{-3}$. With this alteration, the filters yield a perfect reconstruction filter bank with delay $m = 3$ in equation (7.32). It is also convenient to multiply both high-pass filters by -1. These filters are known as the "Le Gall 5/3" filters and are used in the JPEG 2000 compression standard.

Another way we might allocate the factors in equation (7.43) is as

$$L_a(z) = \frac{(\sqrt{3}+1)\sqrt{2}}{8}(1+z^{-1})^2(z-r_1),$$
$$L_s(z) = -\frac{(\sqrt{3}-1)\sqrt{2}}{8}(1+z^{-1})^2(z-r_2)$$

(the product of the constants in front is in fact $-1/16$). When expanded, these yield

$$L_a(z) = \frac{1}{4\sqrt{2}}((1+\sqrt{3})z + (3+\sqrt{3}) + (3-\sqrt{3})z^{-1} + (1-\sqrt{3})z^{-2}),$$
$$L_s(z) = \frac{1}{4\sqrt{2}}((1-\sqrt{3})z + (3-\sqrt{3}) + (3+\sqrt{3})z^{-1} + (1+\sqrt{3})z^{-2}).$$

Table 7.3 Filter coefficients for Le Gall 5/3 filters.

Index	−2	−1	0	1	2
$\boldsymbol{\ell}_a$	−1/8	1/4	3/4	1/4	−1/8
$\mathbf{h}_a$	0	0	1/2	−1	1/2
$\boldsymbol{\ell}_s$	0	0	1/2	1	1/2
$\mathbf{h}_s$	1/8	1/4	−3/4	1/4	1/8

Table 7.4 Filter coefficients for Daubechies 4-tap filters.

Index	0	1	2	3
$\boldsymbol{\ell}_a$	$\frac{1+\sqrt{3}}{4\sqrt{2}}$	$\frac{3+\sqrt{3}}{4\sqrt{2}}$	$\frac{3-\sqrt{3}}{4\sqrt{2}}$	$\frac{1-\sqrt{3}}{4\sqrt{2}}$
$\mathbf{h}_a$	$\frac{1-\sqrt{3}}{4\sqrt{2}}$	$\frac{-3+\sqrt{3}}{4\sqrt{2}}$	$\frac{3+\sqrt{3}}{4\sqrt{2}}$	$\frac{-1-\sqrt{3}}{4\sqrt{2}}$
$\boldsymbol{\ell}_s$	$\frac{1-\sqrt{3}}{4\sqrt{2}}$	$\frac{3-\sqrt{3}}{4\sqrt{2}}$	$\frac{3+\sqrt{3}}{4\sqrt{2}}$	$\frac{1+\sqrt{3}}{4\sqrt{2}}$
$\mathbf{h}_s$	$\frac{-1-\sqrt{3}}{4\sqrt{2}}$	$\frac{3+\sqrt{3}}{4\sqrt{2}}$	$\frac{-3+\sqrt{3}}{4\sqrt{2}}$	$\frac{1-\sqrt{3}}{4\sqrt{2}}$

We obtain $H_a(z)$ and $H_s(z)$ from equation (7.37). The corresponding filters are not causal, but as in the previous example if causality is desired, we can shift all of the filters. In this case, multiply each z-transform by z^{-1}, which is equivalent to shifting the filter vector one index to the right. Then $P(z)$ satisfies $P(z) - P(-z) = 2z^{-3}$, and the resulting filter bank has delay 3. The filters have the coefficients show in Table 7.4 and yield a perfect reconstruction filter bank with delay $m = 3$ in equation (7.32). These are known as the "Daubechies 4-tap" filters. The resulting analysis and synthesis filters are reversals of each other, if we shift the synthesis filters three indexes to the left (and note equation (7.10)), thus the Daubechies 4-tap filters yield an orthogonal filter bank.

Exercise 24 looks at another filter obtained by an alternate splitting of $P(z)$ in equation (7.43).

7.7.6 Orthogonal Filter Banks

7.7.6.1 Design Equations for an Orthogonal Bank

In an orthogonal filter bank, the synthesis filters $\boldsymbol{\ell}_s$ and $\mathbf{h}_s$ satisfy the relations (7.10). If the analysis low-pass filter $\boldsymbol{\ell}_a$ has nonzero components $(\ell_0, \dots, \ell_{N-1})$, then $\boldsymbol{\ell}_s$ has nonzero components $(\ell_{N-1}, \dots, \ell_0)$ (indexes $-N+1$ to 0) and so from Table 7.2 in Remark 7.6, we have $L_s(z) = L_a(z^{-1})$ and $H_s(z) = H_a(z^{-1})$. In this case, equations (7.32) and (7.33) become $L_a(z)L_a(z^{-1}) + H_a(z)H_a(z^{-1}) = 2z^{-m}$ and $L_a(-z)L_a(z^{-1}) + H_a(-z)H_a(z^{-1}) = 0$. However, the quantity $L_a(z)L_a(z^{-1}) + H_a(z)H_a(z^{-1})$ is invariant under the change of variable $z \to 1/z$, from which we conclude that $2z^{-m} = 2z^m$, so that $m = 0$ in the our only choice. In this case, we have

$$L_a(z)L_a(z^{-1}) + H_a(z)H_a(z^{-1}) = 2, \tag{7.44}$$

$$L_a(-z)L_a(z^{-1}) + H_a(-z)H_a(z^{-1}) = 0. \tag{7.45}$$

Note that we are no longer requiring condition (7.37). Equations (7.44) and (7.45) are the z-transform frequency domain conditions for a perfect reconstruction orthogonal filter bank.

When the filter length N of $\boldsymbol{\ell}_a$ is even (as it must be—see Exercise 16 in Chapter 9), there is an easy choice for the high-pass analysis filter that automatically ensures equation (7.45) is satisfied. Specifically, we choose the filter coefficients of $\mathbf{h}_a$ in terms of $\boldsymbol{\ell}_a$ as

$$(\mathbf{h}_a)_k = (-1)^k(\boldsymbol{\ell}_a)_{N-1-k}. \tag{7.46}$$

In short, $\mathbf{h}_a$ is a reversed version of $\boldsymbol{\ell}_a$ with alternating signs. Since $\boldsymbol{\ell}_a$ is assumed to have nonzero entries $(\boldsymbol{\ell}_a)_k$ only for $0 \leq k \leq N-1$, the same is true for $\mathbf{h}_a$. With this choice for $\mathbf{h}_a$, we have from the Table 7.2 that $H_a(z) = (-z)^{-(N-1)}L_a(-z^{-1}) = -z^{-(N-1)}L_a(-z^{-1})$, since N is even. In this case, equation (7.45) is satisfied identically since

$$L_a(-z)L_a(z^{-1}) - z^{-(N-1)}L_a(z^{-1})(z^{-1})^{-(N-1)}L_a(-z) = 0.$$

The choice (7.46) will work for any $\boldsymbol{\ell}_a$ with an even number of taps. In this case, the filters $\boldsymbol{\ell}_a$ and $\mathbf{h}_a$ are an example of *quadrature mirror filters* (see Exercise 25).

The design of the filter bank now comes down to just $L_a(z)$ and equation (7.44), which then reads

$$L_a(z)L_a(z^{-1}) + L_a(-z^{-1})L_a(-z) = 2 \tag{7.47}$$

(since $(-z)^{-(N-1)}(-z^{-1})^{-(N-1)} = 1$).

7.7.6.2 The Product Filter in the Orthogonal Case

Define $P(z)$ as

$$P(z) = L_a(z)L_a(z^{-1}) \tag{7.48}$$

so that (7.47) becomes

$$P(z) + P(-z) = 2. \tag{7.49}$$

The function $P(z)$ corresponds to the *product filter* obtained by convolving $\boldsymbol{\ell}_a$ with its own time reversal. Filters that come from $P(z)$ and satisfy condition (7.49) are called *halfband filters* (see [28]).

From equation (7.49), we can also conclude that all coefficients of the even powers of z in $P(z)$ must be zero, except for the z^0 (constant) term, which is 1. Equation (7.49) imposes no conditions on the coefficients of the odd powers of z. Now, from the beginning we have assumed that $\boldsymbol{\ell}_a$ is causal, so $L_a(z)$ will be of the form $L_a(z) = \ell_0 + \ell_1 z^{-1} + \cdots + \ell_{N-1}z^{-(N-1)}$. Our goal is thus to produce such an $L_a(z)$ so that $P(z) = L_a(z)L_a(z^{-1})$ satisfies equation (7.49).

We also desire that $L_a(-1) = 0$, so that ℓ_a is a low-pass filter, and this forces $P(-1) = 0$. From equation (7.49), this implies that $P(1) = 2$, and so $L_a(1) = \sqrt{2}$. Equivalently, we require

$$\sum_{k=0}^{N-1} \ell_k = \sqrt{2}. \tag{7.50}$$

The converse is also true; that is, if equation (7.50) holds, then $L(1) = \sqrt{2}$, which forces $P(-1) = 0$ and $L_a(-1) = 0$.

We could seek to impose other constraints, for example, on the length or frequency response of ℓ_a.

7.7.6.3 Restrictions on $P(z)$; Spectral Factorization

One might attempt to simply start with a function

$$P(z) = p_{N-1}z^{-(N-1)} + \cdots + p_1 z^{-1} + 1 + p_1 z + \cdots p_{N-1} z^{N-1} \tag{7.51}$$

of the required form, involving only odd powers of z (e.g., with $P(-1) = 0$) and then obtain ℓ_a by factoring $P(z) = L_a(z)L_a(z^{-1})$ where L_a has real coefficients. However, this is not generally possible; as it turns out, the function P must satisfy a certain condition. Specifically, let $z = e^{i\omega}$ for $\omega \in [-\pi, \pi]$. Observe that $\overline{z} = z^{-1}$ in this case so that

$$L_a(z^{-1}) = L_a(\overline{z}) = \overline{L_a(z)}$$

(the last equality follows since L_a has real coefficients). We thus find that $P(z) = L_a(z)L_a(z^{-1}) = L_a(z)\overline{L_a(z)} = |L_a(z)|^2 \geq 0$ if $z = e^{i\omega}$. In short, $P(e^{i\omega})$ must be real and nonnegative if a factorization $P(z) = L_a(z)L_a(z^{-1})$ exists.

The converse is also true. A function $P(z)$ of the form (7.51) (which may also contain even powers of z) that satisfies $P(e^{i\omega}) \geq 0$ for $\omega \in [-\pi, \pi]$ can be factored as $P(z) = L_a(z)L_a(z^{-1})$ where L_a has real coefficients. We sketch a proof in Exercise 28, and show how to actually perform the factorization constructively for modest values of N (see also [10], p. 172, or [28], sec. 5.4). The task of determining $L_a(z)$ from $P(z)$ is often referred to as *spectral factorization*. The factorization is not unique.

Our overall goal then is to design a function $P(z)$ of the form (7.51) with $P(e^{i\omega}) \geq 0$ and $P(-1) = 0$, and perhaps other desirable properties, and then to obtain L_a by performing a spectral factorization.

7.7.6.4 Daubechies Filters

Rather than look at general techniques for designing orthogonal filters, we will content ourselves with an examination of an important subset of orthogonal filters, the *Daubechies* filters. For these filters we have, for any even $N \geq 2$,

$$P(z) = Q\left(\frac{z + z^{-1}}{2}\right), \tag{7.52}$$

where Q is the polynomial defined by

$$Q(x) = c \int_{-1}^{x} (1 - y^2)^{N/2-1}\, dy \tag{7.53}$$

and c is chosen according to $1/c = \int_{-1}^{0} (1 - y^2)^{N/2-1}\, dy$ (this choice for c yields $Q(0) = 1$). For the derivation of this formula, see Exercise 29. In particular, $P(-1) = Q(-1) = 0$; indeed $Q(x)$ is designed to vanish to the highest order possible at $x = -1$. Such filters are called "maxflat" filters. The function P inherits this property. It is also not hard to check that Q and P possess only odd powers of the input argument (except for a constant term), and that $P(e^{i\omega}) \geq 0$.

Example 7.15 *When $N = 2$, we obtain $Q(x) = x + 1$ (here $c = 1$) from equations (7.52) and (7.53) and*

$$P(z) = \frac{1}{2}z^{-1} + 1 + \frac{1}{2}z.$$

It is simple to see $P(z) + P(-z) = 2$; also $P(-1) = 0$. We can factor $P(z) = L_a(z)L_a(z^{-1})$, where

$$L_a(z) = \frac{1}{\sqrt{2}} + \frac{1}{\sqrt{2}}z^{-1},$$

corresponding to the Haar low-pass analysis filter $\boldsymbol{\ell}_a = (1/\sqrt{2}, 1/\sqrt{2})$. We obtain $\mathbf{h}_a = (1/\sqrt{2}, -1/\sqrt{2})$ from equation (7.46), and the synthesis filters from equation (7.10).

Example 7.16 *When $N = 4$, we obtain from equations (7.52) and (7.53) that $Q(x) = 1 + \frac{3}{2}x - \frac{1}{2}x^3$ $\left(\text{here } c = \frac{3}{2}\right)$ and*

$$P(z) = -\frac{1}{16}z^{-3} + \frac{9}{16}z^{-1} + 1 + \frac{9}{16}z - \frac{1}{16}z^3,$$

previously found in Example 7.14 (but multiplied by z^{-1} there). We can then compute $P(z) = L_a(z)L_a(z^{-1})$, where

$$\begin{aligned} L_a(z) &= \frac{1+\sqrt{3}}{4\sqrt{2}}(1 + z^{-1})^2(1 - (2 - \sqrt{3})z^{-1}) \\ &\approx 0.483z^{-3} + 0.837z^{-2} + 0.224z^{-1} - 0.129. \end{aligned}$$

The other analysis and synthesis filter coefficients are obtained as in the previous example.

Equations (7.52) and (7.53) yield an infinite family of orthogonal filters, one for each even $N \geq 2$. There are many other families of orthogonal filters as well, many of which are built into Matlab's Wavelet Toolbox.

7.8 Matlab Project

This section contains Matlab explorations involving the use of filter banks/DWTs for analyzing signals and images.

7.8.1 Basics

1. Start Matlab and create an artificial piecewise constant signal with

```
y=zeros(256,1);
y(100:160)=1.0; y(161:256)=4.0;
```

   As mentioned in the text, Matlab can perform DWTs under a variety of extension modes. For the moment we will use the default extension mode, symmetric periodic extension according to equation (7.11).
   We can perform a one-stage wavelet decomposition of the signal with

```
[cA,cD] = dwt(y,'bior2.2');
```

   Here "cA" is a vector of the approximation coefficients (what we called $\mathbf{X}_\ell$ in the text) and "cD" the detail coefficients (what we called $\mathbf{X}_h$). The vectors will be slightly longer than half the length of the input signal, since Matlab keeps some extra coefficients, but it does not matter. The "bior2.2" argument specifies the wavelets/filters to be used, which in this case correspond to the Le Gall 5/3 wavelet from the text. You can execute `waveinfo('bior')` to see other biorthogonal wavelets that are available, or `waveinfo()` to see information on all available wavelets/filters.
   Plot the vectors cA and cD. Note that cA is the low pass filtered portion of the transform. Since the signal is mostly constant, cA looks like an approximately half-length version of the original signal with some distortion at the jumps. The vector cD should be mostly zeros except at the discontinuity, which contains high-frequency energy that makes it through the high-pass filter. It is also worth noting that since the signal is implicitly extended at the boundary by symmetric reflection, there are no discontinuities there.
2. We can reconstruct an approximation to the original full length signal using only the coefficients cA by executing

```
y2 = upcoef('a', cA, 'bior2.2', 1);
```

   The first "a" indicates that we are reconstructing from approximation coefficients only and the final "1" indicates cA comes from a one-stage transform. In this case, y2 is computed by applying the synthesis filter bank to cA alone, with cD implicitly set to zero. Of course, you should plot y2.

3. Change the extension mode to periodic extension ($\tilde{x}_k = x_{k \bmod N}$) with the command

```
dwtmode('per');
```

Repeat steps 1 and 2 above. Periodic extension introduces discontinuities at the signal boundaries, and this should be obvious in the plots of `cA`, `cD`, and the reconstruction `y2` because the values from each end of the signal now "pollute" the values from the other end.
4. Repeat the steps 1 through 3 for a few other wavelets.

7.8.2 Audio Signals

1. First, set the extension mode back to symmetric extension with the command

```
dwtmode('sym');
```

Load in the "splat" signal with the command `load('splat');` Recall that the audio signal is loaded into a variable "`y`" and the sampling rate into "`Fs`." You can play the sound with `sound(y)`.
2. Use the `dwt` command and the "`bior2.2`" wavelets as above to perform a one-stage transform of the audio signal, and then reconstruct from the vector `cA` alone using the `upcoef` command. Play the sound through the speaker.
3. Let us try some thresholding for signal compression. The `wavedec` and `waverec` commands perform multistage wavelet analysis and reconstructions, but the format of the output of `wavedec` is not so easy to interpret if we want to threshold. Instead we can use the `wdencmp` command. Execute

```
[ycomp,cxc,lxc,perfo,perfl2]
  = wdencmp('gbl',y,'bior2.2',5,0.05,'h',1);
```

The input arguments are "`gbl`," which is global thresholding ("all" portions of the DWT coefficients will be subject to thresholding), "`y`" is the input signal, "`bior2.2`" dictates the wavelet/filters, "`5`" is the number of stages in the transform, and "`0.05`" is the threshold parameter. The "`h`" is for "hard" thresholding (any value not exceeding the threshold in magnitude is zeroed out). The final "`1`" indicates that the lowest level approximation coefficients are not subject to thresholding (change it to "0" to subject all coefficients to the threshold). The outputs are "`ycomp`," which is the signal reconstructed from the thresholded DWT, "`cxc`" and "`lxc`" are certain bookkeeping details we do not need to worry about, "`perfo`" specifies what percentage of the wavelet coefficients from the analysis bank output were zeroed out (a measure of how much the signal was compressed), and "perfl2" is the L^2 distortion of the thresholded image, as a percentage. Play around with the thresholding parameter; try the range 0.0–1.0. Make a table that shows in each case the compression and distortion parameters, and your subjective rating of the sound quality.

4. Experiment with other wavelets!

7.8.3 Images

1. Load an image into Matlab with

```
z=imread('myimage.jpg');
```

Construct an artificial grayscale image zg as in previous chapters, along with the colormap

```
L = 255;
colormap([(0:L)/L; (0:L)/L; (0:L)/L]');
```

View the image with image(zg).

2. Execute the command

```
[cA,cH,cV,cD] = dwt2(zg, 'bior2.2');
```

that performs a two-dimensional DWT. The arrays cA,cH,cV,cD now hold the low- and high-frequency components of the transform; cA contains the low/low elements in the horizontal/vertical directions, the approximation coefficients. The array cH is the horizontal detail, corresponding to high/low in the horizontal/vertical directions, while cV is vertical detail (low/high) and cD is the "diagonal" detail, corresponding to high/high. Each of these arrays is $M/2$ by $N/2$, if the original image is M by N, though they might be a bit larger since Matlab extends the image by a certain amount.

3. You can display cA or the other pieces directly, although they are only half-sized. For a fairer picture, execute

```
zA = upcoef2('a', cA, 'bior2.2', 1);
```

to upsample cA to a full-sized array zA (reconstructing with cH,cV,cD implicitly set to zero). Display zA (you might need to scale it a bit if you want to display on a 0–255 grayscale, by adding a suitable constant and then multiplying by a scalar). Compare to the original image.

Do the same with the other three pieces of the transform, with the commands

```
zH = upcoef2('h', cH, 'bior2.2', 1);
zV = upcoef2('v', cV, 'bior2.2', 1);
zD = upcoef2('d', cD, 'bior2.2', 1);
```

In each case, display the output. Again, rescaling may be helpful.

4. As with audio signals, we can perform multistage decompositions and thresholding. In particular, execute

```
[yz,cxc,lxc,perfo,perfl2]
  = wdencmp('gbl',zg,'bior2.2',5,10.0,'h',1);
```

(input and output arguments as described in the audio signal section; the command is still called wdencmp, even if the argument is 2D). We have begun with the threshold parameter at 10.0. Record the L^2 distortion parameter `perfl2`, and the percentage of zeroed coefficients `perfo`.
Try threshold parameters in the range 0–500. Make a table of your results, showing in each case the quantities `perfo`, `perfl2`, and your rating of the image quality.
Again, try different wavelets!

7.9 Alternate Matlab Project

This section contains Matlab explorations in which we use filter banks/DWTs for analyzing signals and images, but with various supplied commands instead of Matlab's Wavelet Toolbox. This project closely parallels that of the previous Matlab project.

7.9.1 Basics

1. Start Matlab, and create an artificial piecewise constant signal with

```
y=zeros(256,1);
y(100:160)=1.0; y(161:256)=4.0;
```

The supplied routines for DWTs assume periodic extension of all signals, which is sufficient for our purposes. We can perform a one-stage wavelet decomposition of the signal with

```
Y = wavel(y, 'd4');
```

Here `Y` consists of the approximation coefficients, stored in `Y(1:N/2)` (where `N` is the length of the signal, in this case $N = 256$). The detail coefficients are stored in `Y(N/2+1:N)`. The "`d4`" argument specifies the Daubechies 4-tap orthogonal filters; other options are "`haar`," "`orthhaar`," and "`legall`." You can easily add your own.
Plot the vectors `Y(1:128)` and `Y(129:256)`. Note that `Y(1:128)` is the low-pass filtered portion of the transform, and since the signal is mostly constant, `Y(1:128)` looks like a half-length version of the original signal, with some distortion at the jumps. The vector `Y(129:256)` should be mostly zero except at the discontinuity, which contains high-frequency energy that

makes it through the high-pass filter. It is also worth noting that since the signal is implicitly extended periodically, there is some "crosstalk" between the ends of the signal.

2. We can reconstruct an approximation to the original full length using only the approximation coefficients `Y(1:128)` by executing

```
Y(129:256)=0.0;
y2 = invwavel(Y,'d4');
```

   Of course, you should plot `y2`.
3. Repeat the steps 1 and 2 for the other available wavelets.

7.9.2 Audio Signals

1. Load in the "splat" signal with the command `load('splat');`. Recall that the audio signal is loaded into a variable "y" and the sampling rate into "`Fs`." You can play the sound with `sound(y)`.
2. Use the `wavel` command (you need to truncate the signal to even length; e.g., use $N = 10\,000$ and vector `y(1:10000)`) and the "`d4`" wavelets as above to perform a one-stage transform of the audio signal, and then reconstruct from the vector `Y(1:N/2)` alone. Play the sound through the speaker.
3. Let us try some thresholding for signal compression. Let `y1=y(1:10000);` and perform a four-stage wavelet decomposition with

```
Y = fullwave(y, 'd4', 4);
```

   Then threshold `Y` with threshold parameter 0.01 via

```
Y2 = Y.*(abs(Y)>0.01);
```

   Compute the percentage of wavelet coefficients zeroed out by using the command `sum(Y2>0)/10000`. Inverse transform as

```
y2 = invfullwave(Y2, 'd4', 4);
```

   and play the sound `y2`.
   Play around with the thresholding parameter; try the range 0.0–1.0. Make a table that shows in each case the compression and distortion parameters, and your rating of the sound quality.
4. Experiment with other wavelets!

7.9.3 Images

1. Load an image into Matlab with

```
z=imread('myimage.jpg');
```

If necessary, crop the image so that the number of rows and columns is divisible by 8. Construct an artificial grayscale image `zg`, and set the colormap to

```
L = 255;
colormap([(0:L)/L; (0:L)/L; (0:L)/L]');
```

View the image with `image(zg)`.

2. Execute the command

```
Z = imwavel(zg,'d4');
```

which performs a two-dimensional single-stage DWT. You can display `Z` directly, or logarithmically scaled, for example, `image(log(1+abs(Z))*30)`.

3. You can display the approximation coefficients or the other pieces directly, although they are only half-sized. For a fairer picture, execute

```
[m,n] = size(Z)
Z(m/2+1:m,:)=0.0; Z(:,n/2+1:n)=0.0;
z2 = invimwavel(Z, 'd4');
image(z2);
```

to zero out all but the lowest level approximation coefficients (c_A) and then inverse transform. This yields the level-1 approximation α_1. Compare it to the original image.

Recompute `Z`, and do the same with the other three pieces of the transform, that is, the vertical, horizontal, and diagonal details. Rescaling prior to displaying may be helpful.

4. As with audio signals, we can perform multistage decompositions and thresholding. In particular, execute

```
Z = imwavefull(zg, 3, 'd4');
image(log(1+abs(Z))*30);
```

Try zeroing out all but the lowest $\frac{1}{4}$ of the coefficients in each direction, with `Z(m/4+1:m,:)=0;` and `Z(:,n/4+1,n)=0.0;` then inverse transform with

```
z2 = invimwavefull(Z, 3, 'd4');
```

This will produce the level-2 approximation α_2. Executing the command `Z(m/8+1:m,:)=0;` and `Z(:,n/8+1,n)=0.0;` and inverse transforming yields α_3, and so on.

5. Try recomputing the three-stage transform Z, thresholding, and inverse transforming, for example,

```
Z = imwavefull(zg, 3, 'd4');
Z2 = Z.*(abs(Z)>10);
z2 = invimwavefull(Z2,3,'d4');
```

then display z2. Vary the threshold parameter. In each case, record the percentage of wavelet coefficients zeroed out and the distortion.
Try different wavelets!

Exercises

1. Let $\mathbf{x} \in L^2(\mathbb{Z})$ be a signal with components $x_0 = 1, x_1 = -2, x_2 = 2, x_3 = 4$, and all other components zero. Run $\mathbf{x}$ though the Haar filter bank with filter coefficients $(\boldsymbol{\ell}_a)_0 = (\boldsymbol{\ell}_a)_1 = \frac{1}{2}, (\mathbf{h}_a)_0 = \frac{1}{2}, (\mathbf{h}_a)_1 = -\frac{1}{2}$ to compute $\mathbf{X}_\ell$ and $\mathbf{X}_h$ explicitly. Then use the synthesis filters with coefficients $(\boldsymbol{\ell}_s)_{-1} = (\boldsymbol{\ell}_s)_0 = 1, (\mathbf{h}_s)_{-1} = -1, (\mathbf{h}_s)_0 = 1$ to reconstruct $\mathbf{x}$.

2. Consider a filter bank that uses the Haar analysis filters with coefficients $(\boldsymbol{\ell}_a)_0 = (\boldsymbol{\ell}_a)_1 = \frac{1}{2}, (\mathbf{h}_a)_0 = \frac{1}{2}, (\mathbf{h}_a)_1 = -\frac{1}{2}$, and corresponding causal synthesis filters $(\boldsymbol{\ell}_s)_0 = (\boldsymbol{\ell}_s)_1 = 1, (\mathbf{h}_s)_0 = -1, (\mathbf{h}_s)_1 = 1$ (yes, the $(\mathbf{h}_s)_0$ is -1; the synthesis filters here are shifted from the last problem). Show that this filter bank yields perfect reconstruction with output delay 1 by considering an input vector $\mathbf{x} = (\ldots, x_{-1}, x_0, x_1, \ldots)$ and computing the filter bank output $\tilde{\mathbf{x}}$ directly (filter, downsample, upsample, synthesis filter). In particular, show that $\tilde{x}_k = x_{k-1}$. Hint: Most of the work has already been done in equation (7.5).

3. Let $\mathbf{x}$ be a signal with components

$$x_k = \begin{cases} 5 + 3(-1)^k, & k \geq 0 \\ 0, & \text{otherwise.} \end{cases}$$

The "5" portion is dc and $3(-1)^k$ is at the Nyquist frequency. Use the Haar analysis filters as defined by equation (7.1) to compute the components of the Haar transform vectors $\mathbf{X}_\ell = D(\mathbf{x} * \boldsymbol{\ell})$ and $\mathbf{X}_h = D(\mathbf{x} * \mathbf{h})$ explicitly. In particular, look at the components $(\mathbf{X}_\ell)_k$ and $(\mathbf{X}_h)_k$ and comment on why this makes perfect sense. Note that even though $\mathbf{x}$ is not in $L^2(\mathbb{Z})$, the convolution with the FIR Haar filters is still well-defined.

4. Let $\mathbf{x} \in L^2(\mathbb{Z})$ be a signal with components

$$x_k = \begin{cases} 0, & k < 0, \\ 1, & 0 \le k < 50, \\ 3, & 50 \le k < 100, \\ 0, & 100 \le k. \end{cases}$$

Use the Haar analysis filters as defined by equation (7.1) to compute the components of the Haar transform vectors $\mathbf{X}_\ell = D(\mathbf{x} * \boldsymbol{\ell})$ and $\mathbf{X}_h = D(\mathbf{x} * \mathbf{h})$ explicitly. Comment on what you see!

5. Write simple Matlab routines that implement the Haar analysis and synthesis filter banks (say with analysis filters $\boldsymbol{\ell}_a = \left(\frac{1}{2}, \frac{1}{2}\right), \mathbf{h}_a = \left(\frac{1}{2}, -\frac{1}{2}\right)$, synthesis filters $\boldsymbol{\ell}_s = (1, 1), \mathbf{h}_s = (-1, 1)$) for signals $\mathbf{x} \in L^2(\mathbb{Z})$ with $x_k = 0$ for $k < 0$, $\mathbf{x}$ of finite length. Use Matlab's `conv` command for the convolutions.

6. Consider a filter bank for signals in $L^2(\mathbb{Z})$ with analysis low-pass filter coefficients $(\boldsymbol{\ell}_a)_0 = 1$ and all other coefficients zero, while $(\mathbf{h}_a)_1 = 1$ and all other high-pass coefficients zero. These are not low-pass or high-pass—indeed $\boldsymbol{\ell}_a$ is the "identity filter" and $\mathbf{h}_a$ is simply a delay—but that does not matter.

(a) Explicitly compute the components of $\mathbf{X}_\ell = D(\mathbf{x} * \boldsymbol{\ell}_a)$ and $\mathbf{X}_h = D(\mathbf{x} * \mathbf{h}_a)$ in terms of the components x_j.

(b) Explicitly compute the components of the upsampled vectors $U(\mathbf{X}_\ell)$ and $U(\mathbf{X}_h)$.

(c) Determine, by inspection, suitable synthesis filters $\boldsymbol{\ell}_s$ and $\mathbf{h}_s$ so that

$$\boldsymbol{\ell}_s * (U(\mathbf{X}_\ell)) + \mathbf{h}_s * (U(\mathbf{X}_h)) = \mathbf{x}.$$

(d) Suppose that we take the analysis filter $\mathbf{h}_a$ to have $(\mathbf{h}_a)_2 = 1$ and all other coefficients zero. Repeat parts (a) and (b), and then show in this case that no synthesis filters can give perfect reconstruction.

7. Let $\mathbf{x} \in L^2(\mathbb{Z})$. Let $\mathbf{g} \in L^2(\mathbb{Z})$ be an FIR filter, and S the shift operator defined in Remark 7.1. Show that

$$(S^m(\mathbf{g})) * \mathbf{x} = S^m(\mathbf{g} * \mathbf{x}).$$

8. Suppose that we choose analysis filters $\boldsymbol{\ell}_a$ and $\mathbf{h}_a$ with coefficients $(\boldsymbol{\ell}_a)_0 = \frac{3}{4}, (\boldsymbol{\ell}_a)_1 = \frac{1}{2}, (\mathbf{h}_a)_0 = \frac{2}{3}, (\mathbf{h}_a)_1 = -\frac{2}{3}$, all other coefficients zero ($\boldsymbol{\ell}_a$ is not a low-pass filter, but that does not matter). Find appropriate synthesis filters for perfect reconstruction with zero delay. Hint: Look at Example 7.3. In addition, 2-tap filters will work.

9. Consider a "one-channel" filter bank: A signal $\mathbf{x} \in L^2(\mathbb{Z})$ is convolved with some FIR analysis filter $\mathbf{f}$. Since the filter bank is one-channel, there is no downsampling—we keep every component of the filtered signal (see Remark 7.2 so that $\mathbf{X} = \mathbf{x} * \mathbf{f}$. To reconstruct we then apply a synthesis FIR filter $\mathbf{g}$. We seek perfect reconstruction (possibly with some output delay); that is, we want $\mathbf{g} * (\mathbf{f} * \mathbf{x}) = S^m(\mathbf{x})$ for all inputs $\mathbf{x}$, where S is the shift operator.

(a) Show that the only such one-channel filter banks have filters of the form $f_n = a$ for some fixed n and $f_k = 0$ for all $k \neq n$, where a is some nonzero constant. Similarly $\mathbf{g}$ has components the form $g_r = 1/a$ for some fixed r and $g_k = 0$ for all $k \neq r$. Thus the only appropriate filters are simple delays or advances multiplied by a scalar. Hint: Use the Convolution Theorem 4.6 to show that $F(z)G(z) = z^{-m}$. Why is this impossible for FIR filters of length greater that 1?

(b) Show that if we work with signals in $\mathbb{C}^N$ (instead of $L^2(\mathbb{Z})$) and interpret the shift operator S circularly, then such an appropriate synthesis filter will exist for any analysis filter $\mathbf{f}$ provided that the analysis filter's discrete Fourier transform $\mathbf{F} = \mathrm{DFT}(\mathbf{f})$ has all nonzero components. Hint: Look back at Exercise 16.

10. Write out the matrix $\mathbf{W}_6^a$ in equation (7.19) for the Le Gall 5/3 filters from Example 7.4, with coefficients as tabled below in the indicated positions. (It is easy if you notice that the first three rows of $\mathbf{W}^6$ are just rows 0, 2, and 4 of the circulant matrix for the analysis low-pass filter; a similar observation holds for the other rows.) In addition, write out $\mathbf{W}_6^s$ (equation (7.21)). Compute a few entries of the product $\mathbf{W}_6^a\mathbf{W}_6^s$ to check $\mathbf{W}_6^a\mathbf{W}_6^s = \mathbf{I}_6$.

Index	0	1	2	3	4	5
$\boldsymbol{\ell}_a$	3/4	1/4	−1/8	0	−1/8	1/4
$\mathbf{h}_a$	−1/2	1	−1/2	0	0	0
$\boldsymbol{\ell}_s$	1	1/2	0	0	0	1/2
$\mathbf{h}_s$	−1/4	−1/8	0	−1/8	−1/4	3/4

11. Compute and plot the magnitude of the DTFT on the frequency range $-\frac{1}{2} \leq f \leq \frac{1}{2}$ for each of the Le Gall filters tabled in Example 7.14 (use equation (4.16)).

12. Compute and plot the DTFT for each of the Daubechies 4-tap filters tabled in Section 7.7.5 (use equation (4.16)).

13. (a) Show that the downsampling matrix $\mathbf{D}$ defined by equation (7.18) does in fact downsample vectors $\mathbf{x} \in \mathbb{C}^N$, for example, $(\mathbf{Dx})_k = x_{2k}$.

(b) Show that the upsampling matrix **U** defined by equation (7.20) does in fact upsample vectors $\mathbf{x} \in \mathbb{C}^N$, for example, $(\mathbf{Dx})_k = x_{k/2}$ if k is even and $(\mathbf{Dx})_k = 0$ if k is odd.

(c) Show that $\mathbf{U}^T = \mathbf{D}$. In addition, work out the products **DU** and **UD** explicitly.

14. (a) The matrix $\mathbf{W}_4^a$ for the Haar analysis transform on four-point signals (periodic extension) is given in Example 7.7, more specifically by equation (7.14). Write out the corresponding matrix $\mathbf{W}_2^a$, and then use equation (7.23) to form the matrix $\mathcal{W}_2^a$ that governs the two-stage Haar transform for four-point signals with periodic extension.

(b) Use an analogous procedure and equations (7.17) and (7.24) to find $\mathcal{W}_2^s$, and then verify directly that $\mathcal{W}_2^a \mathcal{W}_2^s = \mathbf{I}_4$.

(c) Let $\mathbf{x} = (3, 4, 3, 2)$. Use the results of parts (a) and (b) above and equations (7.26) and (7.27) to compute the quantities $\alpha_2(\mathbf{x})$, $\delta_1(\mathbf{x})$, $\delta_2(\mathbf{x})$, and also $\alpha_1(\mathbf{x})$. Verify directly that equation (7.27) is satisfied. Comment on why $\alpha_2(\mathbf{x})$ has the appearance it does! (Suggestion: First compute **X** using $\mathcal{W}_2^a$, and note that $\mathbf{X}_{\ell\ell}$ is just X_0 here, while $\mathbf{X}_{\ell h}$ is X_1, etc.)

15. Redo Exercise 2 but use equations (7.32) and (7.33), rather than a "brute force" approach.

16. Solve equations (7.32) and (7.33) for $L_s(z)$ and $H_s(z)$ to obtain equations (7.34) and (7.35).

17. Let $\boldsymbol{\ell}_a = (1, 1), \mathbf{h}_a = (1, 1, -2), \boldsymbol{\ell}_s = \left(-\frac{1}{2}, \frac{1}{2}, 1\right), \mathbf{h}_s = \left(\frac{1}{2}, -\frac{1}{2}\right)$. Use equations (7.32) and (7.33) to show that these filters yield a perfect reconstruction filter bank. What is the output delay?

18. Show that in any biorthogonal filter bank the analysis and synthesis filters can be swapped (low-pass with low-pass, high-pass with high-pass) and the resulting filter bank is still biorthogonal. Hint: Look at the conditions for perfect reconstruction.

19. Prove the four relations in Table 7.2 in Remark 7.6.

20. Suppose that **x** and **y** are elements of $L^2(\mathbb{Z})$ with x_k nonzero only for $M_1 \le k \le N_1$ and y_k nonzero only for $M_2 \le k \le N_2$. Show that $(\mathbf{x} * \mathbf{y})_k$ (the kth component of $\mathbf{x} * \mathbf{y}$) is zero outside the finite range $M_1 + M_2 \le k \le N_1 + N_2$. Hint: Do it via z-transforms.

21. Prove Proposition 7.1.

22. Let the analysis filter $\boldsymbol{\ell}_a$ in a filter bank have coefficients $(\boldsymbol{\ell}_a)_0 = \frac{1}{2}, (\boldsymbol{\ell}_a)_1 = 1, (\boldsymbol{\ell}_a)_2 = -\frac{1}{2}$, all other coefficients zero (not really low-pass, but that does not matter). Let the high-pass filter have coefficients $(\mathbf{h}_a)_0 = \frac{1}{4}, (\mathbf{h}_a)_1 = \frac{1}{2}, (\mathbf{h}_a)_2 = \frac{1}{4}$, all other coefficients zero. Use equations (7.32) and (7.33) to find appropriate causal synthesis filters for perfect reconstruction. What is the delay in the filter bank output?

23. Let $G(z)$ be the z-transform of an FIR filter $\mathbf{g}$. Show that $z^m G(z)$ is a polynomial in z for sufficiently large m.

24. Consider splitting $P(z)$ in equation (7.43) as

$$P(z) = L_a(z)L_s(z),$$

where

$$L_a(z) = \frac{1}{4}(1 + z^{-1})^3, \quad L_s(z) = -\frac{1}{4}(1 + z^{-1})(z^2 - 4z + 1).$$

Work out the coefficients of the analysis and synthesis filters, both low- and high-pass, the latter by making use of equation (7.37). Verify that the low-pass filters really are low-pass and the high-pass filters really are high-pass (recall Remark 7.7).

25. Filters $\boldsymbol{\ell}_a$ and $\mathbf{h}_a$ (not necessarily low- and high-pass) are said to be *quadrature mirror filters* if their z-transforms satisfy

$$|L_a(e^{i\theta})| = |H_a(e^{i(\pi-\theta)})|$$

for $0 \le \theta \le \pi$, so that the frequency response of each filter is a "mirror" of the other about $\theta = \pi/2$. Show that filters related by equation (7.46) are quadrature mirror filters.

26. Suppose that $Q_1(z)$ and $Q_2(z)$ are two Laurent polynomials, say

$$Q_1(z) = \sum_{k=-M}^{M} a_k z^{-k},$$

$$Q_2(z) = \sum_{k=-M}^{M} b_k z^{-k}$$

(not all a_k and b_k in this range need be nonzero). Suppose that Q_1 and Q_2 satisfy

$$X(z)Q_1(z) + X(-z)Q_2(z) = X(z)$$

for every z-transform $X(z) = z^q$, $q \in \mathbb{Z}$. All products above are formally well-defined.

(a) Show that $Q_1(z) = 1$ and $Q_2(z) = 0$. Hint: You really only need to consider the cases $q = 0$ and $q = 1$.

(b) Use part (a) to show that equations (7.32) and (7.33) are necessary if equation (7.31) is to hold.

27. Most filters that are used in biorthogonal filter banks are symmetric or antisymmetric. An FIR filter **g** with coefficients $g_0, \dots, g_{M-1}$ is said to be *symmetric* if $g_k = g_{M-1-k}$; **g** is *antisymmetric* if $g_k = -g_{M-1-k}$.

(a) Show that a filter **g** with M taps is symmetric if and only if it is z-transform satisfies $G(z) = z^{1-M}G(z^{-1})$. Show that **g** is antisymmetric if and only if $G(z) = -z^{1-M}G(z^{-1})$. Hint: Use Table 7.2, though it is easy to write out directly too.

(b) Show that the low-pass filters in a perfect reconstruction filter bank cannot be antisymmetric. Hint: If **g** is one of the filters, compute $G(1)$.

(c) Show that if both low-pass filters $\boldsymbol{\ell}_a$ and $\boldsymbol{\ell}_s$ in a perfect reconstruction filter bank of the form dictated by (7.37) are symmetric, with $\boldsymbol{\ell}_a$ having M taps and $\boldsymbol{\ell}_s$ having N taps, then M and N are both even or both odd. Hint: Suppose, in contradiction, that $M + N$ is odd. Use equation (7.39) to show that

$$P(z) = z^{2-M-N}P(z^{-1})$$

and

$$P(-z) = -z^{2-M-N}P(-z^{-1}).$$

Subtract these two equations, and use equation (7.39) to conclude that $P(z)$ has only a single even power of z, namely a constant term c_0z^0. Conclude that if P is low-pass, then $P(z) = c(1 + z^{-m})$ for some odd integer m. Explicitly write out the solutions to $P(z) = 0$. Why does this contradict $L_a(-1) = 0$ and $L_s(-1) = 0$?

(d) Show that under the conditions of (c), if M and N are both even, then $\mathbf{h}_a$ and $\mathbf{h}_s$ are both antisymmetric. Show that if M and N are both odd, then $\mathbf{h}_a$ and $\mathbf{h}_s$ are both symmetric.

28. Let P be of the form in equation (7.51), where all p_k are real numbers and $p_{N-1} \neq 0$. The goal of this exercise is to show that if $P(e^{i\omega})$ is real and nonnegative for all $\omega \in [-\pi, \pi]$, then P can be factored as $P(z) = L(z)L(z^{-1})$ for a suitable function L of the form

$$L(z) = \ell_0 + \ell_1 z + \cdots + \ell_{N-1}z^{N-1}$$

in which the ℓ_k are real and $\ell_{N-1} \neq 0$. In fact, the approach yields a computational method that works for modest values of N. However, L is not typically unique. A minor discrepancy should also be mentioned: here we take $L(z)$ as a polynomial in nonnegative powers of z, whereas the discussion of Section 7.7.6 involves nonpositive powers of z in $L_a(z)$; that does not matter, for we can take $L_a(z) = L(z^{-1})$.

In what follows, it will be helpful to recall a couple elementary facts from algebra. First, if a polynomial $T(z)$ of degree n has roots $z_1, \ldots, z_m$, then $T(z)$ factors as

$$T(z) = c(z - z_1)^{n_1}(z - z_2)^{n_2} \cdots (z - z_m)^{n_m} \tag{7.54}$$

for some constant c, where $n_1 + \cdots + n_m = n$. The exponent n_k is called the *multiplicity* of the root z_k. Moreover, if the coefficients of T are all real and z_k is a root of T of multiplicity n_k, then $\overline{z_k}$ is also a root of multiplicity n_k. Indeed, $T(z)$ has real coefficients if and only if the nonreal roots of T come in conjugate pairs.

Finally, in working this problem, it might be helpful to keep in mind the specific $P(z)$ in part (f), as an example.

(a) Let $Q(z) = z^{N-1}P(z)$. Why must Q be a polynomial of degree $2N - 2$ rather than a lower degree? Show that $Q(z) = 0$ and $P(z) = 0$ have exactly the same solutions z_k, $1 \leq k \leq 2N - 2$, in the complex plane; the z_k need not be distinct. Conclude that $P(z)$ can be expressed as

$$P(z) = p_{N-1}\frac{(z - z_1)(z - z_2)\cdots(z - z_{2N-2})}{z^{N-1}}. \tag{7.55}$$

(b) Show that if z_k is a solution to $P(z) = 0$, then so are $1/z_k, \overline{z_k}$, and $1/\overline{z_k}$ (but not necessarily distinct).

(c) Let z be a nonzero complex number with $|z| \neq 1$. Show that if z is not real, then $z, \overline{z}, 1/z$, and $1/\overline{z}$ are all distinct, but if z is real, then the set $z, \overline{z}, 1/z$, and $1/\overline{z}$ contains only two distinct elements.

Conclude that if z_k is a nonreal solution to $P(z_k) = 0$ with $|z_k| \neq 1$, then $z_k, 1/z_k, \overline{z_k}$, and $1/\overline{z_k}$ are all distinct solutions to $P(z) = 0$, but if z_k is real the set $z_k, 1/z_k, \overline{z_k}$, and $1/\overline{z_k}$ yields only two distinct solutions.

(d) Suppose that z_k is a solution to $P(z) = 0$ (hence by part (a) also $Q(z_k) = 0$) and $|z_k| = 1$. Show that z_k must have even multiplicity. Some hints: If z_k has multiplicity n, then from (7.55),

$$P(z) = \frac{(z - z_k)^n R(z)}{z^{N-1}},$$

where $R(z)$ a polynomial with $R(z_k) \neq 0$. Suppose that $z_k = e^{i\theta}$ for some real θ, and let $z = e^{i\omega}$. Argue that for ω near θ we must have a series expansion of the form

$$P(e^{i\omega}) = A(\omega - \theta)^n + O((\omega - \theta)^{n+1}),$$

where A is some nonzero constant and $O((\omega-\theta)^{n+1})$ indicates powers $(\omega-\theta)^{n+1}$ and higher. Conclude that if n is odd, then $P(e^{i\omega})$ cannot be of one sign for all ω near θ.

(e) Let

$$S_1 = \bigcup_{k=1}^{K} \left\{ z_k, \frac{1}{z_k}, \overline{z_k}, \frac{1}{\overline{z_k}} \right\}$$

denote the set of nonreal solutions to $P(z)=0$, where $|z_k|<1$ and z_k lies in the upper half plane. Note that $|\overline{z_k}|<1$, while $|1/z_k|>1$ and $|1/\overline{z_k}|>1$. Let

$$S_2 = \bigcup_{m=1}^{M} \left\{ z_m, \frac{1}{z_m} \right\}$$

denote the set of real solutions to $P(z)=0$, arranged so that $|z_m|<1$ for each m. Finally, let

$$S_3 = \bigcup_{r=1}^{R} \{z_r, \overline{z_r}\}$$

denote the set of solutions to $P(z)=0$ with $|z_r|=1$ (if z_r is a solution, so is $\overline{z_r}$). Let n_r denote the multiplicity (even) of z_r. Let $L_0(z)$ be defined as

$$L_0(z) = \left(\prod_{k=1}^{K} (z-z_k)(z-\overline{z_k}) \right) \left(\prod_{m=1}^{M} (z-z_m) \right) \times \left(\prod_{r=1}^{R} (z-z_r)^{n_r/2}(z-\overline{z_r})^{n_r/2} \right)$$

where the first product is over S_1, the second over S_2, and the last over S_3. Show that the polynomial $L_0(z)$ has real coefficients, and that we can take $L(z)$ in the factorization $P(z)=L(z)L(z^{-1})$ as $L(z)=cL_0(z)$ where $c=\sqrt{p_{N-1}/L_0(0)}$. Hint: First show that $z^{N-1}L_0(z)L_0(z^{-1})$ is a polynomial with the same roots and multiplicities as $Q(z)$ so that $z^{N-1}L_0(z)L_0(z^{-1})$ is a multiple of $Q(z)$.

(f) Use this method to find an appropriate $L(z)$ for $P(z)=L(z)L(z^{-1})$, where

$$P(z) = -z^{-3} + \frac{11}{2}z^{-2} - 13z^{-1} + \frac{69}{4} - 13z + \frac{11}{2}z^2 - z^3.$$

Hint: One of the roots of P is $1+i$. Another is $\frac{1}{2}$.

29. The purpose of this exercise is to show one method for obtaining the function $P(z)$ in equations (7.52) and (7.53). Recall that the goal is to find, for any even N, a function

$$P(z) = 1 + \sum_{k=-N+1,\ \text{odd}}^{N-1} p_k z^k \tag{7.56}$$

so that $P(z)$ vanishes to high order at $z = -1$; that is, $P(z)$ is divisible by as high a power of $(1+z)$ as possible (so that P will correspond to a low-pass filter). We would also like $P(e^{i\omega}) \geq 0$ for $\omega \in [-\pi, \pi]$.
If we write P in the form

$$P(z) = \frac{(1+z)^{2M} R(z)}{z^{N-1}}, \tag{7.57}$$

then the goal is to make M as large as possible with $R(z)$ a polynomial of degree $2N - 2 - 2M$ that is not divisible by $(1+z)$. The "2" in the exponent of $1+z$ is just for convenience.

(a) Let $P(z) = Q((z + z^{-1})/2)$, where Q is a polynomial of degree $N-1$, of the form

$$Q(x) = 1 + \sum_{k=1,\ \text{odd}}^{N-1} q_k x^k \tag{7.58}$$

and each q_k is real. Show that P will then be of the form (7.56).

(b) Suppose that $Q(x)$ is a polynomial that has a root of multiplicity M at $x = -1$ so that $Q(x) = (1+x)^M S(x)$ for some polynomial $S(x)$ with $S(-1) \neq 0$. Show that $P(z) = Q((z + z^{-1})/2)$ will be of the form (7.57). Our goal is thus to produce a polynomial $Q(x)$ of the form (7.58) such that Q is divisible by as high a power of $(1+x)$ as possible.

(c) Observe that $Q'(x)$ (a polynomial of degree $N-2$) should consist entirely of even powers of x and (like Q) be divisible by as high as power of $(1+x)^{M-1}$ as possible. Explain why taking $Q'(x)$ as an $N-2$ degree polynomial of the form

$$Q'(x) = c(1 - x^2)^{N/2-1} \tag{7.59}$$

for some constant c is consistent with these requirements. Specifically, show that no polynomial of degree $N-2$ consisting of even powers of x can be divisible by any higher power of $(1+x)$.

(d) According to part (c) we should take

$$Q(x) = c \int_{-1}^{x} (1 - y^2)^{N/2-1}\, dy. \tag{7.60}$$

Show that $Q(x)$ of this form must be divisible by $(1+x)^{N/2}$. Hint: Use (7.60) to show that $|Q(x)| \leq K(1+x)^{N/2}$ for some constant K and all x near -1. Why does this imply that $Q(x)$ is divisible by $(1+x)^{N/2}$? Argue that there must be some value of c so that $Q(0) = 1$.

(e) Show that $P(e^{i\omega})$ is real and nonnegative for all $\omega \in [-\pi, \pi]$. Hint: If $z = e^{i\omega}$, then $(z + z^{-1})/2 = \cos(\omega)$.

(f) Compute $Q(x)$ (with the appropriate value of c) for $N = 2, 4, 6$, and then find the corresponding $P(z)$. Compare to the results given in Examples 7.15 and 7.16.

8

Lifting for Filter Banks and Wavelets

8.1 Overview

Lifting methods were first introduced in the 1990s by Wim Sweldens (see [6, 11, 29, 30] for some of the first developments and applications). In a nutshell, lifting provides a way of implementing filter banks and wavelet transforms that can cut the amount of computation and memory required. Perhaps more importantly, lifting gives additional insight into what makes filter banks and discrete wavelet transforms work, and how to design new and more general types of filter banks, leading to so-called "second generation" wavelets. These techniques allow one to adapt this type of analysis to, for example, irregularly spaced data, higher dimensional regions, or regions with irregular boundaries.

The one-stage filter bank or discrete wavelet transform was developed in Chapter 7 for signals in $L^2(\mathbb{Z})$, using the ideas of filtering and downsampling. We then showed how to adapt the analysis for finite length signals, how to iterate the transform to obtain a multistage transform, and how to implement a two-dimensional transform as a sequence of one-dimensional transforms. In this chapter, we focus solely on signals in $L^2(\mathbb{Z})$, and develop lifting schemes for filter banks that act on such signals. The arguments from Chapter 7 still apply, however, and all results in this chapter extend to finite length signals, multistage transforms, and two (or higher) dimensional transforms.

8.2 Lifting for the Haar Filter Bank

We begin this chapter, like Chapter 7, with a careful examination of the Haar filter bank, but from a point of view amenable to a lifting implementation. This allows us to illustrate the key concepts in an easy setting. We then move on to the general case, and a more detailed examination of lifting and polyphase matrix factorization, as well as topics such as filter bank design and integer-to-integer transforms.

Discrete Fourier Analysis and Wavelets: Applications to Signal and Image Processing, Second Edition.
S. Allen Broughton and Kurt Bryan.

Companion Website: www.wiley.com/go/Broughton/Discrete_Fourier_Analysis_and_Wavelets

8.2.1 The Polyphase Analysis

Recall the Haar filter bank from Section 7.2, with analysis filters $\boldsymbol{\ell}_a = (1/2, 1/2)$ and $\mathbf{h}_a = (1/2, -1/2)$. A general signal $\mathbf{x} \in L^2(\mathbb{Z})$, with components x_k is convolved with the analysis filters to produce $\boldsymbol{\ell}_a * \mathbf{x}$ and $\mathbf{h}_a * \mathbf{x}$ as shown in equation (7.2). In general, $(\boldsymbol{\ell}_a * \mathbf{x})_k = (x_k + x_{k-1})/2$ and $(\mathbf{h}_a * \mathbf{x})_k = (x_k - x_{k-1})/2$. The convolved signals are then downsampled to produce $\mathbf{X}_\ell = D(\boldsymbol{\ell}_a * \mathbf{x})$ and $\mathbf{X}_h = D(\mathbf{h}_a * \mathbf{x})$ as in shown in equations (7.3) and (7.4). The mapping $\mathbf{x} \to (\mathbf{X}_\ell, \mathbf{X}_h)$ constitutes the one-stage Haar analysis transform.

When we downsample in computing the transform, we simply throw out the odd-indexed components in the convolution, an obvious inefficiency. In anticipation of downsampling, we should compute only the even-indexed components of $\boldsymbol{\ell}_a * \mathbf{x}$ and $\mathbf{h}_a * \mathbf{x}$. To this end, consider convolving a general two-tap filter $\mathbf{g} = (g_0, g_1)$ (like $\boldsymbol{\ell}_a$ or $\mathbf{h}_a$) with $\mathbf{x}$ and then selecting an even-indexed component. In general,

$$(\mathbf{g} * \mathbf{x})_j = g_0 x_j + g_1 x_{j-1},$$

so that an even-indexed component, say with $j = 2k$, yields

$$(\mathbf{g} * \mathbf{x})_{2k} = g_0 x_{2k} + g_1 x_{2k-1}.$$

The downsampled version $\mathbf{y} = D(\mathbf{g} * \mathbf{x})$ has $y_k = (\mathbf{g} * \mathbf{x})_{2k}$, so that

$$y_k = g_0 x_{2k} + g_1 x_{2k-1}. \tag{8.1}$$

When we use $\mathbf{g} = \boldsymbol{\ell}_a$ in (8.1), we obtain $\mathbf{y} = \mathbf{X}_\ell$, while $\mathbf{g} = \mathbf{h}_a$ yields $\mathbf{y} = \mathbf{X}_h$. In any case, notice that in computing y_k the filter tap g_0 only multiplies even-indexed components of $\mathbf{x}$, while g_1 multiplies only odd-indexed components of $\mathbf{x}$.

We can implement this filtering and downsampling operation efficiently by using a *polyphase* decomposition of $\mathbf{x}$. We split $\mathbf{x}$ into subsignals $\mathbf{x}_e$ and $\mathbf{x}_o$ with even and odd-indexed components from $\mathbf{x}$, as

$$\mathbf{x}_e = (\dots, x_{-4}, x_{-2}, x_0, x_2, x_4, \dots),$$
$$\mathbf{x}_o = (\dots, x_{-3}, x_{-1}, x_1, x_3, x_5, \dots).$$

In general, the components are given by $(\mathbf{x}_e)_k = x_{2k}$ and $(\mathbf{x}_o)_k = x_{2k+1}$. In fact, it is easy to see that

$$\mathbf{x}_e = D(\mathbf{x}) \text{ and } \mathbf{x}_o = D(S^{-1}(\mathbf{x})) \tag{8.2}$$

where D is the downsampling operator of Definition 7.1 and S is the right shift operator (recall Remark 7.1), defined via $S(\mathbf{x})_k = x_{k-1}$, so that S^{-1} is the left shift with $(S^{-1}(\mathbf{x}))_k = x_{k+1}$. (Defining $\mathbf{x}_o$ with a left shift rather than a right is slightly more convenient later on). The signals $\mathbf{x}_e$ and $\mathbf{x}_o$ as defined by equation (8.2) constitute a *polyphase* decomposition of $\mathbf{x}$.

Equation (8.1) can then be written as

$$\mathbf{y} = g_0\mathbf{x}_e + g_1 S(\mathbf{x}_o).$$

The Haar transform can now be implemented, in a polyphase form, by taking $\mathbf{g} = \boldsymbol{\ell}_a$ to obtain $\mathbf{X}_{\ell} = \frac{1}{2}\mathbf{x}_e + \frac{1}{2}S(\mathbf{x}_o)$ and $\mathbf{g} = \mathbf{h}_a$ to obtain $\mathbf{X}_{\mathbf{h}} = \frac{1}{2}\mathbf{x}_e - \frac{1}{2}S(\mathbf{x}_o)$. In matrix form, this becomes

$$\begin{bmatrix} \mathbf{X}_{\ell} \\ \mathbf{X}_{\mathbf{h}} \end{bmatrix} = \begin{bmatrix} 1/2 & 1/2 \\ 1/2 & -1/2 \end{bmatrix} \begin{bmatrix} \mathbf{x}_e \\ S(\mathbf{x}_o) \end{bmatrix}. \tag{8.3}$$

Note that the components of the "vectors" involved in equation (8.3) are in fact elements of $L^2(\mathbf{Z})$. Equation (8.3) shows how to compute the Haar transform $(\mathbf{X}_{\ell}, \mathbf{X}_{\mathbf{h}})$ from the polyphase decomposition $(\mathbf{x}_e, \mathbf{x}_o)$ of $\mathbf{x}$. The matrix

$$\mathbf{P}_a = \begin{bmatrix} 1/2 & 1/2 \\ 1/2 & -1/2 \end{bmatrix} \tag{8.4}$$

on the right in equation (8.3) is called the ***polyphase analysis matrix*** for the Haar transform.

8.2.2 Inverting the Polyphase Haar Transform

Equation (8.3) makes inverting the transform easy. The inverse $\mathbf{P}_s$ of the Haar polyphase matrix $\mathbf{P}_a$ is given by

$$\mathbf{P}_s = \mathbf{P}_a^{-1} = \begin{bmatrix} 1 & 1 \\ 1 & -1 \end{bmatrix}$$

and is called the Haar ***polyphase synthesis matrix.*** Define signals $\mathbf{y}_e$ and $\mathbf{y}_o$ as

$$\begin{bmatrix} \mathbf{y}_e \\ \mathbf{y}_o \end{bmatrix} = \mathbf{P}_s \begin{bmatrix} \mathbf{X}_{\ell} \\ \mathbf{X}_{\mathbf{h}} \end{bmatrix}. \tag{8.5}$$

Comparison to equation (8.3) shows that $\mathbf{y}_e = \mathbf{x}_e$ and $\mathbf{y}_o = S(\mathbf{x}_o)$. That is, $\mathbf{x}_e = \mathbf{y}_e$ and $\mathbf{x}_o = S^{-1}(\mathbf{y}_o)$, so we can recover the polyphase components $\mathbf{x}_e$ and $\mathbf{x}_o$ from $\mathbf{X}_{\ell}$ and $\mathbf{X}_{\mathbf{h}}$. It is easy to check (Exercise 3) that we can recover any $\mathbf{x}$ from its polyphase components $\mathbf{x}_e$ and $\mathbf{x}_o$ as

$$\mathbf{x} = U(\mathbf{x}_e) + S(U(\mathbf{x}_o)) \tag{8.6}$$

where U is the upsampling operator of Definition 7.2, so that we can recover $\mathbf{x}$ from $\mathbf{y}_e$ and $\mathbf{y}_o$ in (8.5) as

$$\mathbf{x} = U(\mathbf{y}_e) + S(U(S^{-1}(\mathbf{y}_o))). \tag{8.7}$$

Equations (8.5) and (8.7) specify the inverse transform.

Example 8.1 *Let* $\mathbf{x} \in L^2(\mathbb{Z})$ *be the signal*

$$\mathbf{x} = (16, 12, 16, -20, 24, 24, 18, 20)$$

beginning with x_0, so $x_0 = 16, \dots, x_7 = 20$. We will assume that $\mathbf{x}$, and all signals below, are listed starting with index 0 and equal to zero outside the listed range. The polyphase decomposition of $\mathbf{x}$ is

$$\mathbf{x}_e = (16, 16, 24, 18) \text{ and } \mathbf{x}_o = (12, -20, 24, 20).$$

We then have $S(\mathbf{x}_o) = (0, 12, -20, 24, 20)$. From (8.3), we obtain

$$\begin{aligned}\mathbf{X}_\ell &= \frac{1}{2}\mathbf{x}_e + \frac{1}{2}S(\mathbf{x}_o) = (8, 8, 12, 9, 0) + (0, 6, -10, 12, 10)\\ &= (8, 14, 2, 21, 10)\\ \mathbf{X}_\mathbf{h} &= \frac{1}{2}\mathbf{x}_e - \frac{1}{2}S(\mathbf{x}_o) = (8, 8, 12, 9, 0) - (0, 6, -10, 12, 10)\\ &= (8, 2, 22, -3, -10).\end{aligned}$$

To invert the transform, apply the polyphase synthesis matrix to obtain

$$\begin{aligned}\mathbf{y}_e &= \mathbf{X}_\ell + \mathbf{X}_\mathbf{h} = (8, 14, 2, 21, 10) + (8, 2, 22, -3, -10) = (16, 16, 24, 18, 0)\\ \mathbf{y}_o &= \mathbf{X}_\ell - \mathbf{X}_\mathbf{h} = (8, 14, 2, 21, 10) - (8, 2, 22, -3, -10)\\ &= (0, 12, -20, 24, 20).\end{aligned}$$

where we have dropped zero components that do not need to be listed. Then $S^{-1}(\mathbf{y}_o) = (12, -20, 24, 20)$ and upsampling as per equation (8.7) produces

$$\begin{aligned}&U(\mathbf{y}_e) = (16, 0, 16, 0, 24, 0, 18, 0) \text{ and}\\ &U(S^{-1}(\mathbf{y}_o)) = (12, 0, -20, 0, 24, 0, 20, 0).\end{aligned}$$

Finally, from (8.7) find

$$\begin{aligned}\mathbf{x} &= (16, 0, 16, 0, 24, 0, 18, 0) + (0, 12, 0, -20, 0, 24, 0, 20)\\ &= (16, 12, 16, -20, 24, 24, 18, 20),\end{aligned}$$

the signal with which we started.

8.2.3 Lifting Decomposition for the Haar Transform

It is simple to check that for the Haar polyphase analysis matrix as defined in equation (8.4) the factorization

$$\mathbf{P}_a = \mathbf{D}\mathbf{M}_1\mathbf{M}_2 \tag{8.8}$$

holds with

$$\mathbf{D} = \begin{bmatrix} 1 & 0 \\ 0 & -1/2 \end{bmatrix}, \quad \mathbf{M}_1 = \begin{bmatrix} 1 & 1/2 \\ 0 & 1 \end{bmatrix}, \quad \mathbf{M}_2 = \begin{bmatrix} 1 & 0 \\ -1 & 1 \end{bmatrix}. \tag{8.9}$$

Where this factorization comes from is a topic for the next section (but here it is a simple variation on the LU decomposition of $\mathbf{P}_a$.) It is worth noting that $\mathbf{M}_1$ and $\mathbf{M}_2$ have determinants equal to 1, and so are invertible. The "scaling" matrix $\mathbf{D}$ is also invertible.

The factorization (8.8) decomposes the Haar transform into three simpler steps. We first multiply the polyphase decomposition $(\mathbf{x}_e, S(\mathbf{x}_o))$ in equation (8.3) by $\mathbf{M}_2$; this is often called a *prediction* step or *dual lifting* step. We then multiply by $\mathbf{M}_1$; this is often called an *update* or *primal lifting* step. Finally, we multiply by $\mathbf{D}$, a *scaling* step. Let us examine what each of these matrix multiplications accomplishes, and why such a factorization might be useful.

To begin, we decompose the signal $\mathbf{x}$ into polyphase components $\mathbf{x}_e$ and $S(\mathbf{x}_o)$, then multiply by $\mathbf{M}_2$ to find signals $\mathbf{x}'_e$ and $\mathbf{x}'_o$ defined as

$$\begin{bmatrix} \mathbf{x}'_e \\ \mathbf{x}'_o \end{bmatrix} = \mathbf{M}_2 \begin{bmatrix} \mathbf{x}_e \\ S(\mathbf{x}_o) \end{bmatrix} = \begin{bmatrix} \mathbf{x}_e \\ -\mathbf{x}_e + S(\mathbf{x}_o) \end{bmatrix}. \tag{8.10}$$

A few observations about this operation are in order:

1. This operation is reversible since $\mathbf{M}_2$ is invertible with inverse
$$\mathbf{M}_2^{-1} = \begin{bmatrix} 1 & 0 \\ 1 & 1 \end{bmatrix}.$$
We can thus recover $\mathbf{x}_e$ and $S(\mathbf{x}_o)$ (hence $\mathbf{x}_o$) from $\mathbf{x}'_e$ and $\mathbf{x}'_o$.
2. When actually performed in a computer, this computation can be done "in place" with no additional memory. Specifically, in practice $\mathbf{x}_e$ and $\mathbf{x}_o$ are finite dimensional vectors. This lifting step leaves $\mathbf{x}'_e = \mathbf{x}_e$ unchanged; the vector $S(\mathbf{x}_o)$ is simply replaced by $\mathbf{x}'_o = -\mathbf{x}_e + S(\mathbf{x}_o)$, and the result of this computation can be stored in the memory that was used for $S(\mathbf{x}_o)$.
3. Why call this a "predict" step? If the signal $\mathbf{x}$ were slowly varying, we would expect that $\mathbf{x}_e$ would be a good approximation to or "predictor" of $\mathbf{x}_o$ (i.e., $x_{2k+1} \approx x_{2k}$, or what is almost the same thing, $S(\mathbf{x}_o) \approx \mathbf{x}_e$ so $x_{2k-1} \approx x_{2k}$). Of course the approximation is not likely exact, so to keep all of the information present in the signal we need to record the small but nonzero difference $S(\mathbf{x}_o) - \mathbf{x}_e$. This is what the right side of (8.10) keeps track of; we predict $S(\mathbf{x}_o) \approx \mathbf{x}_e$ (or $\mathbf{x}_o \approx S^{-1}(\mathbf{x}_e)$), with the additional information $\mathbf{x}'_o = S(\mathbf{x}_o) - \mathbf{x}_e$, so we can recover $\mathbf{x}_o = S^{-1}(\mathbf{x}_e + \mathbf{x}'_o)$ exactly.
Note that $\mathbf{x}'_o$ is just $\mathbf{X}_h$, the high-pass filtered downsampled version of the signal $\mathbf{x}$, and encodes the detail component of $\mathbf{x}$.

The subsequent multiplication by $\mathbf{M}_1$ yields signals $\mathbf{x}''_e$ and $\mathbf{x}''_o$ defined by

$$\begin{bmatrix} \mathbf{x}''_e \\ \mathbf{x}''_o \end{bmatrix} = \mathbf{M}_1 \begin{bmatrix} \mathbf{x}'_e \\ \mathbf{x}'_o \end{bmatrix} = \begin{bmatrix} \mathbf{x}'_e + \frac{1}{2}\mathbf{x}'_o \\ \mathbf{x}'_o \end{bmatrix} = \begin{bmatrix} \frac{1}{2}\mathbf{x}_e + \frac{1}{2}S(\mathbf{x}_o) \\ -\mathbf{x}_e + S(\mathbf{x}_o) \end{bmatrix}.$$

As above, this operation is easily reversible since $\mathbf{M}_1$ is invertible with inverse

$$\mathbf{M}_1^{-1} = \begin{bmatrix} 1 & -1/2 \\ 0 & 1 \end{bmatrix}.$$

Moreover, as above, this operation can be done in place with no additional memory, once $\mathbf{x}_e'$ and $\mathbf{x}_o'$ have been computed. The multiplication by $\mathbf{M}_1$ updates the signal $\mathbf{x}_e$ to become $\mathbf{x}_e'' = (\mathbf{x}_e + S(\mathbf{x}_o))/2$, which is just $\mathbf{X}_\ell$, the approximation or trend portion of $\mathbf{x}$.

Finally, multiplication by $\mathbf{D}$ scales each component to yield the Haar transform of $\mathbf{x}$. Note that the decomposition of the Haar transform into the dual and primal lifting steps followed by scaling allows the entire transform to be done in place with no additional memory.

The lifting and scaling steps into which we decompose the polyphase matrix, especially for more complex transforms than Haar, are not unique. See Exercises 13 and 14 for other factorizations of the Haar polyphase matrix.

8.2.4 Inverting the Lifted Haar Transform

The lifting factorization of equations (8.8) and (8.9) makes inverting the transform easy: we have, from elementary matrix algebra, that the polyphase synthesis matrix is given by $\mathbf{P}_s = \mathbf{P}_a^{-1} = \mathbf{M}_2^{-1}\mathbf{M}_1^{-1}\mathbf{D}^{-1}$. To invert the transform, we just apply the inverse of each lifting step, in the reverse order. Note that the inverses of $\mathbf{M}_1$ and $\mathbf{M}_2$ are of the same general form—triangular matrices with diagonal elements equal to 1. As a consequence, the inverse transform can also be done in place.

8.3 The Lifting Theorem

Before generalizing the polyphase analysis for the Haar transform, it will be helpful to develop a bit more material on Laurent polynomials and show how new perfect reconstruction filter banks can be obtained from existing filters via a process called "lifting."

Let $L(z)$ and $H(z)$ be the z-transforms for $\boldsymbol{\ell}$ and $\mathbf{h}$, the analysis filters for a perfect reconstruction filter bank (we omit the subscript "a" on the filter vectors and transforms, for convenience.) From Lemma 7.2, we know that $\boldsymbol{\ell}$ and $\mathbf{h}$ are part of a perfect reconstruction filter bank if and only if the quantity $d_0(z)$ defined by

$$d_0(z) = L(z)H(-z) - L(-z)H(z) \tag{8.11}$$

is a monomial, in which case, the synthesis filters are given by equation (7.34) and (7.35). Note that $d_0(z)$ must be an odd power of z.

Are there any other filters $\tilde{H}(z)$ that work with $L(z)$ to form a perfect reconstruction filter bank? In fact, there are infinitely many, but all are related to $H(z)$ in a specific way. Define filter $\tilde{\mathbf{h}}$ with transform $\tilde{H}(z)$ via

$$\tilde{H}(z) = H(z) - s(z^2)L(z), \tag{8.12}$$

where $s(z)$ is any Laurent polynomial. Then

$$\begin{aligned}L(z)\tilde{H}(-z) - L(-z)\tilde{H}(z) &= L(z)H(-z) - L(-z)H(z)\\ &\quad -s(z^2)L(z)L(-z) + s(z^2)L(z)L(-z)\\ &= d_0(z),\end{aligned} \tag{8.13}$$

a monomial, so that by Lemma 7.2 the filters corresponding to $L(z), \tilde{H}(z)$ also form a perfect reconstruction filter bank.

In fact, the converse of the above is true–any other filter $\tilde{\mathbf{h}}$ that forms a perfect reconstruction filter bank with $\boldsymbol{\ell}$ necessarily has a z-transform given by (8.12) for some $s(z)$. This is Theorem 8.1, which we prove shortly.

8.3.1 A Few Facts About Laurent Polynomials

8.3.1.1 The Width of a Laurent Polynomial

Consider a Laurent polynomial

$$p(z) = \sum_{k=m}^{n} a_k z^k, \quad m \leq n$$

We will define the *width* of p (often called the *degree* of p) as

$$\text{wid}(p) = n - m, \tag{8.14}$$

provided a_m and a_n are both nonzero. Note that m and n can be positive or negative. To avoid confusion, we prefer to reserve the term "degree" for traditional polynomials. If $p(z) = 0$, we set $\text{wid}(p) = -\infty$.

A Laurent polynomial is of width 0 exactly when it is a monomial. Indeed, it is worth noting that, for example, a monomial like z^{17} has width 0 (as a Laurent polynomial), but is of degree 17 as a standard polynomial. If $p(z)$ is a standard polynomial, it is easy to see that $\text{wid}(p) \leq \deg(p)$ (see Exercise 5). Note also that

$$\text{wid}(z^k p(z)) = \text{wid}(p(z)), \tag{8.15}$$

for any k, and also that if $p(z)$ and $q(z)$ are Laurent polynomials, then (see Exercise 7)

$$\text{wid}(pq) = \text{wid}(p) + \text{wid}(q). \tag{8.16}$$

8.3.1.2 The Division Algorithm

The *Division Algorithm* for Laurent polynomials states that

Lemma 8.1 *If $a(z)$ and $b(z)$ are Laurent polynomials and $b \neq 0$, then for some (not necessarily unique) Laurent polynomials $q(z), r(z)$, we have*

$$a(z) = q(z)b(z) + r(z), \tag{8.17}$$

where $\text{wid}(r) < \text{wid}(b)$.

Proof: To prove (8.17), let z^{s_0} be the lowest power of z in $a(z)$ and z^{p_0} the lowest power in $b(z)$. Then we can write

$$a(z) = z^{s_0}\underbrace{(a_0 + a_1 z + \cdots + a_n z^n)}_{\tilde{a}(z)},$$

$$b(z) = z^{p_0}\underbrace{(b_0 + b_1 z + \cdots + b_m z^m)}_{\tilde{b}(z)}.$$

for some s_0, p_0 and $m, n \geq 0$, with the additional conditions that a_0, b_0, a_n, b_m are all nonzero. Thus $\tilde{a}$ and $\tilde{b}$ are standard polynomials, of degree n and m, respectively. Moreover, as Laurent polynomials, we also have $\text{wid}(\tilde{a}) = n = \deg(\tilde{a})$ and $\text{wid}(\tilde{b}) = m = \deg(\tilde{b})$. Using the division algorithm for standard polynomials, we can write

$$\tilde{a}(z) = \tilde{q}(z)\tilde{b}(z) + \tilde{r}(z), \tag{8.18}$$

for some polynomials $\tilde{q}(z)$ and $\tilde{r}(z)$, with $\deg(\tilde{r}) < \deg(\tilde{b})$. But since $\text{wid}(\tilde{r}) \leq \deg(\tilde{r})$ and $\text{wid}(\tilde{b}) = \deg(\tilde{b})$, we have

$$\text{wid}(\tilde{r}) < \text{wid}(\tilde{b}). \tag{8.19}$$

Multiply both sides of (8.18) by z^{s_0} (note $a(z) = z^{s_0}\tilde{a}(z)$) to obtain

$$\begin{aligned} a(z) &= z^{s_0}\tilde{q}(z)\tilde{b}(z) + z^{s_0}\tilde{r}(z) \\ &= z^{s_0 - p_0}\tilde{q}(z)z^{p_0}\tilde{b}(z) + z^{s_0}\tilde{r}(z) \\ &= q(z)b(z) + r(z), \end{aligned}$$

where $q(z) = z^{s_0 - p_0}\tilde{q}(z)$ and $r(z) = z^{s_0}\tilde{r}(z)$. This is exactly equation (8.17), and note that

$$\text{wid}(r) = \text{wid}(\tilde{r}) < \text{wid}(\tilde{b}) = \text{wid}(b),$$

where we have made use of (8.15) and (8.19). This completes the proof.

Note, however, that $q(z)$ and $r(z)$ are not necessarily unique, in contrast to the case with standard polynomials. For example, let $a(z) = z^3$ and $b(z) = z^2 - z$. Then we have each of

$$a(z) = (z+1)b(z) + z \text{ and } a(z) = zb(z) + z^2 \text{ and } a(z) = 0b(z) + z^3.$$

In each case, the remainder (z or z^2 or z^3) is of width 0 while b is of width 1.

8.3.2 The Lifting Theorem

The statement of the Lifting Theorem is straightforward, though the proof is not essential to understanding the theorem. The Lifting Theorem states

Theorem 8.1 *Let $L(z), H(z)$ be the z-transforms of analysis filters $\boldsymbol{\ell}$ and $\mathbf{h}$ that are part of a perfect reconstruction filter bank. Then a filter $\tilde{\mathbf{h}}$ with z-transform*

$\tilde{H}(z)$ is part of a perfect reconstruction filter bank with $\boldsymbol{\ell}$ if and only if

$$H(z) - \tilde{H}(z) = s(z^2)L(z), \tag{8.20}$$

for some nonzero Laurent polynomial $s(z)$.

Proof: The "if" was shown above in equation (8.13). The converse takes a little more work. Let us define two sets of Laurent polynomials, $\mathcal{J}$ and $\mathcal{E}$, as

$$\mathcal{J} = \{D(z) : L(z)D(-z) - L(-z)D(z) = 0\},$$
$$\mathcal{E} = \{S(z) : S(z) = S(-z)\}.$$

Note that $\mathcal{J}$ depends on $L(z)$, while the set $\mathcal{E}$ consists of Laurent polynomials with only even powers of z. We are going to show that $\mathcal{J} = L(z)\mathcal{E}$, that is, that each element of $\mathcal{J}$ is the product of $L(z)$ with an element of $\mathcal{E}$ and conversely, the product of any element of $\mathcal{E}$ with $L(z)$ yields an element of $\mathcal{J}$. From this, the Lifting Theorem will follow.

Lemma 8.2 *With $\mathcal{J}$ and $\mathcal{E}$ defined as above,*

$$\mathcal{J} = L(z)\mathcal{E}.$$

Proof: The assertion that $L(z)\mathcal{E} \subseteq \mathcal{J}$ is easy to verify. Specifically, if we define $D(z) = L(z)S(z)$ where $S(z) \in \mathcal{E}$ then $L(z)D(-z) - L(-z)D(z) = L(z)L(-z)S(-z) - L(-z)L(z)S(z) = L(z)L(-z)(S(-z) - S(z)) = 0$ since $S \in \mathcal{E}$. Thus $D(z) \in \mathcal{J}$.

To show the converse, that $\mathcal{J} \subseteq L(z)\mathcal{E}$, let us first examine a few straightforward properties of the sets $\mathcal{J}$ and $\mathcal{E}$. Note that $\mathcal{E}$ contains all scalars and is closed under addition and multiplication. The set $\mathcal{J}$ has the properties

$$L(z) \in \mathcal{J}, \tag{8.21}$$
$$D_1(z), D_2(z) \subset \mathcal{J} \Rightarrow D_1(z) + D_2(z) \in \mathcal{J}, \tag{8.22}$$
$$D(z) \in \mathcal{J}, S(z) \in \mathcal{E} \Rightarrow S(z)D(z) \in \mathcal{J}. \tag{8.23}$$

The proofs are left as Exercise 8. We also have

$$D_1(z), D_2(z) \in \mathcal{J} \Rightarrow D_1(z)D_2(-z) = D_1(-z)D_2(z) \tag{8.24}$$

or equivalently, $D_1(z)D_2(-z) - D_1(-z)D_2(z) = 0$. To demonstrate equation (8.24), observe that if $D_1(z), D_2(z) \in \mathcal{J}$, then each of

$$L(z)D_1(-z) = L(-z)D_1(z),$$
$$L(z)D_2(-z) = L(-z)D_2(z)$$

holds. Combine the above equations (cross multiply left by right and right by left) to obtain

$$L(z)D_1(-z)L(-z)D_2(z) = L(-z)D_1(z)L(z)D_2(-z).$$

Dividing by $L(z)L(-z)$ yields (8.24).

Now the nonzero elements of $\mathcal{J}$ have non-negative width and so there is some nonzero element $b(z) \in \mathcal{J}$ (not necessarily unique) of minimum width. Let $a(z)$ be any element in $\mathcal{J}$. From Lemma 8.1, we can write $a(z) = q(z)b(z) + r(z)$ with $\text{wid}(r(z)) < \text{wid}(b(z))$.

Claim: We have $q(z) \in \mathcal{E}$ and $r(z) = 0$.

Proof: Since $a(z) \in \mathcal{J}$, we have

$$\begin{aligned} 0 &= L(z)a(-z) - L(-z)a(z) \\ &= L(z)(q(-z)b(-z) + r(-z)) - L(-z)(q(z)b(z) + r(z)) \\ &= L(z)b(-z)q(-z) - L(-z)b(z)q(z) + L(z)r(-z) - L(-z)r(z). \end{aligned}$$

Since $b(z) \in \mathcal{J}$, we have $L(z)b(-z) = L(-z)b(z)$ and so from above

$$\begin{aligned} 0 &= L(z)b(-z)q(-z) - L(z)b(-z)q(z) + L(z)r(-z) - L(-z)r(z) \\ &= L(z)b(-z)(q(-z) - q(z)) + L(z)r(-z) - L(-z)r(z). \end{aligned}$$

This last equation can be rearranged to

$$L(z)b(-z)(q(-z) - q(z)) = L(-z)r(z) - L(z)r(-z). \tag{8.25}$$

The left side of equation (8.25) has width

$$\begin{aligned} \text{wid}(L(z)b(-z)(q(-z) - q(z))) &= \text{wid}(L(z)) + \text{wid}(b(-z)) + \text{wid}(q(-z) - q(z)) \\ &= \text{wid}(L(z)) + \text{wid}(b(z)) + \text{wid}(q(-z) - q(z)). \end{aligned} \tag{8.26}$$

The right side of (8.25) consists of the odd power terms of $2L(-z)r(z)$ so that

$$\begin{aligned} \text{wid}(L(-z)r(z) - L(z)r(-z)) &\leq \text{wid}(2L(-z)r(z)) \\ &= \text{wid}(L(-z)) + \text{wid}(r(z)) \\ &= \text{wid}(L(z)) + \text{wid}(r(z)). \end{aligned} \tag{8.27}$$

From equation (8.25) as well as (8.26) and (8.27), we conclude that

$$\text{wid}(L(z)) + \text{wid}(b(z)) + \text{wid}(q(-z) - q(z)) \leq \text{wid}(L(z)) + \text{wid}(r(z))$$

or, with a bit of rearrangement,

$$\text{wid}(q(-z) - q(z)) \leq \text{wid}(r(z)) - \text{wid}(b(z)).$$

But $\text{wid}(r(z)) - \text{wid}(b(z)) < 0$ (since we chose $r(z)$ so that $\text{wid}(r(z)) < \text{wid}(b(z))$, a contradiction unless $q(-z) - q(z) = 0$. We conclude that $q(z) = q(-z)$, or equivalently that $q \in \mathcal{E}$. We then have that $r(z) = a(z) - q(z)b(z) \in \mathcal{J}$ by the properties (8.22) and (8.23). But since $\text{wid}(r(z)) < \text{wid}(b(z))$ and $b(z)$ was chosen to have minimal width, we must conclude that $\text{wid}(r(z)) = -\infty$, that is, $r(z) = 0$. This completes the proof of the Claim.

Note that the claim can also be stated as

$$\mathcal{J} = b(z)\mathcal{E}. \tag{8.28}$$

Next we show that $b(z)$ (the element in $\mathcal{J}$ of minimal width) is a monomial times $L(z)$. To this end, since $L(z) \in \mathcal{J}$ and by virtue of the Claim above, we can write $L(z) = q(z)b(z)$ for some $q(z)$, with no remainder. Since the filter pair with z transforms L, H is perfect reconstruction we have

$$\begin{aligned} z^m &= L(z)H(-z) - L(-z)H(z) \\ &= q(z)b(z)H(-z) - q(-z)b(-z)H(z) \\ &= q(z)(b(z)H(-z) - b(-z)H(z)), \end{aligned}$$

for some m; from Remark 7.5, the integer m is odd. This means that $q(z)$ (and $b(z)H(-z) - b(-z)H(z)$) are themselves monomials, and since $b(z)H(-z) - b(-z)H(z)$ is odd, $q(z)$ must in fact be a monomial of even degree. It thus follows that

$$L(z) = z^n b(z),$$

for some even n. Making use of the Claim in the form of equation (8.28) shows that

$$\mathcal{J} = b(z)\mathcal{E} = z^{-n}L(z)\mathcal{E},$$

The last equality follows from the easily verified fact that $z^{-n}\mathcal{E} = \mathcal{E}$ when n is even. This completes the proof of Lemma 8.2.

We can now finish the proof of the Lifting Theorem. Let pairs $L(z), H(z)$ and $L(z), \tilde{H}(z)$ both correspond to perfect reconstruction filter banks so that (8.20) holds. Define

$$\tilde{D}(z) = H(z) - \tilde{H}(z). \tag{8.29}$$

If equation (8.11) holds for both $H(z)$ and $\tilde{H}(z)$ (with the same monomial $d_0(z)$) then it is easy to check that

$$L(z)\tilde{D}(-z) - L(-z)\tilde{D}(z) = 0,$$

since $s(z^2) = s((-z)^2)$ for any s. That is, $\tilde{D}(z) \in \mathcal{J}$, and so by Lemma 8.2 we have

$$\tilde{D}(z) = S(z)L(z), \tag{8.30}$$

for some $S(z) \in \mathcal{E}$. In this case (since $S(z)$ contains only even powers of z), we may define $s(z)$ so that $S(z) = s(z^2)$ and from (8.30) we have $\tilde{D}(z) = s(z^2)L(z)$, which with (8.29) is exactly equation (8.20) of the Lifting Theorem.

Note that a similar version of the Theorem holds if (L, H) and $(\tilde{L}, H)$ are the z-transforms of perfect reconstruction filter pairs $(\boldsymbol{\ell}, \mathbf{h})$ and $(\tilde{\boldsymbol{\ell}}, \mathbf{h})$. We must have

$$L(z) = \tilde{L}(z) + t(z^2)H(z), \tag{8.31}$$

for some Laurent polynomial t.

Remark 8.1 If $\tilde{H}(1) = H(1)$ (e.g., if both filters are high-pass, so $\tilde{H}(1) = H(1) = 0$), then we have $0 = s(1^2)L(1)$ which implies $s(1) = 0$ if L corresponds to a low-pass filter. Similar reasoning shows that if $\tilde{L}(-1) = L(-1)$ (e.g., both filters are low-pass) and H corresponds to a high-pass filter in (8.31), then we need $t(1) = 0$.

Example 8.2 *The Haar analysis filters have z-transforms*

$$L(z) = \frac{1}{2} + \frac{1}{2}z^{-1}, \quad H(z) = \frac{1}{2} - \frac{1}{2}z^{-1}.$$

Let $s(z) = z + 1$ (an arbitrary choice) so that $s(z^2) = z^2 + 1$. We conclude from equation (8.20) and the Lifting Theorem 8.1 that if

$$\tilde{H}(z) = H(z) - s(z^2)L(z) = -z^{-1} - \frac{1}{2}z - \frac{1}{2}z^2,$$

then the filter pair with z-transform $(L(z), \tilde{H}(z))$ is part of a perfect reconstruction filter bank, as can be verified directly by computing $L(z)\tilde{H}(-z) - L(-z)\tilde{H}(z) = z^{-1}$, a monomial.

Example 8.3 *In Example 8.2, the new filter $\tilde{H}$ had more taps than the original high-pass filter H. But given a perfect reconstruction filter pair, equation (8.20) or (8.31) can be used to find a* simpler *(fewer taps) pair of perfect reconstruction filters, by using the division algorithm. To illustrate, consider the Le Gall 5/3 analysis filters with z-transforms*

$$L(z) = -\frac{1}{8}z^2 + \frac{1}{4}z + \frac{3}{4} + \frac{1}{4}z^{-1} - \frac{1}{8}z^{-2}, \quad H(z) = -\frac{1}{2} + z^{-1} - \frac{1}{2}z^{-2}$$

We can write, for example,

$$L(z) = \frac{7}{8} + \frac{1}{4}z - \frac{1}{8}z^2 + \frac{1}{4}H(z).$$

The constant function $1/4$ is a function of z^2, so that if we define $\tilde{L}(z) = \frac{7}{8} + \frac{1}{4}z - \frac{1}{8}z^2$ then $(\tilde{L}(z), H(z))$ yield the analysis filters in a perfect reconstruction filter bank, but $\tilde{\ell}$ has only three taps instead of five. We will exploit this kind of reasoning later to reduce complicated filters to a sequence of simpler filters.

8.4 Polyphase Analysis for Filter Banks

The analysis done for the Haar filter bank above extends in a conceptually straightforward way to more general filter banks, even if the computations are more complicated. As in Chapter 7, it is much easier to do the analysis using the z-transform to analyze convolution, as well as upsampling and downsampling.

8.4.1 The Polyphase Decomposition and Convolution

In a two-channel filter bank, we convolve the input signal $\mathbf{x} \in L^2(\mathbb{Z})$ with low- and high-pass analysis filters $\boldsymbol{\ell}_a$ and $\mathbf{h}_a$ and then downsample to produce filtered signals $\mathbf{X}_\ell$ and $\mathbf{X}_h$, as per Figure 7.2. Just as in the case of the Haar filters, we need only compute the even-indexed components of the convolution, since these are the only components we retain.

Let $\mathbf{x}$ denote the input signal and suppose $\mathbf{g}$ is a FIR filter to be convolved with $\mathbf{x}$. The convolved signal is $\tilde{\mathbf{y}} = \mathbf{g} * \mathbf{x}$ with components

$$\tilde{y}_k = \sum_j g_j x_{k-j}$$

over an appropriate (finite) range of summation. Since we are only interested in the even-indexed components of $\tilde{\mathbf{y}}$, we note that $\tilde{y}_{2k} = \sum_j g_j x_{2k-j}$. Downsampling produces a signal $\mathbf{y} = D(\tilde{\mathbf{y}}) = D(\mathbf{g} * \mathbf{x})$ with components $y_k = \tilde{y}_{2k}$, so that

$$y_k = \sum_j g_j x_{2k-j}.$$

It is easy to see that $2k - j$ is even exactly when j is even and $2k - j$ is odd exactly when j is odd. As a consequence the even-index components of $\mathbf{g}$ multiply only even-indexed components of $\mathbf{x}$; similar considerations hold for the odd-indexed components, just as in the case of the Haar filter bank. This motivates splitting $\mathbf{x}$ and the filter $\mathbf{g}$ into polyphase components, even and odd indexed, and relating these components to the components of $\mathbf{y}$. These computations are most easily done via the z-transform.

Let $X(z)$ denote the z-transform of the signal $\mathbf{x}$ and $\mathbf{x}_e$, $\mathbf{x}_o$ the polyphase components of $\mathbf{x}$ as defined by equations (8.2). Let $\mathbf{g}$ be a FIR filter with polyphase components $\mathbf{g}_e = D(\mathbf{g})$, $\mathbf{g}_o = D(S^{-1}(\mathbf{g}))$, and corresponding z-transforms $G(z)$, $G_e(z)$, and $G_o(z)$, respectively. In Exercise 6, you are asked to show that

$$X_e(z) = \frac{X(\sqrt{z}) + X(-\sqrt{z})}{2}, \tag{8.32}$$

$$X_o(z) = \frac{\sqrt{z}(X(\sqrt{z}) - X(-\sqrt{z}))}{2}, \tag{8.33}$$

$$X(z) = X_e(z^2) + z^{-1}X_o(z^2), \tag{8.34}$$

for any $\mathbf{x} \in L^2(\mathbb{Z})$.

Lemma 8.3 *If* $\mathbf{y} = D(\mathbf{g} * \mathbf{x})$, *then*

$$Y(z) = G_e(z)X_e(z) + z^{-1}G_o(z)X_o(z) \tag{8.35}$$

Proof: From Theorem 4.6, the z-transform of $\tilde{\mathbf{y}} = \mathbf{g} * \mathbf{x}$ is $\tilde{Y}(z) = G(z)X(z)$. The transform of the downsampled signal $\mathbf{y} = D(\tilde{\mathbf{y}})$ is, from Exercise 6 (or equation (8.32)), given by

$$Y(z) = \frac{\tilde{Y}(\sqrt{z}) + \tilde{Y}(-\sqrt{z})}{2} = \frac{G(\sqrt{z})X(\sqrt{z}) + G(-\sqrt{z})X(-\sqrt{z})}{2}.$$

If we now use the relations (8.32) and (8.33) to write out the various quantities on the right in equation (8.35), we obtain

$$G_e(z)X_e(z) = \frac{G(\sqrt{z}) + G(-\sqrt{z})}{2}\frac{X(\sqrt{z}) + X(-\sqrt{z})}{2}, \tag{8.36}$$

while

$$\begin{aligned} z^{-1}G_o(z)X_o(z) &= z^{-1}\frac{\sqrt{z}G(\sqrt{z}) + (-\sqrt{z})G(-\sqrt{z})}{2} \\ &\quad \times \frac{\sqrt{z}X(\sqrt{z}) + (-\sqrt{z})X(-\sqrt{z})}{2} \\ &= \frac{G(\sqrt{z}) - G(-\sqrt{z})}{2}\frac{X(\sqrt{z}) - X(-\sqrt{z})}{2}. \end{aligned} \tag{8.37}$$

Adding equations (8.36) and (8.37) produces, after expanding and canceling,

$$G_e(z)X_e(z) + z^{-1}G_o(z)X_o(z) = \frac{G(\sqrt{z})X(\sqrt{z}) + G(-\sqrt{z})X(-\sqrt{z})}{2},$$

which is, from above, exactly $Y(z)$.

Equation (8.35) works for any filter **g** and signal **x**.

Example 8.4 *Let* **g** *have nonzero components* $g_0 = -1, g_1 = 2, g_2 = 1$, *and let* **x** *have nonzero components* $x_{-1} = 1, x_0 = 0, x_1 = 3, x_2 = -1$. *Then* $(\mathbf{g}_e)_0 = -1, (\mathbf{g}_e)_1 = 1, (\mathbf{g}_o)_0 = 2$ *and* $(\mathbf{x}_e)_0 = 0, (\mathbf{x}_e)_1 = -1, (\mathbf{x}_o)_{-1} = 1, (\mathbf{x}_o)_0 = 3$; *all other components of both vectors are zero. We find*

$$\begin{aligned} X(z) &= z + 0 + 3z^{-1} - z^{-2} \\ G(z) &= -1 + 2z^{-1} + z^{-2} \\ X_e(z) &= -z^{-1} \\ G_e(z) &= -1 + z^{-1} \\ X_o(z) &= z + 3 \\ G_o(z) &= 2. \end{aligned}$$

Let us compute $Y(z)$ *on the left in (8.35) directly from* $\mathbf{y} = D(\mathbf{g} * \mathbf{x})$. *The convolution* $\mathbf{g} * \mathbf{x}$ *can be computed via the product* $G(z)X(z) = -z + 2 - 2z^{-1} + 7z^{-2} + z^{-3} - z^{-4}$ *so that* $\tilde{\mathbf{y}} = \mathbf{g} * \mathbf{x}$ *has nonzero components* $\tilde{y}_{-1} = -1, \tilde{y}_0 = 2, \tilde{y}_1 = -2, \tilde{y}_2 = 7, \tilde{y}_3 = 1, \tilde{y}_4 = -1$. *Then* $\mathbf{y} = D(\tilde{\mathbf{y}})$ *has components* $y_0 = 2, y_1 = 7, y_2 = -1$, *and so*

$$Y(z) = 2 + 7z^{-1} - z^{-2}.$$

If we compute the product $G_e(z)X_e(z) + z^{-1}G_o(z)X_o(z)$ *using the information above, we find*

$$\begin{aligned}G_e(z)X_e(z) + z^{-1}G_o(z)X_o(z) &= (-1 + z^{-1})(-z^{-1}) + (z^{-1})(2)(z+3)\\ &= 2 + 7z^{-1} - z^{-2}\end{aligned}$$

as expected.

8.4.2 The Polyphase Analysis Matrix

Equation (8.35) can be used to implement a polyphase version of a filter bank. Consider a two-channel filter bank with low- and high-pass analysis filters $\boldsymbol{\ell}_a$ and $\mathbf{h}_a$. We will apply (8.35) with each of $\mathbf{g} = \boldsymbol{\ell}_a$ and $\mathbf{g} = \mathbf{h}_a$. Let $L_a(z)$ denote the z-transform of $\boldsymbol{\ell}_a$ and $L_{a,e}(z)$, $L_{a,o}(z)$ the z-transforms of the even- and odd-indexed polyphase components of $\boldsymbol{\ell}_a$. From Lemma 8.3 with filter $\mathbf{g} = \boldsymbol{\ell}_a$ (and then note that $Y(z) = X_\ell(z)$), we have

$$X_\ell(z) = L_{a,e}(z)X_e(z) + z^{-1}L_{a,o}(z)X_o(z).$$

With filter $\mathbf{g} = \mathbf{h}_a$, we find

$$X_h(z) = H_{a,e}(z)X_e(z) + z^{-1}H_{a,o}(z)X_o(z),$$

where $H_{a,e}(z)$ and $H_{a,o}(z)$ denote the z-transforms of the even- and odd-indexed polyphase components of $\mathbf{h}_a$. We can arrange the above relations into a matrix equation

$$\begin{bmatrix} X_\ell(z) \\ X_h(z) \end{bmatrix} = \begin{bmatrix} L_{a,e}(z) & L_{a,o}(z) \\ H_{a,e}(z) & H_{a,o}(z) \end{bmatrix} \begin{bmatrix} X_e(z) \\ z^{-1}X_o(z) \end{bmatrix}. \tag{8.38}$$

The matrix

$$\mathbf{P}_a(z) = \begin{bmatrix} L_{a,e}(z) & L_{a,o}(z) \\ H_{a,e}(z) & H_{a,o}(z) \end{bmatrix} \tag{8.39}$$

is called the *polyphase analysis matrix* for the filter bank $(\boldsymbol{\ell}_a, \mathbf{h}_a)$.

The polyphase matrix is now the vehicle by which the filter bank is implemented, via equation (8.38).

Example 8.5 *Let* $\boldsymbol{\ell}_a$ *and* $\mathbf{h}_a$ *be the Le Gall 5/3 filters with components*

$$(\boldsymbol{\ell}_a)_{-2} = (\boldsymbol{\ell}_a)_2 = -1/8, (\boldsymbol{\ell}_a)_{-1} = (\boldsymbol{\ell}_a)_1 = 1/4, (\boldsymbol{\ell}_a)_0 = 3/4,$$
$$(\mathbf{h}_a)_0 = -1/2, (\mathbf{h}_a)_1 = 1, (\mathbf{h}_a)_2 = -1/2.$$

These are the analysis filters for a perfect reconstruction filter bank. The corresponding z-transforms for the filters, as well as the polyphase components, are

$$L_a(z) = -\frac{1}{8}z^2 + \frac{1}{4}z + \frac{3}{4} + \frac{1}{4}z^{-1} - \frac{1}{8}z^{-2}$$

$$
\begin{aligned}
L_{a,e}(z) &= -\frac{1}{8}z + \frac{3}{4} - \frac{1}{8}z^{-1}\\
L_{a,o}(z) &= \frac{1}{4}z + \frac{1}{4}\\
H_a(z) &= -\frac{1}{2} + z^{-1} - \frac{1}{2}z^{-2}\\
H_{a,e}(z) &= -\frac{1}{2} - \frac{1}{2}z^{-1}\\
H_{a,o}(z) &= 1.
\end{aligned}
$$

The polyphase analysis matrix is

$$
\mathbf{P}_a(z) = \begin{bmatrix} -\frac{1}{8}z + \frac{3}{4} - \frac{1}{8}z^{-1} & \frac{1}{4}z + \frac{1}{4} \\ -\frac{1}{2} - \frac{1}{2}z^{-1} & 1 \end{bmatrix}. \tag{8.40}
$$

A signal $\mathbf{x}$ *with z-transform* $X(z) = \sum_k x_k z^{-k}$ *is then transformed as*

$$
\begin{bmatrix} X_\ell(z) \\ X_h(z) \end{bmatrix} = \mathbf{P}_a(z) \begin{bmatrix} X_e(z) \\ z^{-1}X_o(z) \end{bmatrix},
$$

which yields

$$
\begin{aligned}
X_\ell(z) &= \left(-\frac{1}{8}z + \frac{3}{4} - \frac{1}{8}z^{-1}\right)X_e(z) + \left(\frac{1}{4}z + \frac{1}{4}\right)z^{-1}X_o(z),\\
X_h(z) &= \left(-\frac{1}{2} - \frac{1}{2}z^{-1}\right)X_e(z) + z^{-1}X_o(z).
\end{aligned}
$$

8.4.3 Inverting the Transform

As we saw in Chapter 7, if the filters $\boldsymbol{\ell}_a$ and $\mathbf{h}_a$ are part of a perfect reconstruction filter bank then the original signal $\mathbf{x}$ can be recovered from $\mathbf{X}_\ell$ and $\mathbf{X}_h$ via a process of upsampling and filtering. Let us write out what this looks like in the polyphase framework.

In light of equations (8.38) and (8.39), we will proceed as follows. We will construct a matrix $\mathbf{P}_s(z)$ *with components that are Laurent polynomials* so that $\mathbf{P}_s(z)\mathbf{P}_a(z) = \mathbf{I}$, and then define $Y_e(z)$ and $Y_o(z)$ as

$$
\begin{bmatrix} Y_e(z) \\ Y_o(z) \end{bmatrix} = \mathbf{P}_s(z) \begin{bmatrix} X_\ell(z) \\ X_h(z) \end{bmatrix}. \tag{8.41}
$$

Of course, this means that $Y_e(z) = X_e(z)$ and $Y_o(z) = z^{-1}X_o(z)$, or equivalently, $X_o(z) = zY_o(z)$. We can then reconstruct $X(z)$ (i.e., $\mathbf{x}$) from $Y_e(z)$ and $Y_o(z)$ and equation (8.34) as

$$
X(z) = X_e(z^2) + z^{-1}X_o(z^2) = Y_e(z^2) + zY_o(z^2). \tag{8.42}
$$

The matrix $\mathbf{P}_s(z)$ is called the *polyphase synthesis matrix*. Equations (8.41) and (8.42) comprise the polyphase version of the inverse transform.

Since $\mathbf{P}_s = \mathbf{P}_a^{-1}$, from Cramer's rule for the inverse of a 2×2 matrix, we need

$$\mathbf{P}_s(z) = \begin{bmatrix} H_{a,o}(z)/d(z) & -L_{a,o}(z)/d(z) \\ -H_{a,e}(z)/d(z) & L_{a,e}(z)/d(z) \end{bmatrix}, \tag{8.43}$$

where

$$d(z) = \det(\mathbf{P}_a(z)) = L_{a,e}(z)H_{a,o}(z) - L_{a,o}(z)H_{a,e}(z)$$

is the determinant of $\mathbf{P}_a(z)$. Note that $d(z)$ is itself a Laurent polynomial. Since we will require that the components of $\mathbf{P}_s(z)$ be Laurent polynomials, the determinant of $\mathbf{P}_s(z)$ is also a Laurent polynomial. Moreover, since the determinant is multiplicative and $\mathbf{P}_s(z)\mathbf{P}_a(z) = \mathbf{I}$ we have

$$\det(\mathbf{P}_s(z))\det(\mathbf{P}_a(z)) = 1. \tag{8.44}$$

A consequence of equation (8.44) and Lemma 7.1 is that each determinant on the left must be a monomial in z, that is, of the form cz^r for some $c \neq 0$ and some integer r.

We can summarize the above discussion as

Lemma 8.4 *The polyphase synthesis matrix has Laurent polynomial entries ($\mathbf{P}_a$ is invertible) if and only if $d(z) = L_{a,e}(z)H_{a,o}(z) - L_{a,o}(z)H_{a,e}(z)$ is a monomial in z.*

We then have

Lemma 8.5 *Filters $\boldsymbol{\ell}_a$ and $\mathbf{h}_a$ with z-transforms $L_a(z)$ and $H_a(z)$ are part of a perfect reconstruction filter bank if and only if the quantity $d(z) = L_{a,e}(z)H_{a,o}(z) - L_{a,o}(z)H_{a,e}(z)$ is a monomial in z. In this case, the polyphase synthesis matrix $\mathbf{P}_s = \mathbf{P}_a^{-1}$ exists as a matrix of Laurent polynomials.*

Proof: Suppose that $d(z)$ is a monomial in z. We can use equations (8.32) and (8.33) to write out $L_{a,e}(z^2) = (L_a(z) + L_a(-z))/2$ and similarly for $L_{a,o}(z)$, $H_{a,e}(z), H_{a,o}(z)$. Inserting these into $d(z^2)$ and simplifying shows that

$$d(z^2) = -\frac{1}{2}z(L_a(z)H_a(-z) - L_a(-z)H_a(z)). \tag{8.45}$$

If $d(z)$ is a monomial, then so is $d(z^2)$ and then clearly so is $L_a(z)H_a(-z) - L_a(-z)H_a(z)$. From (7.34) and (7.35), we conclude that appropriate synthesis filters exist for perfect reconstruction.

Conversely, if $L_a(z), H_a(z)$ are part of a perfect reconstruction filter bank, then (recall equation (7.34) and (7.35), and Remark 7.5) the quantity $L_a(z)H_a(-z) - L_a(-z)H_a(z)$ is a monomial of odd degree, say $2q - 1$, so that from (8.45), we conclude that $d(z) = \det(\mathbf{P}_a(z)) = cz^{2q}$ is a monomial of even degree. This means that $d(z) = cz^q$ is a monomial and completes the proof of the Lemma.

Example 8.6 *Recall the Le Gall 5/3 filters and polyphase analysis matrix from Example 8.5. A bit of computation shows that*

$$L_a(z)H_a(-z) - L_a(-z)H_a(z) = -2z^{-1}$$
$$L_{a,e}(z)H_{a,o}(z) - L_{a,o}(z)H_{a,e}(z) = 1$$

in accordance with (8.45). The polyphase synthesis matrix is given by

$$\mathbf{P}_s(z) = \mathbf{P}_a(z)^{-1} = \begin{bmatrix} 1 & -\frac{1}{4}z - \frac{1}{4} \\ \frac{1}{2} + \frac{1}{2}z^{-1} & -\frac{1}{8}z + \frac{3}{4} - \frac{1}{8}z^{-1} \end{bmatrix}. \tag{8.46}$$

We can invert the transform by first using equation (8.41) to find

$$Y_e(z) = X_\ell(z) - (z/4 + 1/4)X_h(z),$$
$$Y_o(z) = \left(\frac{1}{2} + \frac{1}{2}z^{-1}\right)X_\ell(z) + \left(-\frac{1}{8}z + \frac{3}{4} - \frac{1}{8}z^{-1}\right)X_h(z)$$

and recovering $X(z) = Y_e(z^2) + zY_o(z^2)$.

Given analysis filters with z-transforms $L_a(z), H_a(z)$, the polyphase analysis matrix can be expressed in terms the polyphase component z-transforms $L_{a,e}(z)$, and so on, and from $\mathbf{P}_a(z)$ we can compute $\mathbf{P}_s(z)$ by inverting. But it should not be surprising that $\mathbf{P}_s(z)$ can be computed from the synthesis filters $\boldsymbol{\ell}_s$ and $\mathbf{h}_s$ directly.

Lemma 8.6 *Let $\boldsymbol{\ell}_s$ and $\mathbf{h}_s$ be synthesis filters computed via equation (7.34) and (7.35) (thus satisfying (7.32) and (7.33) with an output delay of m time steps) in which $\boldsymbol{\ell}_a$ and $\mathbf{h}_a$ are part of a perfect reconstruction filter bank. Then the polyphase synthesis matrix is given by*

$$\mathbf{P}_s(z) = z^{m/2} \begin{bmatrix} L_{s,e}(z) & H_{s,e}(z) \\ z^{-1}L_{s,o}(z) & z^{-1}H_{s,o}(z) \end{bmatrix} \tag{8.47}$$

when m is even, and

$$\mathbf{P}_s(z) = z^{(m-1)/2} \begin{bmatrix} L_{s,o}(z) & H_{s,o}(z) \\ L_{s,e}(z) & H_{s,e}(z) \end{bmatrix} \tag{8.48}$$

when m is odd.

Proof: This is a straightforward, if tedious, computation. If $L_a(z)$ and $H_a(z)$ are part of a perfect reconstruction filter bank we have, from (7.34), that the synthesis low-pass filter is given by

$$L_s(z) = \frac{2z^{-m}H_a(-z)}{L_a(z)H_a(-z) - L_a(-z)H_a(z)}$$

$$= -\frac{z^{-(m-1)}(H_{a,e}(z^2) - z^{-1}H_{a,o}(z^2))}{d(z^2)}, \tag{8.49}$$

where $d(z) = L_{a,e}(z)H_{a,o}(z) - L_{a,o}(z)H_{a,e}(z)$ and where we have made use of equation (8.45) and (8.34), applied to $H_a(z)$; note the denominator in (8.49) is a monomial. We can compute $L_{s,e}(z)$ and $L_{s,o}(z)$ from (8.49) and either (8.32) or (8.33), though the answer depends on whether m is even or odd. If m is even, we find

$$L_{s,e}(z) = \frac{L_s(\sqrt{z}) + L_s(-\sqrt{z})}{2} = z^{-m/2}\frac{H_{a,o}(z)}{d(z)}, \tag{8.50}$$

$$L_{s,o}(z) = \sqrt{z}\frac{L_s(\sqrt{z}) - L_s(-\sqrt{z})}{2} = -z^{-m/2+1}\frac{H_{a,e}(z)}{d(z)}. \tag{8.51}$$

If m is odd, we find

$$L_{s,e}(z) = \frac{L_s(\sqrt{z}) + L_s(-\sqrt{z})}{2} = -z^{-(m-1)/2}\frac{H_{a,e}(z)}{d(z)}, \tag{8.52}$$

$$L_{s,o}(z) = \sqrt{z}\frac{L_s(\sqrt{z}) - L_s(-\sqrt{z})}{2} = z^{-(m-1)/2}\frac{H_{a,o}(z)}{d(z)}. \tag{8.53}$$

A similar computation with $H_s(z)$ from equation (7.35) shows that

$$H_s(z) = \frac{-2z^{-m}L_a(-z)}{L_a(z)H_a(-z) - L_a(-z)H_a(z)}$$

$$= \frac{z^{-(m-1)}(L_{a,e}(z^2) - z^{-1}L_{a,o}(z^2))}{d(z^2)}.$$

If m is even, we find

$$H_{s,e}(z) = \frac{H_s(\sqrt{z}) + H_s(-\sqrt{z})}{2} = -z^{-m/2}\frac{L_{a,o}(z)}{d(z)}, \tag{8.54}$$

$$H_{s,o}(z) = \sqrt{z}\frac{H_s(\sqrt{z}) - H_s(-\sqrt{z})}{2} = z^{-m/2+1}\frac{L_{a,e}(z)}{d(z)}. \tag{8.55}$$

If m is odd, we find

$$H_{s,e}(z) = \frac{H_s(\sqrt{z}) + H_s(-\sqrt{z})}{2} = z^{-(m-1)/2}\frac{L_{a,e}(z)}{d(z)}, \tag{8.56}$$

$$H_{s,o}(z) = \sqrt{z}\frac{H_s(\sqrt{z}) - H_s(-\sqrt{z})}{2} = -z^{-(m-1)/2}\frac{L_{a,o}(z)}{d(z)}. \tag{8.57}$$

In the case, that m is even equations (8.50),(8.51),(8.54), and (8.55) show that

$$H_{a,o}(z)/d(z) = z^{m/2}L_{s,e}(z), \qquad H_{a,e}(z)/d(z) = z^{m/2-1}L_{s,o}(z)$$
$$L_{a,o}(z)/d(z) = z^{m/2}H_{s,e}(z), \qquad L_{a,e}(z)/d(z) = z^{m/2-1}H_{s,o}(z)$$

Using this in equation (8.43) for $\mathbf{P}_s(z)$ yields equation (8.47). A similar computation using equations (8.52), (8.53), (8.56), and (8.57) shows that

$$H_{a,e}(z)/d(z) = z^{(m-1)/2}L_{s,e}(z), \qquad H_{a,o}(z)/d(z) = z^{(m-1)/2}L_{s,o}(z),$$
$$L_{a,e}(z)/d(z) = z^{(m-1)/2}H_{s,e}(z), \qquad L_{a,o}(z)/d(z) = z^{(m-1)/2}H_{s,o}(z).$$

Using this in equation (8.43) for $\mathbf{P}_s(z)$ yields equation (8.48) and completes the proof of the Lemma.

Example 8.7 *The z-transforms for the Le Gall 5/3 synthesis filters with delay $m = 1$ are, from equation (7.34) and (7.35),*

$$L_s(z) = \frac{1}{2} + z^{-1} + \frac{1}{2}z^{-2} \text{ and } H_s(z) = -\frac{1}{8}z^2 - \frac{1}{4}z + \frac{3}{4} - \frac{1}{4}z^{-1} - \frac{1}{8}z^{-2}.$$

Thus $L_{s,e}(z) = 1/2 + (1/2)z^{-1}$, $L_{s,o}(z) = 1$, $H_{s,e}(z) = -z/8 + 3/4 - (1/8)z^{-1}$, $H_{s,o}(z) = -z/4 - 1/4$. *Lemma 8.6 (in particular, equation (8.48) with $m = 1$) gives exactly $\mathbf{P}_s(z)$ from equation (8.46).*

8.4.4 Orthogonal Filters

Recall that the choice $(\boldsymbol{\ell}_s)_k = (\boldsymbol{\ell}_a)_{-k}$ and $(\mathbf{h}_s)_k = (\mathbf{h}_a)_{-k}$ of equation (7.10) for obtaining the synthesis filters from the analysis filters leads to an orthogonal filter bank with delay $m = 0$ in equation (7.44) and (7.45). Equivalently, $L_s(z) = L_a(z^{-1})$ and $H_s(z) = H_a(z^{-1})$. One can use equation (8.32) to check that

$$\begin{aligned} L_{s,e}(z) &= \frac{L_s(\sqrt{z}) + L_s(-\sqrt{z})}{2} \\ &= \frac{L_a(z^{-1/2}) + L_a(-z^{-1/2})}{2} \\ &= L_{a,e}(z^{-1}). \end{aligned}$$

A similar computation shows that $L_{s,o}(z) = zL_{a,o}(z^{-1}), H_{a,e}(z) = H_{s,e}(z^{-1})$, and $H_{s,o}(z) = zH_{a,o}(z^{-1})$. If we use this in equation (8.47) with $m = 0$ to replace all synthesis polyphase components with the analysis equivalents, we obtain

$$\mathbf{P}_s(z) = \begin{bmatrix} L_{a,e}(z^{-1}) & H_{a,e}(z^{-1}) \\ L_{a,o}(z^{-1}) & H_{a,o}(z^{-1}) \end{bmatrix}.$$

Comparison to equation (8.39) shows that the polyphase synthesis matrix can be computed simply as $\mathbf{P}_s(z) = \mathbf{P}_a(z^{-1})^T$. We have proved

Lemma 8.7 *For an orthogonal filter bank with analysis/synthesis filters that satisfy $(\boldsymbol{\ell}_s)_k = (\boldsymbol{\ell}_a)_{-k}$ and $(\mathbf{h}_s)_k = (\mathbf{h}_a)_{-k}$, we have*

$$\mathbf{P}_s(z) = \mathbf{P}_a(z^{-1})^T.$$

If in addition equation (7.46) holds then $\mathbf{P}_s(z)$ can be written entirely in terms of $L_{a,e}(z)$ and $L_{a,o}(z)$ (see Exercise 20).

Example 8.8 *For the Daubechies 4-tap filters listed in Table 7.4 (these are part of an orthogonal filter bank), we find*

$$L_{a,e}(z) \approx 0.483 + 0.224z^{-1}, \quad L_{a,o}(z) \approx 0.837 - 0.129z^{-1},$$
$$H_{a,e}(z) \approx -0.129 + 0.837z^{-1}, \quad H_{a,o}(z) \approx -0.224 - 0.483z^{-1}$$

to three significant figures. Then

$$\mathbf{P}_a(z) \approx \begin{bmatrix} 0.483 + 0.224z^{-1} & 0.837 - 0.129z^{-1} \\ -0.129 + 0.837z^{-1} & -0.224 - 0.483z^{-1} \end{bmatrix}$$

and so by Lemma 8.7 (replace z^{-1} by z, transpose)

$$\mathbf{P}_s(z) \approx \begin{bmatrix} 0.224z + 0.483 & 0.837z - 0.129 \\ -0.129z + 0.837 & -0.483z - 0.224 \end{bmatrix}.$$

8.5 Lifting

8.5.1 Relation Between the Polyphase Matrices

Suppose that $\boldsymbol{\ell}_a$ and $\mathbf{h}_a$ are the low- and high-pass analysis filters in a perfect reconstruction filter bank. According to the Lifting Theorem 8.1, a second filter pair $\boldsymbol{\ell}_a, \tilde{\mathbf{h}}_a$ also constitutes the analysis filters in a perfect reconstruction filter bank if and only if the various z-transforms are related as

$$H_a(z) = \tilde{H}_a(z) + s(z^2)L_a(z) \tag{8.58}$$

for some Laurent polynomial $s(z)$. Similarly, a filter pair $(\tilde{\boldsymbol{\ell}}_a, \mathbf{h}_a)$ constitutes the analysis filters in a perfect reconstruction filter bank if and only if the various z-transforms are related as

$$L_a(z) = \tilde{L}_a(z) + t(z^2)H_a(z) \tag{8.59}$$

for some Laurent polynomial $t(z)$.

In either case above, it turns out the polyphase matrices are also related in a simple way.

Lemma 8.8 *Equation (8.58) holds if and only*

$$P_a(z) = \begin{bmatrix} 1 & 0 \\ s(z) & 1 \end{bmatrix} \tilde{P}_a(z), \tag{8.60}$$

where P_a is the polyphase matrix for the filter pair $(\boldsymbol{\ell}_a, \mathbf{h}_a)$ and $\tilde{P}_a$ is the polyphase matrix for $(\boldsymbol{\ell}_a, \tilde{\mathbf{h}}_a)$. Similarly, (8.59) holds if and only

$$P_a(z) = \begin{bmatrix} 1 & t(z) \\ 0 & 1 \end{bmatrix} \hat{P}_a(z). \tag{8.61}$$

where $\hat{\mathbf{P}}_a$ is the polyphase matrix for $(\tilde{\boldsymbol{\ell}}_a, \mathbf{h}_a)$.

Proof: Suppose that (8.58) holds for some fixed $s(z)$. Then

$$\begin{aligned} H_{a,e}(z) &= (H_a(\sqrt{z}) + H_a(-\sqrt{z}))/2 \\ &= (\tilde{H}_a(\sqrt{(z)}) + \tilde{H}_a(-\sqrt{z}))/2 + s(z)(L_a(\sqrt{z}) + L_a(-\sqrt{z})) \\ &= \tilde{H}_{a,e}(z) + s(z)L_{a,e}(z). \end{aligned} \tag{8.62}$$

Similarly,

$$\begin{aligned} H_{a,o}(z) &= \sqrt{z}(H_a(\sqrt{z}) - H_a(-\sqrt{z}))/2 \\ &= \sqrt{z}(\tilde{H}_a(\sqrt{(z)}) - \tilde{H}_a(-\sqrt{z}))/2 + s(z)\sqrt{z}(L_a(\sqrt{z}) - L_a(-\sqrt{z})) \\ &= \tilde{H}_{a,o}(z) + s(z)L_{a,o}(z). \end{aligned} \tag{8.63}$$

Equations (8.62) and (8.63) can be amalgamated into the single matrix equation

$$\begin{bmatrix} L_{a,e}(z) & L_{a,o}(z) \\ H_{a,e}(z) & H_{a,o}(z) \end{bmatrix} = \begin{bmatrix} 1 & 0 \\ s(z) & 1 \end{bmatrix} \begin{bmatrix} L_{a,e}(z) & L_{a,o}(z) \\ \tilde{H}_{a,e}(z) & \tilde{H}_{a,o}(z) \end{bmatrix},$$

which is exactly equation (8.60). Virtually the same argument shows that (8.59) implies (8.61).

For the converses—that (8.60) implies (8.58), and that (8.61) implies (8.59)—see Exercise 15.

As with the Haar lifting scheme, the matrix on the right in (8.60) embodies the *dual lifting* step, or *prediction*, while the matrix in (8.61) embodies the *primal lifting* step or *update*.

We can use equations (8.60) and (8.61) in two different ways. Given a perfect reconstruction filter bank with filters $\boldsymbol{\ell}_a, \mathbf{h}_a$, we can compute $\mathbf{P}_a(z)$ and then try to factor $\mathbf{P}_a(z)$ into simpler lifting steps, that is, matrices of the form in (8.58) and (8.59), possibly with scaling steps (diagonal matrices with constant or monomial entries). Alternatively, we can start with a relatively simple filter pair (even the "lazy wavelet" transform, which simply splits the signal into even and odd polyphase components with no filtering; see Exercise 9) or the corresponding polyphase analysis matrix $\mathbf{P}_a$ and then multiply $\mathbf{P}_a(z)$ by matrices of the form in (8.58) or (8.59), to build up a more complex transform. By choosing $s(z)$ or $t(z)$ at each stage we can obtain various desirable properties in the transform.

We first focus on the former case—factoring a polyphase matrix into lifting steps, possibly with a scaling step. That is, we seek a factorization

$$\mathbf{P}_a = \mathbf{D}\mathbf{M}_1 \cdots \mathbf{M}_k,$$

where the $\mathbf{M}_i$ are lifting matrices and $\mathbf{D}$ is a diagonal scaling matrix.

Before proceeding, it is worth noting that these lifting matrices all satisfy $\det(\mathbf{M}_i) = 1$ and are easily invertible, as

$$\begin{bmatrix} 1 & 0 \\ s(z) & 1 \end{bmatrix}^{-1} = \begin{bmatrix} 1 & 0 \\ -s(z) & 1 \end{bmatrix} \text{ and } \begin{bmatrix} 1 & t(z) \\ 0 & 1 \end{bmatrix}^{-1} = \begin{bmatrix} 1 & -t(z) \\ 0 & 1 \end{bmatrix}. \tag{8.64}$$

The scaling matrix will be of the form

$$\mathbf{D} = \begin{bmatrix} a(z) & 0 \\ 0 & d(z) \end{bmatrix}$$

for some monomials $a(z), d(z)$.

We will start by looking at a couple of examples.

8.5.2 Factoring the Le Gall 5/3 Polyphase Matrix

Let us start with a specific example, the Le Gall 5/3 filters. We showed in equation (8.40) that these filters have polyphase analysis matrix

$$\mathbf{P}_a(z) = \begin{bmatrix} -\frac{1}{8}z + \frac{3}{4} - \frac{1}{8}z^{-1} & \frac{1}{4}z + \frac{1}{4} \\ -\frac{1}{2} - \frac{1}{2}z^{-1} & 1 \end{bmatrix}.$$

Our goal is to factor $\mathbf{P}_a(z)$ in the form

$$\mathbf{P}_a(z) = \mathbf{D}\mathbf{M}_1 \cdots \mathbf{M}_k,$$

where $\mathbf{D}$ is a diagonal scaling matrix and the $\mathbf{M}_i$ are lifting matrices of the form shown in (8.60) or (8.61). This is similar to the factorization for the Haar filter bank, given in (8.8) and (8.9). Many factorizations of the form above are possible; we will choose a particular one.

Remark 8.2 We typically alternate primal and dual liftings, for it is easy to see that two successive primal lifts can be amalgamated into a single (more complicated) primal lift, since the product

$$\begin{bmatrix} 1 & t_1(z) \\ 0 & 1 \end{bmatrix} \begin{bmatrix} 1 & t_2(z) \\ 0 & 1 \end{bmatrix} = \begin{bmatrix} 1 & t_1(z) + t_2(z) \\ 0 & 1 \end{bmatrix}$$

is of the form shown in (8.61). A similar observation holds for dual lifts.

We will begin with a primal lifting step as per equation (8.61), which modifies the low-pass filters and leaves the high-pass unchanged. Indeed, if the new filters have z-transforms $\tilde{L}(z)$ and $\tilde{H}(z)$ then (8.61) implies that $\tilde{H}_{a,e}(z) = H_{a,e}(z)$ and $\tilde{H}_{a,o}(z) = H_{a,o}(z)$ (this follows from the dot product of the bottom row of the lifting matrix with $\tilde{\mathbf{P}}_a$), so that $\tilde{H}(z) = H(z)$. However, the dot product of the top row of the lifting matrix in (8.61) shows that, for any given $t(z)$, we have

$$L_{a,e}(z) = t(z)H_{a,e}(z) + \tilde{L}_{a,e}(z) \tag{8.65}$$

and

$$L_{a,o}(z) = t(z)H_{a,o}(z) + \tilde{L}_{a,o}(z). \tag{8.66}$$

By choosing $t(z)$ appropriately, we can make the filter corresponding to $\tilde{L}$ simpler (fewer taps).

The key is to interpret (8.65) and/or (8.66) as a Laurent polynomial division; for equation (8.65), $L_{a,e}(z)$ is divided by $H_{a,e}(z)$, with quotient $t(z)$ and $\tilde{L}_{a,e}(z)$ as the remainder. This will result in $\text{wid}(\tilde{L}_{a,e}) < \text{wid}(H_{a,e})$, so the new low-pass filter will have fewer taps. Alternatively, we can work with equation (8.66) to obtain $t(z)$ as the quotient of $L_{a,o}(z)$ divided by $H_{a,o}(z)$, with $\tilde{L}_{a,o}(z)$ as the remainder. Note that we will obtain a different choice for $t(z)$ by working with (8.65) versus (8.66), though in principle either choice will work (but lead to different factorizations of $\mathbf{P}_a$). Since $H_{a,o}(z) = 1$ here, this division is very easy, and so we will use (8.66) for this first step.

To use (8.66), we require

$$\frac{1}{4}z + \frac{1}{4} = t(z)(1) + \tilde{L}_{a,o}(z)$$

with $\text{wid}(\tilde{L}_{a,o}(z)) < \text{wid}(1) = 0$. The only choice here is to take $t(z) = z/4 + 1/4$ and $\tilde{L}_{a,o}(z) = 0$ (so $\text{wid}(\tilde{L}_{a,o}(z)) = -\infty$). With this choice for $t(z)$, equation (8.61) becomes

$$\begin{bmatrix} -\frac{1}{8}z + \frac{3}{4} - \frac{1}{8}z^{-1} & \frac{1}{4}z + \frac{1}{4} \\ -\frac{1}{2} - \frac{1}{2}z^{-1} & 1 \end{bmatrix} = \begin{bmatrix} 1 & \frac{1}{4}z + \frac{1}{4} \\ 0 & 1 \end{bmatrix} \tilde{\mathbf{P}}_a(z).$$

Let $\mathbf{M}_1$ denote the first matrix on the right above, so that a bit of matrix algebra shows

$$\begin{aligned} \tilde{\mathbf{P}}_a(z) &= \mathbf{M}_1^{-1}\mathbf{P}_a(z) \\ &= \begin{bmatrix} 1 & -\frac{1}{4}z - \frac{1}{4} \\ 0 & 1 \end{bmatrix} \begin{bmatrix} -\frac{1}{8}z + \frac{3}{4} - \frac{1}{8}z^{-1} & \frac{1}{4}z + \frac{1}{4} \\ -\frac{1}{2} - \frac{1}{2}z^{-1} & 1 \end{bmatrix} \\ &= \begin{bmatrix} 1 & 0 \\ -\frac{1}{2} - \frac{1}{2}z^{-1} & 1 \end{bmatrix} \end{aligned} \tag{8.67}$$

(the inverse of $\mathbf{M}_1$ is easy to compute from (8.64).)

Equation (8.67) provides the lifting factorization we want. If we let $\mathbf{M}_2$ denote the matrix on the right-hand side of (8.67), we see that $\mathbf{P}_a(z) = \mathbf{M}_1\mathbf{M}_2$, or

$$\begin{bmatrix} -\frac{1}{8}z + \frac{3}{4} - \frac{1}{8}z^{-1} & \frac{1}{4}z + \frac{1}{4} \\ -\frac{1}{2} - \frac{1}{2}z^{-1} & 1 \end{bmatrix} = \begin{bmatrix} 1 & \frac{1}{4}z + \frac{1}{4} \\ 0 & 1 \end{bmatrix} \begin{bmatrix} 1 & 0 \\ -\frac{1}{2} - \frac{1}{2}z^{-1} & 1 \end{bmatrix}. \tag{8.68}$$

In this case, we can take $\mathbf{D} = \mathbf{I}$, so no scaling is needed. The factorization is complete after just one step!

It should be clear why we started with a lifting step of the form (8.61) instead of (8.60)—the Le Gall low-pass filter has greater width than the high-pass filter, so the division knocks the width of the new low-pass filter down, lower than the width of the high-pass filter.

To invert the transform, we simple invert each lifting and scaling step, in reverse order. Thus, in light of equation (8.68) we have

$$\mathbf{P}_s(z) = \begin{bmatrix} 1 & -\frac{1}{4}z - \frac{1}{4} \\ \frac{1}{2} + \frac{1}{2}z^{-1} & -\frac{1}{8}z + \frac{3}{4} - \frac{1}{8}z^{-1} \end{bmatrix} = \begin{bmatrix} 1 & 0 \\ \frac{1}{2} + \frac{1}{2}z^{-1} & 1 \end{bmatrix} \begin{bmatrix} 1 & -\frac{1}{4}z - \frac{1}{4} \\ 0 & 1 \end{bmatrix}.$$

8.5.3 Factoring the Haar Polyphase Matrix

The Haar polyphase matrix is

$$\mathbf{P}_a(z) = \begin{bmatrix} 1/2 & 1/2 \\ 1/2 & -1/2 \end{bmatrix}.$$

Let us again start with a primal lifting of the form of equation (8.61). Again, equations (8.65) and (8.66) must hold. If we use the latter, we need

$$-\frac{1}{2} = \frac{1}{2}t(z) + \tilde{L}_{a,e}(z)$$

with $\text{wid}(\tilde{L}_{a,e}) < \text{wid}(1/2) = 0$, so that $\text{wid}(\tilde{L}_{a,e}) = -\infty$. That is, $\tilde{L}_{a,e}(z) = 0$, and clearly $t(z) = -1$. From equation (8.61), we have

$$\begin{bmatrix} 1/2 & 1/2 \\ 1/2 & -1/2 \end{bmatrix} = \begin{bmatrix} 1 & -1 \\ 0 & 1 \end{bmatrix} \tilde{\mathbf{P}}_a(z).$$

We find that

$$\tilde{\mathbf{P}}_a(z) = \begin{bmatrix} 1 & 0 \\ 1/2 & -1/2 \end{bmatrix}.$$

We now apply a dual lifting step of the form of equation (8.60), but with $\tilde{P}_a$ above playing the role. In the present case, we need

$$\tilde{\mathbf{P}}_a(z) = \begin{bmatrix} 1 & 0 \\ 1/2 & -1/2 \end{bmatrix} = \begin{bmatrix} 1 & 0 \\ s(z) & 1 \end{bmatrix} \begin{bmatrix} a(z) & b(z) \\ c(z) & d(z) \end{bmatrix},$$

where we are now using a, b, c, d for the polyphase components of the low- and high-pass filters. The above equation implies that $a(z) = 1$ and $b(z) = 0$ (the low-pass filters are unchanged) and also each of $1/2 = s(z)a(z) + c(z)$ and $-1/2 = s(z)b(z) + d(z)$. Since $b(z) = 0$ the division algorithm does not apply to this last equation (it degenerates to $-1/2 = d(z)$, and leaves s undetermined). But from the previous equation, we have

$$1/2 = s(z)(1) + c(z),$$

where we need $\text{wid}(d) < \text{wid}(1) = 0$, so $\text{wid}(c) = -\infty$ and c is zero. This means $s(z) = 1/2$. We then have

$$\tilde{\mathbf{P}}_a(z) = \begin{bmatrix} 1 & 0 \\ 1/2 & -1/2 \end{bmatrix} = \begin{bmatrix} 1 & 0 \\ 1/2 & 1 \end{bmatrix} \begin{bmatrix} 1 & 0 \\ 0 & -1/2 \end{bmatrix}$$

($d(z) = -1/2$ is forced). All in all then we have a factorization

$$\mathbf{P}_a(z) = \begin{bmatrix} 1 & -1 \\ 0 & 1 \end{bmatrix} \begin{bmatrix} 1 & 0 \\ 1/2 & 1 \end{bmatrix} \begin{bmatrix} 1 & 0 \\ 0 & -1/2 \end{bmatrix}. \tag{8.69}$$

The first matrix operation here is a scaling (the diagonal matrix), followed by a prediction, then an update.

This is not the only order in which the steps can be done, however. A diagonal matrix does not quite commute with lifting matrices, but one can verify that

$$\begin{bmatrix} 1 & f \\ 0 & 1 \end{bmatrix} \begin{bmatrix} K_1 & 0 \\ 0 & K_2 \end{bmatrix} = \begin{bmatrix} K_1 & 0 \\ 0 & K_2 \end{bmatrix} \begin{bmatrix} 1 & \frac{K_2}{K_1} f \\ 0 & 1 \end{bmatrix} \tag{8.70}$$

and

$$\begin{bmatrix} 1 & 0 \\ f & 1 \end{bmatrix} \begin{bmatrix} K_1 & 0 \\ 0 & K_2 \end{bmatrix} = \begin{bmatrix} K_1 & 0 \\ 0 & K_2 \end{bmatrix} \begin{bmatrix} 1 & 0 \\ \frac{K_1}{K_2} f & 1 \end{bmatrix}. \tag{8.71}$$

These two equations can be use to move the scaling matrix in (8.69) to the front. By (8.71) (use $K_1 = 1, K_2 = -1/2, f = 1/2$), we have

$$\begin{bmatrix} 1 & 0 \\ 1/2 & 1 \end{bmatrix} \begin{bmatrix} 1 & 0 \\ 0 & -1/2 \end{bmatrix} = \begin{bmatrix} 1 & 0 \\ 0 & -1/2 \end{bmatrix} \begin{bmatrix} 1 & 0 \\ -1 & 1 \end{bmatrix}$$

and from (8.70) (here $f = -1$)

$$\begin{bmatrix} 1 & -1 \\ 0 & 1 \end{bmatrix} \begin{bmatrix} 1 & 0 \\ 0 & -1/2 \end{bmatrix} = \begin{bmatrix} 1 & 0 \\ 0 & -1/2 \end{bmatrix} \begin{bmatrix} 1 & 1/2 \\ 0 & 1 \end{bmatrix}$$

All in all we find

$$\mathbf{P}_a(z) = \begin{bmatrix} 1 & 0 \\ 0 & -1/2 \end{bmatrix} \begin{bmatrix} 1 & -1 \\ 0 & 1 \end{bmatrix} \begin{bmatrix} 1 & 0 \\ 1/2 & 1 \end{bmatrix}$$

which is the factorization given in equations (8.8) and (8.9), with scaling as the final operation applied.

The procedure above works in general. In Ref. [11], the authors prove the following theorem:

Theorem 8.2 *If $\mathbf{P}_a(z)$ is the polyphase matrix for a filter pair $\boldsymbol{\ell}_a, \mathbf{h}_a$ then $\mathbf{P}_a$ has a factorization (not necessarily unique) of the form*

$$\mathbf{P}_a(z) = \mathbf{D}\mathbf{M}_1 \cdots \mathbf{M}_k,$$

where the $\mathbf{M}_i$ are lifting matrices of the form in (8.60) and (8.61) and $\mathbf{D}$ is a diagonal matrix with monomial entries.

Remark 8.3 It is important to note that the lifting factorization is not cut-and-dried; many factorizations are possible, and the order—primal and dual lifting, and scaling—can be changed. We could, for example, have started the Le Gall 5/3 factorization by using equation (8.65), involving the even polyphase components, to obtain $t(z)$ for the first lifting step. In this case, we would require

$$-\frac{1}{8}z + \frac{3}{4} - \frac{1}{8}z^{-1} = t(z)\left(-\frac{1}{2} - \frac{1}{2}z^{-1}\right) + \tilde{L}_{a,e}(z)$$

with $\text{wid}(\tilde{L}_{a,e}(z)) < \text{wid}(-\frac{1}{2} - \frac{1}{2}z^{-1}) = 1$. One possibility is

$$t(z) = \frac{z}{4} - \frac{7}{4} \text{ with } \tilde{L}_{a,e}(z) = z^{-1}.$$

Continuing this process (with other choices) will lead to a different factorization of the polyphase matrix. Or, we could have started with a dual lifting step based on equation (8.60). Any of these will lead to a factorization of the polyphase matrix that may or may not have advantages or desirable properties.

8.5.4 Efficiency

Consider implementing the Le Gall 5/3 transform using $\mathbf{P}_a(z)$ directly. We need to compute the matrix vector product

$$\begin{bmatrix} -\frac{1}{8}z + \frac{3}{4} - \frac{1}{8}z^{-1} & \frac{1}{4}z + \frac{1}{4} \\ -\frac{1}{2} - \frac{1}{2}z^{-1} & 1 \end{bmatrix} \begin{bmatrix} X_e(z) \\ z^{-1}X_o(z) \end{bmatrix}$$

$$= \begin{bmatrix} \left(-\frac{1}{8}z + \frac{3}{4} - \frac{1}{8}z^{-1}\right)X_e(z) + \left(\frac{1}{4}z + \frac{1}{4}\right)z^{-1}X_o(z) \\ \left(-\frac{1}{2} - \frac{1}{2}z^{-1}\right)X_e(z) + z^{-1}X_o(z) \end{bmatrix}.$$

The powers of z correspond to shifts, which we will not count. We will focus on the coefficient multiplications and additions, and simply count multiplying $X_e(z)$ or $X_o(z)$ by a constant c as "one operation," and similarly for the additions. In this case, then the quantity $\left(-\frac{1}{8}z + \frac{3}{4} - \frac{1}{8}z^{-1}\right) X_e(z)$ requires five operations (three multiplies, two adds), while $\left(\frac{1}{4}z + \frac{1}{4}\right) X_o(z)$ requires three operations; adding these to produce $X_\ell(z)$ requires one more operation, a total of nine operations. The second component of the product requires 4 operations, so a total of 13 operations are needed.

Contrast this to computing the lifted/factored version of the transform, as

$$\begin{bmatrix} 1 & \frac{1}{4}z + \frac{1}{4} \\ 0 & 1 \end{bmatrix} \begin{bmatrix} 1 & 0 \\ -\frac{1}{2} - \frac{1}{2}z^{-1} & 1 \end{bmatrix} \begin{bmatrix} X_e(z) \\ z^{-1}X_o(z) \end{bmatrix}.$$

The product

$$\begin{bmatrix} \tilde{X}_e(z) \\ \tilde{X}_o(z) \end{bmatrix} = \begin{bmatrix} 1 & 0 \\ -\frac{1}{2} - \frac{1}{2}z^{-1} & 1 \end{bmatrix} \begin{bmatrix} X_e(z) \\ z^{-1}X_o(z) \end{bmatrix} = \begin{bmatrix} X_e(z) \\ \left(-\frac{1}{2} - \frac{1}{2}z^{-1}\right) X_e(z) + z^{-1}X_o(z) \end{bmatrix}$$

requires four operations, and

$$\begin{bmatrix} X_\ell(z) \\ X_h(z) \end{bmatrix} = \begin{bmatrix} 1 & \frac{1}{4}z + \frac{1}{4} \\ 0 & 1 \end{bmatrix} \begin{bmatrix} \tilde{X}_e(z) \\ \tilde{X}_o(z) \end{bmatrix} = \begin{bmatrix} \tilde{X}_e(z) + \left(\frac{1}{4}z + \frac{1}{4}\right) \tilde{X}_o(z) \\ \tilde{X}_o(z) \end{bmatrix}$$

requires four operations. The lifted version of the transform can be computed in about $8/13 \approx 0.62$ or 62% of the time needed for a more straightforward implementation.

In Ref. [11], the authors compute the potential speed-up for a variety of discrete wavelet transforms via lifting, and note that asymptotically this approach can double the speed of the computation. As they also note, however, there are many other tricks used to improve the performance of the discrete wavelet transform.

8.5.5 Lifting to Design Transforms

As remarked above, we can use lifting to take an existing filter bank and then factor its polyphase matrix into a number of simpler lifting steps. But we can also go the other way. Equations (8.60) and (8.61) can be used to lift a given filter bank polyphase matrix to a new polyphase matrix corresponding to a (possibly) "improved" filter bank, one with more desirable properties for the purpose at hand. In this section, our goal is not an exhaustive treatment of the possibilities, but rather a few simple examples that illustrate the general idea. More applications can be found in the exercises and references (see, e.g., [30]).

Example 8.9 *Let us begin with the "lazy filter bank," with polyphase matrix $\mathbf{P}_a = \mathbf{I}$ (see Exercise 9 for the corresponding filters). The goal is to produce a new perfect reconstruction filter bank with low- and high-pass analysis filters $\boldsymbol{\ell}_a$ and $\mathbf{h}_a$. Recall our characterization of a low-pass filter $\boldsymbol{\ell}_a$ was that $L_a(-1) = 0$, and $L_a(1) \neq 0$, where $L_a(z)$ denotes the z-transform of $\boldsymbol{\ell}_a$. Similarly, our characterization of a high-pass filter $\mathbf{h}_a$ was that $H_a(1) = 0$ and $H_a(-1) \neq 0$. In terms of the polyphase decomposition and equation (8.34) applied with $X(z) = L_a(z)$, the low-pass condition is equivalent to $L_{a,e}((-1)^2) - (1)L_{a,o}((-1)^2) = 0$ or*

$$L_{a,e}(1) - L_{a,o}(1) = 0, \tag{8.72}$$

while $X(z) = H_a(z)$ yields the condition

$$H_{a,e}(1) + H_{a,o}(1) = 0. \tag{8.73}$$

If we begin with the lazy filter bank and apply equations (8.60) and (8.61) to lift, we obtain a polyphase matrix

$$\mathbf{P}_a(z) = \begin{bmatrix} 1 & t(z) \\ 0 & 1 \end{bmatrix} \begin{bmatrix} 1 & 0 \\ s(z) & 1 \end{bmatrix} \mathbf{I} = \begin{bmatrix} s(z)t(z)+1 & t(z) \\ s(z) & 1 \end{bmatrix}$$

for the new filter bank. We can always add a scaling step later. For the new filter bank, we have $L_{a,e}(z) = s(z)t(z) + 1, L_{a,o}(z) = t(z), H_{a,e}(z) = s(z)$, and $H_{a,o}(z) = 1$. Equation (8.73) then yields $s(1) + 1 = 0$ so that $s(1) = -1$. Equation (8.72) yields $s(1)t(1) + 1 - t(1) = 0$ so that $t(1) = 1/2$. All in all the functions $s(z), t(z)$ must satisfy

$$t(1) = 1/2 \tag{8.74}$$

$$s(1) = -1. \tag{8.75}$$

Note also that $L_a(1) = L_{a,e}(1) + L_{a,o}(1) = s(1)t(1) + 1 + t(1) = 1 \neq 0$ and $H_a(-1) = H_{a,e}(1) - H_{a,o}(1) = s(1) - 1 = -2 \neq 0$. With the choices (8.74) and (8.75), these resulting analysis filters will be low and high pass. The factorizations (8.68) for the Le Gall 5/3 filters and (8.8) and (8.9) for the Haar filter bank are in accord with this.

As a specific example, consider the choice $s(z) = -1$, $t(z) = 1/2$, constants. We obtain

$$\mathbf{P}_a = \begin{bmatrix} 1/2 & 1/2 \\ -1 & 1 \end{bmatrix}.$$

It is easy to see this is the polyphase matrix for a rescaled versions of the Haar filters, since $L_{a,e}(z) = 1/2, L_{a,o}(z) = 1/2, H_{a,e}(z) = -1, H_{a,o}(z) = 1$. Thus $L_a(z) = L_{a,e}(z^2) + z^{-1}L_{a,o}(z^2) = \frac{1}{2} + \frac{1}{2}z^{-1}$ and $H_a(z) = H_{a,e}(z^2) + z^{-1}H_{a,o}(z^2) = -1 + z^{-1}$, corresponding to the filters $\boldsymbol{\ell}_a = (1/2, 1/2)$ and $\mathbf{h}_a = (-1, 1)$.

The choice $s(z) = -\frac{1}{2} - \frac{1}{2}z^{-1}$, $t(z) = \frac{1}{4}z + \frac{1}{4}$ yields the Le Gall 5/3 filters.

Example 8.10 *Suppose we already have a filter bank with low- and high-pass analysis filters $\tilde{\boldsymbol{\ell}}_a$ and $\tilde{\mathbf{h}}_a$ and want to construct a "new and improved" filter bank via the polyphase matrix equation (8.60), perhaps to increase the efficacy of the low and high pass performance. Let $\boldsymbol{\ell}_a$ and $\mathbf{h}_a$ denote the new filters. From (8.60), it is easy to see that $L_{a,e}(z) = \tilde{L}_{a,e}(z)$ and $L_{a,o}(z) = \tilde{L}_{a,o}(z)$, so the low-pass filter is unchanged. The new high-pass polyphase components satisfy*

$$H_{a,e}(z) = s(z)\tilde{L}_{a,e}(z) + \tilde{H}_{a,e}(z) \text{ and } H_{a,o}(z) = s(z)\tilde{L}_{a,o}(z) + \tilde{H}_{a,o}(z). \quad (8.76)$$

If $H_a(z)$ is to correspond to a high-pass filter, then we need $H_a(1) = 0$, which means equation (8.73) must hold. Adding both equations in (8.76) and setting $z = 1$ yields

$$s(1)(\tilde{L}_{a,e}(1) + \tilde{L}_{a,o}(1)) + \tilde{H}_{a,e}(1) + \tilde{H}_{a,o}(1) = 0.$$

But since $\tilde{\mathbf{h}}$ is a high-pass filter, we have $\tilde{H}_{a,e}(1) + \tilde{H}_{a,o}(1) = 0$ and moreover, $L_a(1) = \tilde{L}_{a,e}(1) + \tilde{L}_{a,o}(1) \neq 0$. We conclude that $s(1) = 0$.

A very similar analysis shows that if we use equation (8.61) to lift a pair of low- and high-pass filters to a new low- and high-pass filters then we must also have $t(1) = 0$. Note that this analysis assumes that the original filters $\tilde{\boldsymbol{\ell}}$ and $\tilde{\mathbf{h}}$ are already low and high pass (and so does not apply to the previous Example 8.9 in which $\tilde{\mathbf{P}}_a = \mathbf{I}$.)

As an example, take $\tilde{\boldsymbol{\ell}}_a$ and $\tilde{\mathbf{h}}_a$ as the Haar filter bank, with $s(z) = a(z-1)$ and $t(z) = b(z-1)$ for some constants a, b. Our goal is to adjust a and b so that $L_a'(-1) = 0$ and $H_a'(1) = 0$ (improving the frequency response of the filters, by increasing the order to which H_a vanishes at dc and L_a vanishes at the Nyquist frequency). In this case,

$$\mathbf{P}_a = \begin{bmatrix} 1 & 0 \\ a(z-1) & 1 \end{bmatrix} \begin{bmatrix} 1 & b(z-1) \\ 0 & 1 \end{bmatrix} \begin{bmatrix} \frac{1}{2} & \frac{1}{2} \\ \frac{1}{2} & -\frac{1}{2} \end{bmatrix}.$$

A little algebra yields

$$\begin{aligned}\mathbf{P}_a &= \begin{bmatrix} L_{a,e}(z) & L_{a,o}(z) \\ H_{a,e}(z) & H_{a,o}(z) \end{bmatrix} \\ &= \frac{1}{2}\begin{bmatrix} ab(z-1)^2 + 1 - b + bz & ab(z-1)^2 + 1 + b - bz \\ az - a + 1 & az - a - 1 \end{bmatrix}. \end{aligned} \quad (8.77)$$

From (8.34), we have $L_a(z) = L_{a,e}(z^2) + z^{-1}L_{a,o}(z^2)$ and $H_a(z) = H_{a,e}(z^2) + z^{-1}H_{a,o}(z^2)$ from which we can compute

$$L_a'(z) = 2zL_{a,e}'(z^2) + 2L_{a,o}'(z^2) - z^{-2}L_{a,o}(z^2)$$
$$H_a'(z) = 2zH_{a,e}'(z^2) + 2H_{a,o}'(z^2) - z^{-2}H_{a,o}(z^2).$$

In particular,

$$L_a'(-1) = -2L_{a,e}'(1) + 2L_{a,o}'(1) - L_{a,o}(1)$$
$$H_a'(1) = 2H_{a,e}'(1) + 2H_{a,o}'(1) - H_{a,o}(1).$$

With the expressions from (8.77), we find $L_a'(-1) = -2b - 1/2$, *so we take* $b = -1/4$ *to obtain* $L_a'(-1) = 0$. *Similarly, we have* $H'(1) = 2a + 1/2$, *so we need* $a = -1/4$. *This yields*

$$\mathbf{P}_a = \begin{bmatrix} 1 & 0 \\ -\frac{1}{4}(z-1) & 1 \end{bmatrix} \begin{bmatrix} 1 & -\frac{1}{4}(z-1) \\ 0 & 1 \end{bmatrix} \begin{bmatrix} \frac{1}{2} & \frac{1}{2} \\ \frac{1}{2} & -\frac{1}{2} \end{bmatrix}$$
$$= \begin{bmatrix} \frac{z^2}{32} - \frac{3}{16}z + \frac{21}{32} & \frac{z^2}{32} - \frac{1}{16}z + \frac{13}{32} \\ -\frac{1}{8}z + \frac{5}{8} & -\frac{1}{8}z - \frac{3}{8} \end{bmatrix}.$$

This corresponds to filters with z-transforms

$$L_a(z) = \frac{1}{32}z^4 + \frac{1}{32}z^3 - \frac{3}{16}z^2 + \frac{1}{16}z + \frac{21}{32} + \frac{13}{32}z^{-1}$$
$$H_a(z) = -\frac{1}{8}z^2 - \frac{1}{8}z + \frac{5}{8} - \frac{3}{8}z^{-1}.$$

The low- and high-pass frequency response of the original Haar filters and lifted filters is shown in Figure 8.1.

Example 8.11 *In some applications, we know that the signals with which we will be working will have integer components or some other discrete values. The images that we compressed in previous chapters are good examples—each pixel was an integer in the range 0–255. It can be helpful for storage or computational purposes if the filter bank transform outputs* $\mathbf{X}_\ell$ *and* $\mathbf{X}_h$ *also have integer entries. In other words, we want a filter bank transform that maps integer-valued signals to transforms with integer components, while of course giving us controllable frequency and spatial localization and perfect invertibility. A tall order!*

The lifting scheme gives us the ability to design and implement such a transform. Consider an integer-valued signal $\mathbf{x}$ *with z-transform* $X(z)$. *Suppose we have a filter bank with polyphase matrix* $\mathbf{P}_a(z)$ *factored into lifting steps of the form in equation (8.60) and (8.61). Consider a single lifting step of the form*

$$\begin{bmatrix} Y_1(z) \\ Y_2(z) \end{bmatrix} = \begin{bmatrix} 1 & 0 \\ s(z) & 1 \end{bmatrix} \begin{bmatrix} X_e(z) \\ z^{-1}X_o(z) \end{bmatrix}. \tag{8.78}$$

so that $Y_1(z) = X_e(z)$ *and* $Y_2(z) = s(z)X_e(z) + z^{-1}X_o(z)$. *Note that both* X_e *and* X_o *have integer components. Thus* Y_1 *has integer components, but there is no reason* Y_2 *should (since* $s(z)$ *may not have integer components). We can assure that*

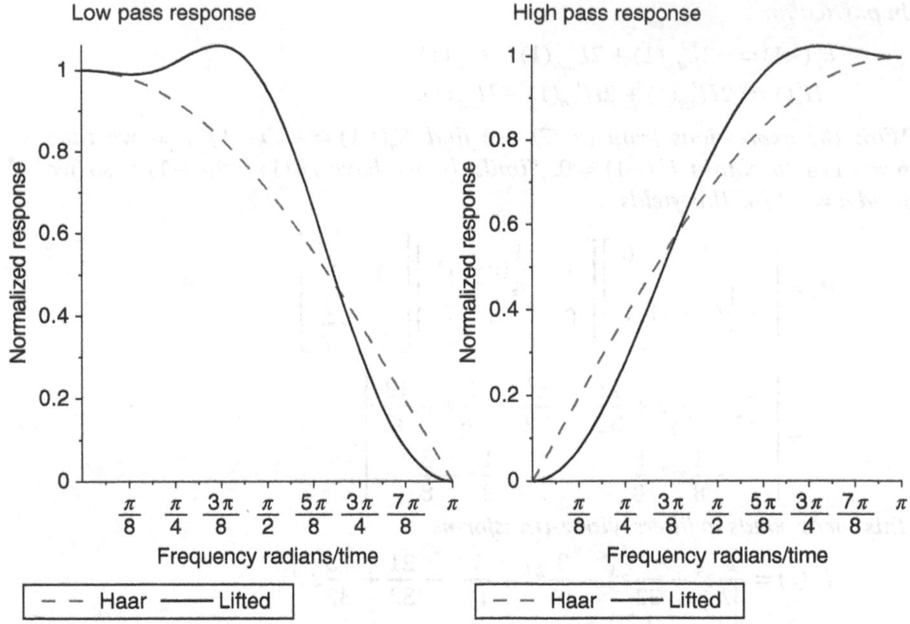

Figure 8.1 Low- and high-pass response of original and lifted filters.

Y_2 does in fact have integer components, however, by modifying the transform embodied by (8.78) as

$$Y_1(z) = X_e(z) \text{ and } Y_2(z) = \lfloor s(z)X_e(z) + z^{-1}X_o(z)\rfloor, \tag{8.79}$$

where $\lfloor a\rfloor$ denotes the greatest integer less than or equal to a; it is understood that this operation is applied to a vector or z-transform component-by-component. It may seem like the transformation of (8.79) loses information due to the rounding, but if we know that **x** *has integer entries, this lifting operation is invertible! Specifically, given Y_1 and Y_2, we obviously recover $X_e = Y_1$ exactly. Moreover, since X_o is known to have integer components, $Y_2(z) = \lfloor s(z)X_e(z) + z^{-1}X_o(z)\rfloor = \lfloor s(z)X_e(z)\rfloor + z^{-1}X_o(z)$ from which we can recover*

$$X_o(z) = zY_2(z) - z\lfloor s(z)X_e(z)\rfloor, \tag{8.80}$$

since $X_e = Y_1$ is known.

To illustrate, suppose $s(z) = -\frac{1}{2} - \frac{1}{2}z^{-1}$ and **x** *has components $x_0 = 14, x_1 = 18, x_2 = 7, x_3 = 28$, so $X(z) = 14 + 18z^{-1} + 7z^{-2} + 28z^{-3}$ and $X_e(z) = 14 + 7z^{-1}$, $X_o(z) = 18 + 28z^{-1}$. Then*

$$\begin{bmatrix} 1 & 0 \\ s(z) & 1 \end{bmatrix} \begin{bmatrix} X_e(z) \\ z^{-1}X_o(z) \end{bmatrix} = \begin{bmatrix} 14 + 7z^{-1} \\ -7 + \frac{15}{2}z^{-1} + \frac{49}{2}z^{-2} \end{bmatrix}.$$

Equation (8.79) then yields $Y_1(z) = 14 + 7z^{-1}$ *and*

$$Y_2(z) = \left\lfloor -7 + \frac{15}{2}z^{-1} + \frac{49}{2}z^{-2} \right\rfloor = -7 + 7z^{-1} + 24z^{-2}.$$

Thus the original signal $\mathbf{x} = (14, 18, 7, 28)$ *has been transformed to* $\mathbf{y}_1 = (14, 7)$ *and* $\mathbf{y}_2 = (-7, 7, 24)$ *(indexing starting at 0). To recover* $\mathbf{x}$ *from the pair* $(\mathbf{y}_1, \mathbf{y}_2)$, *we first recover* $X_e(z) = Y_1(z) = 14 + 7z^{-1}$ *and then from (8.80), we have*

$$\begin{aligned} X_o(z) &= z(-7 + 7z^{-1} + 24z^{-2}) - z\left\lfloor -7 + \frac{21}{2}z^{-1} + \frac{7}{2}z^{-2} \right\rfloor \\ &= -7z + 7 + 24z^{-1} - (-7z - 11 - 4z^{-1}) \\ &= 18 + 28z^{-1}. \end{aligned}$$

Once we have X_e *and* X_o *we recover* $X(z)$ *via equation (8.34) and so reconstruct* $\mathbf{x}$.

Similar observations apply to lifting via equation (8.61). Moreover, the greatest integer function used in (8.79) is not at all essential; any type of quantization could be used, as long as it leaves integers unchanged. If such an operation is performed at each lifting step, we obtain a wavelet transform that maps integers to integers.

Lifting techniques can also be used in a variety of other ways, for example, to develop transforms for irregularly space data, or to develop transforms for finite length or higher dimensional data in a more "customized" form (see [30] for more information).

8.6 Matlab Project

In this section, we focus primarily on the construction and lifting factorization of the polyphase matrices for some standard filter banks, in particular, the CDF(2,2) and Daubechies D4 filter banks, and not so much on the actual transformation of signals (that was the emphasis of the Chapter 7 Matlab project). Most of the work involves manipulating Laurent polynomials. The Matlab Wavelet Toolbox has built-in data structures and routines for handling Laurent polynomials, but since our needs are modest (and for the benefit of the reader without the Wavelet Toolbox), we supply our own routines. They can also be useful for the Exercise Section that follows this Matlab project.

8.6.1 Laurent Polynomials

Consider the Laurent polynomials

$$P(z) = z - 6 + 12z^{-1} - 5z^{-2} \text{ and } Q(z) = z^2 - 7z + 18 - 18z^{-1} + 6z^{-2}$$

the z-transform of signals $\mathbf{p} = (1, -6, 12, -5)$ (first component index -1) and $\mathbf{q} = (1, -7, 18, -18, 6)$ (first component index -2); recall that, for example, the component p_k is paired with z^{-k}. We can define these using the supplied command `lpoly`, as

```
P = lpoly(1, [1,-6,12,-5]);
Q = lpoly(2, [1,-7,18,-18,6]);
```

In general, the command `P = lpoly(n, [p_1,p_2,...,p_m]);` defines a Laurent polynomial

$$P(z) = \sum_{k=1}^{m} p_k z^{n+1-k}$$

with highest power term $p_1 z^n$ and lowest power term $p_m z^{n-m+1}$. This is similar to the corresponding Matlab command "`laurpoly`." You can display the Laurent polynomial $P(z)$ with

```
disp_laurent(P);
```

and similarly for Q.

The Laurent polynomials can be added with

```
S = ladd(P,Q);
```

and followed by `disp_laurent(S)` to see the result. The commands `S = lsub(P,Q);` and `S = lmul(P,Q);` produce $P(z) - Q(z)$ and $P(z)Q(z)$, respectively. The command

```
[D,R] = lquo(P,Q);
```

produces Laurent polynomials $D(z), R(z)$ so that $P(z) = D(z)Q(z) + R(z)$ with $\text{wid}(R) < \text{wid}(Q)$. However, as noted just after the proof of Lemma 8.1, D and R are not unique! Our algorithm works as in the proof of Lemma 8.1: we shift $P(z)$ and $Q(z)$ to be ordinary polynomials with lowest degree term a constant, perform ordinary polynomial division, then shift the quotient and remainder appropriately. But we could have performed the computation differently and so obtained a different result for $D(z)$ and $R(z)$, as per the remarks after the proof of Lemma 8.1.

A few additional routines for manipulating Laurent polynomials are available. One can compute the width of P (equation (8.14)) with

```
lpwid(P)
```

This command returns `-inf` if $P(z) = 0$. The even- and odd-indexed polyphase components of P can be extracted with

```
Pe = extract_even(P); Po = extract_odd(P);
```

Finally, we can reconstruct P from the polyphase components, via equation (8.34), with

```
Prec = reconX(Pe,Po);
```

There are also routines for handling 2×2 matrices of Laurent polynomials, useful in performing lifting computations. Define two additional Laurent polynomials, say

```
V = lpoly(1,[1,-5,7]);
W = lpoly(2,[1,-6,12,-8]);
```

(so $V(z) = z - 5 + 7z^{-1}$ and $W(z) = z^2 - 6z + 12 - 8z^{-1}$). One can then define a 2×2 matrix

$$\mathbf{M}(z) = \begin{bmatrix} P(z) & Q(z) \\ V(z) & W(z) \end{bmatrix}$$

with the command

```
M = lpmat(P,Q,V,W);
```

The matrix can be displayed with

```
disp_lpmat(M);
```

The inverse matrix $\mathbf{M}^{-1}$ is computed with

```
Minv = lpmatinv(M);
```

but only if the inverse exists! If not then the routine returns a warning that the matrix is not invertible, and "NaN" for the entries of $\mathbf{M}^{-1}$. The determinant can be computed with

```
lpmatdet(M)
```

Two matrices, $\mathbf{M}$ and $\mathbf{M}^{-1}$, for example, can be added, subtracted, and multiplied with

```
S = lpmatadd(M,Minv); D = lpmatsub(M,Minv);
Pr = lpmatmul(M,Minv);.
```

This last matrix should of course be the identity, which can be verified with `disp_lpmat(Pr);`.

Given a pair of perfect reconstruction filters, say $\boldsymbol{\ell}_a = (1/2, 1/2)$ and $\mathbf{h}_a = (1/2, -1/2)$, with z-transforms $L_a(z) = \frac{1}{2} + \frac{1}{2}z^{-1}$ and $H_a(z) = \frac{1}{2} - \frac{1}{2}z^{-1}$, defined as

```
La = lpoly(0, [1/2,1/2]);
Ha = lpoly(0, [1/2,-1/2]);
```

we can form the polyphase analysis matrix with

```
Pa = lpmat(extract_even(La),extract_odd(La),
extract_even(Ha),extract_odd(Ha));
```

Alternatively, use the supplied command

```
Pa = polyphase_analysis(La,Ha);
```

which does the polyphase decomposition and installs the components into the matrix in one step.

8.6.2 Lifting for CDF(2,2)

Let us look at how these routines can be used for perform a lifting factorization for the polyphase analysis matrix for a typical filter bank. The CDF(2,2) analysis filters have coefficients

$$\boldsymbol{\ell}_a = \frac{\sqrt{2}}{8}(-1, 2, 6, 2, -1) \approx (-0.177, 0.354, 1.06, 0.354, -0.177), \quad (8.81)$$

$$\mathbf{h}_a = \frac{\sqrt{2}}{4}(-1, 2, -1) \approx (-0.354, 0.707, -0.354), \quad (8.82)$$

where the low-pass coefficients run from index −2 to 2 and the high-pass from 0 to 2.

1. Start Matlab. The filter coefficients can be defined via the commands

   ```
   la = sqrt(2)/8*[-1 2 6 2 -1];
   ha = sqrt(2)/4*[-1 2 -1];
   ```

   The filter z-transforms are defined in Matlab as

   ```
   La = lpoly(2,la);
   Ha = lpoly(2,ha);
   ```

2. The polyphase analysis matrix can be found from the command

   ```
   Pa = polyphase_analysis(La,Ha);
   ```

   and displayed with `disp_lpmat(Pa);`.
3. Let us begin with a primal lift as in equation (8.61). We can compute $t(z)$ using either equation (8.65) or equation (8.66) and the division algorithm (recall they can yield different choices for t). If we use (8.65), we can perform

   ```
   [t,~] = lquo(Pa(1,1),Pa(2,1));
   ```

   We do not need the actual remainder here; hence, we use a "~" to ignore that returned argument. This should yield $t(z) = \frac{1}{2} - \frac{7}{2}z^{-1}$, which you can check with `disp_laurent(t);` Alternatively, we can use equation (8.66) and compute

   ```
   [t,~] = lquo(Pa(1,2),Pa(2,2));
   ```

   which should yield $t(z) = \frac{1}{2} + \frac{1}{2}z^{-1}$.

The computation of t and the associated matrix above is in fact automated into the routine `primal_lift.m`. To use equation (8.65), execute

```
[Pa2,M1] = primal_lift(Pa,0);
```

The second input argument "0" indicates that the odd polyphase components should used to compute $t(z)$ (equation (8.65)), while a "1" indicates the even polyphase components should be used (equation (8.66)). The routine `primal_lift` as called returns arguments `Pa2` and `M1` so that $\mathbf{P}_a(z) = \mathbf{M}_1(z)\mathbf{P}_{a2}(z)$ where

$$\mathbf{M}_1(z) = \begin{bmatrix} 1 & t(z) \\ 0 & 1 \end{bmatrix}. \tag{8.83}$$

Check that equation (8.83) holds, by executing

```
disp_lpmat(Pa);
disp_lpmat(lpmatmul(M1,Pa2));
```

Note that we now have (to three significant figures)

$$\mathbf{P}_a(z) = \begin{bmatrix} 1.0 & 0.500 + 0.500z^{-1} \\ 0 & 1.0 \end{bmatrix} \begin{bmatrix} 1.414 & 0 \\ -0.354z - 0.354 & 0.707z \end{bmatrix}.$$

The first matrix on the right is $\mathbf{M}_1(z)$, the second we will call $\mathbf{P}_{a2}(z)$.

4. We can perform a dual lift on $\mathbf{P}_{a2}$ using `dual_lift.m`. It is clear that we want to work with the first column, that is, the even polyphase components. Execute

```
[Pa3,M2] = dual_lift(Pa2,1);
```

You should find, with `disp_lpmat(M2);` and `disp_lpmat(Pa3);` that

$$\mathbf{M}_2 = \begin{bmatrix} 1.0 & 0.0 \\ -0.250z - 0.250 & 1.0 \end{bmatrix} \quad \mathbf{P}_{a3}(z) = \begin{bmatrix} 1.414 & 0.0 \\ 0.0 & 0.707z \end{bmatrix}.$$

Note that $\mathbf{P}_{a3}$ is diagonal.

5. We now have a factorization $\mathbf{P}_a = \mathbf{M}_1\mathbf{M}_2\mathbf{P}_{a3}$, where $\mathbf{P}_{a3}$ can be considered a scaling matrix, since it is diagonal with monomials on the diagonal. If we want to move the scaling up front, we can use equations (8.70) and (8.71) to commute the scaling with each lifting matrix. This is embodied in the supplied command `flip_scaling.m`. To commute the scaling with $\mathbf{M}_2$ (by altering $\mathbf{M}_2$ appropriately), execute

```
M2 = flip_scaling(M2,Pa3(1,1),Pa3(2,2));
```

In general, `B=flip_scaling(A,K1,K2)` produces a matrix $\mathbf{B}$ so that $\mathbf{AK} = \mathbf{KB}$, where $\mathbf{K}$ is a diagonal matrix with nonzero diagonal entries

K_1, K_2. The matrix $\mathbf{A}$ must be a lifting matrix, that is, 1 on the diagonal and either $A_{1,2} = 0$ or $A_{2,1} = 0$. The matrix $\mathbf{M}_1$ can be altered appropriately with

```
M1 = flip_scaling(M1,Pa3(1,1),Pa3(2,2));
```

You can check that the factorization $\mathbf{P}_a = \mathbf{P}_{a3}\mathbf{M}_1\mathbf{M}_2$ is correct with

```
Pr = lpmatmul(lpmatmul(Pa3,M1),M2);
```

and `disp_lpmat(Pr);`.

8.6.3 Lifting the D4 Filter Bank

The Daubechies D4 filters can be defined with

```
r24 = 4*sqrt(2.0); r3 = sqrt(3.0);
La = lpoly(0,[(1+r3)  (3+r3)  (3-r3)  (1-r3)]/r24);
Ha = lpoly(0,[(1-r3)  (-3+r3)  (3+r3)  (-1-r3)]/r24);
```

1. Define the polyphase matrix $\mathbf{P}_a$ using the `Pa = polyphase_analysis (La,Ha);` command from above.
2. Perform three lifting steps—a primal, a dual, and a primal—using the `primal_lift.m` and `dual_lift.m` commands, and construct the relevant lifting matrices. For the first primal and dual lift, use the even polyphase components (flag 1 to the lifting commands) and for the last, use the odd polyphase components (flag 0). This should reduce $\mathbf{P}_a$ to a diagonal scaling matrix
$$\begin{bmatrix} 3.346z^{-1} & 0.0 \\ 0.0 & -0.299 \end{bmatrix}.$$
3. Use the `flip_scaling.m` command to move the scaling step up front. This should produce a factorization of the D4 polyphase analysis matrix of the form

$$\mathbf{P}_a = \begin{bmatrix} 3.346z^{-1} & 0.0 \\ 0.0 & -0.299 \end{bmatrix} \begin{bmatrix} 1.0 & 0.333z \\ 0.0 & 1.0 \end{bmatrix} \times \begin{bmatrix} 1.0 & 0.0 \\ 0.433 - 2.799z^{-1} & 1.0 \end{bmatrix} \begin{bmatrix} 1.0 & -0.577 \\ 0.0 & 1.0 \end{bmatrix}. \tag{8.84}$$

Exercises

1. Let $\mathbf{x}$ be the signal with components $x_{-2} = 1, x_{-1} = 3, x_0 = 2, x_1 = -2, x_2 = 3$ and all other $x_k = 0$. Compute $\mathbf{x}_e$ and $\mathbf{x}_o$ and verify equation (8.6) by computing the right side explicitly.

2. Let $\mathbf{g}$ be the filter with $g_0 = g_1 = 1/2$ (the Haar low-pass filter). With $\mathbf{x}$ as defined in a previous exercise, compute $\mathbf{X}_\ell$ and $\mathbf{X}_h$ from equation (8.3). Then use equation (8.5) to compute $\mathbf{x}_e = \mathbf{y}_e$ and $\mathbf{x}_o = S^{-1}(\mathbf{y}_o)$ and recover $\mathbf{x}$. Reference to Example 8.1 may be helpful.

3. Verify that equation (8.6) holds for any signal $\mathbf{x}$.

4. Prove equation (8.15).

5. Show that if $p(z)$ is a standard polynomial then $\text{wid}(p) \leq \deg(p)$.

6. Let $\mathbf{w} \in L^2(\mathbb{Z})$ be a signal or filter with z-transform $W(z)$.
   (a) Show that the z-transform of $D(\mathbf{w})$ is given formally by $(W(\sqrt{z}) + W(-\sqrt{z}))/2$.
   (b) Show that the z-transform of $S(\mathbf{w})$ is given by $z^{-1}W(z)$.
   (c) Use parts (a) and (b) above to demonstrate equations (8.32)–(8.34).

7. Prove that if $p(z)$ and $q(z)$ are Laurent polynomials then $\text{wid}(pq) = \text{wid}(p) + \text{wid}(q)$. Hint: Generalize the proof of Lemma 7.1.

8. Prove the assertions of equations (8.21)–(8.23).

9. The "lazy" wavelet transform has polyphase analysis matrix $\mathbf{P}_a(z) = \mathbf{I}$, the identity matrix. What are the corresponding filters $\boldsymbol{\ell}_a$ and $\mathbf{h}_a$? Use the analysis procedure of equation (8.38) to compute $X_\ell(z)$ and $X_h(z)$ if $\mathbf{x} = (2, 4, -3, 5)$ (first component $x_0 = 2$). Then use the polyphase synthesis matrix and equations (8.41) and (8.42) to reconstruct $\mathbf{x}$ (or $X(z)$) from $X_\ell(z)$ and $X_h(z)$.

10. Write out $X_e(z)$ and $X_o(z)$ for the signal $\mathbf{x} = (2, -2, 1, 5, 7)$ by inspection, assuming the first entry is at index position 0, and then by using the formula (8.32) and (8.33).

11. Use the polyphase factorization (8.8) and (8.9) for the Haar transform to compute $\mathbf{X}_\ell$ and $\mathbf{X}_h$ for the signal $\mathbf{x}$ with nonzero components $x_0 = 10, x_1 = -4, x_2 = 5, x_3 = 7$.

12. Use the factorization (8.8) and (8.9) for the Haar polyphase analysis matrix to find a factorization for the Haar polyphase synthesis matrix. Then use this to inverse Haar transform the data $\mathbf{X}_\ell = (1, 2, 1)$, $\mathbf{X}_h = (-1, 1)$, both starting with index zero and all other components zero.

13. Verify that the Haar polyphase analysis matrix $\mathbf{P}_a$ in equation (8.4) can also be factored into scaling and lifting steps as

$$\mathbf{P}_a = \begin{bmatrix} 1 & 0 \\ a & 1 \end{bmatrix} \begin{bmatrix} 1 & b \\ 0 & 1 \end{bmatrix} \begin{bmatrix} c & 0 \\ 0 & d \end{bmatrix}$$

by finding appropriate choices for a, b, c, d. Thus we can implement a polyphase version of the transform that involves scaling first.

14. Find a lifting factorization for the Haar polyphase matrix of the form (8.8) but in which $\mathbf{M}_1$ corresponds to a dual lift (lower triangular) and $\mathbf{M}_2$ to a primal lift (upper triangular).

15. Show that if equation (8.60) holds then equation (8.58) also holds. Equation (8.34) is helpful.

16. Emulate the computations of Example 8.3 with equation (8.20) to find a filter $\tilde{H}(z)$ so that the filter pair $(\tilde{L}(z), \tilde{H}(z))$ from Example 8.3 forms a perfect reconstruction filter bank, with $\mathrm{wid}(\tilde{H}) < \mathrm{wid}(\tilde{L}) = 2$. Hint: Try writing $H(z) = az^m\tilde{L}(z) + \tilde{H}(z)$ for some choice of a, m with m even. Arrange a and m so that $\mathrm{wid}(\tilde{H}) < 2$.

17. Suppose an analysis filter $\boldsymbol{\ell}_a$ is part of a perfect reconstruction filter bank with high-pass filter $\mathbf{h}_a$, and the $\boldsymbol{\ell}_a$ also forms a perfect reconstruction filter pair with filter $\tilde{\mathbf{h}}_a$. Let $L_a(z), H_a(z)$, and $\tilde{H}_a(z)$ denote the corresponding z-transforms, so that $H_a(z) - \tilde{H}_a(z) = s(z^2)L_a(z)$ for some s. Let $L_s(z), H_s(z)$ be synthesis filter transforms for the appropriate synthesis filters for $L_a(z), H_a(z)$ as computed using equation (7.34) and (7.35) with delay $m = 0$, and $\tilde{L}_s(z), \tilde{H}_s(z)$ be synthesis filter transforms for the appropriate synthesis filters for $L_a(z), \tilde{H}_a(z)$ as computed using equation (7.34) and (7.35) with delay $m = 0$. Use (7.34) and (7.35) to show that $\tilde{L}_s(z) = L_s(z) + s(z^2)H_s(z)$ while $\tilde{H}_s(z) = H_s(z)$. (i.e., changing the high-pass analysis filter changes the low-pass synthesis filter, but leaves the high-pass synthesis filter unchanged.)

18. Derive the analog of equations (8.65) and (8.66), but for dual lifting.

19. Verify that if $q(z) = cz^m$ with $c \neq 0$ (so $1/q(z) = z^{-m}/c$ is a Laurent polynomial) then

$$\begin{bmatrix} q(z) & 0 \\ 0 & 1/q(z) \end{bmatrix} = \begin{bmatrix} 1 & q(z) - q^2(z) \\ 0 & 1 \end{bmatrix} \begin{bmatrix} 1 & 0 \\ -1/q(z) & 1 \end{bmatrix} \begin{bmatrix} 1 & q(z) - 1 \\ 0 & 1 \end{bmatrix} \begin{bmatrix} 1 & 0 \\ 1 & 1 \end{bmatrix}.$$

Thus any matrix of the form on the left can be decomposed into four lifting steps.

20. Show that if in addition to equations (7.10), equation (7.46) also holds (so $H_a(z) = -z^{-(N-1)}L_a(-z^{-1})$ where N, the number of filter taps, is even) then

$$H_{a,e}(z) = z^{-N/2+1}L_{a,o}(z^{-1}) \tag{8.85}$$

$$H_{a,o}(z) = -z^{-N/2+1}L_{a,e}(z^{-1}). \tag{8.86}$$

and so by Lemma 8.7 the polyphase synthesis matrix can be expressed as

$$\mathbf{P}_s(z) = \begin{bmatrix} L_{a,e}(z^{-1}) & z^{N/2-1}L_{a,o}(z) \\ L_{a,o}(z^{-1}) & -z^{N/2-1}L_{a,e}(z) \end{bmatrix}.$$

21. Mimic the computations of Section 8.5.4 to count the number of operations needed to apply the D4 filter bank to a signal using the polyphase matrix, and then the number of operations to apply the factored version of equation (8.84).

22. Use the lifting factorization of the Le Gall 5/3 polyphase analysis matrix in equation (8.68) and the floor function as in Example 8.11 at each stage to compute the quantities $\mathbf{X}_\ell$ and $\mathbf{X}_h$ for the signal $\mathbf{x} = (14, 18, 7, 28)$ ($x_0 = 14$). Then show how to recover the signal $\mathbf{x}$ from the transform data.

9

Wavelets

9.1 Overview

9.1.1 Chapter Outline

In Chapter 7, we defined filter banks and started referring to "discrete wavelet transforms" in the finite-dimensional case. In this chapter, we will make clear exactly what wavelets are and what they have to do with filter banks. This chapter contains a certain amount of unavoidable but elementary analysis. Our goal is to provide a basic understanding of wavelets, how they relate to filter banks, and how they can be useful. We state and provide examples concerning the essential truths about wavelets, and some rigorous proofs. But we provide only references for other more technical facts concerning wavelets, such as the existence of scaling functions, convergence of the cascade algorithm for computing scaling functions, and some of the properties of wavelets [12, 20].

9.1.2 Continuous from Discrete

Suppose that by some miracle we had developed the discrete Fourier transform for sampled signals but had no notion of the underlying theory for the continuous case, that is, Fourier series. We know from Chapter 1 that any vector in $\mathbb{C}^N$ can be constructed as a superposition of the basic discrete waveforms $\mathbf{E}_{N,k}$ defined in equation (1.22). If we plot the components of $\mathbf{E}_{N,k}$ for some fixed k and increasing N, the vectors $\mathbf{E}_{N,k}$ seem to stabilize on some "mysterious" underlying function. Of course, this is none other than $e^{2\pi ikt}$ sampled at times $t = m/N$ as illustrated in Figure 9.1 for $k = 3$. Note that the $\mathbf{E}_{N,k}$ are simply the columns of the inverse discrete Fourier transform (IDFT) matrix $\mathbf{F}_N^{-1}$, or equivalently, are obtained by applying the IDFT to the standard basis vectors $\mathbf{e}_k$ in $\mathbf{C}^N$ (recall Example 1.18).

Discrete Fourier Analysis and Wavelets: Applications to Signal and Image Processing, Second Edition.
S. Allen Broughton and Kurt Bryan.

Companion Website: www.wiley.com/go/Broughton/Discrete_Fourier_Analysis_and_Wavelets

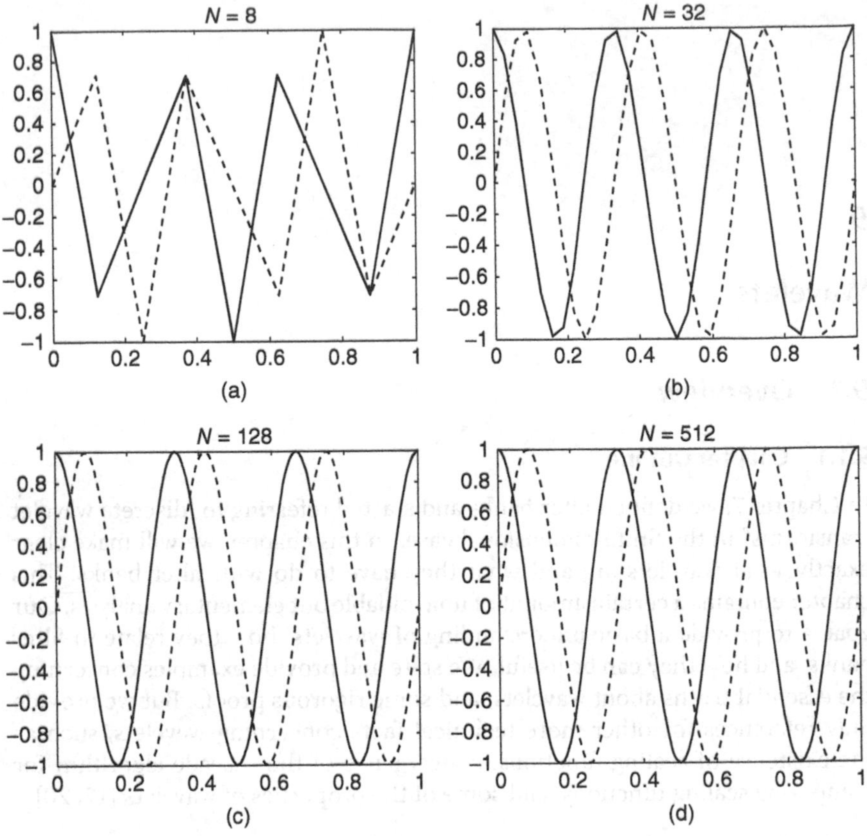

Figure 9.1 Real and imaginary parts of discrete waveforms $\mathbf{E}_{N,3}$ for $N = 8, 32, 128$, and 512.

Even if we had no knowledge of the basis functions $e^{2\pi ikt}$, the plots in Figure 9.1 might lead us to suspect that *some* function underlies the discrete quantity $\mathbf{E}_{N,k}$. We might posit that some more general continuous theory underlies our discrete computations.

Now the idea that we could develop the discrete Fourier transform (DFT) with no notion of the underlying waveforms $e^{2\pi ikt}$ seems farfetched, but we are in precisely this position with regard to wavelets! To illustrate, let us use the multistage orthogonal discrete Haar transform (filters as in Example 7.5, periodic extension of signals) in place of the DFT to perform again the experiment above. We will apply the matrix $\mathcal{W}_n^s$ governing the n-stage inverse transform to the standard basis vector $\mathbf{e}_7$, for the cases $N = 8, 32, 128$, and 512, using $n = 3, 5, 7$, and 9 stages, respectively (the maximum number of stages we can do here is $n = \log_2(N)$). The results are shown in Figure 9.2. In each case we plot the components $(\mathcal{W}_n^s(\mathbf{e}_7))_m$ versus m/N. Aside from a normalization in

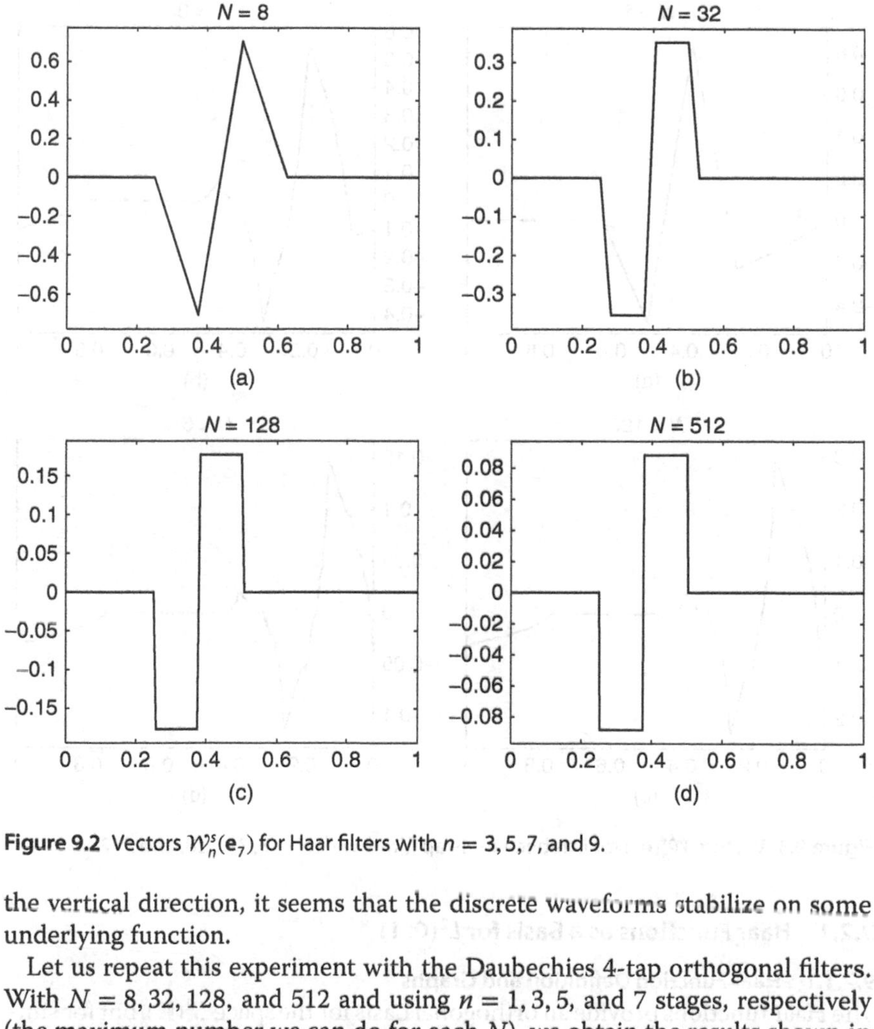

Figure 9.2 Vectors $\mathcal{W}_n^s(\mathbf{e}_7)$ for Haar filters with $n = 3, 5, 7$, and 9.

the vertical direction, it seems that the discrete waveforms stabilize on some underlying function.

Let us repeat this experiment with the Daubechies 4-tap orthogonal filters. With $N = 8, 32, 128$, and 512 and using $n = 1, 3, 5$, and 7 stages, respectively (the maximum number we can do for each N), we obtain the results shown in Figure 9.3. Again, apart from normalization, it appears that the discrete waveforms are converging to some underlying function. It seems that there may be a continuous theory lurking behind filter banks!

9.2 The Haar Basis

We will start by examining the simplest kind of wavelets, the *Haar functions*, which predate modern wavelet theory by about 70 years. See [15] for the original treatment of these functions.

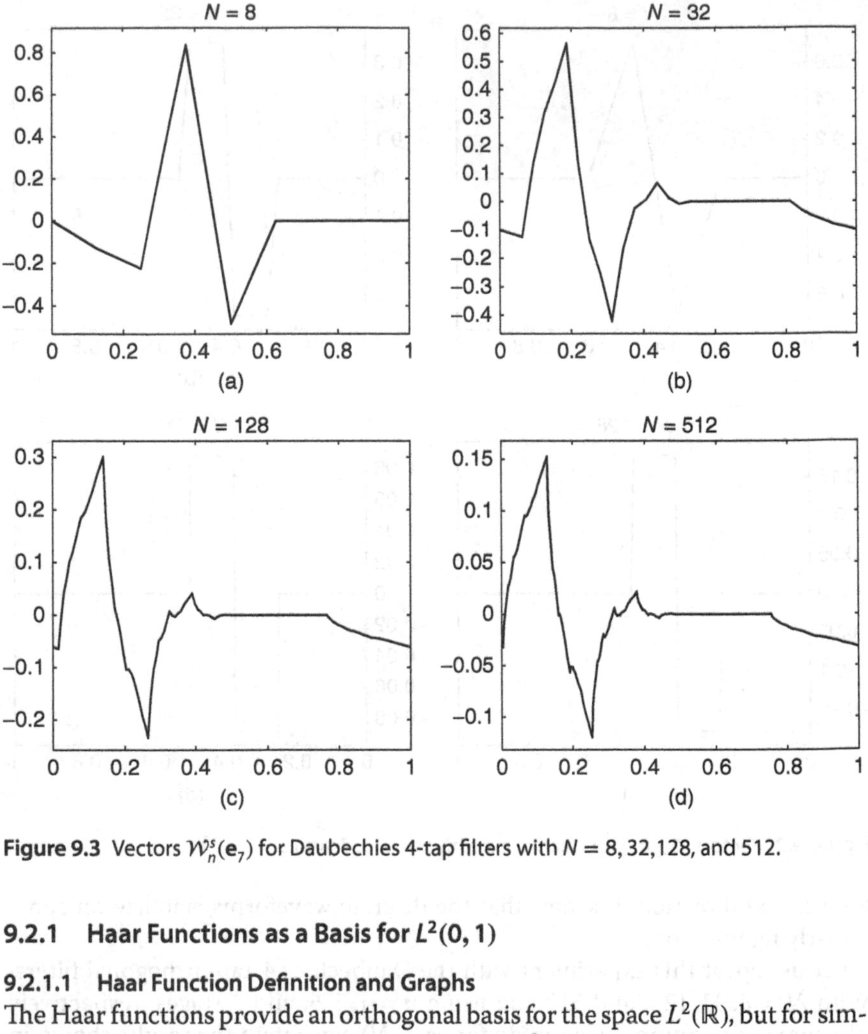

Figure 9.3 Vectors $\mathcal{W}_n^s(\mathbf{e}_7)$ for Daubechies 4-tap filters with $N = 8, 32, 128$, and 512.

9.2.1 Haar Functions as a Basis for $L^2(0, 1)$

9.2.1.1 Haar Function Definition and Graphs

The Haar functions provide an orthogonal basis for the space $L^2(\mathbb{R})$, but for simplicity, we will initially confine our attention to $L^2(0, 1)$. Let $\psi(t)$ be the function defined on $\mathbb{R}$ as

$$\psi(t) = \begin{cases} 1, & 0 \le t < \frac{1}{2}, \\ -1, & \frac{1}{2} \le t < 1, \\ 0, & \text{else.} \end{cases} \tag{9.1}$$

The definition outside of $[0, 1]$ is for later convenience. We call ψ the *mother Haar wavelet*. In Figure 9.4, we provide a graph of $\psi(t)$. The similarity of $\psi(t)$ to that of the graphs in Figure 9.2 is no coincidence!

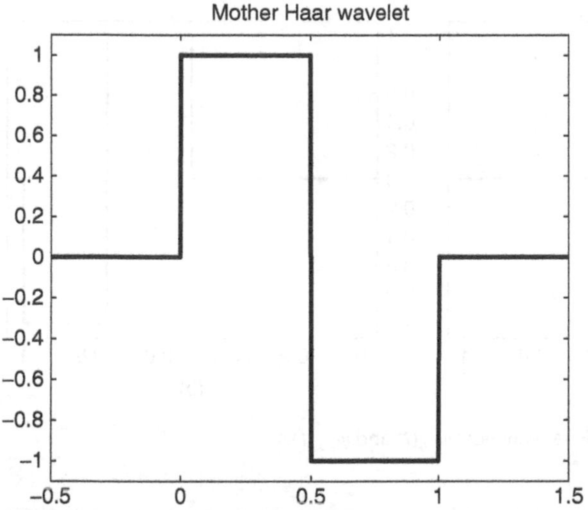

Figure 9.4 Mother Haar wavelet.

From $\psi(t)$ we can build an infinite set of functions $\psi_{k,n}$, the *Haar wavelets*, defined on the interval $[0, 1]$ by

$$\psi_{k,n}(t) = 2^{k/2}\psi(2^k t - n), \tag{9.2}$$

where $k \geq 0$ and $0 \leq n \leq 2^k - 1$. The $2^{k/2}$ in front is simply to normalize so that $\| \psi_{k,n} \|_{L^2(0,1)} = 1$, where we use the notation $\| \cdot \|_{L^2(a,b)}$ to denote the L^2 norm on (a, b), to avoid confusion when we consider functions on multiple intervals. We will also throw in the function

$$\phi(t) = \begin{cases} 1, & 0 \leq t < 1, \\ 0, & \text{else}, \end{cases} \tag{9.3}$$

called the *Haar scaling function*; again, we define ϕ outside $[0, 1]$ merely for later convenience. It turns out that the infinite family $\psi_{k,n}$, together with ϕ, forms an orthonormal basis for $L^2(0, 1)$.

To build up some intuition, let us graph some of these basis functions. The scaling function ϕ is easy to visualize because it is constant, and $\psi_{0,0} = \psi$ is plotted in Figure 9.4. When $k = 1$ we obtain functions $\psi_{1,0}$ and $\psi_{1,1}$, which are graphed in Figure 9.5. These are half-length versions of the mother wavelet, rescaled vertically and translated horizontally. Taking $k = 2$ gives wavelets $\psi_{2,0}, \psi_{2,1}, \psi_{2,2}$, and $\psi_{2,3}$, illustrated in Figure 9.6. These are quarter-scale translations of the mother wavelet. In general, $\psi_{k,n}$ in equation (9.2) can be written explicitly as

$$\psi_{k,n}(t) = \begin{cases} 2^{k/2}, & n/2^k \leq t < (n + 1/2)/2^k, \\ -2^{k/2}, & (n + 1/2)/2^k \leq t < (n + 1)/2^k, \\ 0, & \text{else}. \end{cases} \tag{9.4}$$

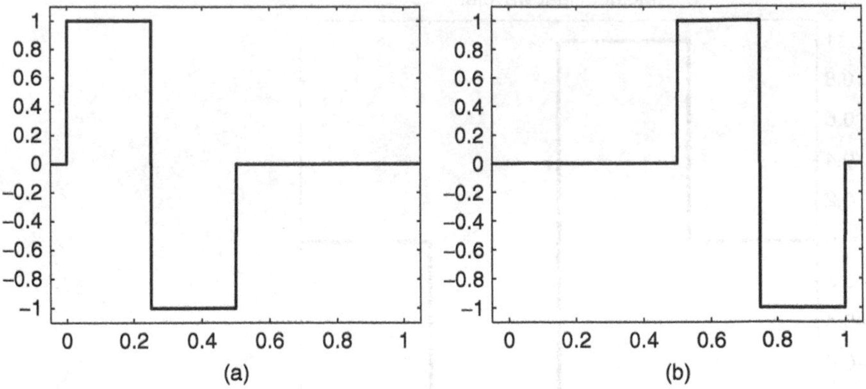

Figure 9.5 Scaled, translated Haar wavelets $\psi_{1,0}(t)$ and $\psi_{1,1}(t)$.

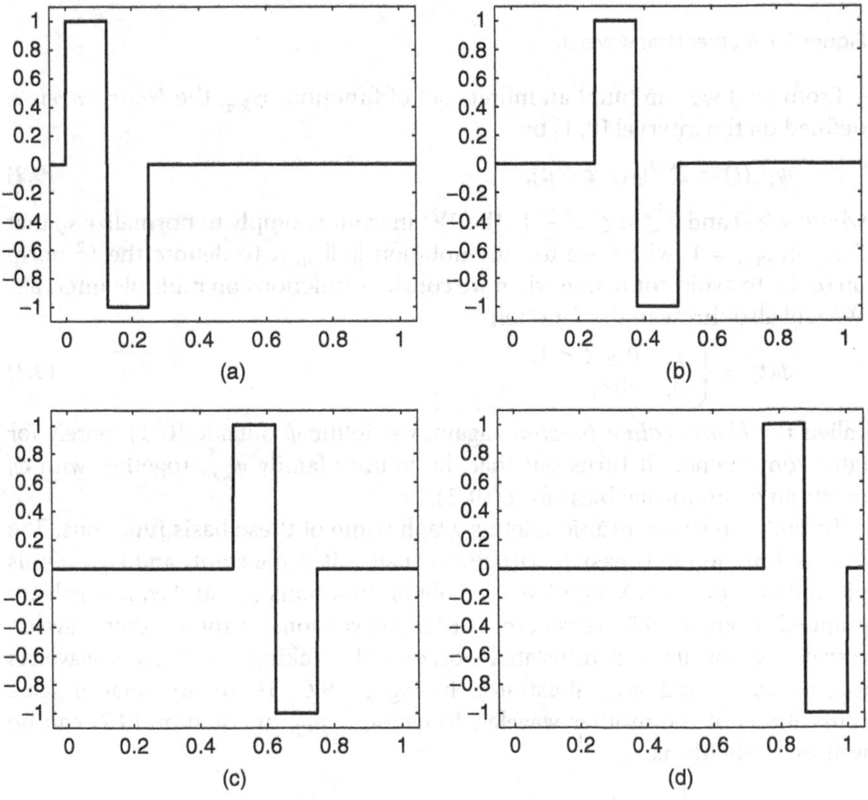

Figure 9.6 Scaled, translated Haar wavelets $\psi_{2,0}, \psi_{2,1}, \psi_{2,2}$, and $\psi_{2,3}$.

In particular, $\psi_{k,n}$ is nonzero only on the interval $[n/2^k, (n+1)/2^k)$, of length $1/2^k$, and for any fixed k the function $\psi_{k,n}$ is simply $\psi_{k,0}$ translated $n/2^{k+1}$ units to the right. The first index k in $\psi_{k,n}$ dictates the scale and n controls position.

Before proceeding, it will be helpful to define some terminology. Recall that the *closure* of a set $S \subset \mathbb{R}$ consists of all $x \in \mathbb{R}$ that can be obtained as a limit of elements in S; the closure of S contains S itself, of course.

Definition 9.1 The *support* of a function g defined on $\mathbb{R}$ is the closure of the set on which g is nonzero. A function g is said to be *supported* in a set $A \subseteq \mathbb{R}$ if the support of g is contained in A. If a function g is supported in a bounded interval, then g is said to have *compact support.*

The word "compact" stems from the fact that closed, bounded subsets of $\mathbb{R}$ are examples of *compact sets*; see [24] or any introductory text on real analysis.

It is easy to see that if a function g has support contained in some set A then g is identically zero (or undefined) outside of A.

The support of the Haar scaling function ϕ is the interval $[0, 1]$. The support of the Haar wavelet $\psi_{k,n}$ is the closed interval $[n/2^k, (n+1)/2^k]$.

9.2.1.2 Orthogonality

Recall the notation

$$(f,g) := \int_a^b f(t)g(t)\,dt$$

for the inner product of two functions in $L^2(a, b)$.

Proposition 9.1 *The set*

$$S = \{\phi\} \cup \{\psi_{k,n} : k \geq 0, 0 \leq n \leq 2^k\}$$

is orthonormal in $L^2(0, 1)$.

Proof: It is easy to see that

$$(\phi, \psi_{k,n}) = \int_0^1 \phi(t)\psi_{k,n}(t)\,dt = \int_0^1 \psi_{k,n}(t)\,dt = 0,$$

since $\phi \equiv 1$; thus ϕ is orthogonal to every $\psi_{k,n}$. It is also easy to see that if $m \neq n$, then the product $\psi_{k,m}(t)\psi_{k,n}(t) \equiv 0$, since $\psi_{k,m}$ and $\psi_{k,n}$ have disjoint support except possibly at a single point (see Figure 9.6). As a result $(\psi_{k,m}, \psi_{k,n}) = 0$.

Now consider the product $\psi_{j,m}(t)\psi_{k,n}(t)$ where $j \neq k$, say $j < k$. One possibility is $\psi_{j,m}\psi_{k,n} \equiv 0$, if $\psi_{j,m}$ and $\psi_{k,n}$ have disjoint support, in which case orthogonality is obvious. Alternatively, we must have one of $\psi_{j,m}\psi_{k,n} = 2^{j/2}\psi_{k,n}$ (e.g., $\psi_{1,0}$ and $\psi_{2,0}$) or $\psi_{j,m}\psi_{k,n} = -2^{j/2}\psi_{k,n}$ (e.g., $\psi_{1,0}$ and $\psi_{2,1}$), because the support of $\psi_{k,n}$ is contained in either the region where $\psi_{j,m} \equiv 2^{j/2}$ or the region where $\psi_{j,m} \equiv -2^{j/2}$. We then have

$$(\psi_{j,m}, \psi_{k,n}) = \int_0^1 \psi_{j,m}(t)\psi_{k,n}(t)\, dt = \pm 2^{j/2} \int_0^1 \psi_{k,n}(t)\, dt = 0,$$

which completes the proof of Proposition 9.1.

Of course, the proposition implies that the set S is linearly independent.

9.2.1.3 Completeness in $L^2(0, 1)$

The set S in Proposition 9.1 is not merely orthonormal in $L^2(0, 1)$, but complete: any function $f \in L^2(0, 1)$ can be approximated to arbitrary precision in the L^2 norm by using linear combinations of elements of S. Specifically, for each $K \geq 0$ let

$$f_{K+1}(t) = c_0\phi(t) + \sum_{k=0}^{K} \sum_{n=0}^{2^K-1} c_{k,n}\psi_{k,n}(t), \tag{9.5}$$

where $c_0 = (f, \phi)/(\phi, \phi) = (f, \phi) = \int_0^1 f(t)\, dt$ and

$$c_{k,n} = (f, \psi_{k,n}) = \int_0^1 f(t)\psi_{k,n}(t)\, dt \tag{9.6}$$

as dictated by Theorem 1.4. Because $(\psi_{k,n}, \psi_{k,n}) = 1$, it is not explicitly needed in the denominator for $c_{k,n}$. For convenience let us also define $f_0(t) \equiv c_0$. We can make $\| f - f_K \|_{L^2(0,1)}$ arbitrarily small by taking K sufficiently large. We will prove this below, but let us pause to look at an example.

Example 9.1 *Let $f(t) = t(1 - t)(2 - t)$ on $(0, 1)$. The crudest possible approximation is obtained by simply using the Haar scaling function. We obtain $f_0(t) = c_0$, where $c_0 = \int_0^1 f(t)\phi(t)\, dt = \frac{1}{4}$, which merely approximates f by its own mean value on the interval $[0, 1]$.*

If we take $K = 0$ in equation (9.5), then $\psi_{0,0}$ is included, and we obtain the approximation f_1 shown in Figure 9.7a. In the terminology we will define shortly, f_1 is the "projection" of f onto the subspace spanned by ϕ and $\psi_{0,0}$ (which consists of functions that are constant on each of $\left[0, \frac{1}{2}\right)$ and $\left[\frac{1}{2}, 1\right)$). In Figure 9.7b, we have the approximation $f_2(t)$ obtained from equation (9.5) with $K = 1$, which throws $\psi_{1,0}$ and $\psi_{1,1}$ into the mix. This is the projection of f onto the subspace of functions which are constant on each interval $\left[0, \frac{1}{4}\right)$, $\left[\frac{1}{4}, \frac{1}{2}\right)$, $\left[\frac{1}{2}, \frac{3}{4}\right)$, $\left[\frac{3}{4}, 1\right)$. In Figure 9.8, we show the approximation $f_5(t)$ obtained by using all $\psi_{k,n}$ up to $k = 4$. It certainly looks like we can approximate any continuous function in

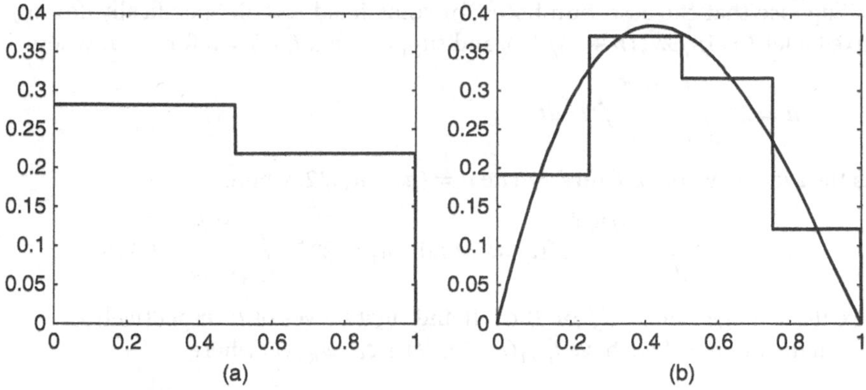

Figure 9.7 Haar expansions using all $\psi_{k,n}$ with $k \leq 0$ (a) and $k \leq 1$ (b).

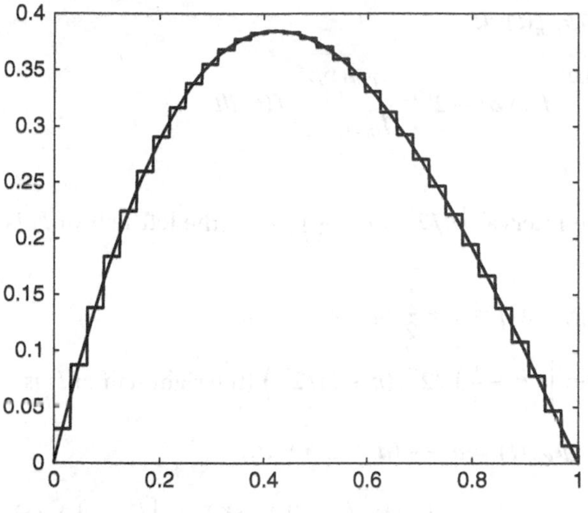

Figure 9.8 Haar expansion using all $\psi_{k,n}$ up to $k = 4$.

this manner! In each case notice that $f_K(t)$ is constant on intervals of the form $[n/2^K, (n+1)/2^K)$, where $0 \leq n \leq 2^K - 1$.

The following proposition will be useful.

Proposition 9.2 *For any $f \in L^2(0,1)$ the (constant) value of the approximation $f_K(t)$ on any interval $[n/2^K, (n+1)/2^K)$ is the average value of f on that interval.*

Proof: This is easy to prove by induction. First note that $f_0(t) \equiv c_0 = \int_0^1 f(t)\, dt$ is indeed the average value of f on the interval $[0, 1]$.

Suppose that the assertion holds for some fixed $K \geq 0$. Specifically, for any fixed n let $I = [n/2^K, (n+1)/2^K)$, and suppose that $f_K(t) = a$ for $t \in I$, where

$$a = 2^K \int_{n/2^K}^{(n+1)/2^K} f(t)\, dt$$

is the average value of f on I. Write $a = (a_0 + a_1)/2$, where

$$a_0 = 2^{K+1} \int_{n/2^K}^{(n+1/2)/2^K} f(t)\, dt \quad \text{and} \quad a_1 = 2^{K+1} \int_{(n+1/2)/2^K}^{(n+1)/2^K} f(t)\, dt$$

are the average values of f on the left and right halves of I, respectively.

On the interval I we have $f_{K+1}(t) = f_K(t) + c_{K,n}\psi_{K,n}(t)$ where

$$\begin{aligned} c_{K,n} &= \int_0^1 f(t)\psi_{K,n}(t)\, dt \\ &= \int_{n/2^K}^{(n+1)/2^K} f(t)\psi_{K,n}(t)\, dt \\ &= 2^{K/2} \int_{n/2^K}^{(n+1/2)/2^K} f(t)\, dt - 2^{K/2} \int_{(n+1/2)/2^K}^{(n+1)/2^K} f(t)\, dt \\ &= 2^{-K/2-1}(a_0 - a_1). \end{aligned}$$

The value of $f_{K+1}(t)$ on the interval $\left[n/2^K, \left(n+\frac{1}{2}\right)/2^K\right)$ (the left half of I) is then

$$f_{K+1}(t) = f_K(t) + c_{K,n}\psi_{K,n}(t) = a + \frac{1}{2}(a_0 - a_1) = a_0,$$

while the value of $f_{K+1}(t)$ on $\left[\left(n+\frac{1}{2}\right)/2^K, (n+1)/2^K\right)$ (the right half of I) is

$$f_{K+1}(t) = f_K(t) + c_{K,n}\psi_{K,n}(t) = a - \frac{1}{2}(a_0 - a_1) = a_1.$$

Thus, f_{K+1} is constant on any interval $\left[n/2^K, \left(n+\frac{1}{2}\right)/2^K\right)$ or $\left[\left(n+\frac{1}{2}\right)/2^K, (n+1)/2^K\right)$, and equal to the average value of f on that interval. Of course, any interval $[m/2^{K+1}, (m+1)/2^{K+1})$ is of the form $\left[n/2^K, \left(n+\frac{1}{2}\right)/2^K\right)$ or $\left[\left(n+\frac{1}{2}\right)/2^K, (n+1)/2^K\right)$, which completes the induction step and proves the proposition.

Remark 9.1 A simple corollary to Proposition 9.2 is that if f is a function that is constant on each interval $[n/2^K, (n+1)/2^K)$, then f can be built exactly from the scaling function ϕ and a finite superposition of Haar wavelets $\psi_{k,n}$ for $k \leq K-1$.

We will use Proposition 9.2 to show that the Haar wavelets form a basis for $L^2(0,1)$. As remarked in Chapter 1, the space $L^2(0,1)$ consists of much more than just continuous functions. Nonetheless, we will just show that any continuous function on $[0,1]$ can be approximated arbitrarily well in $L^2(0,1)$ using the Haar wavelets. Indeed, we can make the stronger statement that f can be approximated to any precision in the supremum norm on $[0,1]$; that is, we can make

$$\sup_{t\in[0,1]} |f_K(t)-f(t)|$$

arbitrarily small by taking K sufficiently large.

Theorem 9.1 *Any continuous function f on $[0,1]$ can be approximated to arbitrary precision in the supremum norm using equations (9.5) and (9.6).*

Proof: The proof requires an elementary fact from real analysis. Specifically, since f is continuous on a closed, bounded (i.e., compact) interval, f is uniformly continuous on this interval. This means that given any $\epsilon > 0$ there is some $\delta > 0$ so that $|f(t)-f(s)| < \epsilon$ whenever $|t-s| < \delta$. Fix any $\epsilon > 0$, and choose K large enough so that $1/2^K < \delta$ for the corresponding δ.

From Proposition 9.2, we know that on any interval of the form $[n/2^K, (n+1)/2^K)$ the function $f_K(t) \equiv a$, where a is the average value of f on this interval. The function f is continuous, so by the integral mean value theorem, we have $a = f(t^*)$ for some $t^* \in [n/2^K, (n+1)/2^K)$. For any t in this interval, then

$$|f(t)-f_K(t)| = |f(t)-a| = |f(t)-f(t^*)| < \epsilon,$$

since $|t-t^*| \leq 1/2^K < \delta$. Thus, on any interval $[n/2^K, (n+1)/2^K)$ contained in $[0,1]$ the functions f and f_K differ by no more than ϵ, and we conclude that

$$\sup_{t\in[0,1]} |f_K(t)-f(t)| \leq \epsilon,$$

which completes the proof.

Remark 9.2 Of course, this means that for any $\epsilon > 0$ we can obtain $|f_K(t) - f(t)|^2 \leq \epsilon^2$ for any continuous f and suitably large K, and then we also have

$$\| f_K - f \|_{L^2(0,1)} \equiv \left(\int_0^1 (f_K(t)-f(t))^2 \, dt \right)^{1/2} \leq \epsilon.$$

The set S of Proposition 9.1 is thus complete in $L^2(0,1)$.

9.2.2 Haar Functions as an Orthonormal Basis for $L^2(\mathbb{R})$

We can use the scaling function and wavelets to define an orthonormal basis for $L^2(\mathbb{R})$. The set

$$\{\phi(t-n) : n \in \mathbb{Z}\}$$

is orthonormal in $L^2(\mathbb{R})$. This set will become part of an orthonormal basis for $L^2(\mathbb{R})$. We will also continue to use the functions $\psi_{k,n}(t) = 2^{k/2}\psi(2^k t - n)$ (equivalently, equation (9.4)), with $k \geq 0$, but now allow n to range over all of $\mathbb{Z}$. This means that all translates are allowed, up and down the real line. We still have $\| \psi_{k,n}(t)\|_{L^2(\mathbb{R})} = 1$.

Let $\tilde{S}$ be the set

$$\tilde{S} = \{\phi(t-n) : n \in \mathbb{Z}\} \cup \{\psi_{k,n} : k \geq 0, n \in \mathbb{Z}\} \tag{9.7}$$

The same reasoning as in Proposition 9.1 shows that $\tilde{S}$ is an orthonormal set in $L^2(\mathbb{R})$. Also, for each $m \in \mathbb{Z}$, some subset of $\tilde{S}$ provides an orthonormal basis for $L^2(m, m+1)$. For example, $\phi(t)$ together with all $\psi_{k,n}$ such that $k \geq 0$ and $0 \leq n < 2^k$ is a basis for $L^2(0,1)$, for this set is simply the basis $\phi(t)$ and $\psi_{k,n}$ for $L^2(0,1)$ considered above. Similarly $\phi(t-1)$ and $\psi_{k,n}$ with $k \geq 0$ and $2^k \leq n < 2^{k+1}$ is a basis for $L^2(1,2)$, for these are just translates one unit to the right of the basis elements for $L^2(0,1)$. More generally, $\phi(t-m)$ together with the set of all $\psi_{k,n}$ such that $k \geq 0$ and $m2^k \leq n < (m+1)2^{k+1}$ is a basis for $L^2(m, m+1)$.

As a consequence, for any $\epsilon > 0$, any $m \in \mathbb{Z}$, and any function $f \in L^2(m, m+1)$, we can obtain

$$\|f - f_m\|_{L^2(m,m+1)} < \epsilon,$$

where f_m is a linear combination of elements of $\tilde{S}$.

Theorem 9.2 *The set $\tilde{S}$ in equation (9.7) is an orthonormal basis for $L^2(\mathbb{R})$.*

Proof: We will again prove the theorem only for continuous functions. Since any function in $L^2(\mathbb{R})$ can be approximated to arbitrary precision by continuous functions, this will suffice. Let $f \in L^2(\mathbb{R})$, so that

$$\int_{-\infty}^{\infty} f^2(t)\, dt < \infty.$$

Write this as

$$\sum_{m=-\infty}^{\infty} \int_m^{m+1} f^2(t)\, dt < \infty.$$

Thus, the series $\sum_m a_m$ with positive terms $a_m = \int_m^{m+1} f^2(t)\,dt$ converges. For any $\epsilon > 0$ we can then choose some M large enough so that

$$\sum_{|m|>M} \int_m^{m+1} f^2(t)\,dt < \epsilon^2/2.$$

From Theorem 9.1 (and its extension to any unit interval $[m, m+1)$ as discussed above) we know that on any given interval $[m, m+1)$ we can approximate f with some function f_m so that $\| f - f_m \|^2_{L^2(m,m+1)} \leq \epsilon^2/(4M+2)$. Define the function $\tilde{f} \in L^2(\mathbb{R})$ as

$$\tilde{f}(t) = \begin{cases} f_m(t), & t \in [m, m+1) \quad \text{and} \quad |m| \leq M, \\ 0, & \text{else}, \end{cases}$$

The function $\tilde{f}(t)$ is a finite linear combination of elements of the set $\tilde{S}$. Then

$$\begin{aligned} \| f - \tilde{f} \|^2_{L^2(\mathbb{R})} &= \int_{-\infty}^{\infty} (f(t) - \tilde{f}(t))^2\,dt \\ &= \sum_{|m|\leq M} \int_m^{m+1} (f(t) - \tilde{f}(t))^2\,dt \\ &\quad + \sum_{|m|>M} \int_m^{m+1} (f(t) - \tilde{f}(t))^2\,dt \\ &= \sum_{|m|\leq M} \| f - \tilde{f} \|^2_{L^2(m,m+1)} + \sum_{|m|>M} \int_m^{m+1} f^2(t)\,dt \\ &\leq \sum_{|m|\leq M} \frac{\epsilon^2}{4M+2} + \frac{\epsilon^2}{2} \\ &\leq \frac{\epsilon^2}{2} + \frac{\epsilon^2}{2} = \epsilon^2. \end{aligned}$$

Thus, for any $\epsilon > 0$ we can obtain $\| f - \tilde{f} \|_{L^2(\mathbb{R})} \leq \epsilon$, where $\tilde{f}$ is a finite linear combination of the elements of $\tilde{S}$. This proves the theorem.

Since $\tilde{S}$ is an orthonormal basis, Theorem 1.4 dictates that if

$$f(t) = \sum_{m=-\infty}^{\infty} c_m \phi(t-m) + \sum_{k=0}^{\infty} \sum_{n=-\infty}^{\infty} c_{k,n} \psi_{k,n}(t), \tag{9.8}$$

then

$$c_m = (f, \phi(t-m)) = \int_m^{m+1} f(t)\,dt, \tag{9.9}$$

$$c_{k,n} = (f, \psi_{k,n}) = 2^{k/2} \left(\int_{n/2^k}^{(n+1/2)/2^k} f(t)\,dt - \int_{(n+1/2)/2^k}^{(n+1)/2^k} f(t)\,dt \right), \tag{9.10}$$

which exactly parallel equations (9.5) and (9.6), except that the interval of integration is not restricted to (0, 1).

9.2.3 Projections and Approximations

Before continuing, it will be useful to look at the notion of "orthogonal projection" in an inner product space. The concept of projection plays a huge role in approximation theory, numerical analysis, and applied mathematics in general. To avoid a long detour, we will develop only what we need, specific to the situation at hand. We will be working in the inner product space $L^2(a,b)$ ($(a,b) = \mathbb{R}$ is allowed), though the results hold for more general Hilbert spaces. In addition, although we assume that the spaces are infinite-dimensional, everything below works in $\mathbb{R}^N$; indeed it is even easier since the sums are then finite and no convergence issues arise.

Let V be a subspace of $L^2(a,b)$ with an orthonormal basis $S = \{\cup_{k\in\mathbb{Z}}\ \eta_k\}$ (alternatively the basis can be finite). We assume that all linear combinations

$$g = \sum_{k\in\mathbb{Z}} a_k \eta_k, \tag{9.11}$$

where $\sum_k a_k^2 < \infty$ are elements of V, so that every square-summable sequence $\{a_k\}$ corresponds to an element of V, and vice versa (recall Theorem 1.5 and the remarks preceding that theorem). Recall also that the precise meaning of the infinite sum on the right-hand side in equation (9.11) is that

$$\lim_{m,n\to\infty} \left\| g - \sum_{k=-m}^{n} a_k \eta_k \right\| = 0. \tag{9.12}$$

as discussed in Chapter 1, for example, equation (1.42).

Definition 9.2 Let V be a subspace of $L^2(a,b)$ and $S = \{\cup_{k\in\mathbb{Z}}\eta_k\}$ an orthonormal basis for V as described above. If $f \in L^2(a,b)$ then the "orthogonal projection of f onto V" is the element $P_V(f)$ of V defined by

$$P_V(f) = \sum_{k\in\mathbb{Z}} (f, \eta_k)\eta_k. \tag{9.13}$$

By Bessel's inequality (equation (1.55), Exercise 36), we have $\sum_k (f,\eta_k)^2 \le \|f\|^2 < \infty$, so the element defined by equation (9.13) really is in V.

What is the relation between $P_V(f)$ and f? First, if $f \in V$, then $P_V(f) = f$, since S is a basis for V. If f is not in V, then it turns out that $P_V(f)$ is the best approximation to f that can be constructed as a superposition of the η_k. Put another way, $P_V(f)$ is the element of V that is "closest" to f, as quantified by the following proposition:

Proposition 9.3 *Let S and $V \subseteq L^2(a,b)$ be as in Definition 9.2. For $f \in L^2(a,b)$, let $P_V(f)$ be the orthogonal projection of f onto V given by equation (9.13). Then $\| P_V(f) - f \| \leq \| g - f \|$ for all $g \in V$. Equality is attained only if $g = P_V(f)$.*

Proof: Consider a function α_n of the form

$$\alpha_n = \sum_{k=-n}^{n} b_k \eta_k \tag{9.14}$$

for scalars b_k; the function α_n clearly lies in V. Let us first consider the problem of minimizing $\| \alpha_n - f \|$, or equivalently, $\| \alpha_n - f \|^2$, by choosing the b_k appropriately. This leads us to the task of minimizing

$$\begin{aligned}(\alpha_n - f, \alpha_n - f) &= \sum_{k=-n}^{n} b_k^2 - 2\sum_{k=-n}^{n} b_k (f, \eta_k) + (f,f)\\ &= \sum_{k=-n}^{n} (b_k - (f,\eta_k))^2 + (f,f) - \sum_{k=-n}^{n} (f,\eta_k)^2 \end{aligned} \tag{9.15}$$

as a function of the b_k, where we have made use of the elementary properties of the inner product, real-valued here. It's obvious that the right-hand side in (9.15) is minimized by taking $b_k = (f, \eta_k)$, since the last two terms don't even depend on the b_k.

Let

$$\tilde{f}_n = \sum_{k=-n}^{n} (f, \eta_k) \eta_k \tag{9.16}$$

denote the minimizer of $\| f - \alpha_n \|$, where α_n is of the form (9.14). Clearly, $\tilde{f}_n$ is unique, for any other choice for the b_k on the right-hand side in (9.15) will increase $\| \alpha_n - f \|^2$. The sequence $\tilde{f}_n$ converges to the function $P_V(f) \in V$, since

$$\| P_V(f) - \tilde{f}_n \|^2 = \sum_{|k|>n} (P_V(f), \eta_k)^2$$

converges to zero as $n \to \infty$.

To see that $\| f - P_V(f) \| \leq \| g - f \|$ for any other $g \in V$, let $g = \sum_k d_k \eta_k$ and define

$$g_n = \sum_{k=-n}^{n} d_k \eta_k.$$

From the discussion above, we know that $\| \tilde{f}_n - f \| \leq \| g_n - f \|$ for all n. We then have

$$\| P_V(f) - f \| = \lim_{n\to\infty} \| \tilde{f}_n - f \| \leq \lim_{n\to\infty} \| g_n - f \| = \| g - f \|$$

by making use of the result of Exercise 37. This shows that $\| P_V(f) - f \| \leq \| g - f \|$ for any $g \in V$. The assertion that equality is attained only when $g = P_V(f)$ is left to Exercise 11.

The projection of an element of an inner product space H onto a subspace V can be defined more generally than we have done. Indeed, *under suitable conditions* the projection of an element $f \in H$ onto V may be defined as the element of V that is closest to f. The phrase "suitable conditions" typically means that the subspace V should be closed (i.e., if a sequence with all elements in V has a limit in $L^2(a, b)$, then the limit is in fact in V). In the present case this property is ensured by the requirement that all functions of the form (9.11) with $\sum_k a_k^2 < \infty$ lie in V, but this may not always be true (see Exercise 10). Our definition of projection is sufficient for our purposes.

Example 9.2 *Let $H = L^2(0, 1)$, and for any fixed integer $N \geq 1$ let*

$$S_N = \{\phi\} \cup \{\psi_{k,n} : 0 \leq k \leq N - 1, 0 \leq n < 2^k\},$$

similar to the set S defined in Proposition 9.1 but truncated at $k \leq N - 1$. The fact that the basis elements $\psi_{k,n}$ are doubly indexed makes no difference at all in the application of the projection formulas (we could re-index them with a single subscript). The set S_N is orthonormal in $L^2(0, 1)$. The subspace $V_N = \text{span}(S_N)$ of $L^2(0, 1)$ consists of exactly those $L^2(0, 1)$ functions that are constant on each interval of the form $[n/2^N, (n + 1)/2^N)$ where $0 \leq n < 2^N$: that any $f \in V_N$ is constant on $[n/2^N, (n + 1)/2^N)$ is clear. The converse, that any function constant on $[n/2^N, (n + 1)/2^N)$ is in V_N (i.e., can be built from elements of S_N) follows from Remark 9.1.

The projection of an arbitrary function $f \in L^2(0, 1)$ onto V_N is given by equations (9.5) and (9.6). The graphs in Figure 9.7 show the projection of the function $f(t) = t(t - 1)(t - 2)$ onto the subspaces V_1 and V_2, while Figure 9.8 shows the projection of f onto V_5.

In this example note that since $S_N \subset S_{N+1}$, we also have $V_N \subset V_{N+1}$. As indicated by Figures 9.7 and 9.8, for any function f the projection of f onto V_N becomes more and more detailed as N increases. Indeed, if we let $\tilde{f}_N$ denote the projection of f onto V_N, then Remark 9.2 shows that $\lim_{N\to\infty} \tilde{f}_N = f$. We will later encounter this idea—a nested sequence of subspaces that embodies more and more detail—in a more general context.

9.3 Haar Wavelets Versus the Haar Filter Bank

Before continuing with our analysis of the Haar basis for $L^2(\mathbb{R})$, it may be helpful to pause and examine some parallels between the Haar decomposition of a function as expressed in equations (9.8) through (9.10) and the Haar filter bank.

9.3.1 Single-stage Case

9.3.1.1 Functions from Sequences

Let $\mathbf{x} \in L^2(\mathbb{Z})$. To illustrate parallels between the Haar filter bank and the Haar wavelets, let us use $\mathbf{x}$ to manufacture a piecewise constant function $x(t)$ on the real line by setting

$$x(t) \equiv x_{k-1} \quad \text{for} \quad t \in [k/2, (k+1)/2) \tag{9.17}$$

at each integer k. For example, $x(t) \equiv x_{-1}$ for $0 \leq t < \frac{1}{2}$ and $x(t) \equiv x_0$ for $\frac{1}{2} \leq t < 1$. The function $x(t)$ is in $L^2(\mathbb{R})$ (Exercise 6). We use x_{k-1} instead of x_k on the right-hand side in equation (9.17) solely to make formulas below prettier; it is not essential. Any $\mathbf{x} \in L^2(\mathbb{Z})$ generates such a function $x(t)$, and conversely, any function $x(t)$ that is constant on half-integer intervals corresponds to such an $\mathbf{x} \in L^2(\mathbb{Z})$. A typical example is illustrated in Figure 9.9.

9.3.1.2 Filter Bank Analysis/Synthesis

We will pass $\mathbf{x}$ through a one-stage Haar analysis filter bank with filters $\boldsymbol{\ell}_a$, $\mathbf{h}_a$, but with one minor change. The astute reader may notice that the scaled Haar wavelet in Figure 9.2 obtained from the Haar filter bank by plotting the vector $\mathcal{W}_n^s(\mathbf{e}_7)$ is "upside down" (drops first, then rises), in opposition to the various Haar wavelets shown in Figures 9.4–9.6. A more perfect correspondence between the Haar filter bank and Haar wavelet framework is obtained if we use the usual low-pass coefficients $(\boldsymbol{\ell}_a)_0 = (\boldsymbol{\ell}_a)_1 = \frac{1}{2}$, but reverse the sign of $\mathbf{h}_a$ and take $(\mathbf{h}_a)_0 = -\frac{1}{2}$, $(\mathbf{h}_a)_1 = \frac{1}{2}$ in the Haar filter bank. This is still a high-pass filter, and if we also negate the synthesis filter $\mathbf{h}_s$, then the filter bank provides perfect reconstruction. We obtain filtered, down-sampled vectors

$$\mathbf{X}_\ell = D(\mathbf{x} * \boldsymbol{\ell}_a) = \left(\ldots, \frac{x_{-2} + x_{-3}}{2}, \frac{x_0 + x_{-1}}{2}, \frac{x_2 + x_1}{2}, \ldots\right),$$

$$\mathbf{X}_h = D(\mathbf{x} * \boldsymbol{\ell}_a) = \left(\ldots, \frac{x_{-3} - x_{-2}}{2}, \frac{x_{-1} - x_0}{2}, \frac{x_1 - x_2}{2}, \ldots\right),$$

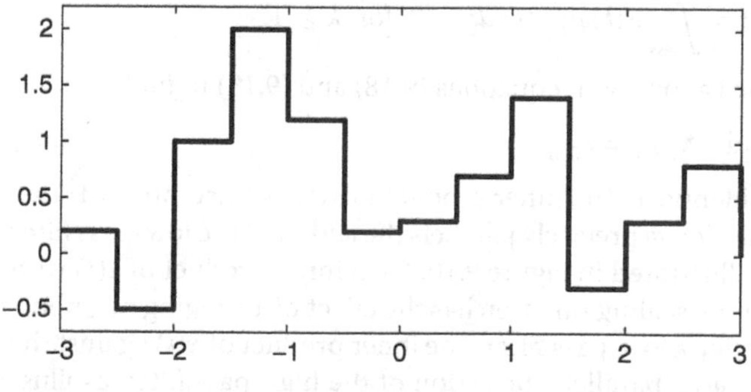

Figure 9.9 Piecewise constant function on half-integer intervals.

similar to those previously computed in equations (7.3) and (7.4). In general, the components of these vectors are given by

$$(\mathbf{X}_\ell)_m = \frac{x_{2m} + x_{2m-1}}{2}, \tag{9.18}$$

$$(\mathbf{X}_h)_m = \frac{x_{2m-1} - x_{2m}}{2}. \tag{9.19}$$

To reconstruct $\mathbf{x}$, we first upsample. The upsampled vectors have components

$$U(\mathbf{X}_\ell)_k = (\mathbf{X}_\ell)_{k/2}, \quad U(\mathbf{X}_h)_k = (\mathbf{X}_h)_{k/2},$$

for k even and 0 for k odd. We then apply the synthesis filters $\boldsymbol{\ell}_s$, $\mathbf{h}_s$ with coefficients $(\boldsymbol{\ell}_s)_0 = (\boldsymbol{\ell}_a)_{-1} = 1$ and $(\mathbf{h}_s)_0 = -1, (\mathbf{h}_s)_{-1} = 1$ (note that the coefficients of $\mathbf{h}_s$ are negated). This yields

$$(\boldsymbol{\ell}_s * (U(\mathbf{X}_\ell)))_k = \begin{cases} (\mathbf{X}_\ell)_{k/2}, & k \text{ even}, \\ (\mathbf{X}_\ell)_{(k+1)/2}, & k \text{ odd}, \end{cases}$$

$$(\mathbf{h}_s * (U(\mathbf{X}_h)))_k = \begin{cases} -(\mathbf{X}_h)_{k/2}, & k \text{ even}, \\ (\mathbf{X}_h)_{(k+1)/2}, & k \text{ odd}. \end{cases}$$

Finally, we add $\boldsymbol{\ell}_s * (U(\mathbf{X}_\ell)) + \mathbf{h}_s * (U(\mathbf{X}_h))$ to reconstruct $\mathbf{x}$. Specifically,

$$x_k = \begin{cases} (\mathbf{X}_\ell)_{k/2} - (\mathbf{X}_h)_{k/2}, & k \text{ even}, \\ (\mathbf{X}_\ell)_{(k+1)/2} + (\mathbf{X}_h)_{(k+1)/2}, & k \text{ odd}. \end{cases} \tag{9.20}$$

9.3.1.3 Haar Expansion and Filter Bank Parallels

The filter bank operation above has a precise parallel in equations (9.8) through (9.10). If we expand $x(t)$ with respect to the Haar basis with equations (9.9) and (9.10), we obtain coefficients

$$c_m = (x, \phi(t-m)) = \int_{-\infty}^{\infty} x(t)\phi(t-m)\,dt = \frac{x_{2m} + x_{2m-1}}{2}, \tag{9.21}$$

$$\begin{aligned} c_{0,m} &= (x, \psi_{0,m}) = \int_{-\infty}^{\infty} x(t)\psi_{0,m}(t)\,dx = \frac{x_{2m-1} - x_{2m}}{2}, \\ c_{k,m} &= (x, \psi_{k,m}) = \int_{-\infty}^{\infty} x(t)\psi_{k,m}(t)\,dx = 0 \text{ for } k \geq 1. \end{aligned} \tag{9.22}$$

Compare the equations above with equations (9.18) and (9.19) to find

$$(\mathbf{X}_\ell)_m = c_m \text{ and } (\mathbf{X}_h)_m = c_{0,m}. \tag{9.23}$$

These sequences are identical! The inner product of $x(t)$ against integer translates of the scaling function ϕ precisely parallels the action of the low-pass filter on $\mathbf{x}$. The situation is illustrated in Figure 9.10. Each inner product of $x(t)$ with an integer translate of the scaling function has the effect of averaging neighboring values of $x(t)$, just like a low-pass filter. The inner product of $x(t)$ against the longest scale wavelets $\psi_{0,m}$ parallels the action of the high-pass filter, as illustrated in Figure 9.11. Each inner product of $x(t)$ with an integer translate of the

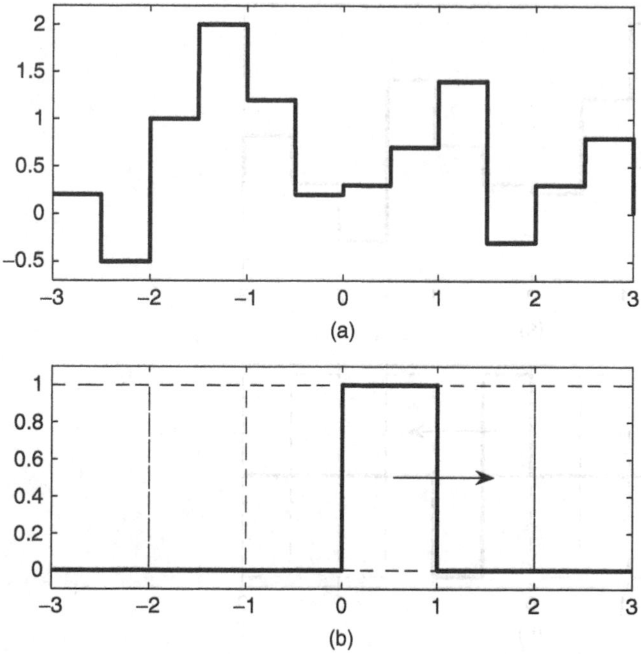

Figure 9.10 Action of scaling function translates on $x(t)$.

wavelet has the effect of differencing neighboring values, just like a high-pass filter. Because the scaling function and wavelet are translated one unit at a time $\left(\text{rather than } \frac{1}{2}\right)$, the "downsampling" is built into the process.

The inner product of $x(t)$ with any shorter scale wavelet $\psi_{k,m}$ with $k \geq 1$ is zero, since $x(t)$ was constructed to be constant on half-integer intervals, hence variation on such shorter scales is zero. Of course, this will not be true for more general functions. That is where a multistage transform comes into play, which we will discuss shortly.

Now consider the process of synthesizing $x(t)$ from the c_m and $c_{0,m}$ using equation (9.8). This precisely mirrors the reconstruction of $\mathbf{x}$ from $\mathbf{X}_\ell$ and $\mathbf{X}_h$ using the synthesis filter bank. Specifically, we have from equation (9.8),

$$x(t) = \sum_{n=-\infty}^{\infty} c_n \phi(t-n) + \sum_{n=-\infty}^{\infty} c_{0,n} \psi_{0,n}(t). \tag{9.24}$$

For any $t \in \mathbb{R}$ only two of the terms on the right-hand side in (9.24) are nonzero: namely, if $t \in [m, m+1)$, then (9.24) becomes

$$x(t) = c_m + c_{0,m}\psi_{0,m}(t), \tag{9.25}$$

since $\phi(t-m) \equiv 1$ on $[m, m+1)$. If t is in the right half of the interval $[m, m+1)$, then $(k+1)/2 \leq t < k/2+1$, where $k = 2m$. Here $x(t) \equiv x_k$ and

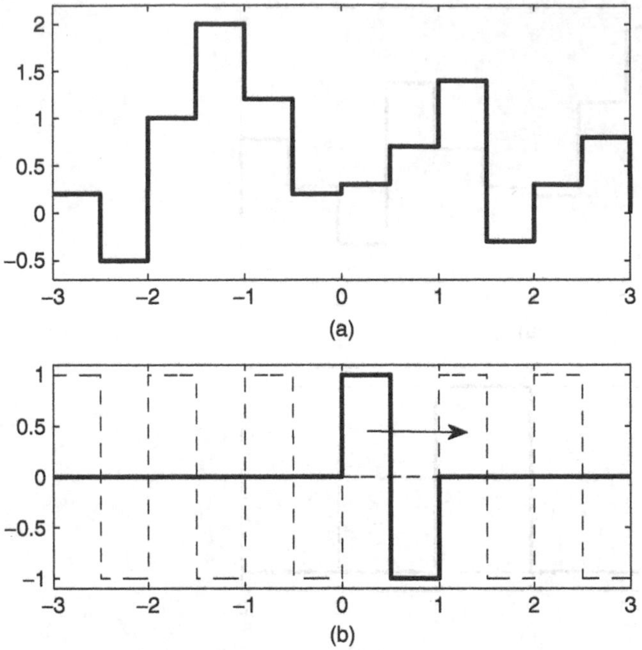

Figure 9.11 Action of wavelet translates on $x(t)$.

$\psi_{0,m} \equiv -1$. In this case equation (9.25) becomes

$$x(t) = c_m - c_{0,m} = (\mathbf{X}_\ell)_m - (\mathbf{X}_h)_m.$$

In the light of equation (9.23) this is just the statement that $x_k = (\mathbf{X}_\ell)_{k/2} - (\mathbf{X}_h)_{k/2}$, in accordance with equation (9.20).

If t is in the left half of the interval $[m, m + 1)$, then $(k + 1)/2 \leq t < k/2 + 1$, where $k = 2m - 1$. Here $x(t) \equiv x_k$ and $\psi_{0,m} \equiv 1$. In this case equation (9.25) becomes

$$x(t) = c_m + c_{0,m} = (\mathbf{X}_\ell)_m + (\mathbf{X}_h)_m.$$

This is just the statement that $x_k = (\mathbf{X}_\ell)_{(k+1)/2} + (\mathbf{X}_h)_{(k+1)/2}$, in accordance with equation (9.20). The synthesis of the signal $x(t)$ from its Haar basis coefficients precisely mirrors the reconstruction $\mathbf{x}$ from the synthesis filter bank.

9.3.2 Multistage Haar Filter Bank and Multiresolution

Multistage filter banks in which we iteratively pass portions of the output back through the filter bank also have a parallel in the functional setting. It is helpful to lay a little groundwork first.

9.3.2.1 Some Subspaces and Bases

For $N \geq 1$ define the set

$$S_N = \{\phi(t-n) : n \in \mathbb{Z}\} \cup \{\psi_{k,n} : 0 \leq k \leq N-1, n \in \mathbb{Z}\},$$

similar to the notation of Example 9.2 except that now we are working over the entire real line. Define the subspace

$$V_N = \text{span}(S_N) \tag{9.26}$$

in $L^2(\mathbb{R})$ for $N \geq 1$. The set S_N is an orthonormal basis for V_N. In the same vein let us also define the subspace $V_0 \subset L^2(\mathbb{R})$ consisting of functions constant on integer intervals $[k, k+1)$. The scaling function translates $\phi(t-n)$ with $n \in \mathbb{Z}$ forms an orthonormal basis for V_0.

It is worth noting that V_N consists precisely of those functions in $L^2(\mathbb{R})$ that are constant on intervals for the form $[k/2^N, (k+1)/2^N)$ for $k \in \mathbb{Z}$. To see this, notice that any element of V_N is, on any interval $[k/2^N, (k+1)/2^N)$, a finite linear combination of elements of S_N, each of which is constant on intervals $[k/2^N, (k+1)/2^N)$. Thus, elements of V_N are also constant on these intervals. Conversely, by Proposition 9.2, it follows that any function that is constant on intervals $[k/2^N, (k+1)/2^N)$ can be built as a linear combination of the elements of S_N.

There is another "obvious" basis for V_N. Define functions

$$\phi_{N,n}(t) = 2^{N/2}\phi(2^N t - n), \tag{9.27}$$

where $n \in \mathbb{Z}$ and ϕ is the Haar scaling function defined in equation (9.3). The function $\phi_{N,n}$ is just ϕ scaled to base width $1/2^N$ and translated $n/2^N$ units; the $2^{N/2}$ factor in front scales the L^2 norm to 1. It is easy to see that the set

$$B_N = \{\phi_{N,n} : n \in \mathbb{Z}\},$$

is also an orthonormal basis for V_N. The basis B_N is simpler than S_N, in that all the basis elements are mere translates of each other. But the basis S_N has certain merits of its own, which we detail below.

9.3.2.2 Multiresolution and Orthogonal Decomposition

When we expand a function f with respect to an orthogonal basis for V_N, whether that basis is B_N, S_N, or any other orthogonal basis, we are projecting f onto V_N as per equation (9.13). The projection $f_N = P_{V_N}(f)$ is unique and is the best approximation to f available in the subspace V_N. In the present case, Theorem 9.2 assures us that f_N converges to f in $L^2(\mathbb{R})$ as $N \to \infty$.

Of course, in practice we settle for a value of N so that f_N is "good enough" as an approximation. Suppose that we have the approximation f_N in hand but want a better approximation to f. Obviously, we have to compute f_{N+1}, f_{N+2}, or higher. In the present setting, there are a couple of ways to do this. We can recompute f_{N+1} "directly" by using the basis B_{N+1}, that is, forming the inner products

$(f, \phi_{N+1,m})$ and then computing $f_{N+1} = \sum_m (f, \phi_{N+1,m})\phi_{N+1,m}$. Alternatively, we can compute f_{N+1} as a "refinement" of f_N. Specifically, we have

$$f_{N+1} = f_N + \delta_N, \tag{9.28}$$

where

$$\delta_N(t) = \sum_{m \in \mathbb{Z}} (f, \psi_{N,m})\psi_{N,m}(t). \tag{9.29}$$

This follows from the fact that $S_{N+1} = S_N \cup \{\psi_{N,m} : m \in \mathbb{Z}\}$. Equations (9.28) and (9.29) show how to hop from the coarser approximation f_N to the finer approximation f_{N+1}; δ_N is the required correction.

Of course, this process can be iterated. Starting with the approximation f_N, we have

$$\begin{aligned} f_N &= f_{N-1} + \delta_{N-1} \\ &= f_{N-2} + \delta_{N-2} + \delta_{N-1} \\ &= f_{N-3} + \delta_{N-3} + \delta_{N-2} + \delta_{N-1} \\ \vdots &= \vdots \\ &= f_0 + \delta_0 + \delta_1 + \delta_2 + \cdots + \delta_{N-1}. \end{aligned} \tag{9.30}$$

Equation (9.30) is very similar to equation (7.28) from Chapter 7, with one small difference. In equation (7.28) as the index r increases the approximations α_r become coarser, while in equation (9.30) increasing indexes correspond to increasingly detailed features.

In either case, though, the point is that we have a quantitative way to relate coarser and finer approximations, and to step from one to the other using the wavelets. This is the central idea of multiresolution. In the present case, we have a nested sequence of subspaces

$$V_0 \subset V_1 \subset V_2 \subset \cdots \subset L^2(\mathbb{R}).$$

The wavelets $\psi_{N,m}$ give us a way to refine f_N to f_{N+1} via equations (9.28) and (9.29). Including the shorter scale wavelets $\psi_{N+1,m}$ in the computation allows us to refine f_{N+1} to f_{N+2}, and so on. Since

$$\lim_{N\to\infty} f_N = f,$$

in the sense that $\lim_{N\to\infty} \| f_N - f \| = 0$, we can iteratively improve the approximation to any desired degree of accuracy (as quantified by the L^2 norm).

9.3.2.3 Direct Sums

For any fixed $N \in \mathbb{Z}$ let $W_N \subset L^2(\mathbb{R})$ be the subspace defined as

$$W_N = \text{span}(\{\psi_{N,m} : m \in \mathbb{Z}\}) \tag{9.31}$$

formed from the $1/2^N$ scale wavelets $\psi_{N,m}$ alone. That is, W_N consists of functions w of the form

$$w = \sum_{m=-\infty}^{\infty} a_m \psi_{N,m},$$

where $\sum_m a_m^2 < \infty$; the infinite sum for $w(t)$ is understood in the sense of equations (9.11) and (9.12). It is easy to see that $W_N \cap V_N = \{0\}$, since every element in each subspace is orthogonal all of the elements of the other subspace.

According to equation (9.28), we can express $f_{N+1} = f_N + \delta_N$, where $f_N \in V_N$ and $\delta_N \in W_N$. In fact both f_N and δ_N are unique (see Exercise 9); δ_N is given by equation (9.29). Since each element of V_{N+1} can be written as the sum of a unique element of V_N and a unique element of W_N, we write

$$V_{N+1} = V_N \oplus W_N.$$

We say that the subspace V_{N+1} is the "direct sum" of the subspaces V_N and W_N. Let us make the more general definition

Definition 9.3 If $U_0, \ldots, U_n$ and U are subspaces of a vector space V, we write

$$U = U_0 \oplus U_1 \oplus \cdots \oplus U_n$$

provided that for each element $u \in U$ there are unique vectors $u_k \in U_k$ such that

$$u = u_0 + u_1 + \cdots + u_n.$$

With the terminology of Definition 9.3, equation (9.30) shows that

$$V_N = V_0 \oplus W_0 \oplus W_1 \oplus \cdots \oplus W_{N-1}.$$

The projection f_N of f onto V_N in equation (9.30) can be written as

$$f_N = f_0 + P_{W_0}(f) + \cdots + P_{W_{N-1}}(f),$$

where $P_{W_k}(f)$ denotes the projection of f onto the wavelet space W_k (and f_0 is the projection of f onto V_0). The addition of each wavelet space projection $P_{W_k}(f)$ adds a layer of increased resolution. Moreover, by Theorem 9.2, we have $f_N \to f$ as $N \to \infty$, so we can reasonably write

$$L^2(\mathbb{R}) = V_0 \oplus W_0 \oplus W_1 \oplus \cdots .$$

In all of the above, we need not use V_0 as the base space; we can just as well write

$$V_N = V_M \oplus W_M \oplus W_{M+1} \oplus \cdots \oplus W_{N-1}$$

for any $M < N$, and so

$$L^2(\mathbb{R}) = V_M \oplus W_M \oplus W_{M+1} \oplus \cdots$$

Indeed, we can even take $M < 0$. In this case V_M consists of functions constant on each interval $[k/2^M, (k+1)/2^M)$, where $k \in \mathbb{Z}$, intervals of length $2^{-M} > 1$ if $M < 0$. The functions $\phi_{M,n}$ defined by equation (9.27) provide an orthonormal basis for the space V_M; for $M < 0$ the scaling functions get wider and lower. The wavelets $\psi_{k,n}$ are also defined for any $k \in \mathbb{Z}$.

Example 9.3 *Look back to Example 9.1 where we compute approximations to a function $f(t) = t(t-1)(t-2)$ for $0 \le t \le 1$. The function f was defined on $[0, 1]$, but let us consider f as a member of $L^2(\mathbb{R})$ via zero extension. The approximation using only the Haar scaling function ϕ_0 yields the approximation*

$$f_0(t) = \begin{cases} \frac{1}{4}, & 0 \le t < 1, \\ 0, & \textit{else}, \end{cases}$$

in V_0. The inclusion of the wavelets $\psi_{0,m}$ (only the translate $m = 0$ matters here) allows us to refine $f_0(t)$ to $f_1(t)$ by adding appropriate variation on the left and right halves of the interval $[0, 1]$, as shown in Figure 9.7a. Inclusion of the wavelets $\psi_{1,m}$ yields $f_2(t)$ as shown in Figure 9.7b. The functions f_3 and f_4 are not shown, but $f_5(t)$ is shown in Figure 9.8.

9.3.2.4 Connection to Multistage Haar Filter Banks

Let us consider again the computations of Section 9.3. In particular, consider what happens if we project the function $x(t)$ defined by equation (9.17) onto the subspace V_N for various values of N. Using $N \ge 2$ is pointless, since $x(t)$ does not vary at this level of detail. Indeed, if $x_N(t)$ denotes the projection of $x(t)$ onto V_N, then $x_N(t) \equiv x(t)$ for all $N \ge 1$.

We computed the coefficients c_m necessary to project $x(t)$ onto V_0 (functions constant on integer intervals $[m, m+1)$) using equation (9.21), and these coefficients turned out to be simply the low-pass filtered/downsampled version of the associated vector $\mathbf{x} \in L^2(\mathbb{Z})$. The coefficients $c_{0,m}$ of the wavelets yield the adjustment necessary to obtain the projection of $x(t)$ onto V_1; moreover $c_{0,m}$ corresponds to the high-pass filtered/downsampled version of $\mathbf{x}$.

Let us instead compute the projection of $x(t)$ onto the subspace V_{-1} consisting of functions constant on intervals $[2k, 2k+2)$ of length 2. The scaling function basis consists of functions

$$\phi_{-1,m}(t) = \frac{1}{\sqrt{2}}\phi\left(\frac{t}{2} - m\right),$$

where ϕ is the Haar scaling function. These functions have base width two units; changing m to $m+1$ translates the function 2 units to the right. An easy computation shows that we obtain coefficients

$$c_m = \int_{-\infty}^{\infty} x(t)\phi_{-1,m}(t)\,dt = \frac{1}{2\sqrt{2}}(x_{4m-1} + x_{4m} + x_{4m+1} + x_{4m+2}).$$

The resulting projection onto V_{-1} is given by $x_{-1}(t) = \sum_m c_m \phi_{-1,m}(t)$.

We can refine our approximation by tossing the wavelets $\psi_{-1,m}$ into the basis. The corresponding coefficients are given by

$$c_{-1,m} = \int_{-\infty}^{\infty} x(t)\psi_{-1,m}(t)\,dt = \frac{1}{2\sqrt{2}}(x_{4m-1} + x_{4m} - x_{4m+1} - x_{4m+2}).$$

The projection of $x(t)$ onto V_0 can then be computed as

$$x_0(t) = x_{-1}(t) + \sum_m c_{-1,m}\psi_{-1,m}(t).$$

We can further refine the approximation $x_0(t)$ by making use of the $\psi_{0,m}(t)$ wavelets. The coefficients are given by equation (9.22), and we find that

$$x_1(t) = x_0(t) + \sum_m c_{0,m}(t)\psi_{0,m}(t).$$

As noted, $x_1(t) \equiv x(t)$, so there is no point in further refinement.

We already saw in equation (9.23) that $c_{0,m} = (\mathbf{X}_h)_m$, that is, the coefficients of the shortest scale wavelets correspond to the high-pass filtered, downsampled version of $\mathbf{x}$. To what do c_m and $c_{-1,m}$ correspond? To the vectors $\mathbf{X}_{\ell\ell}$ and $\mathbf{X}_{\ell h}$, or minor variations thereof! In the present case (with Haar filters $(\boldsymbol{\ell}_a)_0 = (\boldsymbol{\ell}_a)_1 = \frac{1}{2}$ and $(\mathbf{h}_a)_0 = -\frac{1}{2}, (\mathbf{h}_a)_1 = \frac{1}{2}$), it is easy to check that

$$(\mathbf{X}_{\ell\ell})_m = \frac{1}{4}(x_{4m-3} + x_{4m-2} + x_{4m-1} + x_{4m}),$$

$$(\mathbf{X}_{\ell h})_m = \frac{1}{4}(x_{4m-3} + x_{4m-2} - x_{4m-1} - x_{4m}).$$

In short, though not identical, both c_m and $(\mathbf{X}_{\ell\ell})_m$ consist of moving averages of four successive values, while $c_{-1,m}$ and $(\mathbf{X}_{\ell h})_m$ consist of a mix of averaged/differenced successive values. (With enough clever indexing and rescaling, we could make the corresponding quantities identical.)

A similar parallel holds for the corresponding synthesis filter bank and the synthesis of $x(t)$ from the functions $\phi_{-1,m}$, $\psi_{-1,m}$, and $\psi_{0,m}$. Moreover, the parallels exists for three and higher stage banks. As with signals in $L^2(\mathbb{Z})$, we can produce coarser and coarser approximations, to any degree, for any function $f \in L^2(\mathbb{R})$ (analogous to repeatedly low-pass filtering a signal in $L^2(\mathbb{Z})$). However, in the functional setting we can go the other way, to produce finer and finer approximations to the original function. In the discrete signal case, this process is limited by the sampling rate.

9.4 Orthogonal Wavelets

9.4.1 Essential Ingredients

The essential ingredients for the multiresolution framework in the Haar wavelet case are

1. A nested chain of subspaces

$$\cdots V_{-1} \subset V_0 \subset V_1 \subset \cdots$$

   all contained in $L^2(\mathbb{R})$, with the properties that

$$\bigcup_{k=-\infty}^{\infty} V_k \text{ is dense in } L^2(\mathbb{R}) \text{ and } \bigcap_{k=-\infty}^{\infty} V_k = \{0\}. \tag{9.32}$$

   Here *dense* means that any function in $L^2(\mathbb{R})$ can be approximated to arbitrary precision (as measured by the L^2 norm) by using functions in the union of the V_k. The intersection property above was not spelled out in the Haar setting, but it is easy to see only a constant function can be an element of all the V_k in this case, and the only constant function in $L^2(\mathbb{R})$ is 0. That the union of the V_k is dense in $L^2(\mathbb{R})$ is a consequence of Theorem 9.2.
2. A function $f(t) \in V_k$ if and only if $f(2t) \in V_{k+1}$. Also, if $f(t) \in V_0$, then $f(t - n) \in V_0$ for any $n \in \mathbb{Z}$.
3. The space V_0 possesses an orthonormal basis of the form $\phi_n(t) = \phi(t - n)$, $n \in \mathbb{Z}$. The function ϕ is called the *scaling function*.

The wavelets themselves are not essential (yet).

Our goal is to generalize this framework to other than the piecewise constant Haar setting. We begin with a definition.

Definition 9.4 A set of subspaces V_k and function ϕ that satisfy properties 1 through 3 above is called a "multiresolution analysis."

Of course, it is not at all clear that there are any multiresolution analyses other than the one constructed via the Haar functions.

Remark 9.3 It is not absolutely essential that properties 2 and 3 involve integer translates; this is merely for convenience and convention. Also from property 2 it is easy to see that $f(t) \in V_0$ if and only if $f(2^k t) \in V_k$ for any $k \in \mathbb{Z}$. One can easily deduce that for fixed k the set $\phi(2^k t - n)$, $n \in \mathbb{Z}$ is an orthogonal basis for V_k.

9.4.2 Constructing a Multiresolution Analysis: The Dilation Equation

Suppose that we have a multiresolution analysis with scaling function $\phi \in V_0$. The integer translates $\phi(t-n)$ form an orthonormal basis for V_0. From property 2 for a multiresolution analysis, we see that the half-integer translates $\phi(2t-n)$ all lie in V_1 are orthogonal, and in fact $\sqrt{2}\phi(2t-n)$ is an orthonormal basis for V_1. Since $V_0 \subset V_1$ the function ϕ itself lies in V_1, and so it must be the case that

$$\phi(t) = \sqrt{2} \sum_{k=-\infty}^{\infty} c_k \phi(2t-k) \tag{9.33}$$

for some constants c_k. Equation (9.33) is called the *dilation equation.* The $\sqrt{2}$ factor out front could be absorbed into the definition of the c_k, but we leave it in order to later emphasize certain parallels to filter banks.

For the Haar scaling function, we have $c_0 = c_1 = 1/\sqrt{2}$ with all other $c_k = 0$ (see Example 9.4). The fact that only finitely many c_k are nonzero is what makes the Haar multiresolution analysis parallel the filter bank analysis with FIR filters. Can we find other solutions to equation (9.33)—both the c_k and scaling function ϕ—in which only finitely many c_k are nonzero?

Let us see what we can deduce under the assumption that the only nonzero c_k lie in the index range $0 \le k \le N-1$ (but we will assume c_k is defined for all $k \in \mathbb{Z}$, as 0 outside this range). In this case the dilation equation becomes

$$\phi(t) = \sqrt{2} \sum_{k=0}^{N-1} c_k \phi(2t-k). \tag{9.34}$$

We also normalize $\|\phi\|^2 = 1$. Compute the inner product in $(\phi, \phi) = 1$ using the expansion on the right-hand side in equation (9.34) to obtain

$$\begin{aligned}(\phi, \phi) &= \int_{\mathbb{R}} \phi^2(t)\, dt \\ &= 2 \sum_{m=0}^{N-1} \sum_{k=0}^{N-1} \int_{\mathbb{R}} c_m c_k \phi(2t-m)\phi(2t-k)\, dt \\ &= \sum_{k=0}^{N-1} c_k^2,\end{aligned}$$

since the last integral equals 0 if $m \neq k$ and equals $\frac{1}{2}$ if $m = k$. The computation above shows that we must have

$$\sum_{k=0}^{N-1} c_k^2 = 1. \tag{9.35}$$

Now consider computing the inner product $(\phi(t), \phi(t-1)) = 0$ by using the expansion on the right-hand side in equation (9.34). We have

$$\phi(t-1) = \sqrt{2}\sum_{k=0}^{N-1} c_k \phi(2t-k-2),$$

and so

$$\begin{aligned}(\phi(t), \phi(t-1)) &= \int_{\mathbb{R}} \phi(t)\phi(t-1)\,dt \\ &= 2\sum_{m=0}^{N-1}\sum_{k=0}^{N-1}\int_{\mathbb{R}} c_m c_k \phi(2t-m)\phi(2t-k-2)\,dt \\ &= \sum_{k=0}^{N-1} c_k c_{k+2} = 0,\end{aligned}$$

since the last integral equals 0 if $m \neq k+2$ and equals $\frac{1}{2}$ if $m = k+2$. In the last sum the upper limit can be changed to $N-3$, since $c_{k+2} = 0$ for $k > N-3$, but it is not essential.

Generally, if we compute the inner product $(\phi(t), \phi(t-n)) = 0$ for $0 < n < (N-1)/2$ in this fashion, we obtain

$$\sum_{k=0}^{N-1-2n} c_k c_{k+2n} = 0. \tag{9.36}$$

The equation above is invariant under the substitution $n \to -n$, so taking $n < 0$ does not add anything. As it turns out, N will have to be even. So equation (9.36) embodies $N/2 - 1$ equations, since n can assume values $1, 2, \ldots, (N-2)/2$.

One more condition on the c_k can be deduced. If we integrate both sides of equation (9.34) in t over the whole real line, we obtain

$$\begin{aligned}\int_{-\infty}^{\infty} \phi(t)\,dt &= \sqrt{2}\sum_{k=0}^{N-1} c_k \int_{-\infty}^{\infty} \phi(2t-k)\,dt \\ &= \frac{\sqrt{2}}{2}\sum_{k=0}^{N-1} c_k \int_{-\infty}^{\infty} \phi(t)\,dt,\end{aligned}$$

since the integral of $\phi(2t-k)$ is half the integral of $\phi(t)$. If the integral of ϕ converges and is nonzero, we can conclude that

$$\sum_{k=0}^{N-1} c_k = \sqrt{2}. \tag{9.37}$$

Since N is even equations (9.35)–(9.37) yield $1 + N/2$ equations for N unknowns for the filter coefficients, at least if the equations are independent.

For $N > 2$, we have more variables than equations, so we expect many solutions.

Example 9.4 *Consider $N = 2$, in which case equations (9.35) and (9.37) yield*

$$c_0^2 + c_1^2 = 1, \quad c_0 + c_1 = \sqrt{2},$$

while equation (9.36) yields the empty statement $0 = 0$. The only solution is $c_0 = c_1 = 1/\sqrt{2}$, which corresponds to the Haar scaling function. It is easy to see that this yields a solution to equation (9.34). Of course, even if we were given $c_0 = c_1 = 1/\sqrt{2}$, it is not clear how we would have come up with the Haar scaling function without already knowing it. In fact it is not clear that the Haar scaling function is the only possibility, even with these values of c_0 and c_1. We will look at how to obtain ϕ a bit later.

Example 9.5 *Consider $N = 4$, in which case equations (9.35) and (9.37) yield*

$$c_0^2 + c_1^2 + c_2^2 + c_3^2 = 1, \quad c_0 + c_1 + c_2 + c_3 = \sqrt{2},$$

while equation (9.36) yields

$$c_0c_2 + c_1c_3 = 0.$$

With three equations in four unknowns we expect a free variable and a whole family of solutions. Indeed, if we choose $c_3 = t$, then we obtain two families of solutions, one of which looks like

$$c_0 = \frac{\sqrt{2}}{4} + \frac{\sqrt{2 + 8\sqrt{2}t - 16t^2}}{4}, \quad c_1 = \frac{\sqrt{2}}{2} - t,$$

$$c_2 = \frac{\sqrt{2}}{4} - \frac{\sqrt{2 + 8\sqrt{2}t - 16t^2}}{4},$$

where t is a free parameter (a computer algebra system is helpful here, or at least perseverance). The other solution family is similar, but with c_0 and c_2 interchanged. Only in the range $\sqrt{2}/4 - 1/2 \le t \le \sqrt{2}/4 + 1/2$ do all c_k remain real.

The choice $t = 0$ above yields the Haar filters, while $t = \sqrt{2}/2$ yields the Haar filters shifted two indexes. The choice $t = (1 - \sqrt{3})/4\sqrt{2}$ yields the Daubechies 4-tap filters from Chapter 7. Similar results are obtained for the second family of solutions.

9.4.3 Connection to Orthogonal Filters

Recall our design strategy from Chapter 7 for the low-pass analysis filter in an orthogonal filter bank. The FIR filter $\ell = (\ell_0, \ell_1, \dots, \ell_{N-1})$ with

z-transform $L_a(z)$ had to satisfy $P(z) + P(-z) = 2$ (this is equation (7.49)), where $P(z) = L_a(z)L_a(z^{-1})$. This requires that $P(z)$ should have coefficient 1 for z^0 and all even index components equal to zero. A straightforward multiplication shows that the coefficients p_m of $P(z)$ are given by

$$p_m = \sum_{k=0}^{N-1-m} \ell_k \ell_{k-m} \tag{9.38}$$

for $0 \leq m < N-1$ and $p_{-m} = p_m$. The condition that $p_0 = 1$ is, from equation (9.38), just

$$\sum_{k=0}^{N-1} \ell_k^2 = 1. \tag{9.39}$$

Equation (9.38) shows that the condition $p_m = 0$ when m is even becomes (let $m = -2n$; the sign does not matter)

$$p_m = \sum_{k=0}^{N-1-2n} \ell_k \ell_{k+2n} = 0. \tag{9.40}$$

Compare the filter design equations (9.39) and (9.40) to equations (9.35) and (9.36) derived from the dilation equation—aside from the variable names, they are identical! Moreover the condition that the low-pass analysis filter satisfy $L_a(-1) = 0$ was equivalent to equation (7.50), which is exactly equation (9.37). Thus, the methods we used to find appropriate low-pass filter coefficients will yield appropriate constants c_k for the dilation equation.

9.4.4 Computing the Scaling Function

It can be shown that under appropriate conditions there is a unique function ϕ that satisfies equation (9.34). We explore this in Section 9.4.5. But let us first pause and look at an algorithm for computing the scaling function, under the assumption that this function actually exists and is continuous.

We will compute $\phi(t)$ at points of the form $t = m/2^n$ for integers m and n, the so-called *dyadic* rational numbers. One obstacle is that ϕ is defined on the entire line, so we will have to truncate our approximation at some lower and upper bounds for t. What should these limits be?

The Haar scaling function has compact support. It seems reasonable to seek scaling functions in the general case that possess these properties, that is, the scaling functions should be identically zero outside certain lower and upper bounds. What should these bounds be? It is not hard to see that the choice $\phi(t) \equiv 0$ for $t < 0$ is always consistent with equation (9.34) no matter what the behavior of ϕ for $t \geq 0$. This does not prove that $\phi(t)$ must be zero for $t < 0$, but it gives us someplace to start. This property turns out to be correct (see Theorem 9.4). For an upper bound, suppose that $\phi(t) \equiv 0$ for $t > a$ for some constant a. We can determine a as follows: the function $\phi(2t-k)$ will be zero for $2t-k>a$, that is, $t>(a+k)/2$. For the largest value of k, namely $k=N-1$, the right-hand side of equation (9.34) will then be identically zero if $t>(a+N-1)/2$. Choosing a so that $a=(a+N-1)/2$ will at least be consistent, for then both sides of (9.34) will be identically zero for $t>a$. The condition $a=(a+N-1)/2$ yields $a=N-1$.

Thus, we assume that ϕ is supported on the interval $[0, N-1]$ and approximate ϕ only on this interval. If ϕ is continuous (which it will be), then $\phi(0)=\phi(N-1)=0$. From the dilation equation (9.34) with $t=1$ we obtain

$$\phi(1) = \sqrt{2}c_0\phi(2) + \sqrt{2}c_1\phi(1).$$

When $t=2$, we find that

$$\phi(2) = \sqrt{2}c_0\phi(4) + \sqrt{2}c_1\phi(3) + \sqrt{2}c_2\phi(2) + \sqrt{2}c_3\phi(1).$$

More generally, if $t=n\in\mathbb{Z}$, where $1\le n\le N-2$, we find that

$$\phi(n) = \sqrt{2}\sum_k c_k\phi(2n-k), \tag{9.41}$$

where the range of summation in k is such that $1\le 2n-k\le N-2$, since $\phi(2n-k)$ will be zero for integers outside this range. Equation (9.41) provides a set of $N-2$ linear relations among the quantities $\phi(1),\dots,\phi(N-2)$.

In fact these equations can be cast as an eigenvector problem

$$\mathbf{q} = \mathbf{A}\mathbf{q}, \tag{9.42}$$

where $\mathbf{q}=[\phi(1),\phi(2),\dots,\phi(N-2)]^T$ is an eigenvector with eigenvalue 1 for the $N-2\times N-2$ matrix $\mathbf{A}$ with entries

$$a_{n,2n-k+1} = \sqrt{2}c_{k-1}$$

for $1\le n\le N-2$, $1\le k\le N$, and the additional restriction $1\le 2n-k+1\le N-2$ on k. Note that we are indexing matrix columns and rows from 1 now, rather than 0 as was convenient earlier. For example, in the case $N=8$ the matrix $\mathbf{A}$ is 6×6 and looks like

$$\mathbf{A} = \sqrt{2}\begin{bmatrix} c_1 & c_0 & 0 & 0 & 0 & 0 \\ c_3 & c_2 & c_1 & c_0 & 0 & 0 \\ c_5 & c_4 & c_3 & c_2 & c_1 & c_0 \\ c_7 & c_6 & c_5 & c_4 & c_3 & c_2 \\ 0 & 0 & c_7 & c_6 & c_5 & c_4 \\ 0 & 0 & 0 & 0 & c_7 & c_6 \end{bmatrix}.$$

It is not obvious that $\mathbf{A}$ has 1 as an eigenvalue, but it does. We can use any standard method to compute the corresponding eigenvector. We thus obtain

$\phi(n)$ for $1 \le n \le N-2$; indeed we obtain ϕ at all integers, since $\phi(n)$ is zero for n outside this range.

We can then compute ϕ at half-integer points $n/2$ for n odd with $1 \le n \le 2N-3$, for from the dilation equation (9.34) we have

$$\phi(n/2) = \sqrt{2}\sum_{k=0}^{N-1} c_k \phi(n-k).$$

Since $n-k$ is an integer, $\phi(n-k)$ is known. Once we know ϕ at the half-integers, we can compute $\phi(t)$ at points of the form $t = n/4$ where n is odd, $1 \le n \le 4N-5$. We find that

$$\phi(n/4) = \sqrt{2}\sum_{k=0}^{N-1} c_k \phi(n/2-k)$$

and the right-hand side is known because $n/2-k$ is a half-integer. In general, if we have $\phi(n/2^{r-1})$ for some r and $0 < n < 2^{r-1}(N-1)$, we can compute $\phi(t)$ at points $t = n/2^r$ (odd n) as

$$\phi(n/2^r) = \sqrt{2}\sum_{k=0}^{N-1} c_k \phi(n/2^{r-1}-k). \tag{9.43}$$

The right-hand side will be known from previous values of r.

Example 9.6 *Let us compute the scaling function that corresponds to the Daubechies orthogonal 4-tap filter coefficients:*

$$c_0 = \frac{1+\sqrt{3}}{4\sqrt{2}}, \quad c_1 = \frac{3+\sqrt{3}}{4\sqrt{2}}, \quad c_2 = \frac{3-\sqrt{3}}{4\sqrt{2}}, \quad c_3 = \frac{1-\sqrt{3}}{4\sqrt{2}}.$$

In this case the matrix **A** *is*

$$\mathbf{A} = \begin{bmatrix} c_1 & c_0 \\ c_3 & c_2 \end{bmatrix}.$$

This matrix does indeed have 1 as an eigenvalue $\left(\text{also } \frac{1}{2}\right)$*, with eigenvector*

$$\begin{bmatrix} \phi(1) \\ \phi(2) \end{bmatrix} = \begin{bmatrix} 1+\sqrt{3} \\ 1-\sqrt{3} \end{bmatrix}.$$

Of course, the eigenvector can be normalized in any way we wish. We will wait until we have computed ϕ *at a number of other points, then normalize (approximately) to unit* L^2 *norm.*

At the half-integers we find that

$$\phi\left(\frac{1}{2}\right) = \sqrt{2}c_0\phi(1) = 1 + \frac{\sqrt{3}}{2},$$
$$\phi\left(\frac{3}{2}\right) = \sqrt{2}(c_1\phi(2) + c_2\phi(1)) = 0,$$
$$\phi\left(\frac{5}{2}\right) = \sqrt{2}c_3\phi(2) = 1 - \frac{\sqrt{3}}{2}.$$

We can continue the process via equation (9.43). In Figure 9.12, we show the results on the interval $[0, 3]$ *using* $r = 2, 4, 6, 8$ *(dyadic points* $t = n/2^r$*). In each case the vector* $\mathbf{v} \in \mathbb{R}^{2^r(N-1)}$ *that approximates* ϕ *is scaled so that* $\|\mathbf{v}\|^2 = 2^r(N - 1)$*, a discrete analog to* $\|\phi\|^2 = 1$*. The case* $r = 2$ *is at the upper left,* $r = 4$ *the upper right,* $r = 6$ *the lower left,* $r = 8$ *the lower right.*

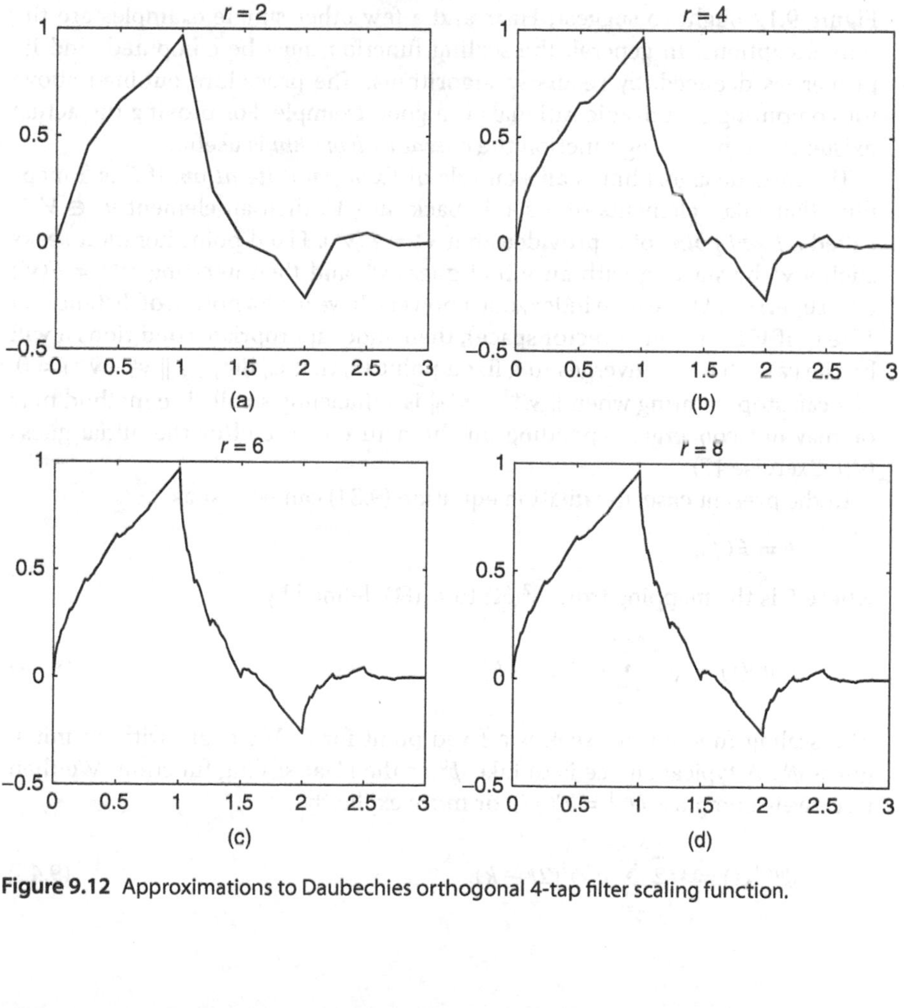

Figure 9.12 Approximations to Daubechies orthogonal 4-tap filter scaling function.

Remark 9.4 The preceding procedure by which we compute the scaling function $\phi(t)$ at dyadic t makes the implicit assumption that ϕ is defined at these points, indeed that $\phi(t)$ is sufficiently "nice" to have well-defined point values anywhere. This is not the case for arbitrary functions in $L^2(\mathbb{R})$ (recall the discussion of $L^2(a, b)$ in Section 1.10.5), but it is true for continuous functions. Almost all of the scaling functions and wavelets that we will encounter are in fact continuous. The exceptions are the Haar scaling function (which is continuous at all but two points) and one other notable exception in Section 9.5.

9.4.5 Scaling Function Existence and Properties

9.4.5.1 Fixed Point Iteration and the Cascade Algorithm

Let $c_0, \ldots, c_{N-1}$ satisfy equations (9.35)–(9.37) in the range $0 < n < (N-1)/2$. From Section 7.7.6, we know how to find such c_k.

There is rarely a closed-form expression for the scaling function, as Figure 9.12 ought to suggest; Haar and a few other simple examples are the only exceptions. In general, the scaling function must be computed, and its properties deduced, by recursive algorithms. The procedure outlined above for computing ϕ at dyadic rationals is a good example. For proving the actual existence of the scaling function the *cascade algorithm* is useful.

The cascade algorithm is an example of *fixed point iteration*. If F is a mapping that takes elements of a set V back into V, then an element $\mathbf{v}^* \in V$ is called a *fixed point* of F provided that $\mathbf{v}^* = F(\mathbf{v}^*)$. Fixed point iteration seeks such a $\mathbf{v}^*$ by starting with an initial guess $\mathbf{v}^0$, and then iterating $\mathbf{v}^{j+1} = F(\mathbf{v}^j)$ (the superscript here is an index, not a power). If we have a notion of distance on V (e.g., if V is a normed vector space), then under appropriate conditions it will be the case that $\mathbf{v}^j$ converges to a fixed point $\mathbf{v}^*$; that is, $\lim_{j\to\infty} \| \mathbf{v}^j - \mathbf{v}^* \| = 0$. We can stop iterating when $\| \mathbf{v}^{j+1} - \mathbf{v}^j \|$ is sufficiently small. The method may or may not converge, depending on the nature of F and/or the initial guess (see Exercise 17).

In the present case, the dilation equation (9.34) can be cast as

$$\phi = F(\phi),$$

where F is the mapping from $L^2(\mathbb{R})$ to $L^2(\mathbb{R})$ defined by

$$F(\phi)(t) = \sqrt{2}\sum_{k=0}^{N-1} c_k\phi(2t-k). \tag{9.44}$$

The scaling function we seek is a fixed point for F. We begin with an initial guess ϕ^0. A typical choice is to take ϕ^0 as the Haar scaling function. We then iteratively compute $\phi^{j+1} = F(\phi^j)$, or more explicitly,

$$\phi^{j+1}(t) = \sqrt{2}\sum_{k=0}^{N-1} c_k\phi^j(2t-k). \tag{9.45}$$

Remember, we treat the c_k as known. Of course, the process is in general hopelessly unwieldy if done symbolically. An exception is the Haar function when $\phi^1 = \phi^0$, and we are done! A few other cases can also be analyzed in closed form, but we are primarily interested in the cascade algorithm as a theoretical tool to show the existence of the scaling function.

9.4.5.2 Existence of the Scaling Function

In equations (9.42) and (9.43) we outlined an algorithm to compute a function ϕ that satisfies the dilation equation (9.34) but never actually showed that ϕ exists. In developing the algorithm, we also made certain unjustified assumptions concerning the properties that ϕ possesses (e.g., that ϕ is zero outside of $[0, N-1]$). In this section, we will examine these issues. However, we will not generally provide detailed proofs, only sketches or references.

First, let us consider the existence of the scaling function that satisfies the dilation equation (9.34). We assume that the c_k have been determined and satisfy equations (9.35)–(9.37). As remarked, equations (9.35) and (9.36) are really the filter bank perfect reconstruction equations (9.39) and (9.40) (which themselves are equivalent to $P(z) + P(-z) = 2$, equation (7.49)). Equation (9.37) is a restatement of the low-pass condition (7.50) on the input filter ℓ_a.

With slightly different notation (and filters scaled differently, by a factor $\sqrt{2}$), the authors of [2] prove the following theorem (Theorem 5.23 in [2]):

Theorem 9.3 *Let $L_a(z) = \sum_{k=0}^{N-1} c_k z^k$ be a polynomial that satisfies the following conditions:*

1. $L_a(1) = \sqrt{2}$,
2. $|L_a(z)|^2 + |L_a(-z)|^2 = 2$ *for* $|z| = 1$,
3. $|L_a(e^{it})| > 0$ *for* $|t| \le \pi/2$.

Let ϕ^0 be the Haar scaling function and the sequence ϕ^j be defined by equation (9.45). Then the sequence ϕ^j converges both pointwise and in $L^2(\mathbb{R})$ to a function ϕ that satisfies the dilation equation (9.34). Moreover. ϕ satisfies $\|\phi\| = 1$ and the orthogonality condition

$$\int_{-\infty}^{\infty} \phi(t-m)\phi(t-n)\,dt = 0$$

for $m \neq n$.

If L_a has real coefficients then $|L_a(e^{it})| = |L_a(e^{-it})|$ (Exercise 15), so we can use $L_a(z) = \sum_{k=0}^{N-1} c_k z^{-k}$ in the statement of the theorem too (thus, matching the conventional z-transform of the filter $\boldsymbol{\ell}_a$). It is also easy to show that the function ϕ so produced satisfies $\int_{\mathbb{R}} \phi(t)\,dt = 1$ (see Exercise 21).

The first condition, $L_a(1) = \sqrt{2}$, is just equation (9.37), or alternatively, equation (7.50). Equation (7.50) is itself equivalent to $L_a(-1) = 0$, which quantifies the fact that the analysis filter ℓ_a is low-pass. Thus, $L_1(1) = \sqrt{2}$ will be satisfied for the filter coefficients c_k (or ℓ_k), which we obtain by the methods of Section 7.7.6.

The second condition, $|L_a(z)|^2 + |L_a(-z)|^2 = 2$ for $|z| = 1$, is also satisfied. To see this, first note that the coefficients $L_a(z)$ are real numbers, and also if $|z| = 1$, then $\bar{z} = z^{-1}$. As a consequence $L_a(z^{-1}) = \overline{L_a(z)}$. Recall also that the product filter satisfies $P(z) = L_a(z)L_a(z^{-1})$ (this is equation (7.48)). As a result for $|z| = 1$ we have

$$P(z) = L_a(z)L_a(z^{-1}) = L_a(z)\overline{L_a(z)} = |L_a(z)|^2. \tag{9.46}$$

Similarly $P(-z) = |L_a(-z)|^2$. This last equation, in conjunction with equation (9.46) and the fact that $P(z) + P(-z) = 2$ (equation (7.49)), then yields $|L_a(z)|^2 + |L_a(-z)|^2 = 2$.

The third condition does not automatically follow from our construction of the c_k or ℓ_k but needs to be checked on a case-by-case or family-by-family basis. Of course, all this condition really says is that $L_a(e^{it}) \neq 0$ for $-\pi/2 \leq t \leq \pi/2$, or equivalently, $P(e^{it}) \neq 0$.

Example 9.7 *Consider the orthogonal Haar filter coefficients $c_0 = c_1 = 1/\sqrt{2}$. In this case $L_a(z) = \frac{1}{\sqrt{2}} + \frac{1}{\sqrt{2}}z^{-1}$. In particular*

$$L_a(e^{it}) = \frac{1}{\sqrt{2}} + \frac{1}{\sqrt{2}}e^{-it}.$$

It is easy to check that

$$|L_a(e^{it})| = \sqrt{1 + \cos(t)}$$

and to see that $|L_a(e^{it})| \geq 1 > 0$ for $|t| \leq \pi/2$. Thus, the cascade algorithm will converge to a scaling function with the required properties. Of course, we already knew this, since $\phi^1 = \phi^0$.

Example 9.8 *Consider the Daubechies filter coefficients, in which case*

$$L_a(z) = \frac{1}{4\sqrt{2}}\left(\left(1+\sqrt{3}\right) + \left(3+\sqrt{3}\right)z^{-1} + \left(3-\sqrt{3}\right)z^{-2} + \left(1-\sqrt{3}\right)z^{-3}\right).$$

We can plot $|L_a(e^{it})|$ for $-\pi/2 \leq t \leq \pi/2$ and easily see that $|L_a(e^{it})| > 0$. However, a bit of algebra yields

$$|L_a(e^{it})| = \left(1 + \frac{3}{2}\cos(t) - \frac{1}{2}\cos^3(t)\right)^{1/2} = (1 + \cos(t))\sqrt{1 - \cos(t)/2},$$

which makes it clear that $|L_a(t)| > 0$ for $t \in [-\pi/2, \pi/2]$. An appropriate scaling function then exists, and the cascade algorithm will converge to this function.

Example 9.9 *Consider the case $N = 4$ with $c_0 = c_3 = 1/\sqrt{2}$ and $c_1 = c_2 = 0$. Then*

$$L_a(z) = \frac{1}{\sqrt{2}} + \frac{1}{\sqrt{2}} z^{-3}.$$

Condition 1 of Theorem 9.3 is satisfied. Condition 2 follows from

$$|L_a(e^{it})|^2 = 1 + \cos(3t),$$

and so $|L_a(-e^{it})|^2 = |L_a(e^{i(t+\pi)})|^2 = 1 + \cos(3(t+\pi)) = 1 - \cos(3t)$. However, the third condition is not satisfied, for $L_a(e^{i\pi/3}) = 0$. What does the cascade algorithm do in this case? The iteration of (9.44) becomes

$$\phi^{j+1}(t) = \phi^j(2t) + \phi^j(2t-3).$$

With ϕ^0 as the Haar scaling function, it is not hard to compute that

$$\phi^1(t) = \begin{cases} 1, & t \in \left[0, \frac{1}{2}\right) \text{ or } t \in \left[\frac{3}{2}, 2\right), \\ 0, & \text{else}, \end{cases}$$

and

$$\phi^2(t) = \begin{cases} 1, & t \in \left[0, \frac{1}{4}\right), t \in \left[\frac{3}{4}, 1\right), t \in \left[\frac{3}{2}, \frac{7}{4}\right), t \in \left[\frac{9}{4}, \frac{5}{2}\right), \\ 0, & \text{else}. \end{cases}$$

A straightforward induction proof (Exercise 22) shows that

$$\phi^j(t) = \begin{cases} 1, & t \in \left[\frac{3k}{2^j}, \frac{3k+1}{2^j}\right), 0 \le k < 2^j, \\ 0, & \text{else}. \end{cases}$$

That is, ϕ^j oscillates between the values 0 and 1, on intervals of length $1/2^j$, and converges to no meaningful limit in the pointwise or L^2 sense. In Figure 9.13 we show a few iterates. Although ϕ^j does not converge to anything in $L^2(\mathbb{R})$, there is a solution to the dilation equation in this instance—see Exercise 23. Moreover the sequence ϕ^j exhibits what is called "weak L^2 convergence"; see [18] for more on weak convergence.

Example 7.9 illustrates one important fact: *not every orthogonal filter bank yields a multiresolution analysis.* Some additional condition on the low-pass filter (e.g., condition 3 of Theorem 9.3) is necessary.

9.4.5.3 The Support of the Scaling Function

Previously we argued informally that the scaling function should be supported in the interval $[0, N-1]$, a fact we can now prove.

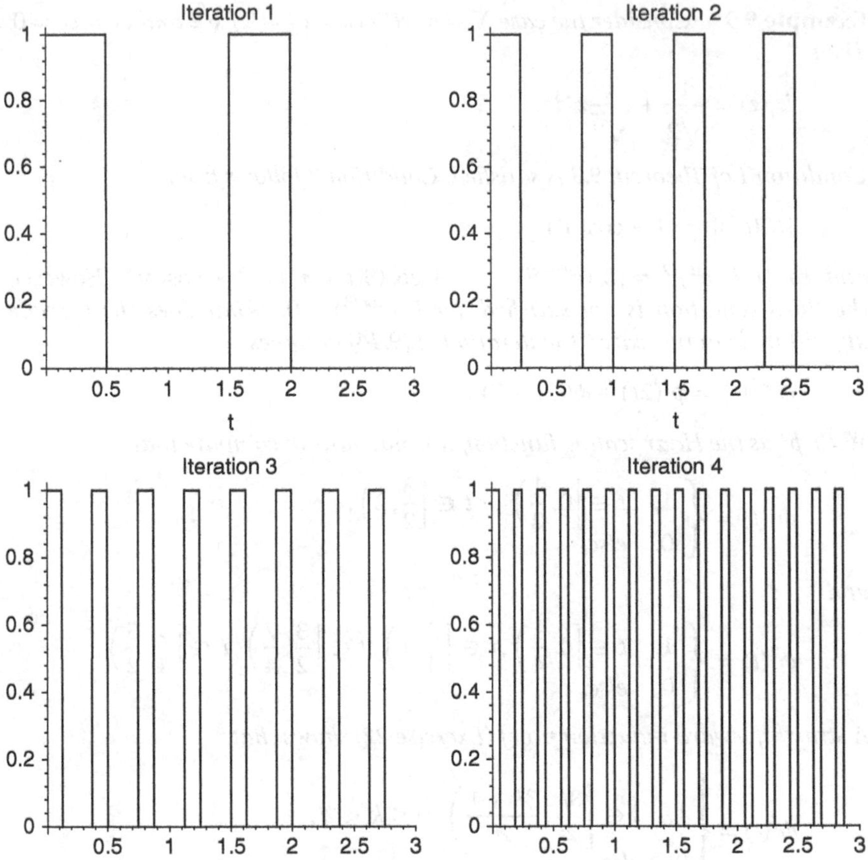

Figure 9.13 Iterates $\phi^1, \phi^2, \phi^3, \phi^4$ of cascade algorithm for filter $(1/2, 0, 0, 1/2)$.

Theorem 9.4 *With the assumptions of Theorem 9.3, the scaling function* ϕ *that satisfies equation (9.34) is supported in the interval* $[0, N-1]$.

Proof: First, we claim that all iterates of the cascade algorithm are supported in $[0, N-1]$. We prove this by induction. The function ϕ^0 is indeed supported on $[0, N-1]$. Suppose that the same holds true for ϕ^j for some fixed j. The function $\phi^j(2t-k)$ is then supported on the interval $[k/2, (N-1)/2 + k/2]$. Since $0 \leq k \leq N-1$, the interval $[k/2, (N-1)/2 + k/2]$ is contained in $[0, N-1]$. Thus, the sum $\sum_{k=0}^{N-1} \phi^j(2t-k)$ is supported in $\cup_{k=0}^{N-1}[k/2, (N-1)/2 + k/2] = [0, N-1]$ (where we use the fact that support of a sum of functions is contained in the union of the individual supports; see Exercise 20). The induction hypothesis is shown, and we conclude that all iterates ϕ^j are supported in $[0, N-1]$.

This means that $\phi^j(t) = 0$ for all t outside of $[0, N-1]$. By Theorem 9.3 the iterates converge pointwise to ϕ, so that

$$\phi(t) = \lim_{j\to\infty} \phi^j(t) = 0$$

for t outside $[0, N-1]$. Thus, ϕ is supported in $[0, N-1]$.

Other properties of the scaling function, such as its differentiability (which depends on N and the c_k), can also be deduced (see Chapter 7 of [28]).

9.4.5.4 Back to Multiresolution

A multiresolution analysis as in Definition 9.4 requires more than just the scaling function—it also requires the subspaces V_k with the appropriate properties. Let ϕ be the scaling function for an appropriate set of c_k, and define

$$V_k = \text{span}\{\phi(2^k t - m) : m \in \mathbb{Z}\} \subset L^2(\mathbb{R}), \tag{9.47}$$

where the span is understood in the sense of equations (9.11) and (9.12). It is easy to see that properties 2 and 3 for a multiresolution analysis hold. The nesting property $\cdots \subset V_{-1} \subset V_0 \subset V_1 \subset \cdots$ holds by virtue of the fact that ϕ satisfies the dilation equation. Only the properties of equation (9.32) need verification (which we did explicitly in the case of the Haar wavelet). These properties are demonstrated in Theorem 5.17 and Appendix A of [2] (the theorem stated in [2] requires $\int \phi(t)\,dt = 1$, which follows in the present case; see Exercise 21).

Let us summarize the results of this section.

Theorem 9.5 *Let coefficients c_k, $0 \le k \le N-1$ satisfy the conditions of Theorem 9.3 and ϕ be the solution to the dilation equation (9.34). Let the V_k be defined via equation (9.47). Then the V_k together with ϕ form a multiresolution analysis.*

9.4.6 Wavelets

For the Haar multiresolution analysis, the Haar wavelets provide a technique for conveniently refining the projection of a function onto V_k to the projection onto V_{k+1}. Since projection onto V_{k+1} embodies greater detail than projection onto V_k, this gives us a way to easily and incrementally improve resolution, to any desired degree of accuracy.

The same approach works more generally. Recall that the mother Haar wavelet in equation (9.1) can be constructed as a linear combination of scaling function dilations and translations, as $\psi(t) = \phi(2t) - \phi(2t-1)$. Let us write this as

$$\psi(t) = \sqrt{2}d_0\phi(2t) + \sqrt{2}d_1\phi(2t-1),$$

where $d_0 = 1/\sqrt{2}, d_1 = -1/\sqrt{2}$. In fact d_0 and d_1 are precisely the high-pass Haar filter coefficients h_0 and h_1. In the light of equation (7.46) of Chapter 7 this suggests that we should define the wavelet in the general case as $\psi = \sqrt{2}\sum_k d_k\phi(2t-k)$, where $d_k = (-1)^k\ell_{N-k-1}$. Since $\ell_k = c_k$ in our normalization, this yields

$$\psi(t) = \sqrt{2}\sum_{k=0}^{N-1} d_k\phi(2t-k), \tag{9.48}$$

where $d_k = (-1)^k c_{N-k-1}$. Precisely the same argument in the proof of Theorem 9.4 shows that the wavelet ψ is supported on the interval $[0, N-1]$. The function $\psi(t)$, like ϕ, satisfies $\|\psi\| = 1$ (using the $L^2(\mathbb{R})$ norm) (see Exercise 24).

Is equation (9.48) the correct definition for the wavelet? What we want is that the set

$$S_1 = \{\phi(t-m) : m \in \mathbb{Z}\} \cup \{\psi(t-m) : m \in \mathbb{Z}\} \tag{9.49}$$

provide an orthogonal basis for the space V_1, an alternative to the basis $B_1 = \{\phi(2t-m) : m \in \mathbb{Z}\}$ of half-width scaling function translates. This is the case, which we demonstrate in the next few propositions. Note that $\psi(t-m)$ is in fact in V_1, for it follows from equation (9.48) that $\psi(t-m)$ is a finite linear combination of translates $\phi(2t-k)$. We also have the following proposition.

Proposition 9.4 *The integer translates $\psi(t-m)$ and $\psi(t-n)$ are orthogonal for $m \neq n$. Also, the function $\psi(t-m)$ is orthogonal to $\phi(t-n)$ for any integer n.*

Proof: Let us first show $\psi(t-m)$ is orthogonal to $\psi(t-n)$. In the equations below all sums can be taken with limits $-\infty$ to ∞ with the understanding that only finitely many terms will be nonzero in any sum (since $c_k = 0$ for $k < 0$ or $k \geq N$). We can compute the inner product $(\psi(t-m), \psi(t-n))$ as

$$\begin{aligned}
(\psi(t-m), \psi(t-n)) &= \int_{-\infty}^{\infty} \psi(t-m)\psi(t-n)\,dt \\
&= \int_{-\infty}^{\infty} \psi(t)\psi(t+r)\,dt \text{ (substitute } r = m-n \neq 0) \\
&= 2\int_{-\infty}^{\infty}\left(\sum_j (-1)^j c_{N-j-1}\phi(2t-j)\right) \\
&\quad \times \left(\sum_k (-1)^k c_{N-k-1}\phi(2t+2r-k)\right)\,dt \\
&= 2\sum_j\sum_k (-1)^{j+k} c_{N-j-1}c_{N-k-1}\left(\int_{-\infty}^{\infty}\phi(2t-j)\phi(2t+2r-k)\,dt\right)
\end{aligned}$$

$$= \sum_k c_{N+2r-k-1} c_{N-k-1}$$

$$= \sum_p c_{p+2r} c_p \qquad (\text{let} \quad p = N - k - 1)$$

$$= 0, \tag{9.50}$$

since the last integral above is zero unless $j = 2r - k$, in which case it equals $\frac{1}{2}$ (and then $(-1)^{j+k} = (-1)^{2r} = 1$). The last line follows from equation (9.36), the "double shift" orthogonality of the c_k.

A computation similar to equation (9.50) shows that $\psi(t-m)$ is orthogonal to $\phi(t-n)$ for any n (see Exercise 13).

Thus, the translates $\psi(t-n)$ are elements of V_1, orthogonal to each other and to the $\phi(t-k)$. But does this collection S_1 of equation (9.49) actually span V_1? This was easy to show explicitly in the Haar case, since each $\phi(2t-m)$ can be built as a superposition of the $\phi(t-k)$ and $\psi(t-k)$. The following result will show this in the general case, and also illustrate the connection between filter banks and multiresolution in the general case, much as equation (9.23) did for the Haar filter bank and multiresolution analysis. Keep in mind that the c_k and d_k here are nothing more than the low- and high-pass coefficients ℓ_k and h_k from the orthogonal filter banks of Chapter 7.

Proposition 9.5 *Let $\psi(t)$ be defined as in equation (9.48) with $d_k = (-1)^k c_{N-k-1}$, where c_k satisfies the conditions of Theorem 9.3. Let*

$$g(t) = \sum_k a_k \phi(2t + k)$$

with $\sum_k a_k^2 < \infty$ be a function in V_1. Then

$$(g, \phi(t+m)) = \frac{1}{\sqrt{2}} (D(\mathbf{a} * \mathbf{c}))_m,$$

$$(g, \psi(t+m)) = \frac{1}{\sqrt{2}} (D(\mathbf{a} * \mathbf{d}))_m,$$

where $\mathbf{a}$, $\mathbf{c}$, and $\mathbf{d}$ denote the vectors in $L^2(\mathbb{Z})$ with components a_k, c_k, and d_k, respectively, and D denotes the downsampling operator.

In short, the process of taking the inner product of g against $\phi(t+m)$ precisely parallels the low-pass filtering and downsampling of the sequence $\mathbf{a}$ with the filter $\mathbf{c}$ (corresponding to $\boldsymbol{\ell}_a$ from Chapter 7). The inner products against $\psi(t+n)$ have a similar interpretation with the high-pass filter $\mathbf{d}$ (which corresponds to $\mathbf{h}_a$).

Proof: Let us first compute the inner product against the scaling function translates, as

$$\begin{aligned}(g, \phi(t+m)) &= \int_{-\infty}^{\infty} g(t)\phi(t+m)\, dt \\ &= \sqrt{2}\int_{-\infty}^{\infty}\left(\sum_j a_j\phi(2t+j)\right)\left(\sum_k c_k\phi(2t+2m-k)\right) \\ &= \sqrt{2}\sum_j\sum_k a_j c_k\left(\int_{-\infty}^{\infty}\phi(2t+j)\phi(2t+2m-k)\, dt\right) \\ &= \frac{1}{\sqrt{2}}\sum_k a_{2m-k}c_k,\end{aligned}$$

where we have used the dilation equation (9.33) and the fact that the last integral is $\frac{1}{2}$ if $j = 2m - k$ and zero otherwise. Note that $(g, \phi(t+m))$ is exactly $(\mathbf{a} * \mathbf{c})_{2m}$, the index $2m$ component of the convolution of $\mathbf{a}$ and $\mathbf{c}$ (considered as element of $L^2(\mathbb{Z})$. Equivalently,

$$(g, \phi(t+m)) = \frac{1}{\sqrt{2}}(D(\mathbf{a} * \mathbf{c}))_m,$$

where D is the downsampling operator. This is the first assertion in the theorem.

The computation of the inner product $(g, \psi(t+m))$ is almost identical, with equation (9.48) in place of the dilation equation, and it yields

$$(g, \psi(t+m)) = \frac{1}{\sqrt{2}}\sum_k a_{2m-k}d_k.$$

This is equivalent to

$$(g, \psi(t+m)) = \frac{1}{\sqrt{2}}(D(\mathbf{a} * \mathbf{d}))_m,$$

where $\mathbf{d}$ is the high-pass filter. This proves the theorem.

Proposition 9.6 *The set S_1 of equation (9.49) spans V_1 defined by equation (9.47).*

Proof: First, by Proposition 9.4, the set S_1 is orthogonal, and indeed orthonormal, since $\|\phi\| = 1$ by design and $\|\psi\| = 1$ follows (Exercise 24). Thus, any function $f \in L^2(\mathbb{R})$ that lies in the span of S_1 can be expressed as

$$f = \sum_k[((f, \phi(t-k))\phi(t-k) + (f, \psi(t-k))\psi(t-k))]. \qquad (9.51)$$

On the other hand, if f is not in the span then the right-hand side above will not equal f.

We will show that $\phi(2t)$ lies in the span of S_1; the other translates $\phi(2t-m)$ are similar. Indeed we will show $\phi(2t)$ is a linear combination of *finitely* many elements of S_1, and then use this to show that any function that can be built as a square-summable superposition of translates $\phi(2t-n)$ (i.e., anything in V_1) can also be built using a square-summable superposition of elements of S_1, so S_1 spans V_1.

Let $f(t) = \phi(2t)$ in equation (9.51). We want to show that both sides are identical, or equivalently, that the function $g(t) \in V_1$ defined by

$$g(t) = \phi(2t) - \sum_k [((\phi(2t), \phi(t-k))\phi(t-k) + (\phi(2t), \psi(t-k))\psi(t-k))] \tag{9.52}$$

is identically zero. In this case the sum in k involves only finitely many terms, since for large k the supports of the functions in the inner products do not overlap. Because $g \in V_1$, we have

$$g(t) = \sum_k a_k \phi(2t+k), \tag{9.53}$$

where $\sum_k a_k^2 < \infty$. Moreover from equation (9.52) it is easy to check that $(g, \phi(t+m)) = 0$ and $(g, \psi(t+m)) = 0$ for all $m \in \mathbb{Z}$. But, by Proposition 9.5, this implies that with input signal $\mathbf{a} \in L^2(\mathbb{Z})$ the output of the corresponding filter bank is identically zero, and since the filter bank is perfect reconstruction (in particular, invertible), we must conclude that $a_k = 0$ for all $k \in \mathbb{Z}$. From equation (9.53) we then have $g \equiv 0$, so $\phi(2t)$ lies in the span of S_1. The same conclusion holds for any other translates: any particular $\phi(2t-m)$ can be built as a superposition of finitely many translates $\phi(t-k)$ and $\psi(t-k)$.

To show that any $f \in V_1$ can be built as a square-summable superposition of the elements of S_1, let

$$f = \sum_m a_m \phi(2t-m) \tag{9.54}$$

be any element of V_1 (so $\sum_k a_k^2 < \infty$) and define

$$\tilde{f} = \sum_{m=M}^{N} a_m \phi(2t-m) \tag{9.55}$$

for fixed M and N. Notice that $\tilde{f}$ is a finite linear combination of the $\phi(2t-m)$ (note $\tilde{f}$ depends on M and N, but we do not explicitly indicate this). For any $\epsilon > 0$ we can obtain $\| f - \tilde{f} \| < \epsilon$ by choosing M and N appropriately. For notational convenience let $p_k^m = (\phi(2t-m), \phi(t-k))$ and $q_k^m = (\phi(2t-m), \psi(t-k))$ so that from above we have

$$\phi(2t-m) = \sum_k p_k^m \phi(t-k) + \sum_k q_k^m \psi(t-k)$$

(because $\phi(2t-m)$ is in the span of S_1). As remarked, the sums in k are in fact finite. For $\tilde{f}$ of equation (9.55) we then have

$$\begin{aligned}
\tilde{f} &= \sum_{m=M}^{N} a_m \phi(2t-m) \\
&= \sum_{m=M}^{N} a_m \sum_k (p_k^m \phi(t-k) + q_k^m \psi(t-k)) \\
&= \sum_k \sum_{m=M}^{N} (a_m p_k^m \phi(t-k) + a_m q_k^m \psi(t-k)) \\
&= \sum_k (\tilde{p}_k \phi(t-k) + \tilde{q}_k \psi(t-k)),
\end{aligned} \tag{9.56}$$

where

$$\tilde{p}_k = \sum_{m=M}^{N} a_m p_k^m, \quad \tilde{q}_k = \sum_{m=M}^{N} a_m q_k^m.$$

The interchange of sums in k and m is justified by the fact that both have a finite index range. Equation (9.56) shows that we can build $\tilde{f}$ exactly from a finite linear combination of elements of S_1. Since we can obtain $\| f - \tilde{f} \| < \epsilon$ for any positive ϵ (by taking appropriate finite values for M and N), the set S_1 spans V_1 and in fact forms an orthonormal basis. This completes the proof.

The function ψ defined by equation (9.48) is called the *mother wavelet* for the corresponding multiresolution analysis.

Example 9.10 *Figure 9.14 is a picture of the wavelet obtained from equation (9.48) and the previously computed scaling function ϕ for the Daubechies 4-tap filters. The wavelet is approximated as follows: Suppose the scaling function ϕ is supported on $[0, N-1]$ at dyadic rational points $t = m/2^r$, $m = 0$ to $m = 2^r(N-1)$. Let $\mathbf{q}$ denote the vector in $\mathbb{R}^{2^r(N-1)+1}$ with components $q_m = \phi(m/2^r)$. If we discretize equation (9.48) in the obvious way we obtain*

$$w_m = \sqrt{2} \sum_{k=0}^{N-1} (-1)^k c_{N-k-1} q_{2m-k2^r},$$

where $w_m = \psi(m/R)$ is the discrete version of the wavelet function (so $\mathbf{w} \in \mathbb{R}^{2^r(N-1)+1}$). Of course, we sum only over those k such that $0 \le 2m - k2^r \le 2^r(N-1)$. In this example we use $r = 7$. Compare the function in Figure 9.14 with those of Figure 9.3.

9.4.7 Wavelets and the Multiresolution Analysis

Let W_0 be the subspace of $L^2(\mathbb{R})$ spanned by the translates $\psi(t-m)$, exactly as in the Haar framework. The set $S_0 = \{\cup_{m\in\mathbb{Z}} \phi(t-m)\}$ provides an orthogonal

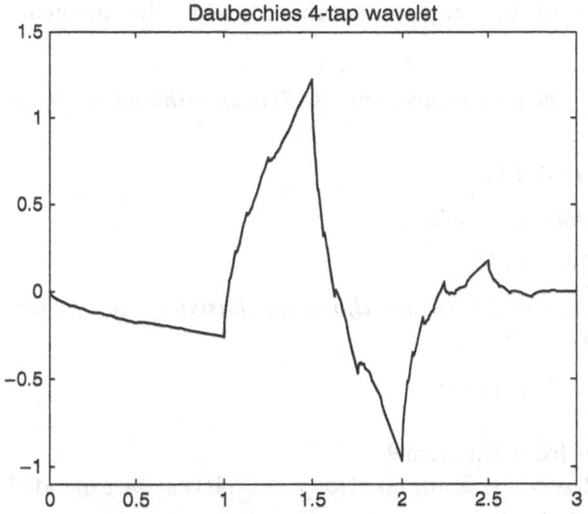

Figure 9.14 Mother wavelet corresponding to Daubechies 4-tap filter.

basis for the space V_0 and $S_1 = \{\cup_{m\in\mathbb{Z}}\phi(t-m)\} \cup \{\cup_{m\in\mathbb{Z}}\psi(t-m)\}$ provides an orthogonal basis for the space V_1. The addition of the wavelets $\psi(t-m)$ to S_0 doubles our resolution. Moreover, the subspace W_0 is orthogonal to V_0 in that, if $f \in V_0$ and $g \in W_0$, then $(f, g) = 0$. As a result we have

$$V_1 = V_0 \oplus W_0.$$

Similar reasoning shows that $V_2 = V_1 \oplus W_1 = V_0 \oplus W_0 \oplus W_1$, where W_1 is the subspace spanned by the wavelets $\psi(2t-n)$ for $n \in \mathbb{Z}$. Each $\psi(2t-m)$ is orthogonal to each basis function $\phi(2t-n) \in V_1$; hence each element of W_2 is orthogonal to each element of V_1. In particular, the function $\psi(2t-n)$ is orthogonal to all of the longer scale wavelets $\psi(t-m)$.

An iteration of the argument above shows that in general for any $N > 0$ we have

$$V_N = V_0 \oplus W_0 \oplus \cdots \oplus W_{N-1},$$

where W_k is spanned by the functions $\psi(2^k t - n)$, $n \in \mathbb{Z}$. Clearly, there is nothing special about using V_0 as the "base space." For any $M < N$ (where M and/or N may be negative) we have

$$V_N = V_M \oplus W_M \oplus \cdots \oplus W_{N-1}.$$

Let us define normalized scaling functions and wavelets

$$\phi_{k,m}(t) = 2^{k/2}\phi(2^k t - m), \quad \psi_{k,m}(t) = 2^{k/2}\psi(2^k t - m)$$

for all $k, m \in \mathbb{Z}$ (the $2^{k/2}$ in front is to obtain L^2 norm 1) and also

$$S_{M,N} = \{\phi_{M,m} : m \in \mathbb{Z}\} \cup \{\psi_{k,m} : M \le k \le N-1, m \in \mathbb{Z}\} \tag{9.57}$$

for $M \leq N$. An iteration of the arguments above yields the following theorem:

Theorem 9.6 *The set $S_{M,N}$ defined in equation (9.57) is an orthonormal basis for the subspace*

$$V_N = span\ (\{\phi_{N,m} : m \in \mathbb{Z}\})$$

of $L^2(\mathbb{R})$, for any $M \leq N$. Moreover, we have

$$V_N = V_M \oplus W_{M+1} \oplus \cdots \oplus W_{N-1},$$

where W_k denotes the subspace of $L^2(\mathbb{R})$ with orthonormal basis $\bigcup_{m\in\mathbb{Z}} \psi_{k,m}$. Each W_k is orthogonal to V_k. For any $M \in \mathbb{Z}$,

$$L^2(\mathbb{R}) = V_M \oplus W_M \oplus W_{M+1} \oplus \cdots .$$

The last assertion follows from Theorem 9.5.

By Theorem 9.6, for any fixed $N \in \mathbb{Z}$ any function $f \in L^2(\mathbb{R})$ can be expanded in this orthonormal basis as

$$f = \sum_{m\in\mathbb{Z}} a_m \phi_{N,m} + \sum_{k\geq N} \sum_{m\in\mathbb{Z}} a_{k,m} \psi_{k,m}, \tag{9.58}$$

where $a_m = (f, \phi_{N,m})$ and $a_{k,m} = (f, \psi_{k,m})$.

As in the Haar case we can let $M \to -\infty$ in Theorem 9.6 to obtain

$$L^2(\mathbb{R}) = \cdots \oplus W_{M-1} \oplus W_M \oplus W_{M+1} \oplus \cdots ,$$

which we interpret to mean that the set

$$S = \bigcup_{k,m\in\mathbb{Z}} \psi_{k,m}$$

of all scalings and translations of the mother wavelet forms an orthonormal basis for $L^2(\mathbb{R})$, with no need for scaling functions (see Exercise 7).

Example 9.11 *In Figure 9.15, panel (a) shows a function $f(t) \in L^2(\mathbb{R})$ supported on the interval $[0, 2]$. The function is discontinuous at $t = 2$. We show the projection of f onto the spaces V_k for various k for the Daubechies 4-tap filters. Since the support of f and the support of all the scaling functions $\phi(t - n)$ are disjoint except when $n = 0$, the projection of f onto V_0 is just a multiple of $\phi(t)$; we do not show this. The panel (b) shows the projection of f onto V_2; panel (c) is the projection onto V_4, and the panel (d) is the projection onto V_6. Note the relative absence of any Gibbs-like phenomena away from the discontinuity at $t = 2$, since the support of the shorter scale wavelets is small and hence most do not intersect the discontinuity.*

9.4.7.1 Final Remarks on Orthogonal Wavelets

We should remark that there are many other types of wavelets and multiresolution analyses beside those of the Daubechies family. The scaling functions

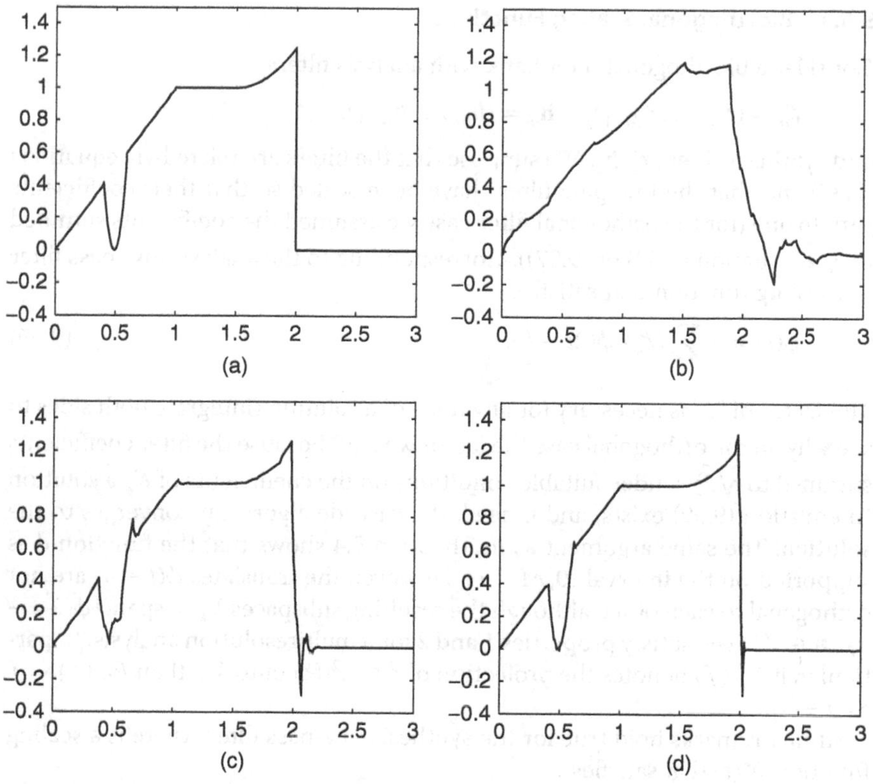

Figure 9.15 Function $f(t)$ (a) and projection onto V_2 (b), V_4 (c), and V_6 (d).

and wavelets do not need to be orthogonal nor compactly supported. The Daubechies scaling function and wavelets are remarkable, however, in that they do possess compact support, yet their simple power-of-two rescalings and translates provide orthogonal bases for $L^2(\mathbb{R})$.

There is a great deal more to say about the properties possessed by various families of wavelets, such as smoothness (continuity, differentiability), ability to approximate various types of functions, and computational issues. We will not pursue these here, although the Matlab project below involves some numerical experimentation related to wavelet smoothness. See, for example, [28] for more on these issues.

9.5 Biorthogonal Wavelets

In this section, we give a very brief synopsis of the essential truths for biorthogonal wavelets, without proof. The properties of biorthogonal wavelets generally mirror those for orthogonal wavelets. See [28] for more details and proofs.

9.5.1 Biorthogonal Scaling Functions

Consider a biorthogonal filter bank with analysis filters

$$\boldsymbol{\ell}_a = (\ell_0, \dots, \ell_{M-1}), \quad \mathbf{h}_a = (h_0, \dots, h_{N-1}),$$

and synthesis filters $\boldsymbol{\ell}_s, \mathbf{h}_s$. We suppose that the filters are related via equations (7.38) and that the low-pass filters have been scaled so that their coefficients sum to one (for the orthogonal filter case we assumed the coefficients summed to $\sqrt{2}$, equation (7.50) or (9.37)). Corresponding to the analysis low-pass filter is a scaling function that satisfies

$$\phi(t) = 2\sum_k (\boldsymbol{\ell}_a)_k \phi(2t-k). \tag{9.59}$$

The factor of "2" is necessary for any hope of a solution (integrate both sides to see why; in the orthogonal case the factor was $\sqrt{2}$ because the filter coefficients summed to $\sqrt{2}$). Under suitable conditions on the coefficients of $\boldsymbol{\ell}_a$ a solution to equation (9.59) exists, and indeed, the cascade algorithm converges to the solution. The same argument as in Theorem 9.4 shows that the function ϕ is supported on the interval $[0, M-1]$. However, the translates $\phi(t-n)$ are not orthogonal to each other, although the resulting subspaces $V_k = \text{span}(\{\phi(2^k t - n) : n \in \mathbb{Z}\})$ will satisfy properties 1 and 2 for a multiresolution analysis. In particular, if $P_{V_n}(f)$ denotes the projection of $f \in L^2(\mathbb{R})$ onto V_n, then $P_{V_n}(f) \to f$ as $n \to \infty$.

Similar remarks hold true for the synthesis low-pass filter. There is a scaling function $\phi^*(t)$ that satisfies

$$\phi^*(t) = 2\sum_k (\boldsymbol{\ell}_s)_k \phi^*(2t-k) = 2\sum_k (-1)^k (\mathbf{h}_a)_k \phi^*(2t-k). \tag{9.60}$$

Again, the integer translates $\phi^*(t-k)$ do not form an orthogonal set, but the subspaces $V_k^* = \text{span}(\{\phi^*(2^k t - n) : n \in \mathbb{Z}\})$ will satisfy properties 1 and 2 for a multiresolution analysis. By Theorem 9.4, the function ϕ^* is supported on the interval $[0, N-1]$.

The scaling functions ϕ and ϕ^* are *biorthogonal* to each other, however, in that there is some integer Q such that after proper rescaling

$$\int_{\mathbb{R}} \phi(t)\phi^*(t-k)\, dt = \begin{cases} 1, & k = Q, \\ 0, & \text{else}. \end{cases} \tag{9.61}$$

That is, the integer translates of ϕ and ϕ^* are orthogonal to each other, except when ϕ^* is translated Q units farther to the right than ϕ. In fact, for the types of biorthogonal filters of interest here, $Q = (M-N)/2$; note that according to Exercise 27 the quantity $(M-N)/2$ is an integer. However, it is much easier to simply redefine ϕ or ϕ^* so that we can assume $Q = 0$; this does not change the multiresolution analysis—the subspaces V_k and/or V_k^* remain the same—hence we will assume this has been done. See Exercise 28.

More generally, if $S_1 = \{\alpha_k\}$ and $S_2 = \{\beta_m\}$ are two bases for an inner product space V, then S_1 and S_2 are said to be *biorthogonal* if $(\alpha_k, \beta_m) = 0$ for $k \neq m$. In this case an element $v \in V$ can be expressed as

$$v = \sum_k b_k \alpha_k, \tag{9.62}$$

where $b_k = (v, \beta_k)/(\alpha_k, \beta_k)$,

9.5.2 Biorthogonal Wavelets

The wavelets are defined as in the orthogonal case. Specifically, let

$$\psi(t) = 2\sum_k (\mathbf{h}_a)_k \phi(2t-k), \quad \psi^*(t) = 2\sum_k (\mathbf{h}_s)_k \phi^*(2t-k). \tag{9.63}$$

Neither ψ nor ψ^* is orthogonal to its own scalings or translates, but

$$\int_{\mathbb{R}} \psi(2^j t - m)\psi^*(2^k t - n)\, dt = 0$$

when $j \neq k$ or $m \neq n$. Thus, the scalings and translates of the wavelets are also biorthogonal to each other. If we define normalized functions

$$\phi_{j,m}(t) = 2^{j/2}\phi(2^j t - m), \quad \phi^*_{k,n}(t) = 2^{k/2}\phi(2^k t - n), \tag{9.64}$$

$$\psi_{j,m}(t) = 2^{j/2}\psi(2^j t - m), \quad \psi^*_{k,n}(t) = 2^{k/2}\psi(2^k t - n), \tag{9.65}$$

then from the scaling on ϕ and ϕ^* we have

$$\int_{\mathbb{R}} \psi_{j,m}(t)\psi^*_{k,n}(t)\, dt = \begin{cases} 1, & \text{if } j = k \text{ and } m = n, \\ 0, & \text{else.} \end{cases}$$

The fact that ϕ is supported on $[0, M-1]$, ϕ^* on $[0, N-1]$, and the reasoning of Theorem 9.4 shows that ψ and ψ^* are both supported on $[0, (M+N)/2 - 1]$; see Exercise 27.

9.5.3 Decomposition of $L^2(\mathbb{R})$

Define subspaces

$$V_k = \text{span}\,(\{\phi_{k,n} : n \in \mathbb{Z}\}), \quad V_k^* = \text{span}\,(\{\phi^*_{k,n} : n \in \mathbb{Z}\}),$$
$$W_k = \text{span}\,(\{\psi_{k,n} : n \in \mathbb{Z}\}), \quad W_k^* = \text{span}\,(\{\psi^*_{k,n} : n \in \mathbb{Z}\}),$$

of $L^2(\mathbb{R})$. We again obtain

$$V_{k+1} = V_k \oplus W_k, \quad V^*_{k+1} = V^*_k \oplus W^*_k,$$

and indeed as by Theorem 9.6 we have

$$L^2(\mathbb{R}) = V_N \oplus W_N \oplus W_{N+1} \oplus \cdots,$$
$$L^2(\mathbb{R}) = V^*_N \oplus W^*_N \oplus W^*_{N+1} \oplus \cdots,$$

for each $N \in \mathbb{Z}$. The space W_k is orthogonal to V_k^* and W_k^* is orthogonal to V_k.

Based on the decompositions above we can write an arbitrary $f \in L^2(\mathbb{R})$ as

$$f = \sum_{n\in\mathbb{Z}} a_n^* \phi_{N,n} + \sum_{k\geq N}\sum_{n\in\mathbb{Z}} a_{k,n}^* \psi_{k,n} \tag{9.66}$$

for constants $a_n^*, a_{k,n}^*$. Indeed, taking the inner product of both sides of (9.66) with $\phi_{N,m}^*$ and using orthogonality shows that

$$a_n^* = (f, \phi_{N,m}^*),$$

while taking the inner product of both sides of (9.66) with $\psi_{j,m}^*$ shows that

$$a_{k,n}^* = (f, \psi_{k,n}^*).$$

Equation (9.66) is the biorthogonal analogue of equation (9.58).

Example 9.12 *Consider the biorthogonal Le Gall 5/3 filters, with coefficients*

$$\boldsymbol{\ell}_a = \left(\frac{1}{2}, 1, \frac{1}{2}\right), \quad \mathbf{h}_a = \left(-\frac{1}{8}, -\frac{1}{4}, \frac{3}{4}, -\frac{1}{4}, -\frac{1}{8}\right),$$

$$\boldsymbol{\ell}_s = \left(-\frac{1}{8}, \frac{1}{4}, \frac{3}{4}, \frac{1}{4}, -\frac{1}{8}\right), \quad \mathbf{h}_s = \left(-\frac{1}{2}, 1, -\frac{1}{2}\right).$$

In Figure 9.16 we show on the left the result of applying the algorithm of equations (9.42) and (9.43) for the low-pass analysis filter (rescaled to coefficients $\left(\frac{1}{4}, \frac{1}{2}, \frac{1}{4}\right)$ *which sum to one). This is the scaling function* $\phi(t)$*, supported on the interval* $[0, 2]$*, and* ϕ *is piecewise linear (see Exercise 26). The algorithm fails on* $\boldsymbol{\ell}_s$*, however, for a somewhat peculiar reason: the function* $\phi^*(t)$ *in this*

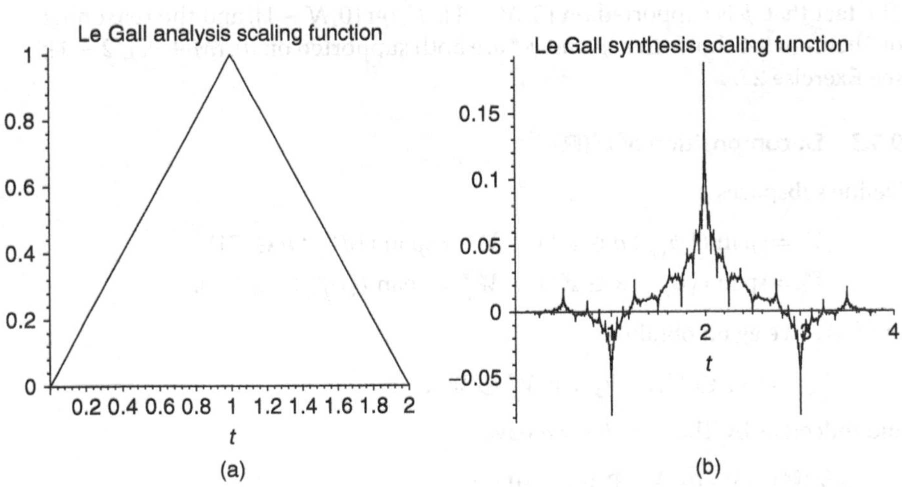

Figure 9.16 Le Gall 5/3 analysis and synthesis scaling functions.

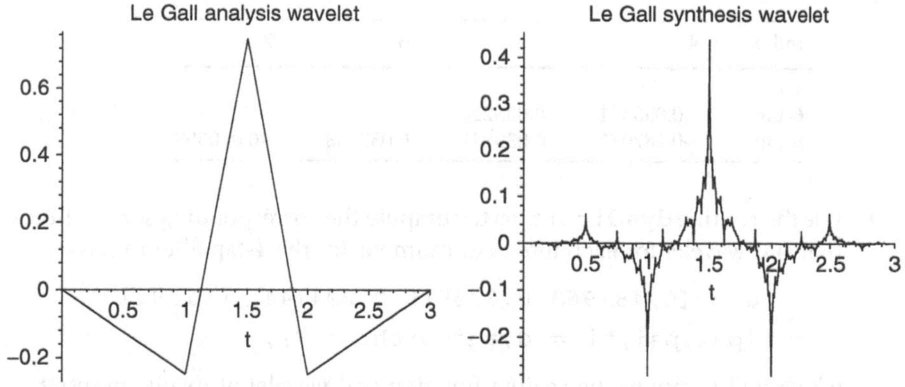

Figure 9.17 Le Gall 5/3 analysis and synthesis wavelets.

case exists as an element of $L^2(\mathbb{R})$, but ϕ^* is singular (unbounded) at each dyadic rational number! (Recall Remark 9.4: this algorithm is predicated on the pointwise existence of the scaling function, which fails in this case—the first step, equation (9.42), never gets off the ground, even though the relevant eigenvector exists.) However, we can construct a somewhat faithful portrait of ϕ^* by implementing the cascade iteration symbolically in a computer algebra system (Maple, in this case). It is actually tractable to do 10 or 12 iterations in closed form, and obtain the piecewise defined approximation to ϕ^* shown in Figure 9.16b. For our particular computation and scaling we find that the translates satisfy the biorthogonality condition $(\phi(t), \phi^*(t-k)) = 0$ for $k \neq -1$ as asserted by equation (9.61), with $Q = (M - N)/2$.

The corresponding wavelets constructed via equation (9.63) are shown in Figure 9.17. We should note that just as the analysis and synthesis filters can be interchanged in a biorthogonal filter bank, so also can the analysis and synthesis scaling functions, or the wavelets.

9.6 Matlab Project

9.6.1 Orthogonal Wavelets

Below are the coefficients for the Daubechies 4-, 6-, and 8-tap orthogonal filters (six significant figures), scaled so coefficients sum to $\sqrt{2}$.

Index	0	1	2	3
4-tap	0.482963	0.836516	0.224144	−0.129410
6-tap	0.332671	0.806892	0.459878	−0.135011
8-tap	0.230378	0.714847	0.630881	−0.027984

Index	4	5	6	7
4-tap				
6-tap	−0.085441	0.035226		
8-tap	−0.187035	0.030841	0.032883	−0.010597

1. Use the routine dyadicortho to compute the corresponding scaling function and wavelet for each filter. For example, for the 4-tap filter execute

```
c = [0.482963 0.836516 0.224144 -0.129410]
[phi,psi,t] = dyadicortho(c,7);
```

which will compute the scaling function and wavelet at dyadic points $t = k/2^7, 0 \le k \le 2^7 \cdot 3$. You can plot the scaling function with `plot(t,phi)` and the wavelet with `plot(t,psi)`.
2. One question of interest (to mathematicians, anyway) is how smooth these functions are, that is, their continuity and differentiability. For example, the 4-tap scaling function and wavelet may look continuous, but have a fairly "jagged" appearance—are they differentiable? Try increasing the second parameter in dyadicortho to 12 or 16 (or whatever your computer can handle). Then zoom in on the graph of ϕ at some point, for example, by plotting a small portion of ϕ, that is, execute

```
c = [0.482963 0.836516 0.224144 -0.129410]
[phi,psi,t] = dyadicortho(c,16);
plot(t(1000:1200),phi(1000:1200))
```

to zoom into the interval $[1000/2^{16}, 1200/2^{16}]$. Does the function appear differentiable?
3. One way to quantify the smoothness of a function f at a point $t = t_0$ is by considering the quantity

$$Q_f(t_0, t, \alpha) = \frac{|f(t) - f(t_0)|}{|t - t_0|^\alpha} \tag{9.67}$$

for t near t_0 and $\alpha \in [0, 1]$. The quantity Q_f may or may not be bounded for all t near t_0, depending on f and the value of α. For example, if f is continuous at $t = t_0$, then $Q_f(t_0, t, 0)$ (the case $\alpha = 0$) will be close to zero for $t \approx t_0$, for the simple reason that $Q_f(t_0, t, 0) = |f(t) - f(t_0)|$. However, if f is merely continuous, then for $\alpha > 0$ it may be the case that $Q_f(t_0, t, \alpha)$ is unbounded near $t = t_0$ for any $\alpha > 0$. If f is in fact differentiable near $t = t_0$, then $Q_f(t_0, t, 1)$

will be bounded and indeed will approach $f'(t_0)$ as $t \to t_0$ (but the converse is not true: Q_f may stay bounded for $\alpha = 1$ but f need not be differentiable). In this case we will also find that $Q_f(t_0, t, \alpha)$ is bounded (in fact close to 0) for any $\alpha < 1$. In brief, the smoother f is, the larger we can take α and keep Q_f bounded for all $t \approx t_0$. The largest value of α such that $Q_f(t_0, t, \alpha)$ remains finite for all t in a neighborhood of t_0 is called the *Hölder exponent* of f at t_0, and it provides a quantitative rating of the smoothness of f at t_0. It can, of course, vary from one value of t_0 to another.

We can use this to study the smoothness of the scaling functions and wavelets numerically by considering a discrete analog of Q_f. Let `phi` denote the vector from part 2 above that approximates the scaling function ϕ; we will assume you used a second argument 16 so that index $m = 2^{16} + 1 = 65,537$ corresponds to $t_0 = 1$. Define `m=65537` and set `alpha=0.2` in Matlab. For any integer p the quantity

```
Qf = abs(phi(m+p)-phi(m))/(abs(p/2^16))^alpha
```

is a discrete approximation to $Q_f(1.0, 1.0 + p/2^{16}, \alpha)$ with $\alpha = 0.2$. If ϕ has Hölder exponent 0.2 or larger at $t_0 = 1$, then `Qf` should remain bounded or even approach 0 as p approaches 0. Compute `Qf` for $p = 128, 64, 32, 16, 8, 4, 2, 1$ (you can use negative p too), and observe the behavior of `Qf`; it should approach zero. Then increase α (e.g., to 0.4) and repeat the computation. Try to home in on a value of α on the razor's edge between growth and decay in Q_f. Compare your value to the published value of approximately 0.55.

Repeat the computation for another value of t_0, such as $t_0 = \frac{15}{32}$ (corresponding to $m = 30,721$). Does the Hölder exponent seem to depend on t_0?

4. If the function ϕ in question is actually k times continuously differentiable (but not $k + 1$), then the Holder continuity is measured by examining

$$Q_f(t_0, t, \alpha) = \frac{|f^{(k)}(t) - f^{(k)}(t_0)|}{|t - t_0|^{\alpha}} \tag{9.68}$$

for t near t_0, where $f^{(k)}$ denotes the kth derivative of f.

The Daubechies 6- and 8-tap filters yield scaling functions that are once but not twice differentiable, so $k = 1$ is appropriate in equation (9.68); that is, we should use ϕ' in place of ϕ. Construct a discrete analog of ϕ' and use the approach of part (3) above to estimate the Hölder exponent for the 6 and 8-tap scaling functions at $t_0 = 1$.

See [28] or [11] for more on the smoothness of wavelets.

9.6.2 Biorthogonal Wavelets

1. The coefficients for the Daubechies 9/7 biorthogonal filters are

Index	$\boldsymbol{\ell}_a$	$\mathbf{h}_a$	$\boldsymbol{\ell}_s$	$\mathbf{h}_s$
0	0.026749	0.091271	0.091271	−0.026749
1	−0.016864	−0.057544	0.057544	−0.016864
2	−0.078223	−0.591272	−0.591272	0.078223
3	0.266864	1.115087	−1.115087	0.266864
4	0.602949	−0.591272	−0.591272	−0.602949
5	0.266864	−0.057544	0.057544	0.266864
6	−0.078223	0.091271	0.091271	0.078223
7	−0.016864			−0.016864
8	0.026749			−0.026749

(these are scaled so the first coefficient is at position 0). Note that $(\boldsymbol{\ell}_s)_k = (-1)^k(\mathbf{h}_a)_k$ and $(\mathbf{h}_s)_k = (-1)^{k+1}(\boldsymbol{\ell}_a)_k$. These are the filters used in the JPEG 2000 compression standard, for lossy compression.

Set up vectors `c` and `d` in Matlab that contain the low- and high-pass analysis filters from the table above (the routine below will generate the synthesis filters). Compute the scaling functions and wavelets with the supplied routine `dyadicbiortho()`, as

```
[phi,phi2,psi,psi2,t,t2,t3]=dyadicbiortho(c,d,8);
```

Here `phi` and `phi2` are the analysis and synthesis scaling functions ϕ and ϕ^*. The t values at which these functions are computed are returned in the arrays `t` (for ϕ) and `t2` (for ϕ^*). The wavelets ψ and ψ^* are contained in `psi` and `psi2`, respectively. The third input argument q controls the resolution, and yields the output functions at points $t = k/2^q$ for an appropriate k range. Thus, you can plot ϕ, for example, with `plot(t,phi)`, ϕ^* with `plot(t2,phi2)`, and the wavelets with `plot(t3,psi)` and `plot(t3,psi2)`.
2. Use Matlab to plot the DFT frequency response of each of the four filters, say by treating them as elements of $\mathbb{R}^{256}$.
3. Find another set of biorthogonal wavelet filter coefficients (e.g., from Matlab's wavelet toolbox, or find some on the internet, or use the methods of Chapter 7 to make your own). Use `dyadicbiortho()` to construct the scaling functions and wavelets.

Exercises

1. Let $\phi_n(t) = \phi(t-n)$ for $n \in \mathbb{Z}$, where ϕ is the Haar scaling function defined by equation (9.3), and let $\psi_{k,n}$ denote the orthonormal Haar

wavelets defined by equation (9.4). Let $f(t)$ be defined

$$f(t) = \begin{cases} t, & 0 \le t \le 2, \\ 0, & \text{else.} \end{cases}$$

(a) Compute the $L^2(\mathbb{R})$ inner products

$$c_n = (f, \phi_n), \quad d_{k,n} = (f, \psi_{k,n}),$$

for $k \ge 0$ explicitly in terms of k and n. *Hint:* For the wavelets only the range $0 \le n \le 2^{k+1} - 1$ is needed.

(b) Verify Parseval's identity, that

$$\int_0^2 f^2(t)\,dt = \sum_{n\in\mathbb{Z}} c_n^2 + \sum_{k\ge 0}\sum_{n\in\mathbb{Z}} d_{k,n}^2.$$

2. Let f and the $\psi_{k,n}$ be as in Exercise 1, but with no restriction on $k \in \mathbb{Z}$ (the set $\psi_{k,n}$ forms an all-wavelet orthonormal basis for $L^2(\mathbb{R})$). Compute the inner products

$$d_{k,n} = (f, \psi_{k,n}),$$

and verify that

$$\int_0^2 f^2(t)\,dt = \sum_{k\in\mathbb{Z}}\sum_{n\in\mathbb{Z}} d_{k,n}^2.$$

3. Justify Remark 9.1.

4. (a) Explicitly compute the projection of the function f from Exercise 1 onto the subspaces V_0 and V_1 of $L^2(\mathbb{R})$ defined by equation (9.26) with the Haar scaling functions/wavelets.

(b) Explicitly compute the projection of the function f from Exercise 1 onto the subspaces W_0 defined by equation (9.31) with the Haar scaling functions/wavelets.

(c) Verify directly that $P_{V_1}(f) = P_{V_0}(f) + P_{W_0}(f)$, where $P_V(f)$ denotes the projection of f onto the subspace V.

5. Show that the Haar wavelets $\psi_{0,n}$, $n \in \mathbb{Z}$ by themselves do not span the subspace V_1 defined by equation (9.26).

6. Show that the function $x(t)$ defined by equation (9.17) is in $L^2(\mathbb{R})$ if $\mathbf{x} \in L^2(\mathbb{Z})$.

7. Let f be a continuous function in $L^2(\mathbb{R})$. In this exercise, we show that f can be approximated arbitrary well using only the wavelets $\psi_{k,n}$, with no need for the scaling function. As a consequence the wavelets $\psi_{k,n}$ form an orthonormal basis for $L^2(\mathbb{R})$.
We will begin with the Haar basis. Recall that for any fixed N we can expand f with respect to the orthonormal Haar basis consisting of $\phi_{N,m}$ and $\psi_{k,m}$ for $m \in \mathbb{Z}, k \geq N$ as

$$f = \sum_{m=-\infty}^{\infty} d_{N,m}\phi_{N,m} + \sum_{k\geq N}\sum_{m=-\infty}^{\infty} c_{k,m}\psi_{k,m},$$

where $d_{N,m} = (f, \phi_{N,m})$ and $c_{k,m} = (f, \psi_{k,m})$. Let $\tilde{f}_N$ denote the expansion

$$\tilde{f}_N = \sum_{k\geq N}\sum_{m=-\infty}^{\infty} c_{k,m}\psi_{k,m},$$

which omits the $\phi_{N,m}$.

(a) Suppose f is supported in a bounded interval $[-a, a]$ for some constant a. Show that

$$\lim_{N\to-\infty} \|f - \tilde{f}_N\|_{L^2(\mathbb{R})} = 0.$$

Hint:

$$|d_{N,m}| = \left|\int_{-\infty}^{\infty} f(t)\phi_{N,m}(t)\, dt\right| \leq \sup_{[-a,a]} |\phi_{N,m}| \int_{-a}^{a} |f(t)|\, dt.$$

Use this to show $\sum_m d_{N,m}^2$ goes to zero as $N \to -\infty$.

(b) Use part (a) and the reasoning in the proof of Theorem 9.2 to extend to the case in which f is not compactly supported.

(c) Extend parts (a) and (b) to a general multiresolution analysis, under the assumption that the scaling function is continuous and compactly supported.

8. Let V be the subspace in $\mathbb{R}^3$ spanned by the orthonormal vectors $\mathbf{v}_1 = (1/\sqrt{2}, 0, 1/\sqrt{2})$ and $\mathbf{v}_2 = (0, 1, 0)$. Compute the projection of $\mathbf{w} = (1, 3, 2)$ onto V using Definition 9.2. Verify that this vector is indeed the closest element of V to $(1, 3, 2)$ by explicitly minimizing

$$\| \mathbf{w} - c_1\mathbf{v}_1 - c_2\mathbf{v}_2\|^2$$

as a function of c_1 and c_2.

9. Suppose that W and $\tilde{W}$ are subspaces of a vector space V, and that $W \cap \tilde{W} = \{\mathbf{0}\}$ (the zero vector in V).

(a) Show that if $\mathbf{w} \in W$ and $\tilde{\mathbf{w}} \in \tilde{W}$ satisfy $\mathbf{w} + \tilde{\mathbf{w}} = \mathbf{0}$, then $\mathbf{w} = \tilde{\mathbf{w}} = \mathbf{0}$.

(b) Suppose that for some $\mathbf{v} \in V$ we have $\mathbf{v} = \mathbf{w} + \tilde{\mathbf{w}}$ for $\mathbf{w} \in W$ and $\tilde{\mathbf{w}} \in \tilde{W}$. Show that $\mathbf{w}$ and $\tilde{\mathbf{w}}$ are unique. *Hint:* Use part (a).

10. Let V denote the subset of $L^2(\mathbb{R})$ consisting of those functions that are continuous.

(a) Show that V is a subspace of $L^2(\mathbb{R})$.

(b) Let ϕ be the Haar scaling function defined in equation (9.3) (though we could use any discontinuous function for what follows). Show that there is no element of V that is "closest" to ϕ, in that

$$\| \phi - f \|_{L^2(\mathbb{R})}$$

with $f \in V$ can be made arbitrarily small but not zero. This shows that projection onto this subspace is not well-defined, at least by any definition we have given.

11. Finish the proof of Proposition 9.3 by showing that if $\alpha \in V$ and $\alpha \neq P_V(f)$, then $\| f - P_V(f) \| < \| \alpha - f \|$. *Hint:* Let $\tilde{f}_n$ be as defined in equation (9.16). Let $\alpha \in V$ have expansion

$$\alpha = \sum_{k=-\infty}^{\infty} b_k \eta_k,$$

and let α_n be as defined by equation (9.14). If $\alpha \neq P_V(f)$, then $b_M \neq (f, \eta_M)$ for some M. Use equation (9.15) to show that $\| f - \alpha_n \|^2 - \| f - P_V(f) \| \geq (b_M - (f, \eta_m))^2$ for $n \geq M$. Conclude that $\| f - P_V(f) \| < \| f - \alpha \|$.

12. Suppose that V is a subspace of an inner product space and V has orthonormal basis $\cup_{k=1}^{\infty} \eta_k$. Let $P_V(f)$ denote the projection of f onto V.

(a) Show that $(f - P_V(f), \phi) = 0$ for any $\phi \in V$. *Hint:* Just show that $(f - P_V(f), \eta_m) = 0$; a careful proof might invoke Exercise 39 in Chapter 1.

(b) Show that

$$\| f \|^2 = \| P_V(f) \|^2 + \| f - P_V(f) \|^2.$$

13. Show that the wavelet defined by equation (9.48) is orthogonal to $\phi(t - n)$ for any n. *Hint:* Mimic the computation of equation (9.50), and make use of the dilation equation (9.34). It should come down to showing that

$$\sum_{j=0}^{N-1} (-1)^j c_j c_{n'-j-1} = 0$$

when n' is any even integer (and take $c_k = 0$ for any k outside the range 0 to $N - 1$, of course). Show that this is always true when N is even, for any numbers $c_0, \ldots, c_{N-1}$.

14. (a) Show that the Haar scaling function defined by equation (9.3) satisfies the dilation equation (9.34) (with $c_0 = c_1 = 1/\sqrt{2}$).
(b) Show that the mother Haar wavelet defined by equation (9.1) satisfies equation (9.48).

15. Show that if $L_a(z) = \sum_k c_k z^k$ where the c_k are real, then $|L_a(e^{it})| = |L_a(e^{-it})|$ for $t \in [0, 2\pi)$.

16. Use equation (9.36) to show that N, the number of nonzero coefficients in the low-pass filter vector $(c_0, \dots, c_{N-1})$ for an orthogonal filter bank, must be even. *Hint:* Consider a simple special case like $N = 3$ first; note that c_0 and c_{N-1} are nonzero by assumption.

17. (a) Use fixed point iteration to find a solution to the equation $x = \cos(x)$. Does the initial guess matter?
(b) Use fixed point iteration to try to find a solution to the equation $x = -10e^x$ (this equation has a unique solution; plot $x + 10e^x$ for $-5 \le x \le 0$ to see this). What does the fixed point iteration do? Can you interpret the behavior graphically?

18. Show that the mapping F defined in equation (9.44) really does take $L^2(\mathbb{R})$ to $L^2(\mathbb{R})$.

19. Show that the approximation algorithm of Section 9.4.4 works for the Haar scaling function if we define $\phi(0) = \phi(1) = \frac{1}{2}$ (and filter coefficients $c_0 = c_1 = 1/\sqrt{2}$, and that ϕ is supported in $[0,1]$).

20. (a) Show that if functions $f_1, \dots, f_N$ are all defined on $\mathbb{R}$ and f_k is supported in a set $S_k \subset \mathbb{R}$, then the sum $f_1 + \cdots + f_N$ is supported in the union $\cup_k S_k$. Show that the product $f_1 \cdots f_N$ is supported in the intersection $\cap_k S_k$.
(b) Show that if each f_k in part (a) has compact support, then the sum $f_1 + \cdots + f_N$ and product $f_1 \cdots f_N$ also have compact support.

21. Show that the function $\phi(t)$ produced by the cascade algorithm with coefficients $c_0, \dots, c_{N-1}$ satisfies

$$\int_0^{N-1} \phi(t)\,dt = 1.$$

Hint: The Haar scaling function ϕ^0 (the initial guess) has integral one, and make use of $L_a(1) = \sqrt{2}$.

22. Show that $\phi^j(t)$ in Example 9.9 is of the form asserted. Also verify that for any fixed j we have $(\phi^j(t), \phi^j(t-n)) = 0$ for any integers n (i.e., ϕ^j is orthogonal to its own integer translates).

23. Even though the cascade algorithm fails to converge in Example 9.9, show nonetheless that the function

$$\phi(t) = \begin{cases} 1/3, & 0 \le t < 3, \\ 0, & \text{else,} \end{cases}$$

satisfies the dilation equation with the relevant filter coefficients and normalization $\int \phi = 1$. Does this function satisfy $(\phi(t), \phi(t-n)) = 0$ for all integers n?

24. Show that the wavelet ψ defined by equation (9.48) satisfies $\|\psi\| = 1$ (L^2 norm), assuming that the scaling function is normalized so that $\|\phi\| = 1$.

25. Let S_1 and S_2 be the bases

$$S_1 = \{(1,2,1),\ \ (0,1,1),\ \ (1,1,1)\},$$
$$S_2 = \{(0,1,-1),\ \ (-1,0,1),\ \ (1,-1,1)\},$$

for $\mathbb{R}^3$. Show that S_1 and S_2 are biorthogonal with respect to the usual inner product. Use equation (9.62) to decompose the vector $(3,1,2)$ into a sum of the basis elements of S_1.

26. Show that the Le Gall analysis scaling function shown in Figure 9.16a satisfies the dilation equation (9.59) with low-pass coefficients $\ell_a = \left(\frac{1}{4}, \frac{1}{2}, \frac{1}{4}\right)$.

27. Show that if biorthogonal scaling functions ϕ and ϕ^* are supported in intervals $[0, M-1]$ and $[0, N-1]$, respectively, then the wavelets ψ and ψ^* defined in equation (9.63) are both supported in the interval $[0, (M+N)/2 - 1]$. *Hint:* Reason as in Theorem 9.4; Exercise 20 may be helpful.

28. The goal of this problem is to show that if ϕ and ϕ^* satisfy the relation (9.61), then $Q = (M-N)/2$. Suppose that ϕ and ϕ^* are the unique solution to equations (9.59) and (9.60) with filter coefficients $((\ell_a)_0, \ldots, (\ell_a)_{M-1})$ and $((\ell_s)_0, \ldots, (\ell_s)_{M-1})$, respectively. We suppose that the filters are both symmetric; that is, $(\ell_a)_k = (\ell_a)_{M-k-1}$ and $(\ell_s)_k = (\ell_s)_{N-k-1}$. Recall from Exercise 27 that the integers M and N are both even or both odd.

(a) Define a function $\tilde{\phi}(x) = \phi(M - 1 - x)$. Show that $\tilde{\phi}$ also satisfies equation (9.59) and so $\phi = \tilde{\phi}$. A similar conclusion holds for ϕ^*.
(b) Use part (a) to show that if ϕ and ϕ^* satisfy equation (9.61), then $Q = (M - N)/2$.

Bibliography

1 Biggs, W. and Henson, V.E. (1995) *The DFT: An Owner's Manual for the Discrete Fourier Transform*, SIAM, Philadelphia, PA.

2 Boggess, A. and Narcowich, F. (2001) *A First Course in Wavelets with Fourier Analysis*, Prentice Hall, New Jersey.

3 Brigham, E.O. (1988) *The Fast Fourier Transform and Its Applications*, Prentice Hall, New Jersey.

4 Brislawn, C. (1995) Fingerprints go digital. *Not. Am. Math. Sci.*, **42**, 1278–1283.

5 Brislawn, C. (1996) Classification of nonexpansive symmetric extension transforms for multirate filter banks. *Appl. Comput. Harmon. Anal.*, **3**, 337–357.

6 Calderbank, A.R. and Daubechies, I. (1998) Wavelet transforms that map integers to integers. *Appl. Comput. Harmon. Anal.*, **5**, 332–369.

7 Casazza, P. and Kutyniok, G. (eds) (2013) *Finite Frames: Theory and Applications*, Springer-Verlag, New York.

8 Cooley, J. and Tukey, J. (1965) An algorithm for the machine calculation of complex Fourier series. *Math. Comput.*, **19** (90), 297.

9 Courant, R. and Hilbert, D. (1989) *Methods of Mathematical Physics*, vol. 1, John Wiley & Sons, Inc., New York.

10 Daubechies, I. (1992) *Ten Lectures on Wavelets, CBMS-NSF Regional Conference Series in Applied Mathematics*, vol. 61, SIAM, Philadelphia, PA.

11 Daubechies, I. and Sweldens, W. (1998) Factoring wavelet transforms into lifting steps. *J. Fourier Anal. Appl.*, **4** (3), 245–267; (1965), 297–301.

12 Frazier, M.W. (1999) *An Introduction to Wavelets Through Linear Algebra, Undergraduate Texts in Mathematics*, Springer-Verlag, New York.

13 Gabor, D. (1946) Theory of communication. *J. Inst. Electr. Eng.*, **93**, 429–457.

14 Gonzales, R., Woods, R., and Eddins, S. (2009) *Digital Image Processing Using MATLAB*, 2nd edn, Gatesmark Publishing, Knoxville, TN.

Discrete Fourier Analysis and Wavelets: Applications to Signal and Image Processing, Second Edition.
S. Allen Broughton and Kurt Bryan.

Companion Website: www.wiley.com/go/Broughton/Discrete_Fourier_Analysis_and_Wavelets

15 Haar, A. (1910) Zur Theorie der orthogonalen Funktionensysteme. *Math. Ann.*, **69** (3), 331–371. doi: 10.1007/BF01456326.

16 Han, D., Kornelson, K., Larson, D., and Weber, E. (2007) *Frames for Undergraduates, Student Mathematical Library*, vol. 40, AMS, Providence, RI.

17 https://jpeg.org/jpeg2000/index.html (accessed 19 March 2018).

18 Kreyszig, E. (1978) *Introductory Functional Analysis with Applications*, John Wiley & Sons, Inc., New York.

19 McClellan, J., Schafer, R., and Yoder, M. (1998) *DSP First, A Multimedia Approach*, Prentice Hall, Upper Saddle River, NJ.

20 Mallat, S. (2009) *A Wavelet Tour of Signal Processing*, Elsevier, New York.

21 Okoudjou, S. (ed.) (2016) *Finite Frame Theory: A Complete Introduction to Overcompleteness, Proceedings of Symposia in Applied Mathematics*, vol. 73, AMS, Providence, RI.

22 Pennebaker, W.B. and Mitchell, J.L. (1993) *JPEG Still Image Data Compression Standard*, Van Nostrand Reinhold, New York.

23 Pratt, W. (2013) *Introduction to Digital Image Processing*, CRC Press, Boca Raton, FL.

24 Rudin, W. (1976) *Principles of Mathematical Analysis*, 3rd edn, McGraw-Hill, New York.

25 Saxe, K. (2002) *Beginning Function Analysis*, Springer-Verlag, New York.

26 Skodras, A., Christopoulos, C., and Ebrahimi, T. (2000) The JPEG 2000 still image compression standard. *IEEE Trans. Consum. Electron.*, **46**, 1103–1127.

27 Steinberg, B. (2011) *Representation Theory of Finite Groups: An Introductory Approach*, Springer-Verlag, NewYork.

28 Strang, G. and Nguyen, T. (1996) *Wavelets and Filter Banks*, Wellesley-Cambridge Press, Wellesley, MA.

29 Sweldens, W. (1996) The lifting scheme: a custom-design construction of biorthogonal wavelet. *Appl. Comput. Harmon. Anal.*, **3** (2), 186–200.

30 Sweldens, W. (1998) The lifting scheme: a construction of second generation wavelets. *SIAM J. Math Anal.*, **29** (2), 511–546.

31 Taubman, D. and Marcellin, M. (2001) *JPEG2000 Image Compression Fundamentals*, Kluwer Academic Publishers, Norwell, MA.

32 Text Resources web page, http://www.wiley.com/go/Broughton/Discrete_Fourier_Analysis_and_Wavelets (accessed 19 March 2018).

33 Vaidyanathan, P.P. (1993) *Multirate Systems and Filter Banks*, Prentice Hall, Englewood Cliffs, NJ.

34 Vetterli, M. and Kovacevic, J. (1995) *Wavelets and Subband Coding*, Prentice Hall, New Jersey.

A

Solutions to Selected Exercises

A.1 Chapter 1 Solutions

Solution 1 Color JPEG compressed images are typically 5–50 times smaller than they would be if stored "naively," so the ratio of naively-stored to JPEG-stored might range from a low of 0.02–0.2.

Solution 3 The eighth roots of unity are the numbers $e^{2\pi ik/8} = \cos(k\pi/4) + i\sin(k\pi/4)$, where $0 \le k \le 7$.

Solution 4 The mth component of $\mathbf{E}_{N,k}$ is given by equation (1.23) and is $\mathbf{E}_{N,k}(m) = \exp(2\pi ikm/N)$. But $(\exp(2\pi ikm/N))^N = \exp(2\pi ikm) = 1$ since k and m are integers (refer to Exercise 2).

Solution 5 From Euler's identity and a bit of algebra

$$x(t) = a\cos(\omega t) + b\sin(\omega t) = \frac{1}{2}(a - ib)e^{i\omega t} + \frac{1}{2}(a + ib)e^{-i\omega t}.$$

Comparison to $x(t) = ce^{i\omega t} + de^{-i\omega t}$ shows that $c = \frac{a-ib}{2}$, $d = \frac{a+ib}{2}$. These last two linear equations are easily solved for a and b to yield $a = c + d, b = i(c - d)$.

Solution 7

(a) You should find values 0.3090169944, 0.5877852525, 0.8090169943, and 0.9510565165.

(b) Rounding to the nearest multiple of 0.25 yields 0.25, 0.50, 0.75, and 1.00. This distortion is 8.534×10^{-3}.

(c) Rounding to the nearest multiple of 0.25 yields 0.3, 0.6, 0.8, and 1.00. This distortion is 1.354×10^{-3}. Rounding to the nearest multiple of 0.05 yields 0.30, 0.60, 0.80, and 0.95. This distortion is 1.56×10^{-4}.

(d) The codebook here consists of some range of multiples of h. The dequantization map is simply $\tilde{q}(k) = kh$.

Discrete Fourier Analysis and Wavelets: Applications to Signal and Image Processing, Second Edition.
S. Allen Broughton and Kurt Bryan.

Companion Website: www.wiley.com/go/Broughton/Discrete_Fourier_Analysis_and_Wavelets

Solution 9 This is not a vector space—it is not closed under addition or scalar multiplication.

Solution 10 $\mathbb{R}^n$ is NOT a subspace of $\mathbb{C}^n$, if $\mathbb{C}^n$ is taken as a vector space over $\mathbb{C}$, for we do not have closure under scalar multiplication.

Solution 12 Use $(p+q)^2 \leq 2p^2 + 2q^2$ to show that

$$\sum_{k=0}^{\infty} (x_k + y_k)^2 \leq 2\left(\sum_{k=0}^{\infty} x_k^2 + \sum_{k=0}^{\infty} y_k^2\right) < \infty$$

so the sum $\mathbf{x}+\mathbf{y}$ is in $L^2(\mathbb{N})$ and we have closure under addition. Clearly $\sum_k (cx_k)^2 = c^2 \sum_k x_k^2$, so we have closure under scalar multiplication. All other properties are verified exactly just as for Exercise 11.

To show $L^2(\mathbb{N})$ is a subset of $L^\infty(\mathbb{N})$ argue that there must be some M so that $|x_k| \leq M$ for all k.

Solution 16 Use Euler's identity:

$$\begin{aligned}\cos(\alpha x)\cos(\beta y) &= \frac{e^{i\alpha x} + e^{-i\alpha x}}{2}\,\frac{e^{i\beta y} + e^{-i\beta y}}{2} \\ &= \frac{1}{4}e^{i\alpha x}e^{i\beta y} + \frac{1}{4}e^{-i\alpha x}e^{i\beta y} + \frac{1}{4}e^{i\alpha x}e^{-i\beta y} + \frac{1}{4}e^{-i\alpha x}e^{-i\beta y}.\end{aligned}$$

The other three cases are similar.

Solution 23

(a) One can easily compute that $\mathbf{v}_1 \cdot \mathbf{v}_2 = \mathbf{v}_1 \cdot \mathbf{v}_3 = \mathbf{v}_2 \cdot \mathbf{v}_3 = 0$. Note also that $\|\mathbf{v}_1\|^2 = 2, \|\mathbf{v}_1\|^2 = 3, \|\mathbf{v}_3\|^2 = 6$.

(b) Here $\mathbf{w} \cdot \mathbf{v}_1 = 7, \mathbf{w} \cdot \mathbf{v}_2 = 6, \mathbf{w} \cdot \mathbf{v}_3 = 9$. Thus (use part (a)) $\mathbf{w} = \frac{7}{2}\mathbf{v}_1 + 2\mathbf{v}_2 + \frac{3}{2}\mathbf{v}_3$.

(c) The rescaled vectors are (in row form) $\mathbf{u}_1 = (1/\sqrt{2}, 1/\sqrt{2}, 0)$, $\mathbf{u}_2 = (-1/\sqrt{3}, 1/\sqrt{3}, 1/\sqrt{3})$, $\mathbf{u}_3 = (1/\sqrt{6}, -1/\sqrt{6}, 2/\sqrt{6})$.

(d) In this case we find $\mathbf{w} = \frac{7\sqrt{2}}{2}\mathbf{v}_1 + 2\sqrt{3}\mathbf{v}_2 + \frac{3\sqrt{6}}{2}\mathbf{v}_3$.

(e) From part (d) $\left(\frac{7\sqrt{2}}{2}\right)^2 + (2\sqrt{3})^2 + \left(\frac{3\sqrt{6}}{2}\right)^2 = 3^2 + 4^2 + 5^2$. Both sides in fact equal 50.

Solution 24

(a) Here $\mathbf{w} = \frac{7}{4}\mathbf{E}_{4,0} + \frac{3+2i}{4}\mathbf{E}_{4,1} - \frac{9}{4}\mathbf{E}_{4,2} + \frac{3-2i}{4}\mathbf{E}_{4,3}$.

(b) Each vector $\mathbf{E}_{4,k}$ has norm 2, so the rescaled vectors are just $\widetilde{\mathbf{E}}_{4,k} = \frac{1}{2}\mathbf{E}_{4,k}$.

(c) Here $\mathbf{w} = \frac{7}{2}\widetilde{\mathbf{E}}_{4,0} + (3/2 + i)\widetilde{\mathbf{E}}_{4,1} - \frac{9}{2}\widetilde{\mathbf{E}}_{4,2} + (3/2 - i)\widetilde{\mathbf{E}}_{4,3}$.

(d) We find $(7/2)^2 + |3/2 + i|^2 + (9/2)^2 + |3/2 - i|^2 = 1^2 + 5^2 + (-2)^2 + 3^2$. Both sides are in fact equal to 39.

Solution 28 That $\|\mathbf{v}\| \geq 0$ follows from $(\mathbf{v}, \mathbf{v}) \geq 0$ as does $\|\mathbf{v}\| = 0$ if and only if $(\mathbf{v}, \mathbf{v}) = 0$. The property $\|a\mathbf{v}\| = |a|\|\mathbf{v}\|$ is also straightforward.

To show the triangle inequality take the hint to find

$$\|\mathbf{v} + \mathbf{w}\|^2 \leq \|\mathbf{v}\|^2 + \|\mathbf{w}\|^2 + 2|(\mathbf{v}, \mathbf{w})|. \tag{A.1}$$

Apply the Cauchy–Schwarz inequality to obtain $\|\mathbf{v} + \mathbf{w}\|^2 \leq \|\mathbf{v}\|^2 + \|\mathbf{w}\|^2 + 2\|\mathbf{v}\|\|\mathbf{w}\|$ and take the square root of both sides above to obtain the triangle inequality.

Solution 41

(a) Working the relevant integral shows

$$\begin{aligned}(e^{i\pi kt}/\sqrt{2}, e^{i\pi mt}/\sqrt{2}) &= \frac{1}{2}\int_{-1}^{1} e^{i\pi kt}e^{-i\pi mt}\,dt \\ &= \frac{1}{2}\int_{-1}^{1} e^{i\pi(k-m)t}\,dt.\end{aligned}$$

If $k \neq m$ we have antiderivative $\frac{e^{i\pi(k-m)t}}{i\pi(k-m)}$ and the integral works out to be zero. If $k = m$ the integral is 1. The set is orthonormal.

(b) Here $\alpha_0 = 0$ and $\alpha_k = \frac{(-1)^{k+1} i\sqrt{2}}{k\pi}$.

(c) From Parseval's identity, we have $\int_{-1}^{1} t^2\,dt = \sum_{k=-\infty}^{\infty} |\alpha_k|^2$ which becomes, in this case, (use $|\alpha_{-k}| = |\alpha_k|$ to write the sum from $k = 1$ to ∞) $\frac{2}{3} = 2\sum_{k=1}^{\infty} \frac{2}{k^2\pi^2}$. A bit of simple algebra (multiply both sides by $\pi^2/4$) yields $\sum_{k=1}^{\infty} \frac{1}{k^2} = \frac{\pi^2}{6}$.

A.2 Chapter 2 Solutions

Solution 1 This follows immediately from the fact that $e^{-2\pi ikm/N}$ is periodic in k with period N (so $e^{-2\pi i(k+jN)m/N} = e^{-2\pi ikm/N}$ for any j), hence

$$X_{k+jN} = \sum_{m=0}^{N-1} x_m e^{-2\pi i(k+jN)m/N} = \sum_{m=0}^{N-1} x_m e^{-2\pi ikm/N} = X_k.$$

Solution 3 The DFT is $\mathbf{F}_4\mathbf{x} = (2, 1 - 3i, 0, 1 + 3i)^T$.

Solution 4 The IDFT is $\frac{1}{4}\mathbf{F}_4^*\mathbf{X} = (3/2, 0, 1/2, 1)^T$.

Solution 5 Use equation (2.8); it is a routine matrix multiplication.

Solution 16 A column-by-column transform followed by a row-by-row transform yields

$$\hat{\mathbf{A}} = \mathbf{F}_2\mathbf{A}\mathbf{F}_2^T = \begin{bmatrix} 2 & 4 \\ -2 & 0 \end{bmatrix}.$$

Solution 17

(a) The row r column p entry of $\mathbf{x}$ is x_r (only one column here); the row p column s entry of $\mathbf{y}^T$ is y_s (only one row). The row r column s entry of $\mathbf{z}$ is, from the definition of matrix multiplication,

$$z_{r,s} = \sum_{p=1}^{1} x_r y_s = x_r y_s.$$

(b) If we use $Z_{k,l}$ to denote the row k column l entry of $\mathbf{Z}$ then from the definition of the 2D DFT

$$\begin{aligned} Z_{k,l} &= \sum_{r=0}^{m-1}\sum_{s=0}^{n-1} z_{r,s} e^{-2\pi i(kr/m+ls/n)} \\ &= \sum_{r=0}^{m-1}\sum_{s=0}^{n-1} x_r y_s e^{-2\pi i(kr/m+ls/n)} \\ &= \left(\sum_{r=0}^{m-1} x_r e^{-2\pi ikr/m}\right)\left(\sum_{s=0}^{n-1} y_s e^{-2\pi ils/n}\right) \\ &= X_k Y_l. \end{aligned}$$

Equivalently, $\mathbf{Z} = \mathbf{X}\mathbf{Y}^T$.

Solution 21

(a) Of course, the image should look the same, but off center.
(b) The DFT magnitudes should be identical!

A.3 Chapter 3 Solutions

Solution 1 The DCT of $\mathbf{x} = (1, 2, -1)$ is $(2\sqrt{3}/3, \sqrt{2}, -2\sqrt{6}/3) \approx (1.15, 1.41, -1.63)$.

Solution 2 The inverse DCT of $\mathbf{X} = (3, 0, 1)$ is $(\sqrt{3} + \sqrt{6}/6, \sqrt{3} - \sqrt{6}/3, \sqrt{3} + \sqrt{6}/6) \approx (2.14, 0.92, 2.14)$.

Solution 4 This is just the mth column of C_N.

Solution 9 For the DCT with $d = 0.01, 0.1, 1.0, 10.0, 100.0$ we obtain percentage errors $6.4 \times 10^{-5}, 0.0053, 0.792, 30.7, 100$ respectively. In the last case the entire transform is quantized to zero.

Solution 10 According to the matrix form of the DCT in equation (3.17) the 2D DCT of the identity matrix is $\hat{\mathbf{I}} = C_N \mathbf{I} C_N^T = C_N C_N^T = \mathbf{I}$ from equation (3.14) (i.e., C_N is an orthogonal matrix).

Solution 11 If C_2 denotes the 2×2 DCT matrix then

$$C_2 \mathbf{A} C_2^T = \begin{bmatrix} 1 & 2 \\ -1 & 0 \end{bmatrix}.$$

Inverse transforming reverses the steps and leads back to **A**.

Solution 12 Take the hint. Let $\mathbf{X} = \mathrm{DCT}(\mathbf{x}) = C_N \mathbf{x}$ so that

$$\|\mathbf{X}\|^2 = \mathbf{X}^* \mathbf{X} = (C_N \mathbf{x})^* (C_N \mathbf{x}) = \mathbf{x}^* C_N^T C_N \mathbf{x} = \mathbf{x}^* \mathbf{x} = \|\mathbf{x}\|^2.$$

Note C_N is real-valued, so $C_N^* = C_N^T$.

Solution 13 This follows straight from equation (3.10):

$$C_0 = \sqrt{\frac{1}{N}} \sum_{m=0}^{N-1} x_m = \sqrt{N} \left(\frac{1}{N} \sum_{m=0}^{N-1} x_m \right) = \sqrt{N}\ \mathrm{ave}(x).$$

The 2D case is similar

A.4 Chapter 4 Solutions

Solution 1 Note that $|w|$ and $|z|$ are just real numbers, and so clearly $(|w| + |z|)^2 \geq 0$. Expand and rearrange to obtain $|w||z| \leq \frac{1}{2}(|w|^2 + |z|^2)$.

It is clear that for any complex number $c = a + bi$ we have $|\mathrm{Re}(c)| = |a| \leq \sqrt{a^2 + b^2} = |c|$. In particular then $|\mathrm{Re}(zw)| \leq |zw| = |z||w|$. Virtually identical reasoning yields $|\mathrm{Im}(zw)| \leq |z||w|$.

Solution 2 You should obtain $\mathbf{g} * \mathbf{x} = [16, 20, 7, 2]^T$.

Solution 3

(a) Let $\mathbf{y} = \mathbf{x} * \mathbf{w}$. From Definition 4.1

$$y_r = \sum_{k=0}^{N-1} x_k w_{(r-k) \bmod N}$$

The only nonzero summand is when $k = r$, which yields $y_r = x_r$, so $\mathbf{y} = \mathbf{x}$.

(b) The DFT $\mathbf{W}$ of $\mathbf{w}$ has

$$W_m = \sum_{k=0}^{N-1} e^{2\pi ikm/N} w_k = 1$$

since the only nonzero summand is when $k = 0$. If $\mathbf{y} = \mathbf{x} * \mathbf{w}$, then $Y_m = X_m W_m = X_m$, so $\mathbf{Y} = \mathbf{X}$. We conclude $\mathbf{y} = \mathbf{x}$.

Solution 7 Just compare the "0th" columns of the circulant matrices!

Solution 9 The row j column k entry of $\mathbf{M_h}$ is h_{j-k}, indices interpreted mod N. The row j column k entry in $\mathbf{M_h^T}$ is thus h_{k-j} (reverse the role of j and k). But h_{k-j} (index mod N) is precisely the row j column k entry in $\mathbf{M_{h'}}$. Hence $\mathbf{M_{h'}} = \mathbf{M_h^T}$.

Solution 11 For $M = N = 256$ the convolved signal is flat, and is of course just the dc component of $\mathbf{f}$. As M gets smaller the convolution $\mathbf{f} * \mathbf{v}$ looks like a progressively less and less blurry version of $\mathbf{f}$. When $M = 1$ the convolution is just $\mathbf{f}$. In essence, convolution with $\mathbf{v}$ is a kind of low-pass filter, and the wider $\mathbf{v}$ gets, the more high-frequencies get cut out.

Solution 15 The DFT $\mathbf{H}$ has coefficients

$$H_k = \frac{1}{2} - \frac{1}{2}e^{-2\pi ik/N}.$$

Then compute that $|H_k| = \sin(k\pi/N)$. The filter zeros out dc and passes the Nyquist frequency unchanged. The argument has a jump through index 0.

Solution 16

(a) The relevant DFT's are $\mathbf{G} = (6, 2+2i, -2, 2-2i)$ and $\mathbf{F} = (48, -16+8i, 0, -16-8i)$. The DFT $\mathbf{X}$ must satisfy $G_k X_k = F_k$, so we can solve for $\mathbf{X}$ component by component and obtain $\mathbf{X} = (8, -2+6i, 0, -2-6i)$. An inverse DFT yields

$$\mathbf{x} = (1, -1, 3, 5).$$

(b) We always have $G_k X_k = F_k$ for $0 \le k \le N-1$, so if $G_k \neq 0$ for all k, $\mathbf{X}$ and hence $\mathbf{x}$ exists and is unique.

(c) If some DFT coefficient $G_m = 0$ then the equation will be solvable if $F_m = 0$ also, but will not be unique (any X_m would work).

Solution 26

(a) A simple ratio test shows that $\sum_{k=0}^{\infty}(e^{-k})^2 < \infty$.
(b) The transform is

$$X(f) = \sum_{k=0}^{\infty} e^{-k}e^{-2\pi ikf} = \sum_{k=0}^{\infty}(e^{-(1+2\pi if)})^k = \frac{1}{1-e^{-(1+2\pi if)}}.$$

if we make use of $1 + z + z^2 + \cdots = 1/(1-z)$ for $|z| < 1$.

Solution 33

(a) The transforms are $X(z) = -1 + 2z^{-1} + 0z^{-2} + 4z^{-3}$ and $Y(z) = 1 + 1z^{-1} + 2z^{-2} - z^{-3}$. The product is $X(z)Y(z) = -1 + z^{-1} + 0z^{-2} + 9z^{-3} + 2z^{-4} + 8z^{-5} - 4z^{-6}$.
(b) The convolution is (starting with the 0th component) $(-1, 1, 0, 9, 2, 8, -4)$, all other components zero.
(c) If we look at $X(z)Y(z)$ modulo z^{-4} (identify 1 and z^{-4}, z^{-1} and z^{-5}, and so on), we obtain $1 + 9z^{-1} - 4z^{-2} + 9z^{-3}$. So the circular convolution is the vector $(1, 9, -4, 9)$.
(d) Virtually the same computation when we identify indices 0 and 4, 1 and 5, and so on. The circular convolution is $(-1 + 2, 1 + 8, 0 - 4, 9 + 0)$, or just $(1, 9, -4, 9)$.

Solution 35 The DTFT of **h** is given by $H(f) = \frac{1}{2}(1 - e^{-2\pi if}) = ie^{-\pi if}\left(\frac{1}{2i}(e^{\pi if} - e^{-\pi if})\right) = ie^{-\pi if}\sin(\pi f)$. Then $|H(f)| = |\sin(\pi f)|$. In particular, $|H(0)| = 0$, so dc is zeroed out by the filter **h**, while $|H(1/2)| = 1$. The Nyquist frequency is passed without attenuation.

Solution 36 Use the z-transform. We need $X^2(z) = 1 - 4z^{-2} + 4z^{-4}$. It is not too hard to see that $X(z) = 1 + 2z^{-2}$, so that $x_0 = 1, x_2 = 2$, all other $x_k = 0$.

A.5 Chapter 5 Solutions

Solution 1 Let $\mathbf{x} = (x_0, x_1, x_2, x_3)$ be a signal and $\mathbf{X} = (X_0, X_1, X_2, X_3)$ the DFT of **x**. Let $\mathbf{w} = (0, 1, 1, 0)$ be the window vector and $\mathbf{y} = \mathbf{w} \circ \mathbf{x}$.

(a) The DFT of **x** is $\mathbf{X} = \mathbf{F}_4\mathbf{x} = [x_0 + x_1 + x_2 + x_3, x_0 - ix_1 - x_2 + ix_3, x_0 - x_1 + x_2 - x_3, x_0 + ix_1 - x_2 - ix_3]^T$.

(b) Find $\mathbf{W} = (2, -1-i, 0, -1+i)^T$ and $Y = [x_1+x_2, -ix_1-x_2, -x_1+x_2, ix_1-x_2]^T$. The quantity $\mathbf{W} * \mathbf{X}$ has components $4x_1+4x_2, -4ix_1-4x_2, -4x_1+4x_2$, and $4ix_1-4x_2$, respectively. Divide by 4 and obtain exactly $\mathbf{Y}$.

(c) The 2-point DFT of (x_1, x_2) yields components $\tilde{X}_0 = x_1+x_2, \tilde{X}_1 = x_1-x_2$. According to Theorem 5.1 we should have (using $m = 1, M = 2, N = 4$, $q = 2$)

$$\tilde{X}_0 = \frac{1}{4}e^{2\pi i\cdot 1\cdot 0/2}(\mathbf{W} * \mathbf{X})_0 = \frac{1}{4}(4x_1+4x_2) = x_1+x_2$$
$$\tilde{X}_1 = \frac{1}{4}e^{2\pi i\cdot 1\cdot 1/2}(\mathbf{W} * \mathbf{X})_2 = -\frac{1}{4}(-4x_1+4x_2) = x_1-x_2.$$

Solution 3 Mimic the proof of the convolution theorem, Theorem 4.2. Let $\mathbf{Y} = \frac{1}{N}\mathbf{X} * \mathbf{W}$ and let $\mathbf{y} = \text{IDFT}(\mathbf{Y})$. Then

$$y_k = \frac{1}{N}\sum_{m=0}^{N-1} e^{2\pi ikm/N} Y_m \tag{A.2}$$

from the definition of the IDFT. Since $\mathbf{Y} = \frac{1}{N}\mathbf{X} * \mathbf{W}$ we have $Y_m = \frac{1}{N}\sum_{j=0}^{N-1} X_j W_{m-j}$. Substitute this into the right side of equation (A.2) to find $y_k = \sum_{j=0}^{N-1}\sum_{m=0}^{N-1} e^{2\pi ikm/N} X_j W_{m-j}$. Note the $1/N$ factor drops out. Make a change of index $m = r+j$ above, split the exponential, and find $y_k = \left(\sum_{j=0}^{N-1} X_j e^{2\pi ikj/N}\right)\left(\sum_{r=0}^{N-1} e^{2\pi ikr/N} W_r\right) = x_k w_k$.

Solution 7

(a) This is just the first-order Taylor polynomial or "tangent line approximation" for the function $t\omega(t)$ about the point $t = t_0$. If $\omega(t)$ is, for example, twice continuously differentiable we in fact expect $\omega(t)t = \omega(t_0)t_0 + (\omega(t_0) + \omega'(t_0)t_0)(t-t_0) + O((t-t_0)^2)$.

(b) Based on part (a) we expect (to order $O((t-t_0)^2)$)

$$f(t) \approx \sin(\omega(t_0)t_0 + (\omega(t_0) + \omega'(t_0)t_0)(t-t_0))$$
$$= \sin((\omega(t_0) + \omega'(t_0)t_0)t - t_0^2\omega'(t_0)),$$

which is just $f(t) \approx \sin(\omega_0 t + a)$ with $a = -t_0^2\omega(t_0)$ and $\omega_0 = \omega(t_0) + \omega'(t_0)t_0$.

(c) For the spectrogram of Figure 5.8, we find that $\omega_0 = 150 + 50\cos(2\pi t) - 100\pi t\sin(2\pi t)$. The graph of ω_0 versus t looks exactly like the frequency variation in Figure 5.8! In particular, in the range $0 \le t \le 1$ it bottoms out at about $t = 0.35$, at 30 Hz, and peaks at $t = 0.8$ at about 400 Hz.

Solution 9 Let $Z = X * W$ and define let $\mathbf{z}$ denote the inverse DTFT of Z. Start with $z_k = \int_{-1/2}^{1/2} Z(f)e^{2\pi ikf}\,df$ then reason as in the proof of the convolution theorem, replacing sums with integrals.

A.6 Chapter 6 Solutions

Solution 4 Let $c = \|\phi_1\|$, then simplify $\sum_{n=1}^{N} \left|\left(\mathbf{x}, \frac{1}{c}\phi_n\right)\right|^2$.

Solution 6 Noting $(\mathbf{z}, \mathbf{w}) = \mathbf{w}^*\mathbf{z}$, manipulate the expression $(A\mathbf{x}, \mathbf{y}) = \mathbf{y}^*A\mathbf{x}$.

Solution 7 Noting $\sum_{n=1}^{N} |(\mathbf{x}, \psi_n)|^2 = (S^{-1}F^*\mathbf{x}, S^{-1}F^*\mathbf{x})$, expand

$$\sum_{n=1}^{N} |(\mathbf{x}, \psi'_n)|^2 = ((S^{-1}F + K)^*\mathbf{x}, (S^{-1}F + K)^*\mathbf{x}).$$

Solution 8 Here is a sample

$$\begin{aligned} P^2 &= VP_0V^*VP_0V^* \\ &= VP_0^2V^* \\ &= VP_0V^* \\ &= P. \end{aligned}$$

Solution 9 If $P\mathbf{x} = \mathbf{x}$, then $F\mathbf{x} = FP\mathbf{x} = 0$, If $P\mathbf{x} \neq \mathbf{x}$ then show $Q_0V^*\mathbf{x} = \begin{bmatrix} \mathbf{y} \\ 0 \end{bmatrix}$ for some $\mathbf{y} \in \mathbb{C}^d$, and then $F\mathbf{x} \neq \mathbf{0}$.

Solution 10 The range of F^* is the orthogonal complement of $\ker F$.

Solution 13 First show $S = DF(DF)^* = Ndiag(|a_1|^2, \dots |a_d|^2)$.

Solution 15 Let $S_1\mathbf{x}_i = \alpha_i\mathbf{x}_i$ and $S_2\mathbf{y}_j = \beta_j\mathbf{y}_j$, then simplify $S_1(\mathbf{x}_i\mathbf{y}_j^t)S_2^t$.

A.7 Chapter 7 Solutions

Solution 1 The convolutions with the low- and high-pass filters yield vectors with components

Index	0	1	2	3	4
$\mathbf{x} * \boldsymbol{\ell}_a$	1/2	−1/2	0	1	2
$\mathbf{x} * \mathbf{h}_a$	1/2	−3/2	2	1	−2

and all other components zero. Downsampling produces $L^2(\mathbb{Z})$ vectors $\mathbf{X}_\ell = (1/2, 0, 2)$ and $\mathbf{X}_h = (1/2, 2, -2)$ (in each case the element 1/2 is the index zero component), with all other components zero.

Upsampling produces $U(\mathbf{X}_\ell) = (1/2, 0, 0, 0, 2)$ and $\mathbf{X}_h = (1/2, 0, 2, 0, -2)$, all other components zero. Convolution with the synthesis filters produces vectors $\mathbf{v}_\ell = \boldsymbol{\ell}_s * (U(\mathbf{X}_\ell))$ and $\mathbf{v}_h = \mathbf{h}_s * (U(\mathbf{X}_h))$ with components

Index	−1	0	1	2	3	4
$\mathbf{v}_\ell$	1/2	1/2	0	0	2	2
$\mathbf{v}_h$	−1/2	1/2	−2	2	2	−2

Adding produces the vector $(0, 1, -2, 2, 4, 0)$, with the first component listed in the -1 index position, all other components zero. This is just $\mathbf{x}$!

Solution 3 The convolution $\mathbf{x} * \boldsymbol{\ell}_a$ looks like $\mathbf{x} * \boldsymbol{\ell}_a = (\dots, 5, 5, 5, \dots)$ while $\mathbf{x} * \mathbf{h}_a = (\dots, -3, 3, -3, \dots)$ (even components are 3, odd components are -3). Downsampling produces $\mathbf{X}_\ell = (\dots, 5, 5, 5, \dots)$, $\mathbf{X}_h = (\dots, 3, 3, 3, \dots)$. This makes sense: the dc passes through the low-pass filter and shows up in $\mathbf{X}_\ell$, the Nyquist frequency passes through the high-pass filter and shows up in $\mathbf{X}_h$.

Solution 6

(a) Find $\mathbf{X}_\ell = (\dots, x_{-2}, x_0, x_2, \dots)$, $\mathbf{X}_h = (\dots, x_{-3}, x_{-1}, x_1, \dots)$ so $(\mathbf{X}_\ell)_j = x_{2j}$ and $(\mathbf{X}_h)_j = x_{2j-1}$.

(b) Here

$$U(\mathbf{X}_\ell)_j = \begin{cases} x_j, & j \text{ even}, \\ 0, & j \text{ odd}, \end{cases} \qquad U(\mathbf{X}_h)_j = \begin{cases} x_{j-1}, & j \text{ even}, \\ 0, & j \text{ odd}, \end{cases}$$

or $U(\mathbf{X}_\ell) = (\dots, x_{-2}, 0, x_0, 0, x_2, \dots)$ and $U(\mathbf{X}_h) = (\dots, x_{-3}, 0, x_{-1}, 0, x_1, \dots)$.

(c) Take $\boldsymbol{\ell}_s = \boldsymbol{\ell}_a$, the identity filter, and $\mathbf{h}_s$ to have $(\mathbf{h}_s)_{-1} = 1$, all other components zero.

(d) In this case, we arrive at

$$U(\mathbf{X}_\ell)_j = \begin{cases} x_j, & j \text{ even}, \\ 0, & j \text{ odd}, \end{cases} \qquad U(\mathbf{X}_h)_j = \begin{cases} x_{j-2}, & j \text{ even}, \\ 0, & j \text{ odd}, \end{cases}$$

or $U(\mathbf{X}_\ell) = (\dots, x_{-2}, 0, x_0, 0, x_2, \dots)$ and $U(\mathbf{X}_h) = (\dots, x_{-4}, 0, x_{-2}, 0, x_0, \dots)$. All trace of the odd-indexed components is gone, so no filtering is going to recover it!

Solution 8 It is easy to compute that $(\mathbf{x} * \boldsymbol{\ell}_a)_k = \frac{3}{4}x_k + \frac{1}{2}x_{k-1}$, $(\mathbf{x} * \mathbf{h}_a)_k = \frac{2}{3}x_k - \frac{2}{3}x_{k-1}$. Downsampling produces $(\mathbf{X}_\ell)_k = \frac{3}{4}x_{2k} + \frac{1}{2}x_{2k-1}$, $(\mathbf{X}_h)_k = \frac{2}{3}x_{2k} - \frac{2}{3}x_{2k-1}$. Immediately upsampling produces $(U(\mathbf{X}_\ell))_k = \frac{3}{4}x_k + \frac{1}{2}x_{k-1}$, $(U(\mathbf{X}_h))_k = \frac{2}{3}x_k - \frac{2}{3}x_{k-1}$ when k is even, 0 for both components when k is odd.

Let us suppose that the synthesis filters have two nonzero coefficients each, at index positions -1 and 0 as in Example 7.3. When k is even we find

$(\boldsymbol{\ell}_s * U(\mathbf{X}_\ell))_k = (\boldsymbol{\ell}_s)_0(\frac{3}{4}x_k + \frac{1}{2}x_{k-1})$ while (again, if k is even) $(\mathbf{h}_s * U(\mathbf{X}_h))_k = (\mathbf{h}_s)_0(\frac{2}{3}x_k - \frac{2}{3}x_{k-1})$. From the above two equations the requirement that $(\boldsymbol{\ell}_s * U(\mathbf{X}_\ell))_k + (\mathbf{h}_s * U(\mathbf{X}_h))_k = x_k$ forces both $\frac{3}{4}(\boldsymbol{\ell}_s)_0 + \frac{2}{3}(\mathbf{h}_s)_0 = 1$, $\frac{1}{2}(\boldsymbol{\ell}_s)_0 - \frac{2}{3}(\mathbf{h}_s)_0 = 0$. This yields $(\boldsymbol{\ell}_s)_0 = 4/5, (\mathbf{h}_s)_0 = 3/5$.

Now if k is odd then $(\boldsymbol{\ell}_s * U(\mathbf{X}_\ell))_k = (\boldsymbol{\ell}_s)_{-1}(\frac{3}{4}x_{k+1} + \frac{1}{2}x_k)$ and $(\mathbf{h}_s * U(\mathbf{X}_h))_k = (\mathbf{h}_s)_{-1}(\frac{2}{3}x_{k+1} - \frac{2}{3}x_k)$. In this case, we are led to equations $\frac{3}{4}(\boldsymbol{\ell}_s)_{-1} + \frac{2}{3}(\mathbf{h}_s)_{-1} = 0$, $\frac{1}{2}(\boldsymbol{\ell}_s)_{-1} - \frac{2}{3}(\mathbf{h}_s)_{-1} = 1$. This yields $(\boldsymbol{\ell}_s)_{-1} = 4/5, (\mathbf{h}_s)_{-1} = -9/10$.

All in all the synthesis filters are $\boldsymbol{\ell}_s = (4/5, 4/5)$ and $\mathbf{h}_s = (-9/10, 3/5)$, for the index -1 and 0 coefficients, all other coefficients zero.

Solution 14 For part (c) you should obtain

$$\alpha_2(\mathbf{x}) = \begin{bmatrix} 3 \\ 3 \\ 3 \\ 3 \end{bmatrix}, \delta_2(\mathbf{x}) = \begin{bmatrix} -1/2 \\ 1/2 \\ 1/2 \\ -1/2 \end{bmatrix}, \delta_1(\mathbf{x}) = \begin{bmatrix} 1/2 \\ 1/2 \\ -1/2 \\ -1/2 \end{bmatrix}.$$

Of course $\alpha_2(\mathbf{x})$ is just the dc portion of the signal. Also, $\alpha_1(\mathbf{x}) = \alpha_2(\mathbf{x}) + \delta_2(\mathbf{x}) = [5/2, 7/2, 7/2, 5/2]^T$.

Solution 18 Equations (7.32) and (7.33) make this very straightforward.

Solution 21 If $\mathbf{y} = U(D(\mathbf{w}))$ then

$$y_k = \begin{cases} w_k, & k \text{ even}, \\ 0, & 0 \text{ odd}. \end{cases}$$

But it is equally easy to see that $\frac{1}{2}W(z) + \frac{1}{2}W(-z)$ has coefficient w_k when k is even and 0 when k is odd, and hence $\frac{1}{2}W(z) + \frac{1}{2}W(-z) = Y(z)$.

Solution 22 The z-transforms of the analysis filters are $L_a(z) = \frac{1}{2} + z^{-1} - \frac{1}{2}z^{-2}, H_a(z) = \frac{1}{4} + \frac{1}{2}z^{-1} + \frac{1}{4}z^{-2}$. Equations (7.32) and (7.33) yield $L_s(z) = \frac{1}{2}z^2 - z + \frac{1}{2}, H_s(z) = -z^2 + 2z + 1$. These correspond to filters that are not causal. However, if we multiply both synthesis z-transforms by z^{-2} we obtain causal filters with coefficients with coefficients $(\boldsymbol{\ell}_s)_0 = 1/2, (\boldsymbol{\ell}_s)_1 = -1, (\boldsymbol{\ell}_s)_0 = 1/2$ and $(\mathbf{h}_s)_0 = -1, (\mathbf{h}_s)_1 = 2, (\mathbf{h}_s)_0 = 1$. In this case, we find $L_a(z)L_s(z) + H_a(z)H_s(z) = 2z^{-3}$, so the filter bank has delay 3.

Solution 24 The z-transforms turn out to be $L_a(z) = \frac{1}{4} + \frac{3}{4}z^{-1} + \frac{3}{4}z^{-2} + \frac{1}{4}z^{-3}, L_s(z) = -\frac{1}{4}z^2 + \frac{3}{4}z + \frac{3}{4} - \frac{1}{4}z^{-1}, H_a(z) = -\frac{1}{4}z^2 - \frac{3}{4}z + \frac{3}{4} + \frac{1}{4}z^{-1}, H_s(z) = -\frac{1}{4} + \frac{3}{4}z^{-1} - \frac{3}{4}z^{-2} + \frac{1}{4}z^{-3}$. Then check that $L_a(-1) = L_s(-1) = 0$ and $H_a(1) = H_s(1) = 0$, while $L_a(1) = 2, L_s(1) = 1, H_a(-1) = 1$, and $H_s(-1) = -2$.

A.8 Chapter 8 Solutions

Solution 1 Here $\mathbf{x}_e = (1, 2, 3)$ (first index -1) and $\mathbf{x}_o = (3, -2)$ (first index -1). Then $U(\mathbf{x}_e) = (1, 0, 2, 0, 3)$ (first index -2), $U(\mathbf{x}_o) = (3, 0, -2)$ (first index -2) and $S(U(\mathbf{x}_o)) = (0, 3, 0, -2)$ (first index -2). Adding produces $U(\mathbf{x}_e) + S(U(\mathbf{x}_o)) = (1, 0, 2, 0, 3) + (0, 3, 0, -2, 0) = (1, 3, 2, -2, 3)$.

Solution 2 From equation (8.3), we have (all signals indexed from index -2, zero-padded when necessary) $\mathbf{X}_\ell = (0, 1/2, 5/2, 1/2, 0)$ and $\mathbf{X}_h = (0, 1/2, -1/2, 5/2, 0)$.

From equation (8.5), we find $\mathbf{y}_e = (0, 1, 2, 3, 0)$ and $\mathbf{y}_o = (0, 0, 3, -2, 0)$, all signals indexed from -2 to 2. Then $S^{-1}(\mathbf{y}_o) = (0, 3, -2, 0, 0)$ and upsampling as per equation (8.7) produces $U(\mathbf{y}_e) = (0, 0, 1, 0, 2, 0, 3, 0, 0)$ and $U(S^{-1}(\mathbf{y}_o)) = (0, 0, 3, 0, -2, 0, 0, 0, 0)$ (indices -4 to 4). Finally, from equation (8.7) find (remember to right shift $U(S^{-1}(\mathbf{y}_o))$) $\mathbf{x} = (0, 0, 1, 0, 2, 0, 3, 0, 0) + (0, 0, 0, 3, 0, -2, 0, 0, 0) = (0, 0, 1, 3, 2, -2, 3, 0, 0)$.

Solution 4 Let $p(z) = \sum_{j=m}^{n} a_j z^j$ with a_m and a_n nonzero, so $\text{wid}(p(z)) = n - m$. Then $z^k p(z) = \sum_{j=m}^{n} a_j z^{j+k} = \sum_{j=m+k}^{n+k} a_{j-k} z^j$ so that $\text{wid}\,(z^k p(z)) = (n + k) - (m + k) = n - m = \text{wid}(p(z))$.

Solution 10 We have $\mathbf{x}_e = (2, 1, 7)$ and $\mathbf{x}_o = (-2, 5)$ (both starting at index 0) so that $X_e(z) = 2 + z^{-1} + 7z^{-2}$ and $X_o(z) = -2 + 5z^{-1}$. Now $X(z) = 2 - 2z^{-1} + z^{-2} + 5z^{-3} + 7z^{-4}$. From equation (8.32), one can check directly that $\frac{X(\sqrt{z})+X(-\sqrt{z})}{2} = X_e(z)$ and $\sqrt{z}\frac{X(\sqrt{z})-X(-\sqrt{z})}{2} = X_o(z)$.

Solution 12 Compute $\mathbf{P}_s = \mathbf{P}_a^{-1} = (\mathbf{D}\mathbf{M}_1\mathbf{M}_2)^{-1} = \mathbf{M}_2^{-1}\mathbf{M}_1^{-1}\mathbf{D}^{-1}$ from which we find

$$\mathbf{P}_s = \begin{bmatrix} 1 & 0 \\ 1 & 1 \end{bmatrix} \begin{bmatrix} 1 & -1/2 \\ 0 & 1 \end{bmatrix} \begin{bmatrix} 1 & 0 \\ 0 & -2 \end{bmatrix}.$$

To apply this to the given data, first apply $\mathbf{D}^{-1}$ to obtain

$$\begin{bmatrix} \mathbf{y}_\ell \\ \mathbf{y}_h \end{bmatrix} = \begin{bmatrix} \mathbf{X}_\ell \\ -2\mathbf{X}_h \end{bmatrix} = \begin{bmatrix} (1, 2, 1) \\ (2, -2, 0) \end{bmatrix}.$$

Then

$$\begin{bmatrix} \mathbf{y}'_\ell \\ \mathbf{y}'_h \end{bmatrix} = \mathbf{M}_1^{-1} \begin{bmatrix} (1, 2, 1) \\ (2, -2, 0) \end{bmatrix} = \begin{bmatrix} (0, 3, 1) \\ (2, -2, 0) \end{bmatrix}.$$

Finally

$$\begin{bmatrix} \mathbf{x}_e \\ S(\mathbf{x}_o) \end{bmatrix} = \mathbf{M}_2^{-1} \begin{bmatrix} (0,3,1) \\ (2,-2,0) \end{bmatrix} = \begin{bmatrix} (0,3,1) \\ (2,1,1) \end{bmatrix}$$

so $\mathbf{x}_o = (2,1,1)$, index starting at -1. Then $\mathbf{x} = U(\mathbf{x}_e) + S(U(\mathbf{x}_o)) = (0,0,0,3,0,1) + (2,0,1,0,1,0) = (2,0,1,3,1,1)$.

Solution 13 We need

$$\begin{bmatrix} 1/2 & 1/2 \\ 1/2 & -1/2 \end{bmatrix} = \begin{bmatrix} 1 & 0 \\ a & 1 \end{bmatrix} \begin{bmatrix} 1 & b \\ 0 & 1 \end{bmatrix} \begin{bmatrix} c & 0 \\ 0 & d \end{bmatrix} = \begin{bmatrix} c & bd \\ ac & (ab+1)d \end{bmatrix}.$$

Matching entries in the upper left yields $c = 1/2$ and then the lower left yields $a = 1$. The upper and lower right sides entries yield $bd = 1/2$ and $bd + d = -1/2$ (using $a = 1$) so that $d = -1$ and $b = -1/2$. The factorization is

$$\mathbf{P}_a = \begin{bmatrix} 1 & 0 \\ 1 & 1 \end{bmatrix} \begin{bmatrix} 1 & -1/2 \\ 0 & 1 \end{bmatrix} \begin{bmatrix} 1/2 & 0 \\ 0 & -1 \end{bmatrix}$$

Solution 16 A little experimentation/inspection shows that $m = -2$ is promising. With the hint we find

$$\tilde{H}(z) = H(z) - az^{-2}\tilde{L}(z) = \frac{a-4}{8} + \left(1 - \frac{a}{4}\right)z^{-1} - \left(\frac{1}{2} + \frac{7}{8}a\right)z^{-2}.$$

The choice $a = 4$ leaves us with $\tilde{H}(z) = -4z^{-2}$ with width 0 ($\tilde{\mathbf{h}}$ has one tap).

Solution 22 Recall that

$$\mathbf{P}_a = \begin{bmatrix} 1 & \frac{1}{4}z + \frac{1}{4} \\ 0 & 1 \end{bmatrix} \begin{bmatrix} 1 & 0 \\ -\frac{1}{2} - \frac{1}{2}z^{-1} & 1 \end{bmatrix}.$$

Apply this to the signal with z-transform $X(z) = 14 + 18z^{-1} + 7z^{-2} + 28z^{-3}$ (so $X_e(z) = 14 + 7z^{-1}, X_o(z) = 18 + 28z^{-1}$), by first computing as in equation (8.79). In fact, it is the same computation already done in Example 8.11. We then apply the primal lift to compute $Y_1'(z) = \lfloor Y_1(z) + t(z)Y_2(z) \rfloor = Y_1(z) + \lfloor t(z)Y_2(z) \rfloor$ and $Y_2'(z) = Y_2(z)$. This yields $Y_1'(z) = \lfloor 14 + \frac{59}{4}z^{-1} - \frac{7}{4}z + 6z^{-2} \rfloor = 14 - 2z + 14z^{-1} + 6z^{-2}$ and of course $Y_2'(z) = Y_2(z) = -7 + 7z^{-1} + 24z^{-2}$. The full forward transform takes signal $(14,18,7,28)$ to vectors $\mathbf{y}_1' = (-2,14,14,6)$ (index starting at -1) and $\mathbf{y}_2' = (-7,7,24)$ (index starting at 0).

To invert the transform, we obviously recover $Y_2(z) = Y_2'(z) = -7 + 7z^{-1} + 24z^{-2}$ and then from above $Y_1(z) = Y_1'(z) - \lfloor t(z)Y_2(z) \rfloor = 14 + 7z^{-1}$. Then apply exactly the same procedure from Example 8.11 to recover $X(z)$ (and so $\mathbf{x}$) from $Y_1(z)$ and $Y_2(z)$.

A.9 Chapter 9 Solutions

Solution 1

(a) Find $c_0 = 1/2, c_1 = 3/2, d_{k,n} = -\frac{2^{-3k/2}}{4}$ $(0 \le k \le 2^{n+1} - 1)$ and $c_n = 0$ for $n \ne 0, 1$, while $d_{k,n} = 0$ for $k < 0$ or $k > 2^{n+1} - 1$ (since the support of f and $\psi_{k,n}$ do not overlap).

(b) Find $c_0^2 + c_1^2 = 5/2$ while

$$\sum_{k\ge 0}\sum_{n\in\mathbb{Z}} d_{k,n}^2 = \frac{1}{16}\sum_{k=0}^{\infty}\sum_{n=0}^{2^{k+1}-1} 2^{-3k} = \frac{1}{8}\sum_{k=0}^{\infty} 2^{-2k} = \frac{1}{6}$$

(the last series is geometric, ratio 1/4). Then

$$\sum_{n\in\mathbb{Z}} c_n^2 + \sum_{n\in\mathbb{Z}}\sum_{k\ge 0} d_{k,n}^2 = 5/2 + 1/6 = 8/3$$

which is precisely $\int_0^2 f^2(t)\, dt$.

Solution 4

(a) The projections of f are the piecewise functions

$$P_{V_0}(f)(t) = \begin{cases} 1/2, & t \in [0,1), \\ 3/2, & t \in [1,2), \\ 0, & \text{else.} \end{cases}$$

and

$$P_{V_0}(f)(t) = \begin{cases} 1/4, & t \in [0,1/2), \\ 3/4, & t \in [1/2,1), \\ 5/4, & t \in [1,3/2), \\ 7/4, & t \in [3/2,2), \\ 0, & \text{else.} \end{cases}$$

(b) The projection of f onto W_0 is

$$P_{W_0}(f)(t) = \begin{cases} -1/4, & t \in [0,1/2), \\ 1/4, & t \in [1/2,1), \\ -1/4, & t \in [1,3/2), \\ 1/4, & t \in [3/2,2), \\ 0, & \text{else,} \end{cases}$$

since $(f, \psi_{0,0}) = (f, \psi_{0,1}) = -1/4$.

(c) This is easy to see—just check each interval $[k/2,(k+1)/2)$ for $k = 0, 1, 2, 3$.

Solution 5 Find any function that lies in V_1 and cannot be built from the $\psi_{0,n}$. The obvious choice is $\phi(t)$! More generally, functions that are a superposition of the $\psi_{0,n}$ must average to zero over any unit interval $[n-1/2, n+1/2]$ for $n \in \mathbb{Z}$.

Solution 6 Compute $\int_{k/2}^{(k+1)/2} x^2(t)\,dt = \frac{1}{2}x_{k-1}^2$ so that

$$\int_{-\infty}^{\infty} x^2(t)\,dt = \sum_{k=-\infty}^{\infty} \int_{k/2}^{(k+1)/2} x^2(t)\,dt = \frac{1}{2}\sum_{k=-\infty}^{\infty} x_{k-1}^2 < \infty.$$

Solution 8 The projection $P_V(\mathbf{w})$ is given by

$$\begin{aligned} P_V(\mathbf{w}) &= (\mathbf{w},\mathbf{v}_1)\mathbf{v}_1 + (\mathbf{w},\mathbf{v}_2)\mathbf{v}_2 = \frac{3}{\sqrt{2}}(1/\sqrt{2}, 0, 1/\sqrt{2}) + 3(0,1,0) \\ &= (3/2, 3, 3/2). \end{aligned}$$

Solution 15 If $L_a(z) = \sum_k c_k z^k$ then (whether or not the c_k are real) then $|L_a(e^{it})|^2 = L_a(e^{it})\overline{L_a(e^{it})}$ and $|L_a(e^{-it})|^2 = L_a(e^{-it})\overline{L_a(e^{-it})}$. But if the c_k are real then $\overline{L_a(e^{it})} = L_a(\overline{e^{it}}) = L_a(e^{-it})$ and of course $\overline{L_a(e^{-it})} = L_a(e^{it})$. Use this in the expressions for $|L_a(e^{it})|^2$ and $|L_a(e^{-it})|^2$ above to see that $|L_a(e^{it})|^2 = |L_a(e^{-it})|^2$.

Solution 17

(a) The iterates should converge to about 0.739, attaining three significant figures in less than 20 iterations.
(b) In this case the iterates get stuck in a two-point loop, oscillating between −9.995440299 and −0.0004560741254 (roots of the system $t_2 = -10e^{t_1}$, $t_1 = -10e^{t_2}$).

Solution 18 Just note that $\int_{-\infty}^{\infty} \phi^2(2t-n)\,dt = \frac{1}{2}\int_{-\infty}^{\infty} \phi^2(u)\,du$ by the simple change of variable $u = 2t - n$. Thus, $\phi(2t-n)$ is in $L^2(\mathbb{R})$. Since $L^2(\mathbb{R})$ is closed under scalar multiplication and addition, $F(\phi)$ is in $L^2(\mathbb{R})$.

Solution 20 Let t_0 be a point in the support of the sum $f_1 + \cdots + f_N$; then at least one of the f_k is nonzero at $t = t_0$, that is, t_0 is in at least one of the S_k, and hence t_0 is in the union $\cup_k S_k$.

Similarly if the product $f_1 \cdots f_N$ is nonzero at $t = t_0$ then each function f_k is non-zero at $t = t_0$, that is, $t_0 \in S_k$ for all $1 \le k \le N$, and hence $t_0 \in \cap_k S_k$.

Solution 24 The wavelet ψ has L^2 norm

$$(\psi, \psi) = \left(\sqrt{2} \sum_k d_k \phi(2t - k), \sqrt{2} \sum_m d_m \phi(2t - m) \right) = \sum_k d_k^2 = \sum_k c_k^2 = 1,$$

if we make use of the fact that $(\phi(2t-k), \phi(2t-m)) = 0$ for $k \neq m$, $(\phi(2t-k), \phi(2t-k)) = 1/2$, and equation (9.35).

Solution 25 It is straightforward to check that the dot product of the jth element of S_1 with the kth element of S_2 is zero, except when $j = k$ (in the order given). All other dot products are in fact equal to one. Then we find $(3,1,2) = c_1(1,2,1) + c_2(0,1,1) + c_3(1,1,1)$ with $c_1 = (3,1,2) \cdot (0,1,-1) = -1, c_2 = (3,1,2) \cdot (-1,0,1) = -1, c_3 = (3,1,2) \cdot (1,-1,1) = 4$. It works.

Solution 27 If ϕ is supported on $[0, M-1]$ then $\phi(2t-k)$ is supported on $[k/2, (M-1+k)/2]$. In particular, if $0 \leq k \leq N-1$ (as it does for ψ in equation (9.63)) then all such $\phi(2t-k)$ are supported in $[0, (M+N)/2 - 1]$ (using the minimum value $k = 0$, the max $k = N-1$). Then ψ must be supported on $[0, (M+N)/2 - 1]$.

Exactly the same argument works for ψ^*, with ϕ^* in place of ϕ and the roles of M and N interchanged.

Index

Discrete Fourier Analysis and Wavelets: Applications to Signal and Image Processing, Second Edition.
S. Allen Broughton and Kurt Bryan.

Companion Website: www.wiley.com/go/Broughton/Discrete_Fourier_Analysis_and_Wavelets